State-of-the-Art Polymer Science and Technology in Italy (2019–2020)

State-of-the-Art Polymer Science and Technology in Italy (2019–2020)

Editors

Ignazio Blanco
Roberto Pantani
Andrea Sorrentino
Antonio Pizzi

MDPI • Basel • Beijing • Wuhan • Barcelona • Belgrade • Manchester • Tokyo • Cluj • Tianjin

Editors

Ignazio Blanco
University of Catania
Italy

Roberto Pantani
University of Salerno
Italy

Andrea Sorrentino
National Research Council of Italy (CNR)
Italy

Antonio Pizzi
Université de Lorraine
France

Editorial Office
MDPI
St. Alban-Anlage 66
4052 Basel, Switzerland

This is a reprint of articles from the Special Issue published online in the open access journal *Polymers* (ISSN 2073-4360) (available at: https://www.mdpi.com/journal/polymers/special_issues/ITA_2019_2020).

For citation purposes, cite each article independently as indicated on the article page online and as indicated below:

LastName, A.A.; LastName, B.B.; LastName, C.C. Article Title. *Journal Name* **Year**, *Article Number*, Page Range.

ISBN 978-3-03943-180-9 (Hbk)
ISBN 978-3-03943-181-6 (PDF)

Contents

About the Editors

Ignazio Blanco is a full professor of Chemical Foundations of Technologies at the Department of Civil Engineering and Architecture, University of Catania. His research activities are focused on the following themes: synthesis and characterization of toughened thermoset blends; process technology of polymeric fiber-reinforced composites; heat capacities, fusion, and solid-to-solid transition of series of organic molecules; comparative kinetic studies of the thermal degradation of model polymers; comparative kinetic studies of the thermal degradation of nanocomposites; synthesis and characterization of aromatic copolymers; thermal characterization of films used in food packaging applications; synthesis and characterization of nanoparticles; POSS synthesis and characterization; and material characterization for rapid prototyping. To date, he has authored and/or co-authored 110 papers indexed on Scopus, 1 book chapter, 3 books and about 80 presentations in national and international conferences.

Roberto Pantani is a full professor at the Department of Industrial Engineering of the University of Salerno. He is a lecturer of Transport Phenomena and of Principles of Chemical Engineering, and is currently Director of the Department. In 1995, he graduated "cum laude" in Chemical Engineering, defending a thesis on shrinkage in injection molded polystyrene. He obtained his Ph.D. at the University of Palermo, Italy, in February 2000, defending a thesis entitled "Analysis of Shrinkage Development in Injection Molded Samples". From 1999 to 2001, he completed a research fellowship at University of Salerno, Dept. of Chemical Engineering, in "Processing and Characterization of Polymeric Materials". From September 2001 to 2007, he was an assistant professor, and from 2007 to 2017, he was an associate professor of Transport Phenomena at University of Salerno, Dept. of Chemical Engineering. His main research interests focus on the analysis and simulation of injection molding of thermoplastics, structure development in polymer processing, volume accuracy and stability in polymer processing, processing and degradation of biodegradable polymers. He collaborates with companies and research centers and is the author of two patents, more than 100 publications on international peer reviewed scientific journals, several book chapters and more than 200 publications on proceedings of international conferences.

Andrea Sorrentino is a senior researcher at the Institute for Polymers, Composites and Biomaterials (IPCB) of the Italian National Research Council (CNR), where he works on polymer processing including product design and process development. His research focuses on the effect of thermo-mechanical history on the final morphology and properties of polymeric materials. Some of his areas of interest include the development of functional nanocomposites materials for barrier and electrical applications. Furthermore, he has experience in renewable and bio-based polymer preparation and bio/photo-degradation analysis. His lab in Lecco currently conducts research on innovative additive manufacturing processing, specifically designed for orthopedic and biomedical applications. He has received over c1.5 million in funding from public bodies, international organizations, and various industrial collaborations. He has over 150 publications and 10 international patents. Since 2012, he has been a member of the Editorial Board of renowned scientific journals and regularly serves as a peer reviewer for various funding agencies.

Antonio Pizzi is a full professor (Emeritus) of Industrial Chemistry at the LERMAB-ENSTIB of the University of Lorraine, France. His activities are in the conception and formulation of synthetic, bio-sourced and natural wood adhesives and resins, fire resistant and insulating foams, and in the assembly of wood composites, in wood welding without adhesives, and in bio-sourced non-isocyanate polyurethanes (NIPU) for a variety of applications, on polymer chemistry in particular in applied and theoretical aspects of polycondensation, and in some other lesser fields. He has conducted innumerable national, European and industrial contract projects with companies in Europe, North America, South America, Asia and Africa. He is the author of more than 800 science journal publications of good standing, 41 patents, 11 books, 70 book chapters and innumerable invited conference proceedings. He has received more than 20 international scientific prizes, including the finalist prize of the European Commission Descartes Prize, being the only person to have achieved this twice.

polymers

MDPI

Editorial

State-of-the-Art Polymer Science and Technology in Italy

Ignazio Blanco [1,*], Roberto Pantani [2,*], Antonio Pizzi [3,*] and Andrea Sorrentino [4,*]

1 Department of Civil Engineering and Architecture and INSTM UdR, University of Catania, V.le A. Doria 6, 95125 Catania, Italy
2 Industrial Engineering Department, University of Salerno, I-84084 Fisciano (Salerno), Italy
3 LERMAB, Laboratoire d'Etude et de Recherche sur le MAteriau Bois, Université de Lorraine, 27 rue Philippe Seguin, CS60036, 88021 Epinal, France
4 Institute of Polymers, Composites and Biomaterials (IPCB), National Research Council of Italy (CNR), Via Previati 1/C, 23900 Lecco, Italy
* Correspondence: iblanco@unict.it (I.B.); rpantani@unisa.it (R.P.); antonio.pizzi@univ-lorraine.fr (A.P.); andrea.sorrentino@cnr.it (A.S.)

Received: 21 July 2020; Accepted: 30 July 2020; Published: 31 July 2020

The history of polymers in Italy certainly begins with Giulio Natta and the discovery of isotactic polypropylene [1,2], culminated in the delivery to Prof. Natta of the Nobel Prize for chemistry in 1963. Before those years, thanks to the development of the hydrocarbons and petrochemical sector by Montecatini (General Society for the Mining and Chemical Industry), the foundations for the development of macromolecular chemistry in Italy had been laid, but the real expansion occurred between 1958 and 1967 with a sector growth of 30% on an annual basis [3]. The research of Natta and his collaborators, on metal-organic catalysis applied to the polymerization reactions of -olefins and other unsaturated monomers, promoted the growth and consolidation, in Italy, of a network of industrial centers and academic institutes operating in the polymer science and technology sector, creating international skills. Since that period until today, Italy has kept pace with other countries and has certainly contributed to the development of the polymeric materials sector. Today, polymer science and technology in Italy represents a long-lasting and interdisciplinary field in which chemistry, physics, and engineering mix together to produce studies, which are considered among the best in the world. Several research groups coming from very different fields often collaborate in the design of the material, of the part, and of the processing technology to obtain innovative products with outstanding, new, and smart properties. Examples of the contributions of Italian research in the field are spread out in the top Journals and Conferences throughout the world. This Special Issue, which consists of 22 articles, including two review articles, aims at collecting an overview of Polymer Science and Technology in Italy, hoping that it can portray the state of the Polymeric Research in Italy and give the rest of the world a bright image of what is carried out in the field in our country.

In the molecular imprinting technique, the use of preformed oligomers instead of functional monomers increases the stability of the non-covalent interactions with the template molecule, providing a sharp gain in terms of binding properties for the resulting imprinted polymer. Based on this theory, Baggiani and co-workers assumed that the delayed addition of template molecules to a polymerization mixture enhances the binding properties of the resulting polymer, concluding that the delayed addition approach could be useful in prepare imprinted polymers with higher binding affinity and increased binding selectivity in cases of difficult imprinting polymerization [4].

Catauro et al. investigated the effectiveness of nanocomposites (composed of ultra-high molecular weight polyethylene (UHMWPE) mixed with carbon nano-filler (CNF) and medical grade paraffin oil (PO), from the biological point of view. They carried out wear measurements without (air) and with lubricant (distilled water, natural, and artificial lubricant) and evaluated the antibacterial activity and cytotoxicity. Their results highlighted that the presence of CNF is important in the nanocomposite formulation because it reduces the wear rate and prevents oxidative degradation during its processing [5].

Pucci et al. investigated the functionalization of polyketone 30 (PK30) with glycyl-glycine (Gly-Gly) via the Paal–Knorr reaction with the aim of homogenously dispersing two types of reduced graphene oxide (rGO, i.e., lrGO and hrGO, the former characterized by a lower degree of reduction in comparison to the latter) by non-covalent interactions. All the composites showed a typical semiconductive behavior and stable electrical response after several heating/cooling cycles from 30 to 115 °C, demonstrating their usability as an ON–OFF temperature sensor and as sensing material in soft robotics applications [6].

Malucelli et al. evaluated the influence of carbonaceous materials, such as biochar (BC) and/or multiwalled carbon nanotubes (MWCNTs), in the rheological and thermal properties of UV-LED curable coatings. They reported on the synthesis and characterization of carbon-reinforced films containing nanometric (MWCNTs) and micrometric (BC) carbon-based materials. They analyzed the rheological properties of the UV-LED curable dispersions, as well as the thermal and optical properties of the resulting films, establishing some correlations between filler dispersion/loading with the main observed properties [7].

Leporini and co-workers, by the means of molecular dynamics simulations, studied two melts of polymer chains with different bond length, resulting in rather different strength of the Johari-Goldstein (JG) relaxation. They found that, even if changing the bond length alters both the strength and the relaxation time of the JG relaxation, it leaves unaffected the correlation between the vibrational dynamics and the primary relaxation [8].

Film blowing technique was used to prepare polymer films based on biodegradable polymers, polyethylene (PE) and modified PE with oxo-degradable additive by La Mantia et al. [9]. The Sicilian researchers investigated mechanical properties, soil burial degradation and surface wettability of the obtained polymers, before and after UV irradiation. They observed that UV irradiation increased surface wettability and encouraged the disintegration in soil of all the samples, due to molar mass reduction and the production of hydrophilic end groups by photo-oxidation, thus increasing surface erosion and weight loss.

Tondi and co-workers produced high homogeneity tannin foams suitable for industrial upscaling. Foams catalyzed with nitric acid showed similar physical properties and more phenolic character, which favors the absorption of ionic pollutants. Conversely, the foams blown with aliphatic solvents and surfactants presented smaller pores, and higher mechanical and insulating properties, without affecting the chemical properties or the heating value. The combined foam produced with nitric acid as a catalyst and petroleum ether as a blowing agent result in sulfur-free and small cell material with overall improved features [10].

Gorrasi et al. studied the effect of ionic liquid (IL) in improving the dispersion of a Zn/Al layered double hydroxides (LDHs) hosting carbon nanotubes (80% of CNTs) into a commercial biodegradable highly amorphous vinyl alcohol polymer [11]. The analysis of thermal, mechanical and electrical properties of the composites, resulted improved compared to the unfilled material, allowed to hypothesize a good dispersion of the LDH-CNTs lamellar filler into the polymer matrix-assisted by the ionic liquid. This was demonstrated comparing electrical conductivity of composite at 5% of LDH-CNTs in the presence and in the absence of IL.

The monitoring of the properties' development of different hybrid epoxy formulations, during low temperature curing and aging, was carried out by Frigione and co-workers [12]. The durability of the produced hybrids was probed in comparison to the corresponding neat epoxies by glass transition temperature (T_g) measurements. Flexural mechanical properties after exposure to different levels of humidity and immersion in water and at temperatures slightly higher than the local ambient temperature were also evaluated, in order to simulate the conditions encountered during summer seasons in very humid environments. Their epoxy-based systems displayed significant advantages over the conventional epoxy resins used as structural adhesives or as matrices for fiber reinforced composites in terms of higher T_g, better mechanical properties and enhanced durability in aqueous environments.

Piras and co-workers investigated the potential application of ammonium–chitosan grafted with 2-methyl-β-cyclodextrin (N^+-rCh-MCD) as the functional macromolecular complexing agent for the oral administration of the neuropeptide dalargin (DAL). The results of NMR characterization, UV and fluorescence techniques, as well as biological in vitro assessments, indicated that N^+-rCh-MCD forms water-soluble complexes with DAL, with a prevalent involvement of Tyr or Phe over Leu and Ala residues. The association constant of DAL with the polymeric derivative is one order of magnitude higher than that with the pristine cyclodextrin (K_a: 2600 M^{-1} and 120 M^{-1}, respectively). Additionally, N^+-rCh-MCD shields DAL from enzymatic degradation in gastrointestinal in vitro models with a three-fold time delay, suggesting a future pharmaceutical exploitation of the polymeric derivative [13].

Aiming at understanding the relationship between the properties of the plastic objects and the process conduction, and thus the evaluation of process parameters, such as the mold temperature, Speranza et al. placed a thin electrical heater below the cavity surface in order to obtain rapid and localized surface heating/cooling cycles during the injection molding process [14]. The modulation of the cavity temperature was found able to control the distribution of relevant morphological characteristics, thus, properties along the sample thickness. The crystalline degree slightly increased with the cavity temperature, and this induced an increase in the elastic modulus when high temperatures were adopted for the cavity surface. The cavity temperature strongly influenced the orientation distribution that, on its turn, determined the highest values of the elastic modulus found in the shear layer. In particular, they observed that, if the macromolecules spent adequate time at temperatures close to the crystallization temperature, it was possible achieving high levels of structuring and, thus, high values of elastic modulus.

With the purpose to fulfill higher standard prerequisites and properties in dental restorative materials, such as high modulus, high hardness, and chemically inertness, Tammaro et al. developed novel composites with multiple biofunctions [15]. They prepared light-cured Bis-GMA/TEGDMA dental resin (RK)/layered double hydroxide intercalated with fluoride ions (LDH-F)/calcium bentonite (Bt) hybrid composites, recording an improvement of the mechanical properties with respect to the pristine resin. Furthermore, they observed, for the synthetized composites, antibacterial and antibiofilm effects against three bacterial strains isolated from healthy volunteers' saliva (*Streptococcus* spp., *Bacteroides fragilis*, and *Staphylococcus epidermidis*) without affecting its ability to induce dental pulp stem cells differentiation into odontoblast-like cells, thus making these materials a promising strategy in preventive and restorative dentistry.

De Marco and Franco worked at the design and production of composite microparticles with a prolonged drug release for oral delivery. The supercritical antisolvent (SAS) process was used to coprecipitate Eudragit L100-55 (EUD) with diclofenac (DICLO) and theophylline (THEOP), obtaining well-defined spherical microspheres with a mean diameter ranging from 3.75 and 5.93 μm. After a wide characterization (scanning electron microscopy, differential scanning calorimetry, X-ray microanalysis, FT-IR, and UV–vis spectroscopy), they carried out dissolution studies, showing the potential of EUD to prolong the drug release, significantly, up to a few days [16].

In order to solve dark aspects related to the Polyfurfuryl alcohol (PFA) molecular structure derived from the polymerization in acid environment, Tondi et al. carried out a wide spectroscopic investigation [17]. Thanks to a multi-technique approach, involving Solid-State ^{13}C-NMR, Attenuated Total Reflectance (ATR), Fourier Transform Infrared (FTIR) spectroscopy, and UV-resonant Raman spectroscopy at different excitation wavelengths, using both an UV laser source and UV synchrotron radiation and simulations with first-principles and semi-empiric methods to evaluate their matching with experimental ones, they concluded that, in addition to the major linear unconjugated polymerization, the PFA structure consists of Diels-Alder rearrangements occurring after the opening of some furanic units, while the terminal moieties of the chain involves γ-lactone arrangements. The occurrence of head-head methylene ether bridges and free hydroxyl groups (from unreacted furfuryl alcohol, FA, or terminal chains) could be excluded, while the conjugated systems could be considered rather limited.

In their work, Roppolo and colleagues presented new 3D printable materials based on the introduction of different commercially available ionic liquids (ILs) in the starting formulations. They evaluated the influence of these additives on the printability of such formulations through light-induced 3D printing (digital light processing—DLP), as well as investigating the effect of ionic liquids with polymerizable groups. The physical chemical properties of such materials were compared, focusing on the permeability towards CO_2 of the different ILs present in the formulations, showing the possibility of 3D printing high complexity structures, which could be the base of new high complexity filters for a more efficient CO_2 capture [18].

Lazzara et al. have developed active, healable, and safely dissolvable alginate-pectin-based biocomposites for possible applications in food packaging [19]. After morphological, mechanical, thermal tests, they observed properties in the range of commercially available packaging films. Furthermore, the prepared biocomposites exhibited higher hydrophobicity properties and antibacterial properties against different bacteria, making them ideal for utilization as packaging material.

Continuing their study on biodegradable polymers, La Mantia and co-workers prepared irrigation tubes by extruding polylactide/poly (butyleneadipate-co-butyleneterephthalate) (PLA/PBAT) blend (Bio-Flex®) and Mater-Bi®. After rheological and mechanical characterization, the obtained pipes were subjected to photoaging with continued exposure to UV radiation up to 22 days together with a soil burial degradation test at 30 °C and 50 °C for up to 70 days. They observed that the degradation rate of irrigation tube samples based on Mater-Bi® was higher at 30 °C and was stimulated after 14 days of UV irradiation, thus concluding that higher temperatures or UV aging encouraged the disintegration in soil of Bio-Flex®-based irrigation tubes [20].

Pizzi and co-workers studied the reaction among Soy protein isolate (SPI) and insoluble soy flour polymeric carbohydrates with sodium periodate for the specific oxidation of vicinal –OH groups aiming at investigate the reactions involved in this approach to soy flour adhesives. They observed the generation of carbohydrate oligomer fractions presenting one, two or multiple aldehyde groups. The reaction of periodate with soy protein isolate has been shown to generate some aldehydes too, whilst when the mix of SPI and soy insoluble carbohydrates is treated with periodate, the majority of the observed aldehyde carrying species appear to be higher molecular weight carbohydrate oligomer fractions [21].

Two series of novel dumbbell-shaped polyhedral oligomeric silsesquioxanes (POSSs), fully functionalized with phenyl groups at the corner of the silicon cages, were synthesized by Blanco et al. to prepare, by the means of in situ polymerization, polystyrene (PS) nanocomposites. Thermal and morphological properties were evaluated and compared among the nanocomposites obtained using the two different series of dumbbell-shaped POSSs and with the net PS showing much higher values when compared with those of neat PS, and highlighting significant differences when an aliphatic or aromatic bridge was used to link the silicon cages [22].

A new formulation based on Diglycidyl ether of bisphenol A (DGEBA) mixed with Diethyltoluene diamine (DETDA) and with a commercial daylight photocurable resin was prepared by Tosto et al. to obtain printable materials in a liquid crystal display (LCD) 3D printer. After the calorimetric, dynamic mechanical and rheology testing, the authors carried out 3D printing trials, thus determining the best printing conditions [23].

In the first review of this Special Issue, Trotta and co-workers gave an overview of cyclodextrin-based nanosponges research, focusing on the origin and key points of the historical development in the last 50 years, progressing from relatively simple crosslinked networks in the 1960s to today's multifunctional polymers. They explored the historical evolution of nanosponges in order to understand their role today, and imagine their future [24].

Another review by Siracusa et al. offered a rundown of life-cycle assessment (LCA) studies on polymers, used in the most important production and commercial sectors, carried out in the last few years by Italians researchers in order to remark, in the role of researchers who have made polymers and their derivatives their main research object, that a sustainable use of polymeric materials is not only possible but is, above all, necessary [25].

Conflicts of Interest: The authors declare no conflict of interest.

References

1. Natta, G.; Pino, P.; Corradini, P.; Danusso, F.; Mantica, E.; Mazzanti, G.; Moraglio, G. Crystalline High Polymers of α-Olefins. *J. Am. Chem. Soc.* **1955**, *77*, 1708–1710. [CrossRef]
2. Natta, G.; Pino, P.; Mazzanti, G. Isotactic Poly Propylene. U.S. Patent 3,112,300, 26 November 1963.
3. Martuscelli, E. *Dalla scoperta di Natta. Lo sviluppo dell'industria e della ricerca sulle plastiche in Italia*; CNR Edizioni: Roma, Italia, 2001; ISBN 8880800205.
4. Anfossi, L.; Cavalera, S.; Di Nardo, F.; Spano, G.; Giovannoli, C.; Baggiani, C. Delayed Addition of Template Molecules Enhances the Binding Properties of Diclofenac-Imprinted Polymers. *Polymers* **2020**, *12*, 1178. [CrossRef] [PubMed]
5. Catauro, M.; Scolaro, C.; Dal Poggetto, G.; Pacifico, S.; Visco, A. Wear Resistant Nanocomposites Based on Biomedical Grade UHMWPE Paraffin Oil and Carbon Nano-Filler: Preliminary Biocompatibility and Antibacterial Activity Investigation. *Polymers* **2020**, *12*, 978. [CrossRef] [PubMed]
6. Araya-Hermosilla, E.A.; Carlotti, M.; Picchioni, F.; Mattoli, V.; Pucci, A. Electrically-Conductive Polyketone Nanocomposites Based on Reduced Graphene Oxide. *Polymers* **2020**, *12*, 923. [CrossRef] [PubMed]
7. Strongone, V.; Bartoli, M.; Jagdale, P.; Arrigo, R.; Tagliaferro, A.; Malucelli, G. Preparation and Characterization of UV-LED Curable Acrylic Films Containing Biochar and/or Multiwalled Carbon Nanotubes: Effect of the Filler Loading on the Rheological, Thermal and Optical Properties. *Polymers* **2020**, *12*, 796. [CrossRef] [PubMed]
8. Tripodo, A.; Puosi, F.; Malvaldi, M.; Capaccioli, S.; Leporini, D. Coincident Correlation between Vibrational Dynamics and Primary Relaxation of Polymers with Strong or Weak Johari-Goldstein Relaxation. *Polymers* **2020**, *12*, 761. [CrossRef]
9. La Mantia, F.P.; Ascione, L.; Mistretta, M.C.; Rapisarda, M.; Rizzarelli, P. Comparative Investigation on the Soil Burial Degradation Behaviour of Polymer Films for Agriculture before and after Photo-Oxidation. *Polymers* **2020**, *12*, 753. [CrossRef]
10. Eckardt, J.; Neubauer, J.; Sepperer, T.; Donato, S.; Zanetti, M.; Cefarin, N.; Vaccari, L.; Lippert, M.; Wind, M.; Schnabel, T.; et al. Synthesis and Characterization of High-Performing Sulfur-Free Tannin Foams. *Polymers* **2020**, *12*, 564. [CrossRef]
11. Bugatti, V.; Viscusi, G.; Di Bartolomeo, A.; Iemmo, L.; Zampino, D.C.; Vittoria, V.; Gorrasi, G. Ionic Liquid as Dispersing Agent of LDH-Carbon Nanotubes into a Biodegradable Vinyl Alcohol Polymer. *Polymers* **2020**, *12*, 495. [CrossRef]
12. Frigione, M.; Lettieri, M.; Lionetto, F.; Mascia, L. Experimental Cold-Cured Nanostructured Epoxy-Based Hybrid Formulations: Properties and Durability Performance. *Polymers* **2020**, *12*, 476. [CrossRef]
13. Cesari, A.; Recchimurzo, A.; Fabiano, A.; Balzano, F.; Rossi, N.; Migone, C.; Uccello-Barretta, G.; Zambito, Y.; Piras, A.M. Improvement of Peptide Affinity and Stability by Complexing to Cyclodextrin-Grafted Ammonium Chitosan. *Polymers* **2020**, *12*, 474. [CrossRef] [PubMed]
14. Liparoti, S.; Speranza, V.; Titomanlio, G.; Pantani, R. Effect of Rapid Mold Heating on the Structure and Performance of Injection-Molded Polypropylene. *Polymers* **2020**, *12*, 341. [CrossRef] [PubMed]
15. Tammaro, L.; Di Salle, A.; Calarco, A.; De Luca, I.; Riccitiello, F.; Peluso, G.; Vittoria, V.; Sorrentino, A. Multifunctional Bioactive Resin for Dental Restorative Materials. *Polymers* **2020**, *12*, 332. [CrossRef] [PubMed]
16. Franco, P.; De Marco, I. Eudragit: A Novel Carrier for Controlled Drug Delivery in Supercritical Antisolvent Coprecipitation. *Polymers* **2020**, *12*, 234. [CrossRef] [PubMed]
17. Tondi, G.; Cefarin, N.; Sepperer, T.; D'Amico, F.; Berger, R.J.F.; Musso, M.; Birarda, G.; Reyer, A.; Schnabel, T.; Vaccari, L. Understanding the Polymerization of Polyfurfuryl Alcohol: Ring Opening and Diels-Alder Reactions. *Polymers* **2019**, *11*, 2126. [CrossRef] [PubMed]

18. Gillono, M.; Chiappone, A.; Mendola, L.; Gomez Gomez, M.; Scaltrito, L.; Pirri, C.F.; Roppolo, I. Study on the Printability through Digital Light Processing Technique of Ionic Liquids for CO_2 Capture. *Polymers* **2019**, *11*, 1932. [CrossRef]
19. Makaremi, M.; Yousefi, H.; Cavallaro, G.; Lazzara, G.; Goh, C.B.S.; Lee, S.M.; Solouk, A.; Pasbakhsh, P. Safely Dissolvable and Healable Active Packaging Films Based on Alginate and Pectin. *Polymers* **2019**, *11*, 1594. [CrossRef]
20. Rapisarda, M.; La Mantia, F.P.; Ceraulo, M.; Mistretta, M.C.; Giuffrè, C.; Pellegrino, R.; Valenti, G.; Rizzarelli, P. Photo-Oxidative and Soil Burial Degradation of Irrigation Tubes Based on Biodegradable Polymer Blends. *Polymers* **2019**, *11*, 1489. [CrossRef]
21. Frihart, C.R.; Pizzi, A.; Xi, X.; Lorenz, L.F. Reactions of Soy Flour and Soy Protein by Non-Volatile Aldehydes Generation by Specific Oxidation. *Polymers* **2019**, *11*, 1478. [CrossRef]
22. Abate, L.; Bottino, F.A.; Cicala, G.; Chiacchio, M.A.; Ognibene, G.; Blanco, I. Polystyrene Nanocomposites Reinforced with Novel Dumbbell-Shaped Phenyl-POSSs: Synthesis and Thermal Characterization. *Polymers* **2019**, *11*, 1475. [CrossRef]
23. Tosto, C.; Pergolizzi, E.; Blanco, I.; Patti, A.; Holt, P.; Karmel, I.S.; Cicala, G. Epoxy based blends for additive manufacturing by Liquid Crystal Display (LCD) printing. *Polymers* **2020**, *12*, 1594. [CrossRef] [PubMed]
24. Krabicová, I.; Appleton, S.L.; Tannous, M.; Hoti, G.; Caldera, F.; Rubin Pedrazzo, A.; Cecone, C.; Cavalli, R.; Trotta, F. History of Cyclodextrin Nanosponges. *Polymers* **2020**, *12*, 1122. [CrossRef] [PubMed]
25. Blanco, I.; Ingrao, C.; Siracusa, V. Life-Cycle Assessment in the Polymeric Sector: A Comprehensive Review of Application Experiences on the Italian Scale. *Polymers* **2020**, *12*, 1212. [CrossRef] [PubMed]

polymers

MDPI

Article

Polystyrene Nanocomposites Reinforced with Novel Dumbbell-Shaped Phenyl-POSSs: Synthesis and Thermal Characterization

Lorenzo Abate [1,*], Francesco Agatino Bottino [2], Gianluca Cicala [1], Maria Assunta Chiacchio [3], Giulia Ognibene [1] and Ignazio Blanco [1,*]

1 Department of Civil Engineering and Architecture and INSTM UdR, University of Catania, V.le A. Doria 6, 95125 Catania, Italy

2 Department of Chemical Science, University of Catania, V.le A. Doria 6, 95125 Catania, Italy

3 Department of Pharmaceutical Sciences University of Catania, V.le A. Doria 6, 95125 Catania, Italy

* Correspondence: labate@dmfci.unict.it (L.A.); iblanco@unict.it (I.B.)

Received: 22 August 2019; Accepted: 4 September 2019; Published: 9 September 2019

Abstract: Two series of novel dumbbell-shaped polyhedral oligomeric silsesquioxanes (POSSs), fully functionalized with phenyl groups at the corner of the silicon cages, were used to prepare polystyrene (PS) nanocomposites through the method of in situ polymerization. The percentage of the molecular filler reinforcement was set as 5% *w/w* of POSS and was checked by ^{1}H-NMR spectroscopy. The obtained nanocomposites were characterized by thermogravimetric analysis (TGA), differential scanning calorimetry (DSC), Fourier transform infrared spectroscopy (FTIR) and scanning electron microscopy (SEM). Thermal and morphological properties were evaluated and compared among the nanocomposites obtained using the two different series of dumbbell-shaped POSSs and with the net PS. The thermal parameters for the prepared nanocomposites were very high when compared with those of neat PS, and they evidenced significant differences when an aliphatic or aromatic bridge was used to link the silicon cages. SEM analysis results allow us to hypothesize a justification for the different resistance to thermal degradation showed by the two series of molecular reinforcement.

Keywords: polyhedral oligomeric silsesquioxanes; nanocomposites; POSS; polystyrene; thermal behavior; SEM analysis

1. Introduction

Since the middle of the last century, from common household goods to the important biomedical devices, by way of food packaging or building materials, polymers have made our lives easier and safer [1–5]. More recently, both for improving their properties and for reducing their environmental impact, eco-friendly inorganic materials have been added to polymeric matrices, thus leading to the production and commercialization of new polymer-based hybrid systems [6–8]. Literature has evidenced that property improvements cannot be obtained by a simple addition of an inorganic filler to the matrix [9], but a nanometric dispersion of the particles [10–12] and a good compatibility, obtainable via filler particle functionalization [13,14] are necessary. In the last few years, following these suggestions and by looking at the use of particular molecules like polyhedral oligomeric silsesquioxanes (POSSs) [15–17], our research group has been engaged in the design, synthesis and utilization of polymers reinforcement (polystyrene in particular) of POSSs [18–20]. This particular filler presents a hybrid nature due to its inorganic cage of silicon and oxygen atoms linked to organic groups by covalent bonds, thus making them molecules with the resulting advantages [21]. This aspect allows researchers to best use their nano-sized cage structures, dimensionally comparable to the most common polymer segments, thus producing an innovative chemical composition [22–25]. Starting from these considerations, we studied in the past several families of POSSs containing isobutyl, cyclopentyl and phenyl groups, with the aim of investigating whether and to what extent the physical properties of POSSs were modified by the introduction of these groups on the silicon cage [20,26,27]. We found that for phenyl-substituted POSSs, a higher resistance to the thermal degradation than cyclopentyl derivatives, and, for these last ones, a higher thermal stability than isobutyl-substituted POSSs. However, it has been well known in literature that the presence of phenyl groups increases the thermal stability of silicon cages [28,29], but their presence in a matrix decreases their compatibility, in comparison with aliphatic groups [30]. Hence, we had the need to verify how our POSSs, differently functionalized, behaved in a polymeric matrix—in particular in the polystyrene (PS) selected by us for this purpose. This need, combined with the aspiration to synthesize innovative POSS systems, which the literature has dealt with in the meantime, called "bridged" or "dumbbell-shaped" POSSs [31–33], induced us to verify the compatibility of this typology of hyper functionalized POSSs with a polystyrene matrix. Dumbbell-shaped POSSs aroused our interest because their change of organic bridge leads to a change of geometry, rigidity and symmetry of the obtained POSS, thus affecting their ability to be dispersed at a nanometric level [12,13,21].

In the present work, we synthesized the PS-based nanocomposites which had two different series of novel dumbbell-shaped POSS molecules (as fillers), fully phenyl functionalized in the periphery of the cage but linked with different aliphatic and aromatic bridges. The prepared PS/POSS nanocomposites were subjected to ^{1}H NMR to check if the filler content fit with the desired one, whilst thermal stability was evaluated by means of thermogravimetric analysis (TGA) and differential scanning calorimetry (DSC). Finally, scanning electron microscopy (SEM) was performed to gain information about the POSSs dispersion within the polystyrene (PS) matrix, whilst Fourier transform infrared spectroscopy (FTIR) was used to analyze the residue obtained from the TGA experiments.

2. Experimental

2.1. Materials and Methods

1,2-bis(triethoxysilyl)ethane and 4,4′-bis(triethoxysilyl)biphenyl have been acquired from Aldrich Co. (St. Gallen, Switzerland) and used without further purification. A Na-benzophenone mixture was employed to distill the tetrahydrofuran (THF). 1,6-bis(triethoxysilyl)hexane, 1,10-bis(triethoxysilyl)decane and 4,4′-bis(triethoxysilyl)phenyl ether were obtained by using an excess of tetraethoxysilane (TEOS) in dry THF and the Grignard reagent according to the description reported elsewhere [34]. Trisilanolphenyl POSS was acquired from Hybrid Plastics Co. (Hattiesburg, MS, USA) and used without further purification. Styrene, 2,2-azobis (isobutyronitrile) (AIBN), methanol and toluene were acquired from Aldrich Co. (St. Gallen, Switzerland) and purified in our laboratory: Styrene was purified in an inhibitor removal column; AIBN was re-crystallised twice from dry ethanol at temperatures less than 40 °C and in a dark light condition; toluene was stirred over calcium hydride for 24 h and distilled in a nitrogen flow.

Dumbbell-shaped POSSs, (Tables 1 and 2), were synthesized according with the procedure reported elsewhere [35]. A 5% *w/w* POSS/styrene mixture, in toluene, was used for in situ free-radical polymerization: After being dissolved in toluene, AIBN was added to the POSS/styrene mixture, and then we proceeded to freeze everything using liquid nitrogen; after degassing by using a vacuum pump, we thawed it. This procedure was repeated three times, and thus the tube was sealed under vacuum and heated at 70 °C for 24 h under stirring. Finally, we obtained a solution that was poured in methanol, from which we collected, by filtration, the PS/POSS nanocomposite. The samples were then dried at 40 °C, under vacuum, obtaining yields of 82%, 83%, 88%, 86% and 79% for samples **1**, **2**, **3**, **4** and **5**, respectively. PS, used for comparison, was prepared by using the same procedure.

Table 1. (**1**), (**2**) and (**3**) dumbbell-shaped octaphenyl polyhedral oligomeric silsesquioxanes (POSSs) structures, with aliphatic bridge, used in polystyrene (PS) nanocomposites.

$C_{86}H_{74}Si_{16}O_{24}$ (**1**)

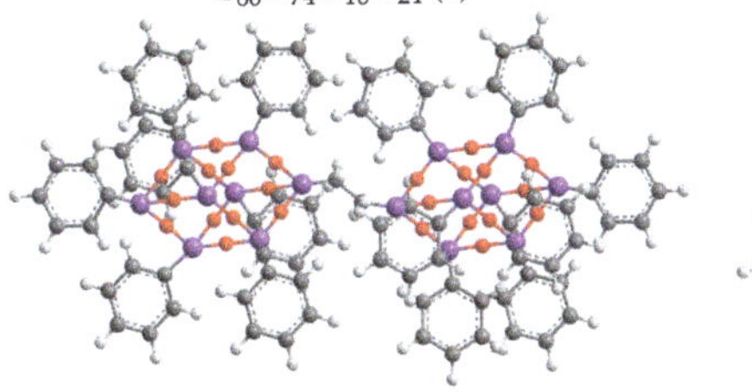

$C_{90}H_{82}$ $Si_{16}O_{24}$ (**2**)

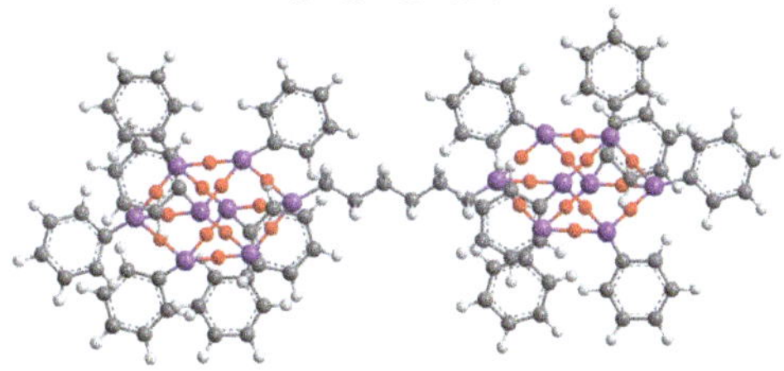

$C_{94}H_{90}$ $Si_{16}O_{2}$ (**3**)

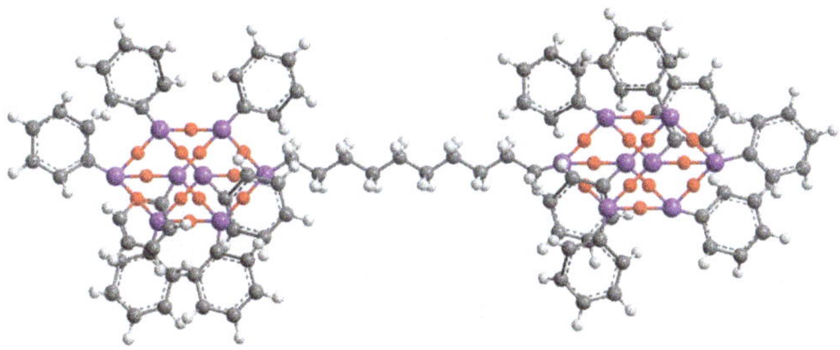

Table 2. (**4**) and (**5**) dumbbell-shaped octaphenyl POSS structures, with aromatic bridge, used in PS nanocomposites.

$C_{96}H_{78}$ $Si_{16}O_{25}$ (**4**)

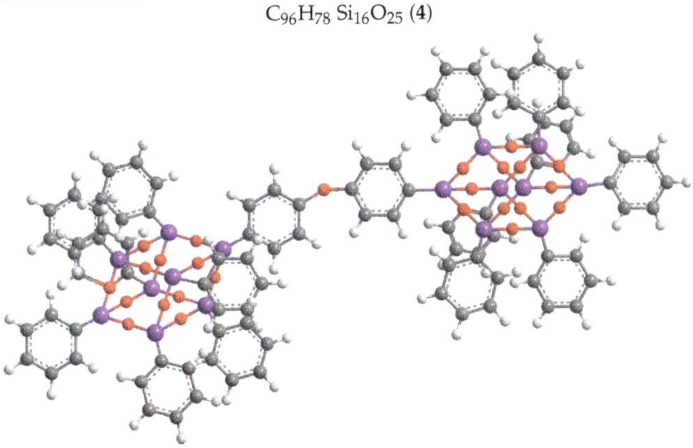

$C_{96}H_{78}$ $Si_{16}O_{24}$ (**5**)

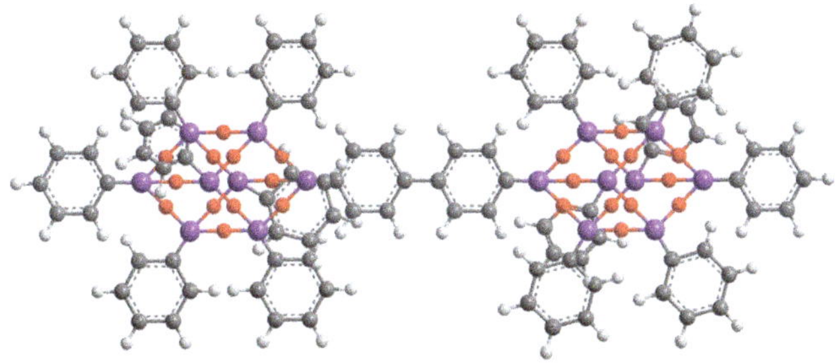

2.2. ^1H NMR Spectroscopy

Varian Unity INOVA equipment (^{1}H 500 MHz, Varian, Palo Alto, CA, USA) was employed to check the exact filler content. Measurements were performed by using $CDCl_3$ as a solvent and tetramethylsilane (TMS) as an internal standard. The evaluation was carried out by taking into account the ratio in the ^{1}H NMR spectra among the POSS aromatic hydrogens and the polystyrene ones.

2.3. Scanning Electron Microscopy (SEM)

The nanocomposites morphology was studied by SEM (ZEISS EVO MA 15, EVO-ZEISS, Cambridge, UK). In order to perform the analysis, an accelerating voltage of 15,000 KV was used. All the samples were treated overnight in an oven at 120 °C to remove any traces of toluene (Figure 1). Scans were carried out at different magnifications ranging from 500× to 25,000×.

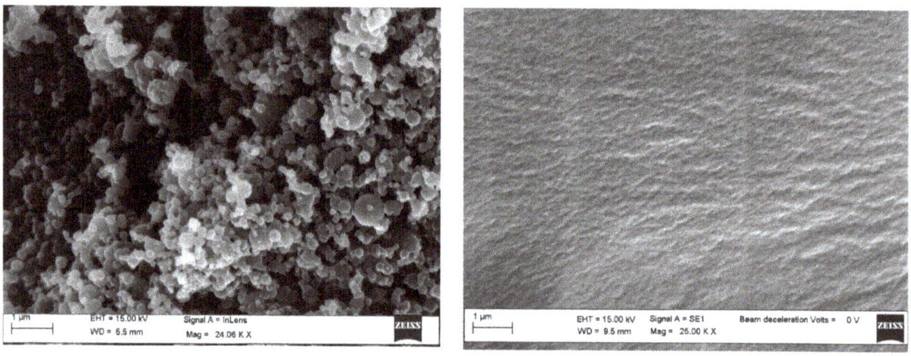

Figure 1. SEM images of polystyrene before (**left**) and after (**right**) drying in a stove.

2.4. Fourier Transform Infrared Spectroscopy (FTIR)

A Perkin Elmer Spectrum 100 spectrometer was employed in the determination of the FTIR spectra of the solid residues obtained from the TGA experiments at 700 °C. Spectra were carried out, without treatments, directly on samples at room temperature. For this purpose, we used an attenuated total reflectance (ATR) sampling accessory. Measurements were performed in the range from 4000 to 650 cm^{-1}, with a resolution of 4.0 cm^{-1}.

2.5. Differential Scanning Calorimetry (DSC)

Calorimetric investigations were performed, by using a Mettler DSC 1 Star System (Greifensee, Switzerland), for the determination of the glass transition temperatures (T_g) of PS and PS/POSSs nanocomposites. The calibration of the DSC apparatus was carried out following a consolidated procedure [36], showing a very good agreement with the literature values of indium and tin [36,37]. The DSC scans were carried out from room temperature up to 200 °C, at a heating rate of 10 °C·min^{-1}, on samples of about 6.0×10^{-3} g, held in sealed aluminum crucibles.

2.6. Thermogravimetric Analysis (TGA)

Thermal degradations were carried out in a Mettler Thermogravimetric Analyzer TGA 1 Star System (Mettler, Greifensee, Swizterland). For the equipment calibration, we considered the magnetic properties change of three metal standards (Isatherm, Nickel-alloy and Trafoperm 86, Mettler, Greifensee, Swizterland) at their Curie points (148, 355 and 750 °C), and we followed a consolidated procedure [38]. TGA scans were carried out by using 5×10^{-3} g of the sample at 10 °C·min^{-1} in the temperature range of 25–700 °C in a nitrogen (flow = 0.02 L·min^{-1}) and a static air atmosphere. The International Confederation for Thermal Analysis and Calorimetry (ICTAC) Kinetics Committee recommendations were followed to correct the error in mass determination due to the reduction of the buoyancy (Archimedes) force with the temperature increase [39]. The percentage of the non-degraded sample, $(1 - D)\%$, was reported as a function of the temperature, where $D = (W_o - W)/W_o$, and W_o and W were the weights at the starting point and during the experiment, respectively. TGA scans were performed in triplicate, and the average value was considered.

3. Results and Discussion

After the in situ polymerization of styrene in the presence of 5% of POSS, the obtained composites were subjected to ^{1}H NMR spectroscopy, with the aim to verify if the POSS percentage corresponded to the desired one.

A difference between the actual POSS contents in the obtained samples and the theoretical content was observed, which could be explained by an oligomers formation in the reactant mixtures [40]. These oligomers, which are soluble in methanol, remained in the solution, thus resulting in a richer POSS. This occurrence justified the yields obtained and the POSS percentage in nanocomposites that are reported in Table 3 together with the thermal parameters.

Table 3. POSS content (*w/w*), glass transition temperatures (T_g), temperatures at 5% mass loss ($T_{5\%}$), residue percentage at 700 °C of PS, and nanocomposites in a static air atmosphere and in flowing nitrogen.

Compounds	POSS%	$T_{g\%}$ °C	Static Air		Nitrogen Flow	
			$T_{5\%}$ °C	Residue %	$T_{5\%}$ °C	Residue %
PS	/	101.2	309.0	0	341.3	0
1	5.96	102.6	375.7	11.6	360.0	7.1
2	5.82	103.5	370.3	7.8	359.3	7.9
3	5.12	101.7	364.7	9.4	355.3	3.1
4	6.01	103.5	363.2	13.3	374.8	10.4
5	5.60	102.5	360.3	17.3	375.7	19.3

Thermogravimetric (TG) experiments on PS and prepared nanocomposites were performed in dynamic heating conditions in the temperature range of 25–700 °C by using a scanning rate of 10 °C min^{-1}, selected among the values normally used in this typology of experiments.

The degradations in inert environments were first performed, and the corresponding TG curves are reported in Figures 2 and 3, whilst the TG curves obtained in oxidative conditions are reported in Figures 4 and 5. Two parameters were considered for the assessment of the resistance to the thermal degradation, namely the temperature at 5% mass loss ($T_{5\%}$) and the solid residue at 700 °C.

Both in the inert atmosphere and in the oxidizing one, the net PS reached complete degradation more quickly in air than in nitrogen, whilst all the nanocomposites evidenced the formation of a stable residue up to 700 °C (Figures 2–5), ranging from about 10% for PS-reinforced with aliphatic bridged POSSs to about 20% for PS-reinforced by using phenyl bridged POSSs. The presence of a residue at 700 °C, a temperature at which PS was completely degraded in a quantity higher than the percentage of POSS in the matrix (Table 3) allowed us to speculate about a good dispersion of the POSS molecules in the matrix that resulted in an active interaction among the molecular filler and the PS.

SEM investigation confirmed this hypothesis, showing a good dispersion of the functionalized silicon cages within the polymer (Figure 6). Probably due to the higher flexibility of the aliphatic bridges in respect to the aromatic ones that joint the POSS cages used to reinforce the polymer, a better dispersion of POSS **1–3** then for POSS **4** and **5** was observed within the matrix (Figure 7).

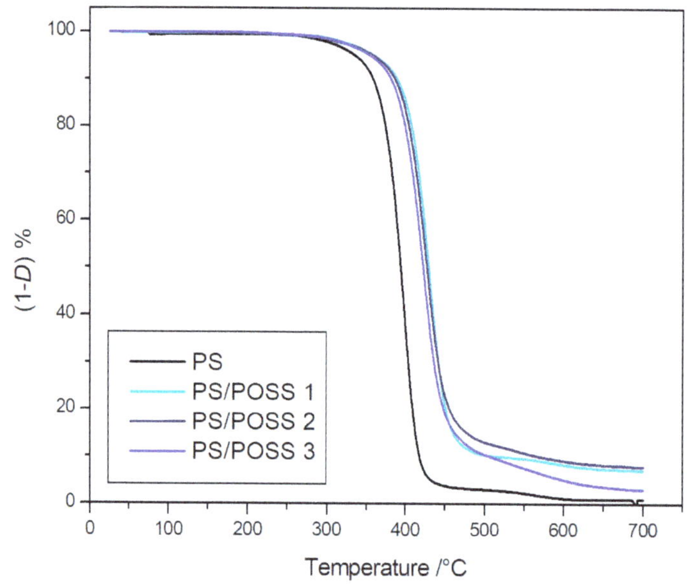

Figure 2. Thermogravimetric curves, in an inert atmosphere, for PS and samples **1**, **2** and **3**.

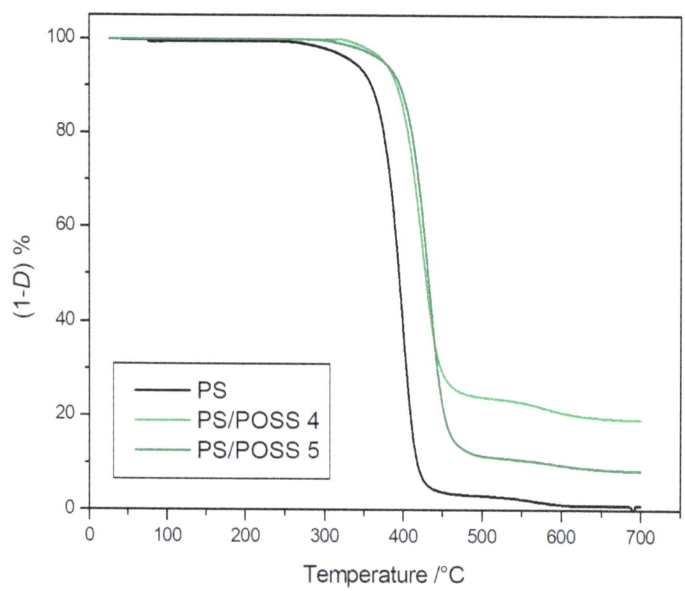

Figure 3. Thermogravimetric curves, in an inert atmosphere, for PS and samples **4** and **5**.

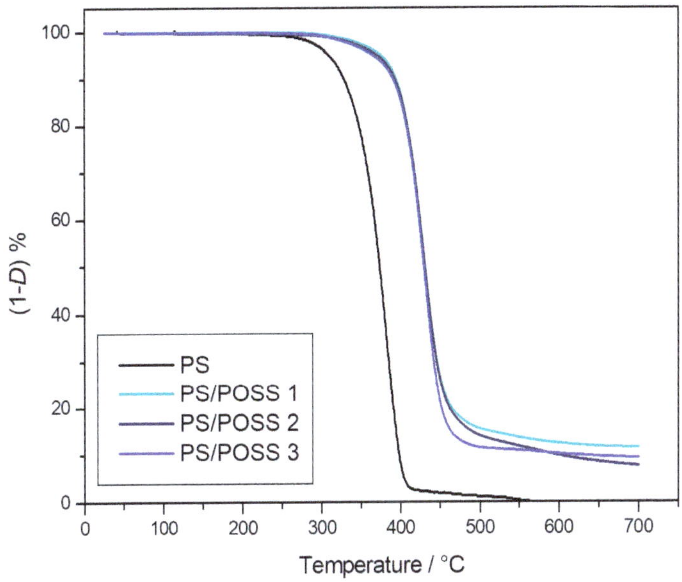

Figure 4. Thermogravimetric curves, in an oxidative atmosphere, for PS and samples **1**, **2** and **3**.

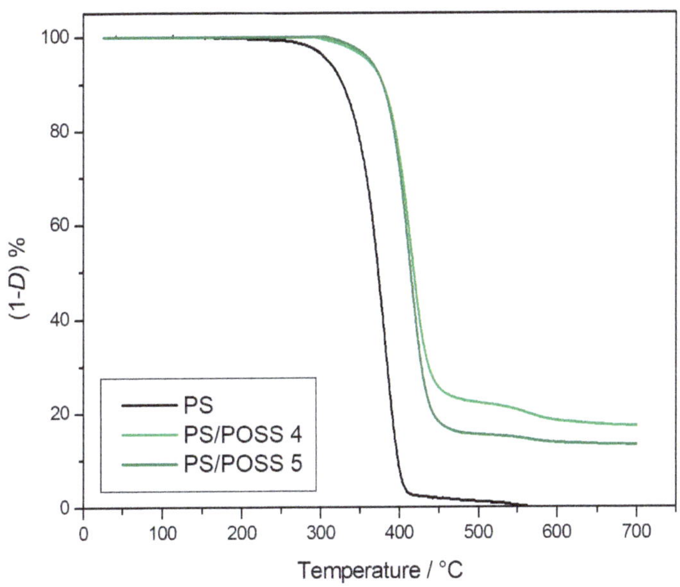

Figure 5. Thermogravimetric curves, in an oxidative atmosphere, for PS and samples **4** and **5**.

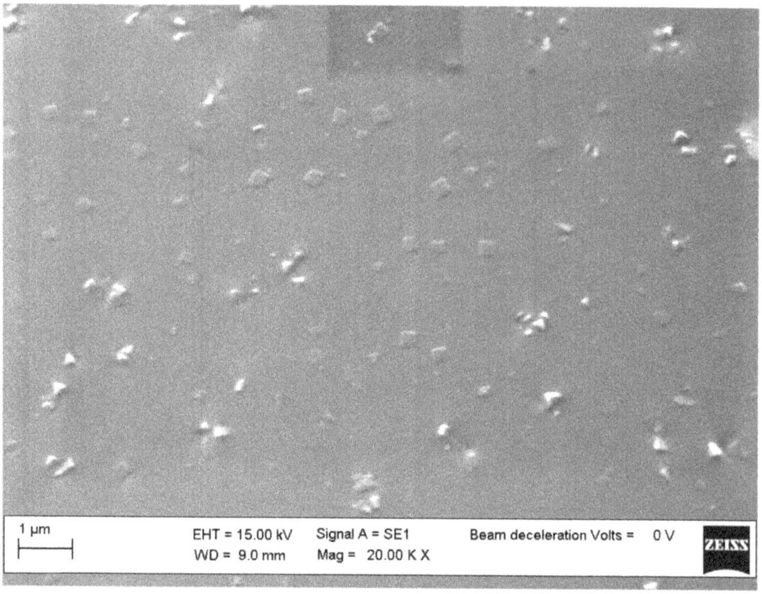

Figure 6. SEM images of PS/POSS **2** nanocomposites.

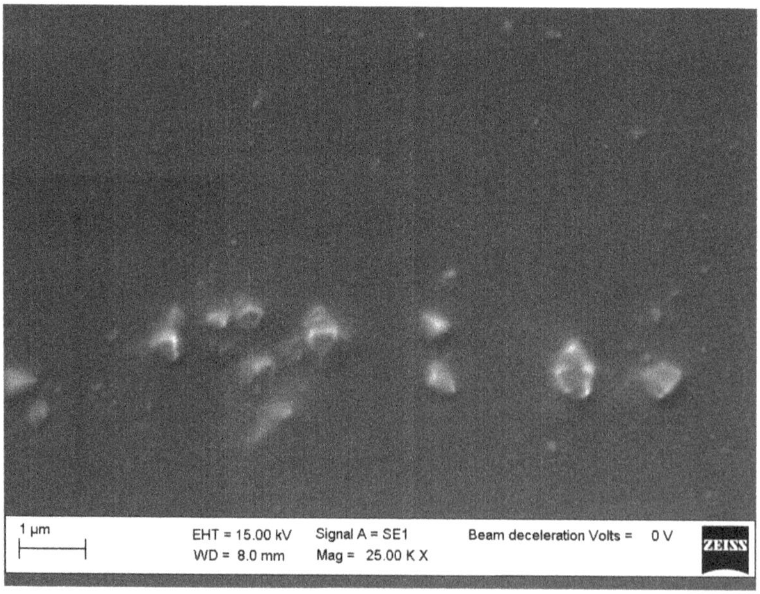

Figure 7. SEM images of PS/POSS **5** nanocomposites.

POSS **4** and **5**, with their spatially blocked aromatic bridges, probably did not permit the same freedom of movement of the silicon cages in the dumbbell-shaped POSSs with aliphatic bridges, thus leading to a different symmetric/asymmetric structure that facilitated POSS auto-aggregation phenomena for samples **4** and **5** (Figure 7), which formed large block aggregates.

The nanocomposites evidenced, both in nitrogen and in air, values of $T_{5\%}$ higher than those of the pristine polymer, $\Delta T_{5\%} \approx +35$ °C in nitrogen and $\Delta T_{5\%} \approx +60$ °C in air (Table 3). These high values of $T_{5\%}$, indicative of a good resistance to the thermal degradation, highlighted an increment of the thermal stability of the various nanocomposites in respect to the PS following the introduction of dumbbell-shaped POSSs in the matrix.

The degradations in oxidative and in inert environments proceeded similarly—both in the shape of TG curves and in the values at a temperature of 5% mass loss—that were comparable each other, meaning that the different atmosphere did not affect the degradation mechanism. Derivative thermogravimetric (DTG) curves are reported in Figures 8–11 and show a main degradation step along with a second, smaller one at higher temperatures (550–600 °C) that was more evident for the nanocomposites reinforced with the phenyl bridged POSS, thus leading us to think that this stage may be due to POSS degradation.

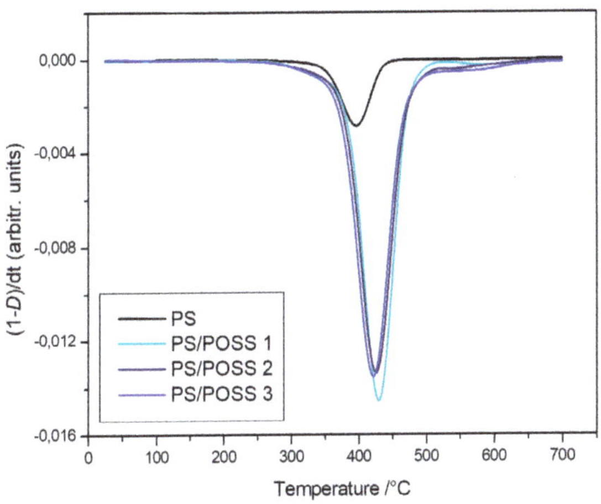

Figure 8. Differential thermogravimetric curves, in an inert atmosphere, for PS and samples **1**, **2** and **3**.

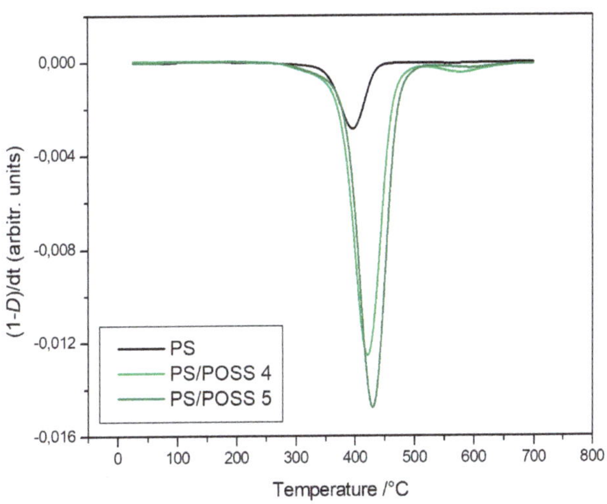

Figure 9. Differential thermogravimetric curves, in an inert atmosphere, for PS and samples **4** and **5**.

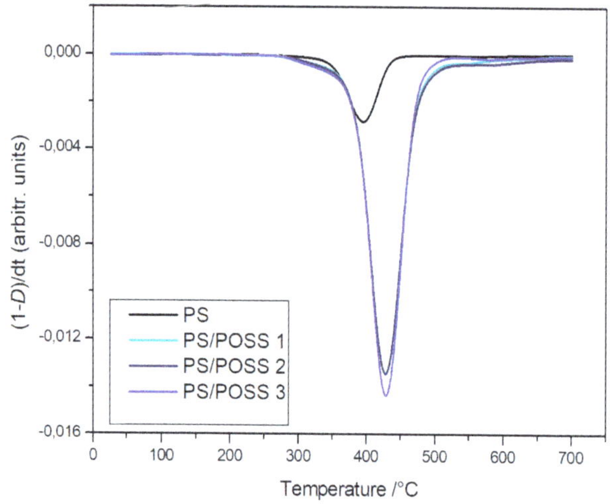

Figure 10. Differential thermogravimetric curves, in an inert atmosphere, for PS and samples **1**, **2** and **3**.

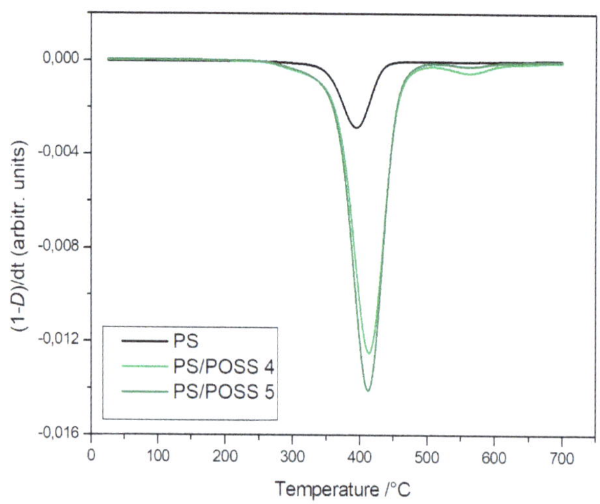

Figure 11. Differential thermogravimetric curves, in an inert atmosphere, for PS and samples **4** and **5**.

In order to gain information on the nature of the solid residue obtained at the end of TGA scans in both used atmospheres, we performed FTIR analysis, which evidenced only the presence of the silica characteristics peaks at about 1100 cm^{-1} (Figures S1 and S2).

Finally, the glass transition temperature was calorimetrically determined to better characterize the obtained compounds. DSC curves are reported in Figure 12, and the T_g values are reported in Table 3, showing a very small increment of the glass transition temperature for the nanocomposites in respect to virgin PS. This behavior could be attributed, in our hypothesis, to the nature of the groups at the POSS cage vertex. On considering that the incorporation of the POSSs on the polymeric chains change the matrix topology and alter its dynamics [41], the use of a less reactive phenyl-substituted POSSs during in situ polymerization may have had a mild effect on the reaction rate, thus not greatly altering the degree of crosslinking [42] and keeping the T_g values of composites quite constant in respect to that of neat polymer.

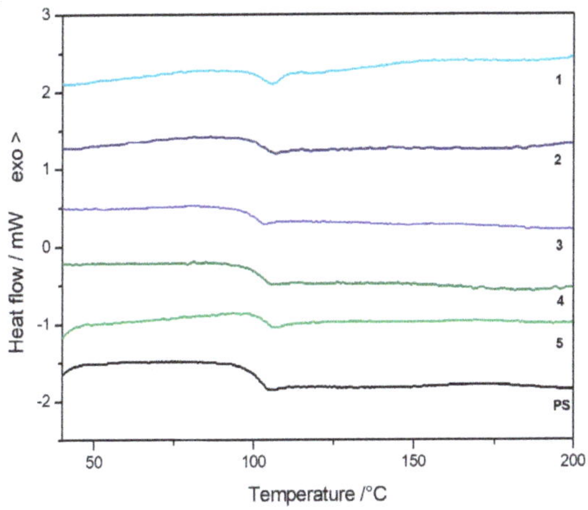

Figure 12. Differential scanning calorimetry curves for PS and PS/POSS nanocomposites.

4. Conclusions

The novel dumbbell-shaped POSSs, easily synthesized by a well-established methodology, have been proven as a valid reinforcement for the polystyrene matrix, showing a considerable improvement of the resistance to the thermal degradation of the prepared nanocomposites. We have shown encouraging SEM images regarding a good dispersion of the POSSs within the matrix, which has not always been reported in the literature for POSSs with a completely phenyl functionalized cage's periphery. Finally, the environment in which degradation was performed (inert or oxidative) seemed to not influence the degradation mechanism that proceeded with two different stages, the first relevant one at a lower temperature than the second little one; this was attributed to the POSS degradation.

Supplementary Materials: The following are available online at http://www.mdpi.com/2073-4360/11/9/1475/s1.

Author Contributions: I.B. and F.A.B. conceived and designed the experiments; F.A.B. synthesized the materials and interpreted HNMR; M.A.C. performed HNMR; I.B. performed the TGA and DSC experiments; G.O. performed SEM experiments; I.B., L.A. and G.C. analyzed the data; I.B. and L.A. wrote the paper.

Funding: This research was funded by the University of Catania within the "Piano della Ricerca Dipartimentale 2016-2018" of the Department of Civil Engineering and Architecture, grant name project: MATErials LIfe foreCAst (MATELICA).

Conflicts of Interest: The authors declare no conflict of interest.

References

1. Langer, R.; Tirrell, D.A. Designing materials for biology and medicine. *Nature* **2004**, *428*, 487. [CrossRef] [PubMed]
2. Siracusa, V.; Blanco, I.; Romani, S.; Tylewicz, U.; Dalla Rosa, M. Gas permeability and thermal behavior of polypropylene films used for packaging minimally processed fresh-Cut potatoes: A case study. *J. Food Sci.* **2012**, *77*, 264. [CrossRef]
3. Thakur, V.K.; Thakur, M.K.; Gupta, R.K. Review: Raw Natural Fiber–Based Polymer Composites. *Int. J. Polym. Anal. Charact.* **2013**, *18*, 495. [CrossRef]
4. Peplow, M. The plastics revolution: How chemists are pushing polymers to new limits. *Nature* **2016**, *536*, 266. [CrossRef] [PubMed]
5. Członka, S.; Strąkowska, A.; Strzelec, K.; Adamus-Włodarczyk, A.; Kairytė, A.; Vaitkus, S. Composites of Rigid Polyurethane Foams Reinforced with POSS. *Polymers* **2019**, *11*, 336. [CrossRef]
6. Cavallaro, G.; Grillo, I.; Gradzielski, M.; Lazzara, G. Structure of Hybrid Materials Based on Halloysite Nanotubes Filled with Anionic Surfactants. *J. Phys. Chem. C* **2016**, *120*, 13492. [CrossRef]

7. Cavallaro, G.; Lazzara, G.; Konnova, S.; Fakhrullin, R.; Lvov, Y. Composite films of natural clay nanotubes with cellulose and chitosan. *Green Mater.* **2014**, *2*, 232–242. [CrossRef]
8. Catauro, M.; Bollino, F.; Veronesi, P.; Lamanna, G. Influence of PCL on mechanical properties and bioactivity of ZrO2-Based hybrid coatings synthesized by sol-Gel dip coating technique. *Mater. Sci. Eng. C* **2014**, *39*, 344. [CrossRef]
9. Rong, M.Z.; Zhang, M.Q.; Zheng, Y.X.; Walter, R.; Friedrich, K. Structure–Property relationships of irradiation grafted nano-Inorganic particle filled polypropylene composites. *Polymer* **2001**, *42*, 167. [CrossRef]
10. Shvartzman-Cohen, R.; Nativ-Roth, E.; Baskaran, E.; Szleifer, I.; Yerushalmi-Rozen, R. Selective dispersion of single-Walled carbon nanotubes in the presence of polymers: The role of molecular and colloidal length scales. *J. Am. Chem. Soc.* **2004**, *126*, 14850. [CrossRef]
11. Monticelli, O.; Fina, A.; Ullah, A.; Waghmare, P. Preparation, Characterization, and Properties of Novel PSMA−POSS Systems by Reactive Blending. *Macromolecules* **2009**, *42*, 6614. [CrossRef]
12. Blanco, I.; Bottino, F.A.; Cicala, G.; Cozzo, G.; Latteri, A.; Recca, A. Synthesis and thermal characterization of new dumbbell shaped POSS/PS nanocomposites: Influence of the symmetrical structure of the nanoparticles on the dispersion/aggregation in the polymer matrix. *Polym. Compos.* **2015**, *36*, 1394. [CrossRef]
13. Blanco, I.; Abate, L.; Bottino, F.A.; Bottino, P. Synthesis, characterization and thermal stability of new dumbbell-shaped isobutyl-Substituted POSSs linked by aromatic bridges. *J. Therm. Anal. Calorim.* **2014**, *117*, 243. [CrossRef]
14. Yue, Y.; Wang, X.; Wu, Q.; Han, J.; Jiang, J. Assembly of Polyacrylamide-Sodium Alginate-Based Organic-Inorganic Hydrogel with Mechanical and Adsorption Properties. *Polymers* **2019**, *11*, 1239. [CrossRef] [PubMed]
15. Li, Y.; Dong, X.H.; Zou, Y.; Wang, Z.; Yue, K.; Huang, M.; Liu, H.; Feng, X.; Lin, Z.; Zhang, W.; et al. Polyhedral oligomeric silsesquioxane meets "click" chemistry: Rational design and facile preparation of functional hybrid materials. *Polymer* **2017**, *125*, 303–329. [CrossRef]
16. Cha, C.G.; Yu, Y.G.; Seo, H.B.; Kim, M.J.; Grubbs, R.H.; Lee, J.S. Experimental Formulation of Photonic Crystal Properties for Hierarchically Self-Assembled POSS–Bottlebrush Block Copolymers. *Macromolecules* **2018**, *51*, 3458–3466. [CrossRef]
17. Li, Z.; Fu, Y.; Li, Z.; Nan, N.; Zhu, Y.; Li, Y. Froth flotation giant surfactants. *Polymer* **2019**, *162*, 58–62. [CrossRef]
18. Blanco, I.; Bottino, F.A.; Cicala, G.; Latteri, A.; Recca, A. Synthesis and characterization of differently substituted phenyl hepta isobutyl-Polyhedral oligomeric silsesquioxane/polystyrene nanocomposites. *Polym. Compos.* **2014**, *35*, 151. [CrossRef]
19. Blanco, I.; Bottino, F.A.; Abate, L. Influence of n-Alkyl substituents on the thermal behaviour of Polyhedral Oligomeric Silsesquioxanes (POSSs) with different cage's periphery. *Thermochim. Acta* **2016**, *623*, 50. [CrossRef]
20. Blanco, I.; Abate, L.; Bottino, F.A. Mono substituted octaphenyl POSSs: The effects of substituents on thermal properties and solubility. *Thermochim. Acta* **2017**, *655*, 117. [CrossRef]
21. Blanco, I. The Rediscovery of POSS: A Molecule Rather than a Filler. *Polymers* **2018**, *10*, 904. [CrossRef] [PubMed]
22. Lichtenhan, J.D. Polyhedral Oligomeric Silsesquioxanes: Building Blocks for Silsesquioxane-Based Polymers and Hybrid Materials. *Comments Inorg. Chem.* **1995**, *17*, 115–130. [CrossRef]
23. Haddad, T.S.; Lichtenhan, J.D. Hybrid organic-Inorganic thermoplastics: Styryl-Based polyhedral oligomeric silsesquioxane polymers. *Macromolecules* **1996**, *29*, 7302–7304. [CrossRef]
24. Kalia, S.; Pielichowski, K. Polymer/POSS Nanocomposites and Hybrid Materials. In *Preparation, Properties, Applications*; Springer Nature: Basel, Switzerland, 2019.
25. Wei, W.; Zhang, Y.; Liu, M.; Zhang, Y.; Yin, Y.; Gutowski, W.S.; Deng, P.; Zheng, C. Improving the Damping Properties of Nanocomposites by Monodispersed Hybrid POSS Nanoparticles: Preparation and Mechanisms. *Polymers* **2019**, *11*, 647. [CrossRef] [PubMed]
26. Blanco, I.; Abate, L.; Bottino, F.A.; Bottino, P.; Chiacchio, M.A. Thermal degradation of differently substituted Cyclopentyl Polyhedral Oligomeric Silsesquioxane (CP-POSS) nanoparticles. *J. Therm. Anal. Calorim.* **2012**, *107*, 1083–1091. [CrossRef]
27. Blanco, I.; Abate, L.; Bottino, F.A.; Bottino, P. Hepta isobutyl polyhedral oligomeric silsesquioxanes (hib-POSS): A thermal degradation study. *J. Therm. Anal. Calorim.* **2012**, *108*, 807–815. [CrossRef]

28. Puglisi, C.; Sturiale, L.; Montaudo, G. Thermal Decomposition Processes in Aromatic Polycarbonates Investigated by Mass Spectrometry. *Macromolecules* **1999**, *327*, 2194–2203. [CrossRef]
29. Deshpande, G.; Rezac, M.E. The effect of phenyl content on the degradation of poly(dimethyl diphenyl) siloxane copolymers. *Polym. Degrad. Stabil.* **2001**, *74*, 363–370. [CrossRef]
30. Fina, A.; Tabuani, D.; Carniato, F.; Frache, A.; Boccaleri, E.; Camino, G. Polyhedral oligomeric silsesquioxanes (POSS) thermal degradation. *Thermochim. Acta* **2006**, *440*, 36. [CrossRef]
31. Shea, K.J.; Loy, D.A. Bridged polysilsesquioxanes. Molecular-Engineered hybrid organic–Inorganic materials. *Chem. Mater.* **2001**, *13*, 3306–3319. [CrossRef]
32. Araki, H.; Naka, K. Syntheses and properties of dumbbell-Shaped POSS derivatives linked by luminescent π-conjugated units. *J. Polym. Sci. Part A* **2012**, *50*, 4170–4181. [CrossRef]
33. Araki, H.; Naka, K. Syntheses and properties of star- and dumbbell-shaped POSS derivatives containing isobutyl groups. *Polym. J.* **2012**, *44*, 340–346. [CrossRef]
34. Hunks, W.J.; Ozin, G.A. Periodic mesoporous phenylenesilicas with ether or sulfide hinge groups—A new class of PMOs with ligand channels. *Chem. Commun.* **2004**, *21*, 2426–2427. [CrossRef] [PubMed]
35. Blanco, I.; Abate, L.; Bottino, F.A. Synthesis and thermal behaviour of phenyl-substituted POSSs linked by aliphatic and aromatic bridges. *J. Therm. Anal. Calorim.* **2018**, *131*, 843. [CrossRef]
36. Badea, E.; Blanco, I.; Della Gatta, G. Fusion and solid-To-Solid transitions of a homologous series of alkane α, ω dinitriles. *J. Chem. Thermodyn.* **2007**, *39*, 1392–1398. [CrossRef]
37. Della Gatta, G.; Richardson, M.J.; Sarge, S.M.; Stølen, S. Standards, calibration, and guidelines in microcalorimetry. Part 2. Calibration standards for differential scanning calorimetry (IUPAC Technical Report). *Pure Appl. Chem.* **2006**, *78*, 1455–1476. [CrossRef]
38. Abate, L.; Blanco, I.; Cicala, G.; Recca, G.; Scamporrino, A. The influence of chain-Ends on the thermal and rheological properties of some 40/60 PES/PEES copolymers. *Polym. Eng. Sci.* **2009**, *49*, 1477. [CrossRef]
39. Vyazovkin, S.; Chrissafis, K.; Di Lorenzo, M.L.; Koga, N.; Pijolat, M.; Roduit, B.; Sbirrazzuoli, N.; Suñol, J.J. ICTAC Kinetics Committee Recommendations for Collecting Experimental Thermal Analysis Data for Kinetic Computations. *Thermochim. Acta* **2014**, *590*, 1. [CrossRef]
40. Blanco, I.; Abate, L.; Bottino, F.A.; Chiacchio, M.A. Synthesis and Thermal Behaviour of Novel Aliphatic/Aromatic Hepta-Cyclopentyl Bridged Polyhedral Oligomeric Silsesquioxanes (POSSs)/Polystyrene (PS) Nanocomposites. *J. Inorg. Organomet. Polym. Mater.* **2015**, *25*, 1456. [CrossRef]
41. Raftopoulos, K.N.; Pielichowski, K. Segmental dynamics in hybrid polymer/POSS nanomaterials. *Prog. Polym. Sci.* **2016**, *52*, 36. [CrossRef]
42. Blanco, I.; Bottino, F.A. Thermal characterization of a series of novel hepta cyclopentyl bridged POSS/PS nanocomposites. *J. Therm. Anal. Calorim.* **2016**, *125*, 637. [CrossRef]

polymers

MDPI

Article

Reactions of Soy Flour and Soy Protein by Non-Volatile Aldehydes Generation by Specific Oxidation

Charles R. Frihart [1], Antonio Pizzi [2,3,*], Xuedong Xi [2] and Linda F. Lorenz [1]

1 Forest Products Laboratory, USDA, 1 Gifford Pinchot Ave., Madison, WI 53726, USA
2 LERMAB-ENSTIB, University of Lorraine, 27 rue Philippe Seguin, 88000 Epinal, France
3 Department of Physics, King Abdulaziz University, Al Ehtifalat St, Jeddah 21589, Saudi Arabia
* Correspondence: antonio.pizzi@univ-lorraine.fr; Tel.: +33-623126940

Received: 30 August 2019; Accepted: 7 September 2019; Published: 10 September 2019

Abstract: Soy protein isolate (SPI) and insoluble soy flour polymeric carbohydrates have been reacted with sodium periodate for the specific oxidation of vicinal –OH groups to investigate the reactions involved in this approach to soy flour adhesives. The reactions have been shown to generate carbohydrate oligomer fractions presenting one, two or multiple aldehyde groups. With the exception of the small molecular weight heptanedial, the smaller molecular weight aldehydes generated from mono- and disaccharides by the same reaction do not appear to form from the insoluble soy flour carbohydrates, or have already reacted. The reaction of periodate with soy protein isolate has been shown to generate some aldehydes too. When the mix of SPI and soy insoluble carbohydrates is treated with periodate, the majority of the observed aldehyde carrying species appear to be higher molecular weight carbohydrate oligomer fractions.

Keywords: soy flour; soy protein isolate; insoluble carbohydrates; periodate oxidation; aldehydes generation; condensation reactions; soy adhesives

---

1. Introduction

Recently, to meet new environmental standards, a number of different approaches to the use of natural materials have been used to prepare wood adhesives without the use of formaldehyde. The use of carbohydrate oxidants is a long well-known practice for a number of different applications. Thus, a number of different oxidation systems have been used to generate aldehydes from polymeric carbohydrate, the literature on this being rather abundant. These include sulfoxide-carbodiimide and related methods [1], nitroxyl radical mediated aqueous oxidation [2], special salts oxidation [3] and more recently nitro-oxidation methods [4,5], this latter being well suited for nanocellulose generation. Among these, the periodate specific oxidation of oligomeric carbohydrates is a well-known system. Thus, such a system is reported from the older periodate oxidation literature on aldehyde generation [6–8] to the more recent literature on nanocellulose generation by such a system followed by ozonization [9]. Among the more recent applications, the use of specific oxidants of monomeric or oligomeric carbohydrates in soy flour have shown particularly interesting results, allowing not only the protein fraction but also the carbohydrate fraction to participate positively in the preparation of wood panel adhesives [10–12]. In this approach, soy flour adhesives have been successfully prepared by treating soy flour with either potassium permanganate or sodium periodate [12]. Sodium periodate is a specific oxidant for carbohydrates reacting with adjacent vicinal hydroxyl groups to form dialdehydes [6,12] according to the reaction:

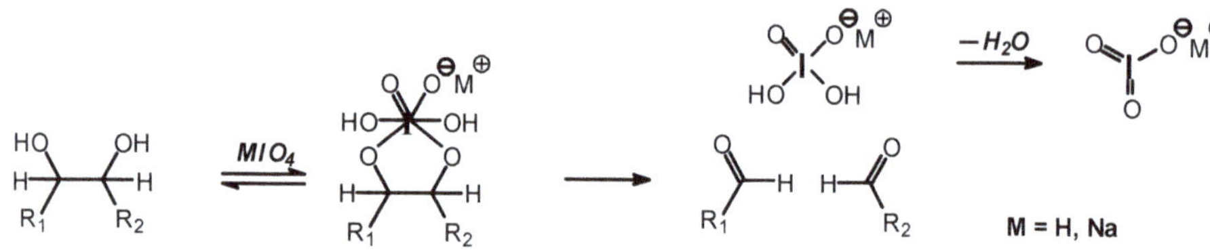

While this reaction and outcome are better known for carbohydrate monomers [6,13] and dimers [13] such as glucose and sucrose, equivalent reactions are known also for higher carbohydrate oligomers up to cellulose itself [12,14,15]. In the case of cellulose and long carbohydrate oligomers alone, the reactions that have been shown to occur are the condensation of the aldehydes formed with other carbohydrate chains to yield crosslinking, leading to solid panels in the case of cellulose [15]. By increasing the level of oxidation with further periodate in the presence of an aldehyde-reactive species such as a soy protein [12], a flavonoid tannin [13] or other reactive species, including lignin, cross-linking can also occur, leading to feasible wood adhesives [12,13].

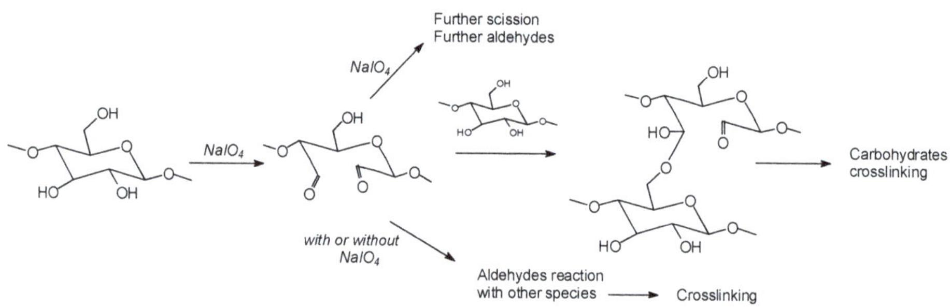

The oxidation with the periodate ion, resulting in a 1,2–glycol scission, is one of the most widely used reactions in carbohydrate chemistry. The mild reaction and the aqueous solvent conditions for periodate oxidation are particularly apt for use with water-soluble carbohydrates. The development and wide application of the reaction are due to its high degree of selectivity [6,14,15].

Aldehydes are well known for reacting with the amino group on proteins [10–12]. The aldehydes generated from the carbohydrate fraction of soy flour appear to react with the active sites of the soy protein, leading to cross-linking and yielding adhesives bonding plywood panels that are very encouraging for shear bond strengths, especially under wet testing conditions [12]. Equally, some of the aldehydes generated by the periodate treatment of glucose and sucrose also react with a polyflavonoid tannin, yielding equally encouraging plywood bond strengths [13].

In the case of Na periodate acting on glucose at 120 °C for 1 h, a number of different aldehydes have been shown to be generated, for example:

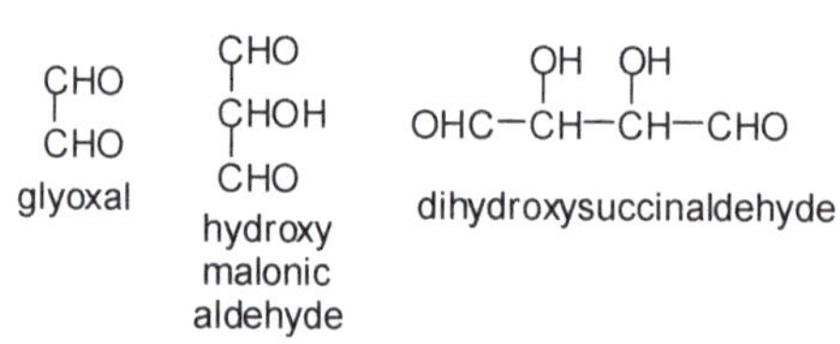

However, other aldehydes, not only from the oxidative cleavage of the carbohydrates but also via different mechanisms following the cleavage, have been shown to form [13]. The first class of these are given by the recombination of the dialdehydes obtained from the cleavage by aldol condensation [13]. Aldol condensation can occur under acid or alkaline conditions [16]. Thus, due possibly to the acid environment induced by periodate, the aldol condensation reaction starting from the glyoxal formed is for example:

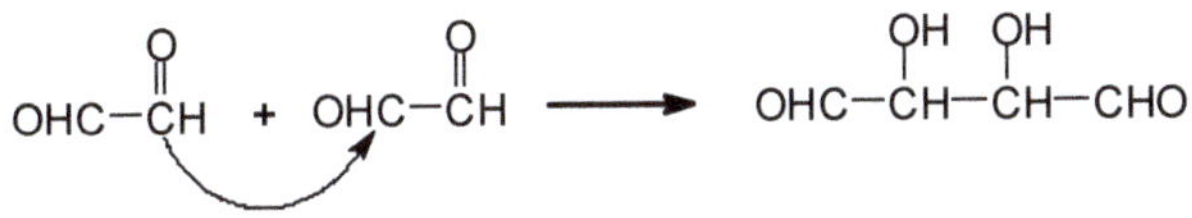

Additionally, the following species have been found to be formed via aldol condensation [13]:

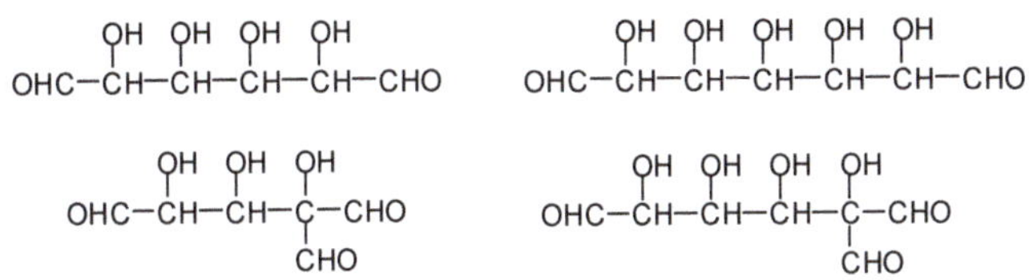

A third type of reaction has also been shown to occur due to water elimination between two aldehyde groups [13]. Aldehydes in water are in general present as hemiacetals, these being due to the reaction of an aldehyde with an alcohol or water [16]. This is the case, for example, of formaldehyde forming hemiformals in water in the preparation of formaldehyde-based resins [17].

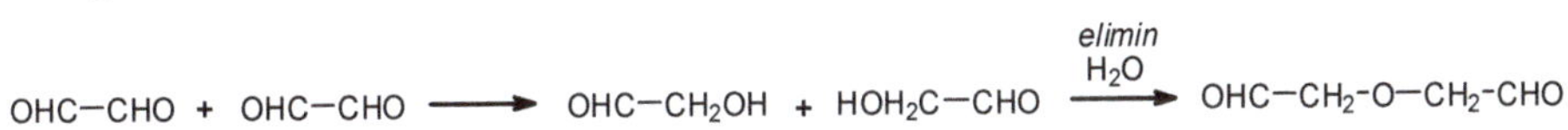

From this reaction of elimination, species such as follows were formed [3].

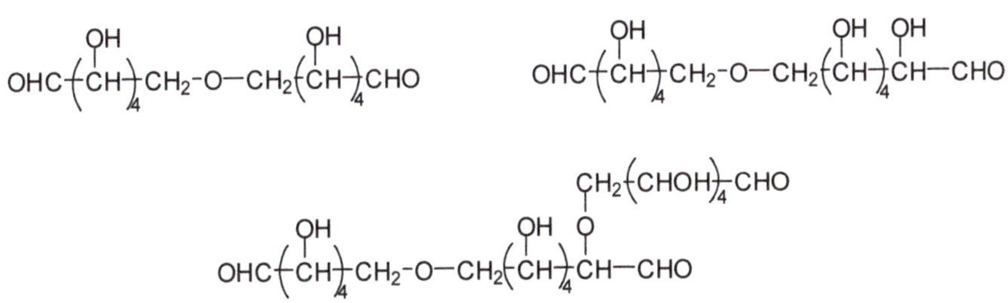

Compounds obtained by the combination of the three reactions outlined above have been shown to occur even at higher molecular weights, but aldehyde groups are always present, thus maintaining the reaction capacity of these compounds with any available reactive sites of other present compounds [13]. The reactivity of the higher molecular weight aldehydes is likely to be relatively lower than that of more active, lower molecular weight aldehydes. It must also be noted that all the aldehydes involved in subsequent reactions also have a very low volatility, eliminating the possibility of VOC emission.

However, in addition to the background given above and the reactions indicated, it has also been deduced from applied bonding results [3] that: (i) the reactions do appear to occur with soy flour, thus on insoluble polymeric carbohydrates and soluble sucrose and its alpha-galactosyl derivatives reacting, after selective oxidation, with its proteic part; and (ii) even in the absence of carbohydrates in soy protein isolates, the protein itself is able, alone, to cross-link once treated with sodium periodate, yielding encouraging plywood results. The research work presented here is then aimed to investigate (i) what aldehydes are formed via the action of periodate on the soy flour insoluble carbohydrates to explain the already experienced cross-linking [12] of periodate-treated soy flour; (ii) what occurs in the reactions of the cleaved polymeric carbohydrates with the soy protein in soy flour; and even more importantly, (iii) what happens when the soy protein isolate is treated with periodate, in the absence of carbohydrates, to make it cross-link.

2. Materials and Methods

2.1. Preparation of Samples for Analysis

In the purification of the soy protein isolate the first step involves dispersing the native flour in water at 10 percent solids, adjusting the pH to 8, centrifuging out the insoluble carbohydrates and then adjusting the pH of the insoluble carbohydrates to 6.5 [12]. (a) 10 g of isolated insoluble soy flour carbohydrates, (b) 10 g of soy protein isolate alone, and (c) 10 g of samples of a $NaIO_4$-treated mix of soy protein isolate + insoluble soy flour carbohydrates + sucrose in relative proportions by a weight of 50:25:25 were all three treated with 1.5 g $NaIO_4$. Controls of 10 g of soy protein isolate + 5% glutaraldehyde and 1% glutaraldehyde by weight were also prepared under the same conditions. The mixtures were placed in an oven at 120 °C for 1 h, then cooled.

2.2. Matrix Assisted Laser Desorption Ionization (MALDI-TOF) Mass Spectrometry

All samples for the matrix assisted laser desorption ionization time-of-flight (MALDI-TOF) analysis were prepared by first dissolving the (a) 5 mg samples of $NaIO_4$-treated insoluble soy flour carbohydrates, (b) 5 mg samples of $NaIO_4$-treated soy protein flour, or (c) 5 mg samples of $NaIO_4$-treated soy protein isolate alone, in 1 mL of a 50:50 *v/v* acetone/water solution. Then, 10 mg of this solution was added to 10 µL of a 2,5–dihydroxy benzoic acid (DHB) matrix. The locations dedicated to the samples on the analysis plaque were first covered with 2 µL of a NaCl solution 0.1 M in 2:1 *v/v* methanol/water, and pre-dried. Then, 1 µL of the sample solution was placed on its dedicated location, and the plaque was dried again. MALDI-TOF spectra were obtained using an Axima-Performance mass spectrometer from Shimadzu Biotech (Kratos Analytical Shimadzu Europe Ltd., Manchester, UK) using a linear polarity-positive tuning mode. The measurements were carried out making 1000 profiles per sample with 2 shots accumulated per profile. The spectrum precision is of ±1 Da.

3. Results and Discussion

In the case of the periodate treatment of the insoluble polymeric carbohydrates of soy flour, the cleavage of C–C bonds with vicinal –OH groups does yield aldehydes, but the peaks of the smaller aldehydes, such as glyoxal, hydroxymalonic dialdehyde and dihydroxy succinaldehyde as obtained for glucose [13] and in some cases for sucrose, are not observed, either because they are not formed (most likely due to the polymeric nature of the insoluble carbohydrates), because they are lost due to their volatile nature, or because they have already reacted in some manner. One peak already observed for glucose, given by the recombination of small dialdehydes species by aldol condensation, is nonetheless present. This sole peak, also present in glucose at 231 Da (calc. 231Da), including the Na+ enhancer, appears to belong to the action of 15% sodium periodate at 120 °C on the insoluble carbohydrates. It has already been reported for monomeric carbohydrates [13], and has the following structure:

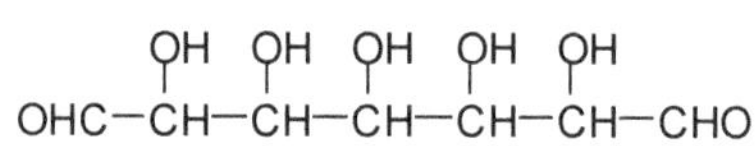

This peak, corresponding to a pentahydroxy heptanedial, is very marked when the insoluble carbohydrates are treated with periodate at 120 °C, while it is almost absent, showing just traces, when the insoluble carbohydrates alone are heated for the same period at 120 °C. It is absent at 20 °C.

The other aldehydes, found when glucose and sucrose are treated with periodate as a consequence of the other reactions outlined in the introduction, are not found for soy flour insoluble polymeric carbohydrates. Different aldehydes of much greater molecular weight are instead formed (Figure 1a–c and Table 1). Table 1 lists all the species formed, whether they carry aldehyde groups or not.

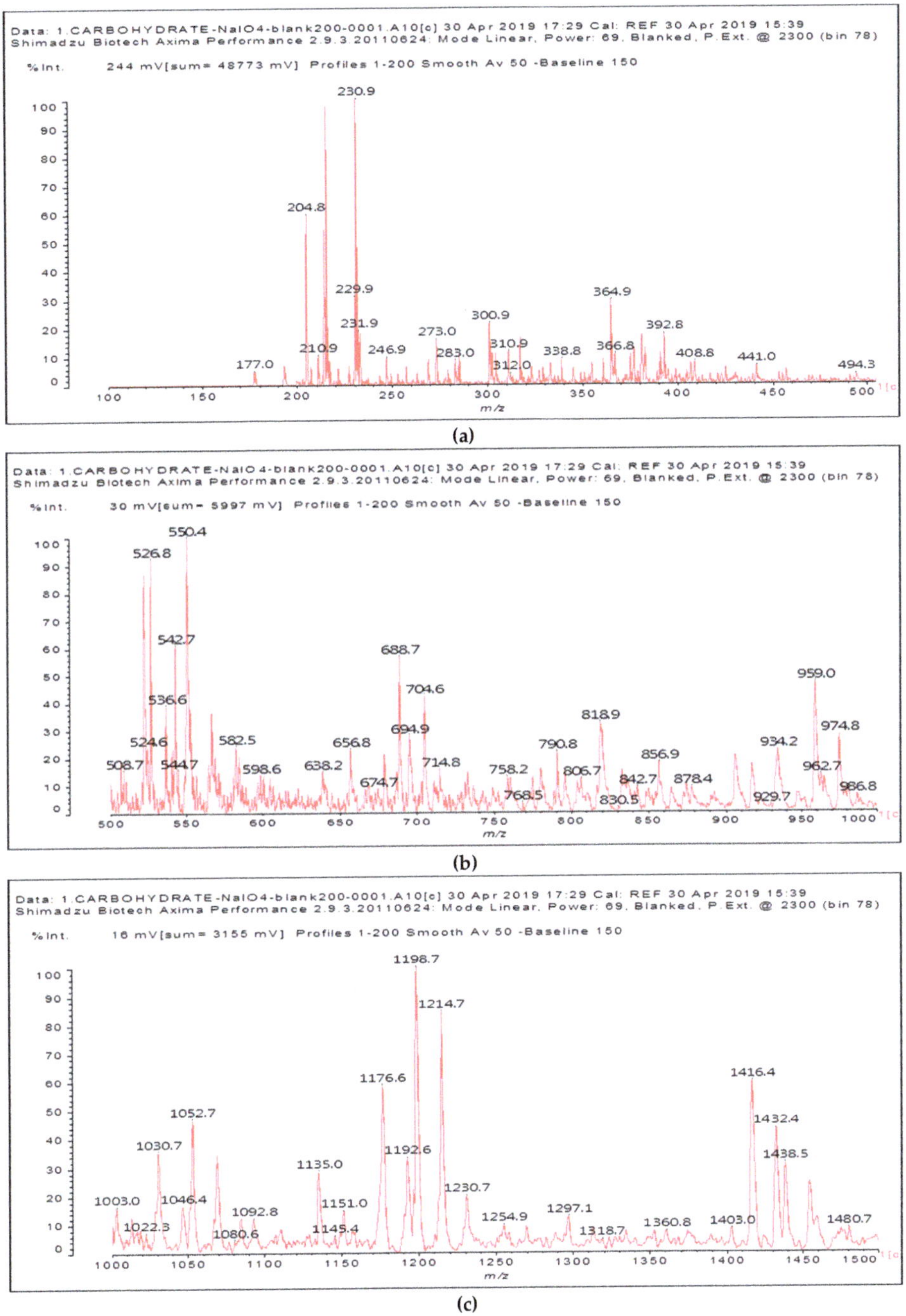

Figure 1. MALDI ToF spectra of insoluble soy flour carbohydrates after a treatment with sodium periodate at 120 °C for 1 h. (**a**) 100 Da to 500 Da range. (**b**) 500 Da to 1000 Da range. (**c**) 1000 Da to 1500 Da range.

In the insoluble carbohydrates treated with sodium peroxide and in the mix proteins + carbohydrates + sucrose + periodate at 120 °C, there are a number of peaks that do not appear in the carbohydrates control at 120 °C without periodate (Supplementary Material). These are listed in Table 1, which lists all the species formed that carry aldehyde groups or not. Among these, several of the species formed after the periodate treatment are noteworthy, as they present one or more aldehyde groups. These are the peaks at 768 Da, 791 Da, 930 Da, 975 Da, 1031 Da, 1046 Da, 1053 Da, 1092 Da, 1135 Da, 1153 Da, 1176 Da, 1192 Da, 1198 Da, 1214 Da, 1230 Da, 1254 Da, 1362 Da, and 1416 Da. All present at least one aldehyde group. Among all of these there are monoaldehyde species obtained by different cleavage sites of one glucose, such as the species represented by the peaks at 768 Da, 791 Da, 1254 Da and 1416 Da:

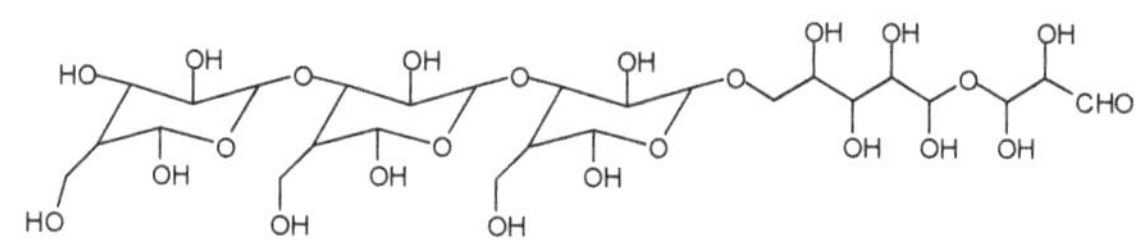

768 Da

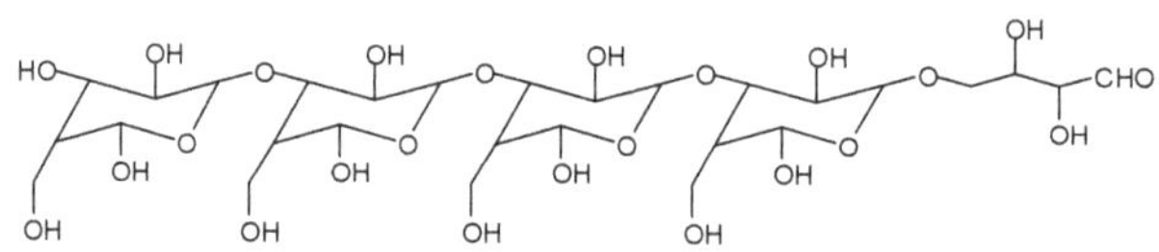

791 Da

There are species presenting two aldehyde groups due to the specific oxidation of two glucoses by the periodate, such as those represented by the peaks at 1031 Da and 1053 Da:

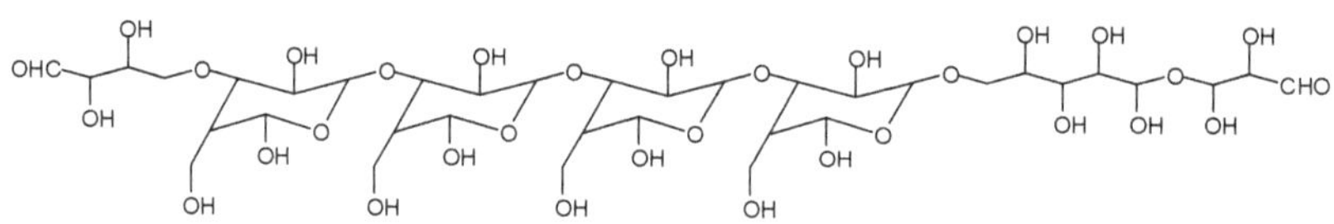

1031 Da

Finally, species presenting a multitude of aldehyde groups, such as several glucose pyran rings in the carbohydrate oligomer, have been cleaved by periodate. Two examples of these are the species represented by the peaks at 1214 Da and 1230 Da:

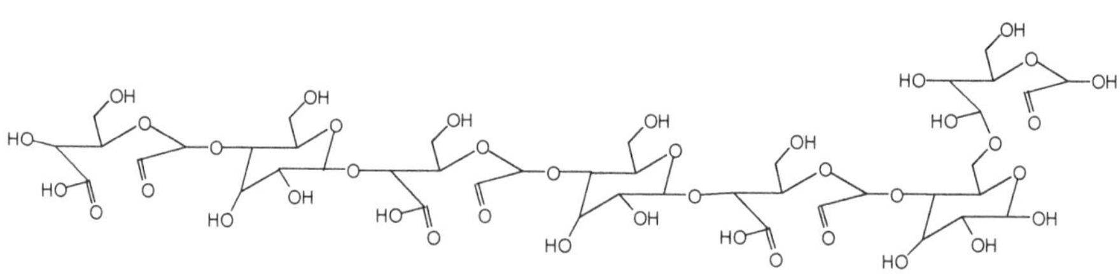

1214 Da

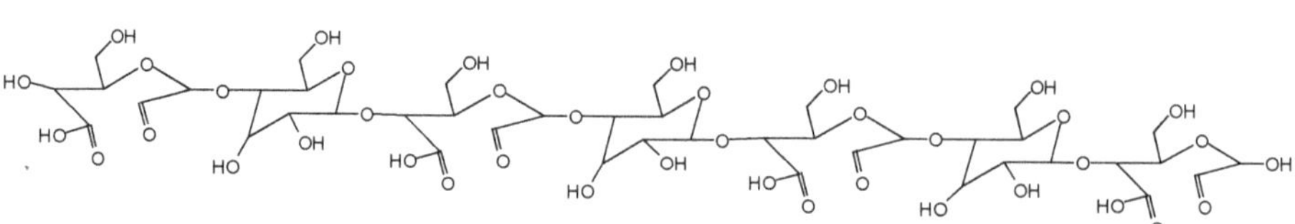

1230 Da

Both species confirm the reactions brought about on cellulose and other carbohydrate oligomers by previous work [6,7]. In the structures in Table 1 for the 1214 Da and 1230 Da species, the open and oxidized forms of glucose can be placed anywhere and alternate in any manner in relation to the non-oxidized glucose closed forms. Thus, the above structures and their structures in Table 1 represent just one of the possible isomers. The 1230 Da peak is separated by 16 Da from the 1214 Da peak and by 32 Da from the 1198 Da peak. This means that these two species have respectively 1 and 2 –O-atoms less than the 1230 Da one. These can have been lost in the MALDI analysis; carbohydrate chains are known to do this [18], but it could also be due to similar but slightly different species.

The marked 1198 Da peak could be a case of 6 open units linked together: 176 × 6 = 1032 + 162 units =1194 Da. It is the nearest to 1198 Da. One could also imagine that it could be a peak obtained by reactions of the small aldehydes such as glyoxal. The reactions of an aldehyde with an alcohol yields reversible hemiacetals under acid conditions. However, better and more accurate explanations are possible. Effectively, the group of peaks 1198 Da, 1214 Da, and 1230 Da are of interest due to their relative height in the high molecular weights part of the spectra in Figure 1. Of these, the 1230 Da and 1214 Da ones are interesting due to their multiple aldehyde groups. The 1198 Da species appears to be the same as the 1214 Da or the 1230 Da ones, having lost just one or two –OH groups in the MALDI analysis.

There are some peculiarities too. The peak at 1092 Da is also present in the SPI treated at 120 °C without periodate. Thus, one cannot be sure that the interpretation as an aldehyde carrying carbohydrate oligomer is correct, and a doubt must logically persist.

Monoaldehydes of higher molecular weights are also noted, such as the one at 1416 Da:

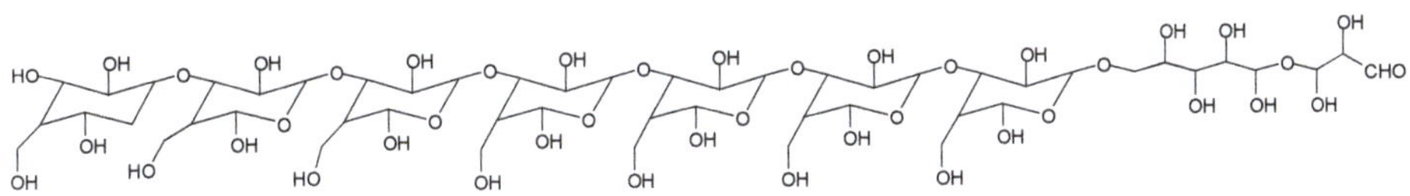

1416 Da

This indicates that even higher molecular weight oligomers containing aldehyde groups are likely to be generated the longer the periodate treatment is.

All of the above information indicates that the specific oxidation of the periodate appears to be able to also occur with polymeric carbohydrates, also generating aldehyde groups in high molecular weight fragments of carbohydrate chains. Consequently, they are these high molecular weight carbohydrate oligomers carrying periodate-generated aldehyde groups and the heptanedial aldehyde, which contribute to the applied bonding results with the proteic part of soy flour already reported [12]. In particular, it must be noted that species presenting a number of active aldehyde groups such as those represented by the 1230 Da and 1214 Da peaks can be a centre of tridimensional networking by a reaction with the protein, but also (as already demonstrated) by a reaction with other present carbohydrate chains [14,15].

Another interesting curiosity are the peaks at 975 Da (Calc. 976 Da) and 1135 Da (Calc. 1137 Da). The species represented by these two peaks are associated with 2 Na^+, the 975 Da corresponding to the 930 Da peak + 2xNa^+, and the 1135 Da peak corresponding to 1092 Da + 2xNa^+. The 959 Da peak is the same as the 975 Da but has lost one –OH group in the MALDI analysis. The species containing two Na^+ in the MALDI analysis are relatively unusual but are not rare and have also been reported in the previous literature [19].

Table 1. The proposed oligomer species for the MALDI ToF spectra peaks in Figure 1a–c for the reaction at 120 °C for 1 h of sodium periodate with insoluble soy flour carbohydrates. Note: the links between the glucose units are alpha-glucosidic. They are represented as being beta glucosidic to shorten the table.

231 Da (with Na^+)
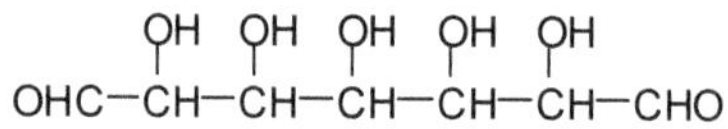
365 Da (calc. 365 Da) (with Na^+)
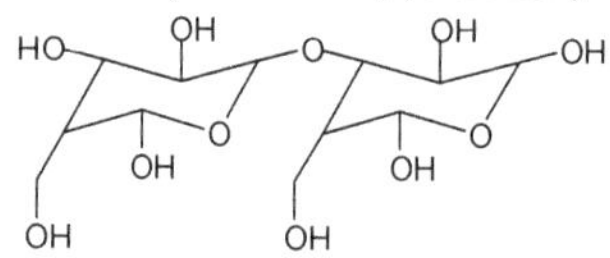
527 Da (calc. 527 Da) (with Na^+)
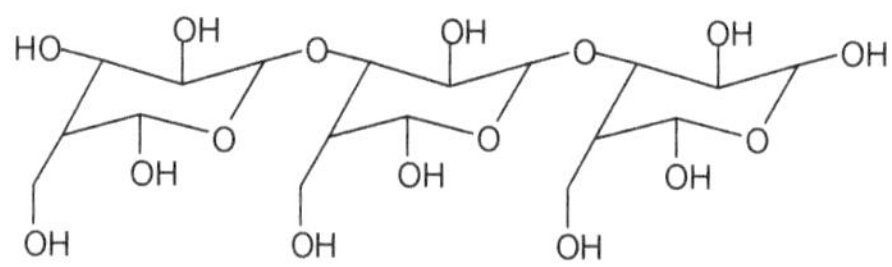
768 Da (calc.766 Da) (with Na^+)
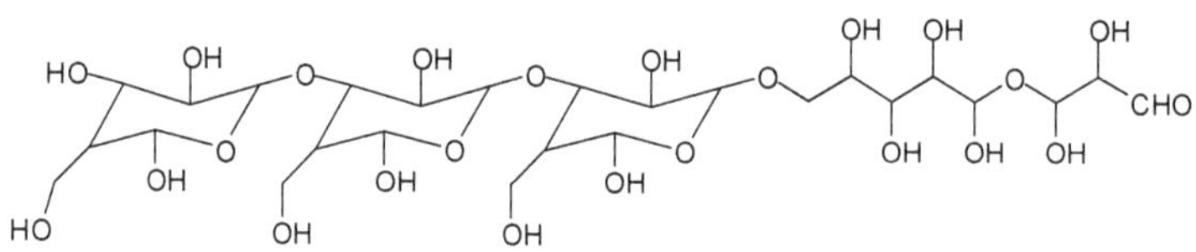
689 Da (calc. 689 Da) (with Na^+)
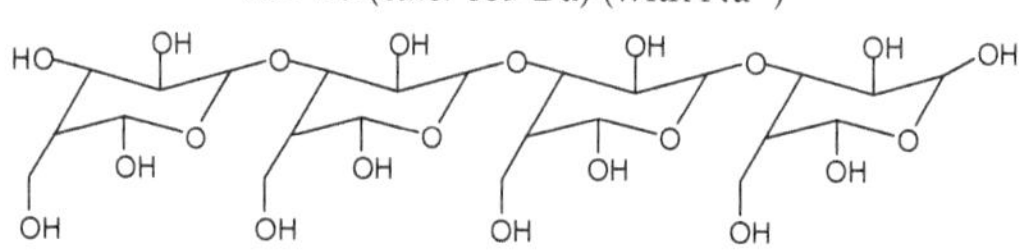
768 Da (calc. 768 Da) (no Na^+), 791 Da (calc. 791 Da) (with Na^+)
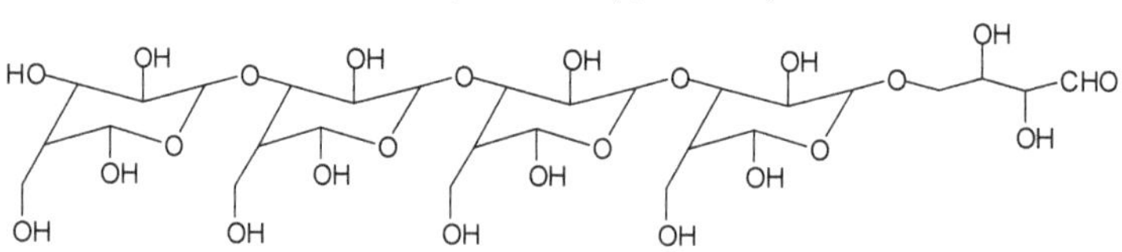
930 Da (calc. 928 Da) (with Na^+)
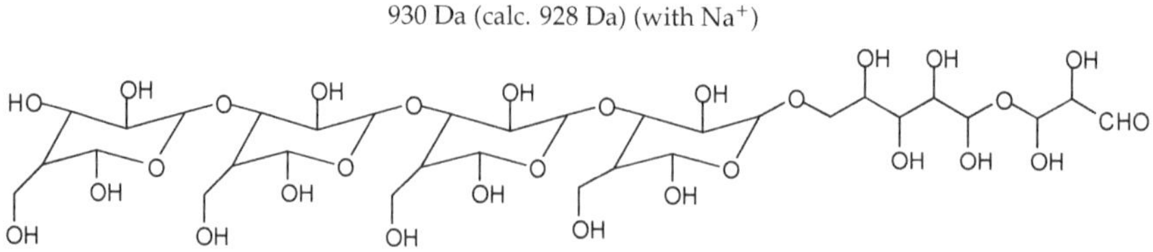
930 Da (calc. 930 Da) (no Na^+)
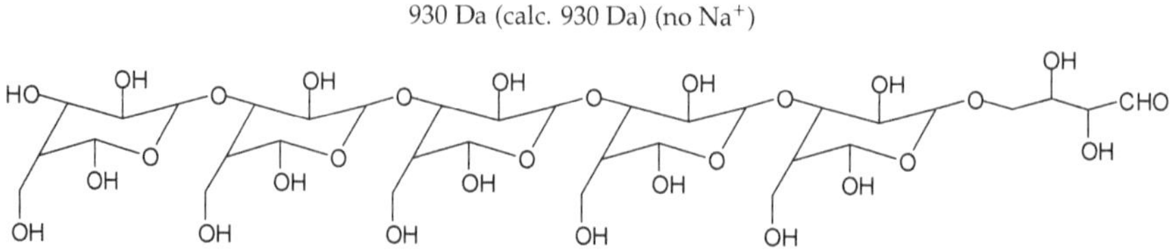

Table 1. *Cont.*

975 Da (calc. 976 Da) (= 930 Da but with 2xNa^+)
1031 Da (calc. 1032 Da) (no Na^+)
1053 Da (calc. 1054 Da) (with Na^+)

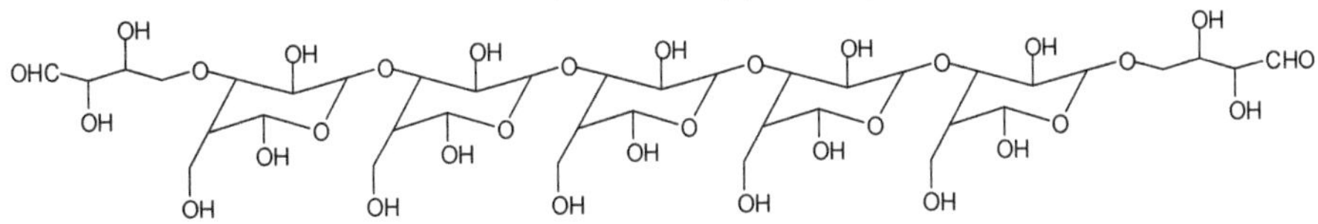

1031 Da (calc. 1029 Da) (with Na^+)

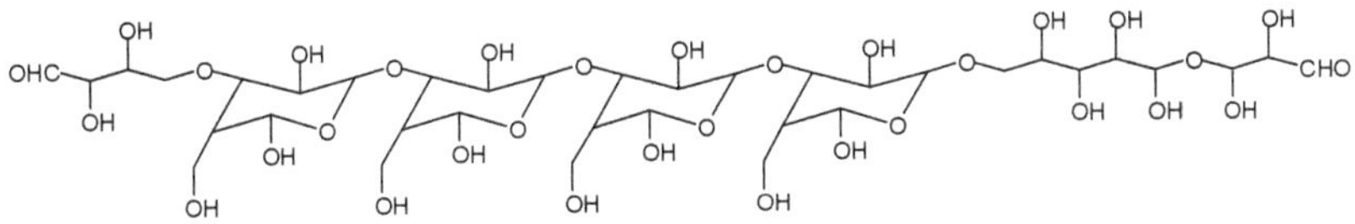

1046 Da (calc. 1043 Da) (with Na^+)

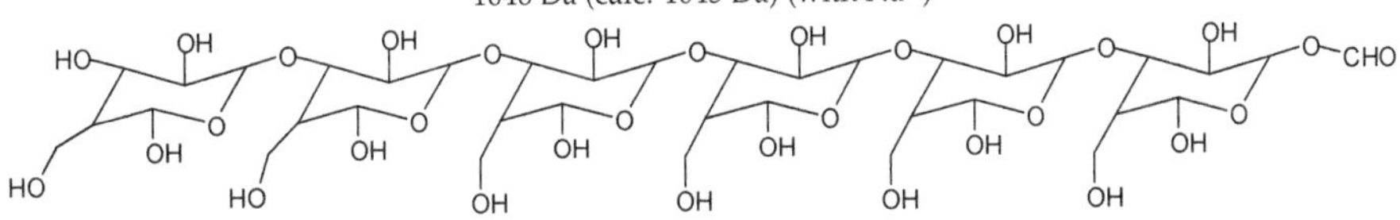

1092 Da (calc.1089 Da) (with Na^+)

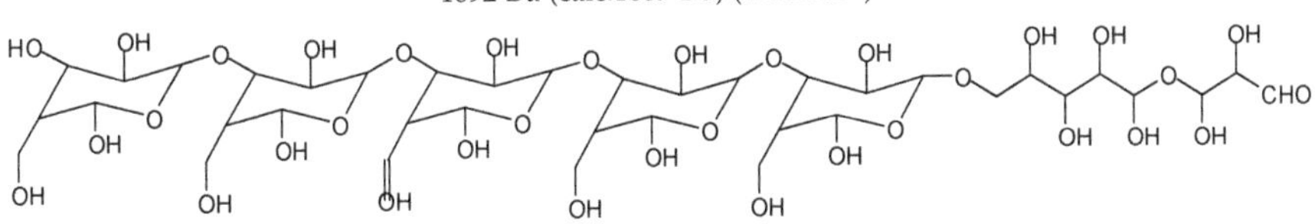

1092 Da (calc.1092 Da) (no Na^+)

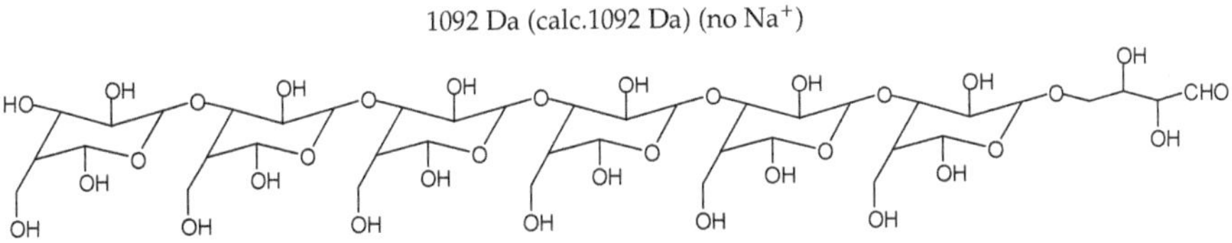

1135 Da (calc. 1136 Da) (= 1092 Da but with 2x Na^+)
1153 Da (calc. 1153 Da) (no Na^+)
1176 Da (calc. 1176 Da) (with Na^+)

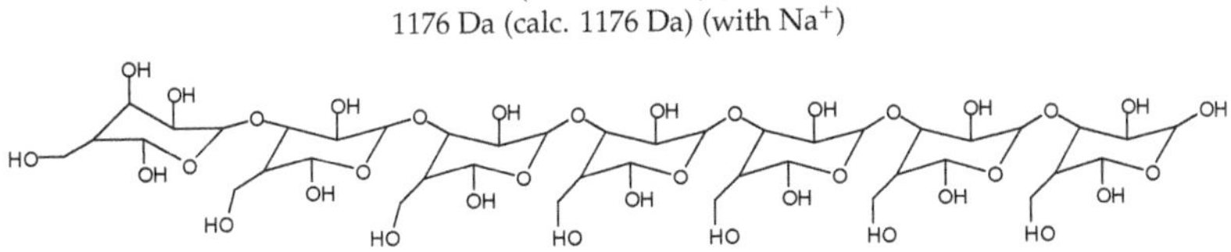

1192 Da (calc. 1192 Da) (with Na^+)

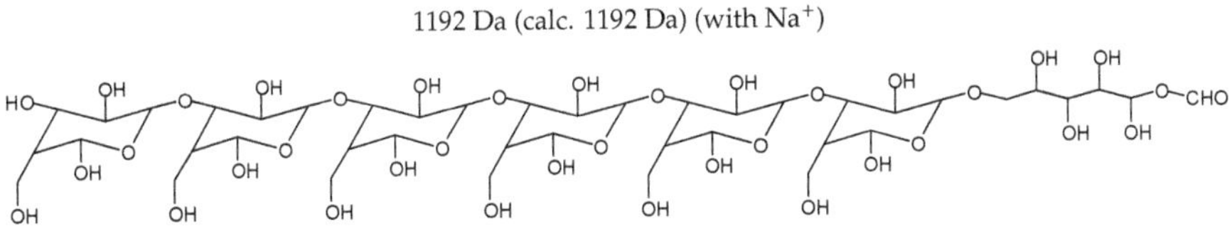

Table 1. *Cont.*

1214 Da (calc. 1212 Da) (no Na$^+$)

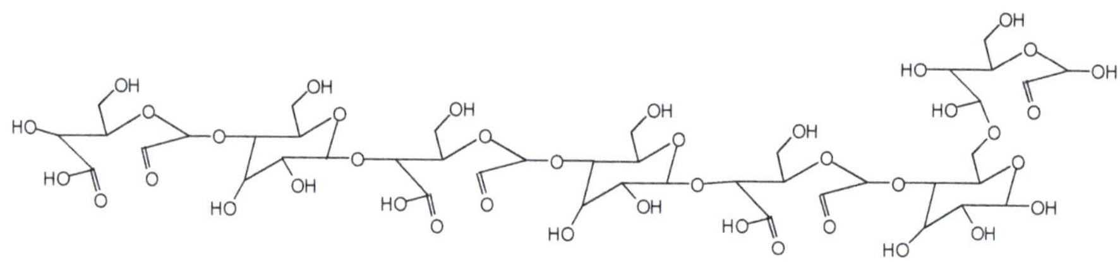

1230 Da (calc. 1231 Da) (with Na$^+$)

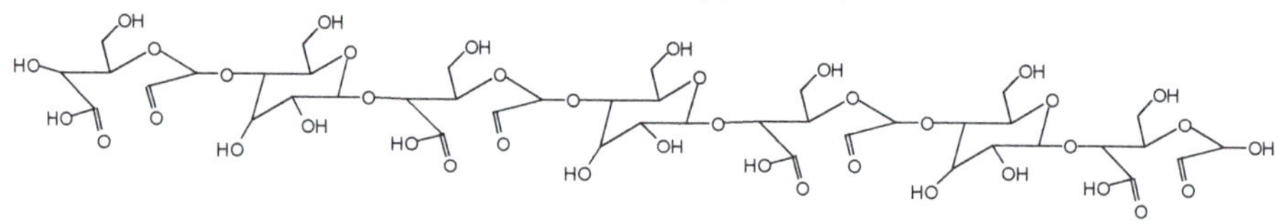

1254 Da (calc. 1253 Da) (with Na$^+$)

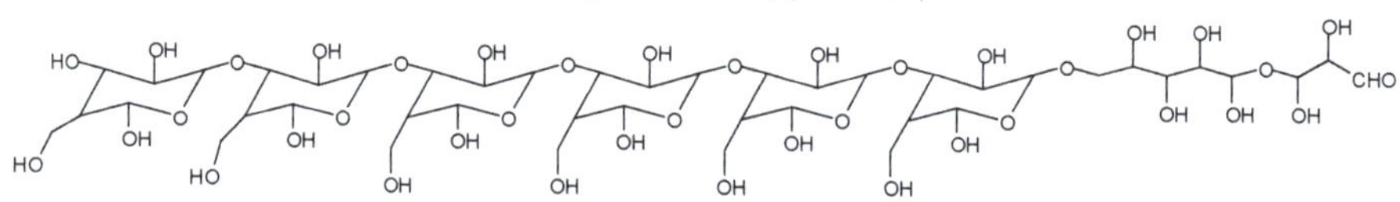

1297 Da = glucose octamer missing one –OH, (no Na$^+$)
1362 Da (calc. 1364 Da) (with Na$^+$)

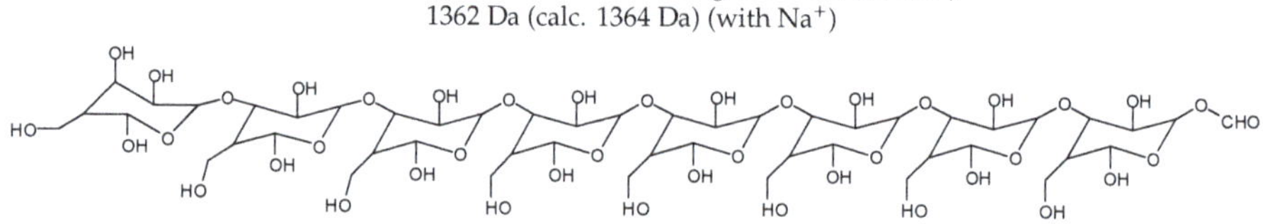

1416 Da (Calc. 1415 Da)

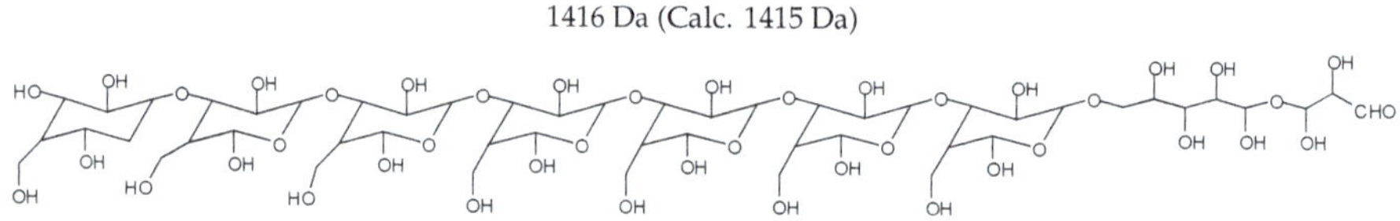

It is more interesting to examine the behavior of the soy protein in the soy protein isolate (SPI) when treated with sodium periodate. Notwithstanding that the amino acids and the peptides involved in these reactions are always rather difficult to identify, the task is facilitated by the fact that the relative abundance of each amino acid in the soy protein is known [20,21]. The results clearly show that the protein is also affected by a periodate treatment. Thus, on observing the peaks of the soy protein isolate + 15% peroxide present in the MALDI analysis as well as of the mix protein + soy insol. carbohydrates + sucrose + periodate, there are peaks clearly coming from the protein itself. This also shows the formation of aldehydes via the periodate oxidation of some sites of the protein itself, thus explaining why the soy protein isolate alone + peroxide also yields good bonding results without the presence of carbohydrates [12]. The main species formed that is coming from the protein are a series of monoaldehydes but also one dialdehyde, as shown in Table 1 and Figure 2a–c, the rest being difficult to assign. The question that must be answered is between which groups does the specific oxidation cleavage occur in the protein primary skeleton, as vicinal –OH carrying carbons are not present. The only possibility is that the peptidic bond itself is cleaved, as its C=O double bond is in reality delocalized with the vicinal –NH, the electronegativity being somewhat similar to that of the vicinal –OH groups.

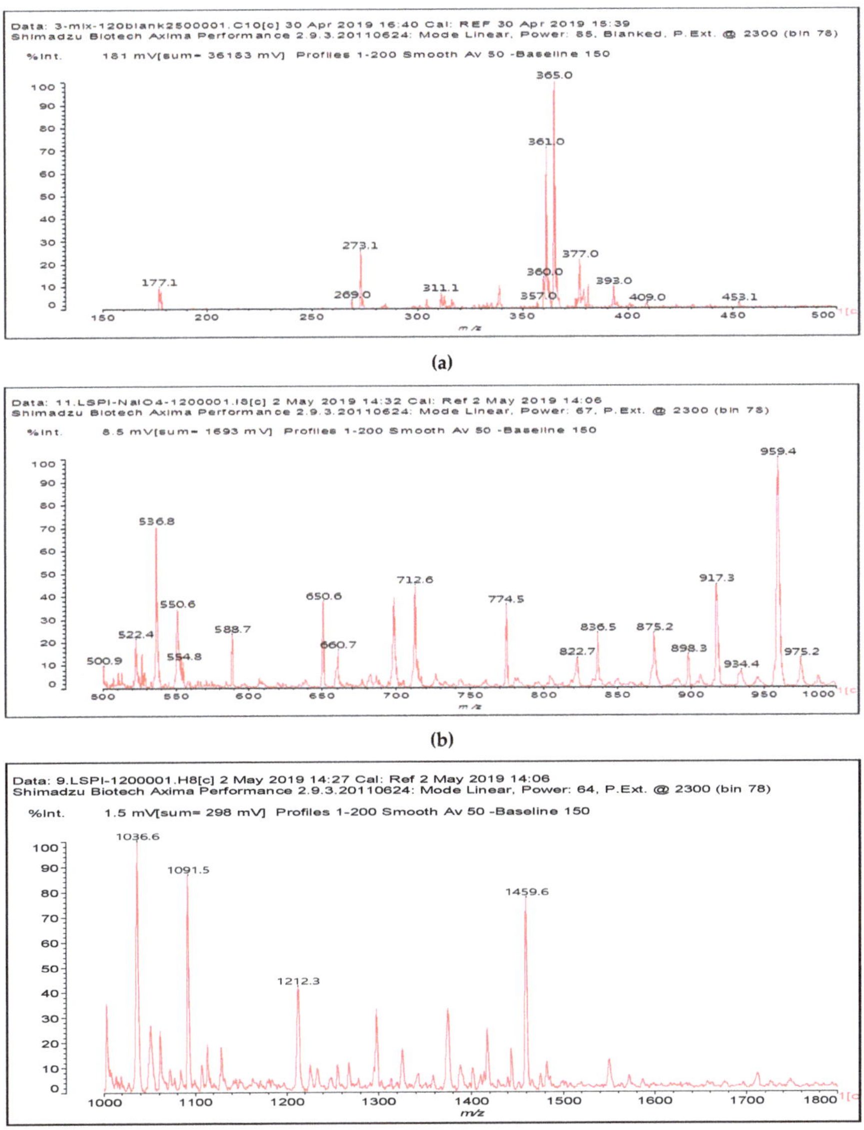

(a)

(b)

(c)

Figure 2. The MALDI ToF spectra of the soy protein isolate after a treatment with sodium periodate at 120 °C for 1 h. (**a**) 100 Da to 500 Da range. (**b**) 500 Da to 1000 Da range. (**c**) 1000 Da to 1500 Da range.

The location of the formed aldehyde groups seems to support such an idea. Table 2 lists the structures of the species identified according to the peaks in Figures 2 and 3. Among these, several monoaldehydes are present, namely those represented by the peaks at 377 Da, 408 Da, 455 Da and the dialdehyde at 659–660 Da, namely a serine–[alanine]5–glutamic acid dialdehyde + Na+ with the following structure:

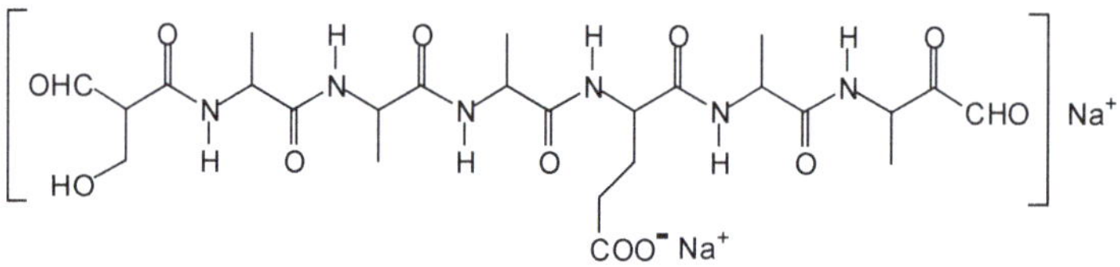

There could be several other aldehydes, but it is difficult to identify from the MALDI peaks the structures formed in the case of a protein.

Table 2. The proposed oligomer species for the MALDI ToF spectra peaks in Figures 2a–c and 3a,b for the reaction at 120 °C for 1 h of sodium periodate with the soy protein isolate.

377 Da = [alanine]$_4$-aspartic acid **monoaldehyde** without Na$^+$
409 Da (calc. 409 Da) = alanine pentamer **monoaldehyde** with Na$^+$

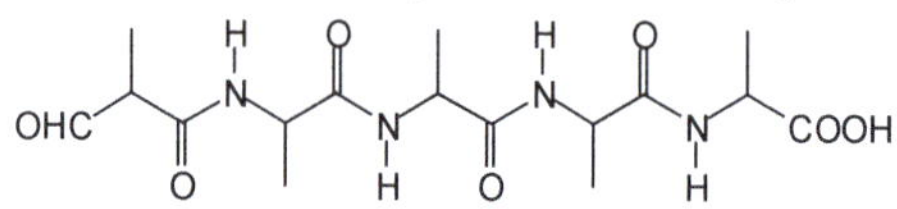

But it could also be (but less likely)
409 Da = tyrosine-triptophane + Na$^+$ (dimer)
455 Da (calc. 457 Da) = alanine hexamer **monoaldehyde** without Na$^+$

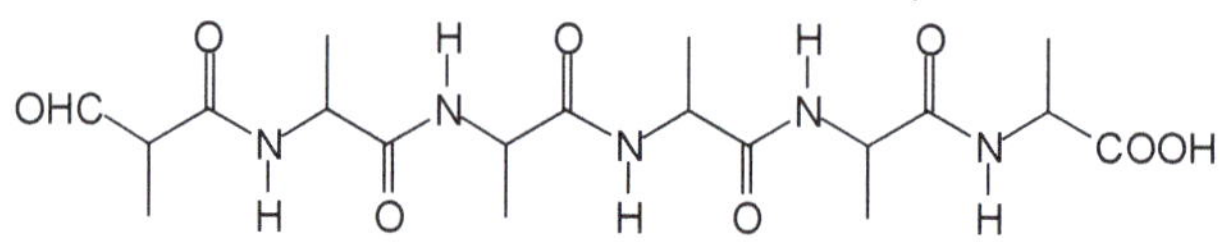

OR455 Da (calc. 457 Da) = [alanine]$_2$-[serine]$_3$ **monoaldehyde** + Na$^+$

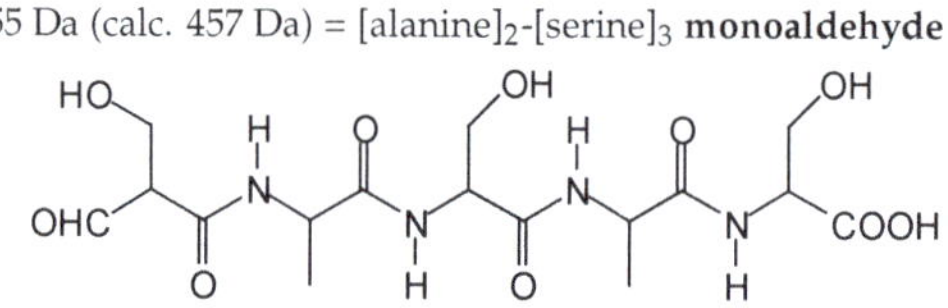

659–660 Da = phenylalanine-triptophan-leucine-histidine + Na$^+$ (tetramer)
659–660 Da = phenylalanine-triptophan-histidine-aspartic acid + Na$^+$ (tetramer)
659–660 Da = serine-[alanine]$_5$-glutamic acid dialdehyde + Na$^+$

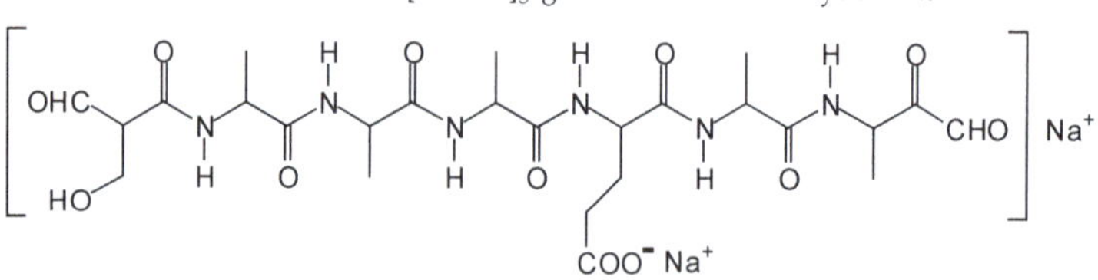

959 Da = phenylalanine-triptophan-leucine-histidine-phenylalanine-arginine + Na$^+$ (hexamer)
959 Da = phenylalanine-triptophan-histidine-aspartic acid-phenylalanine-arginine + Na$^+$ (hexamer)

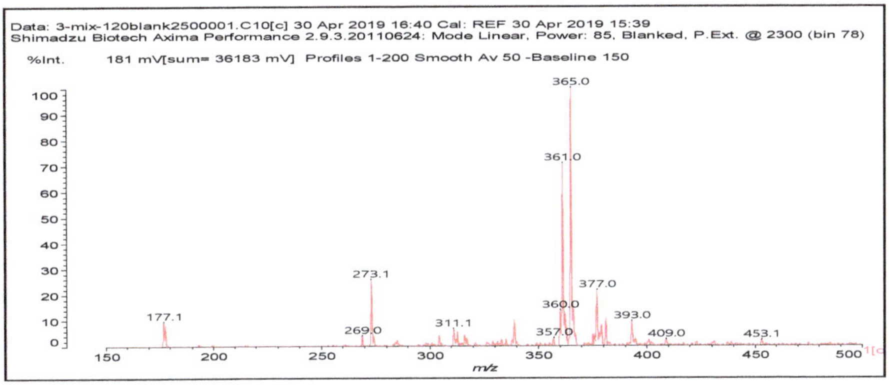

(a)

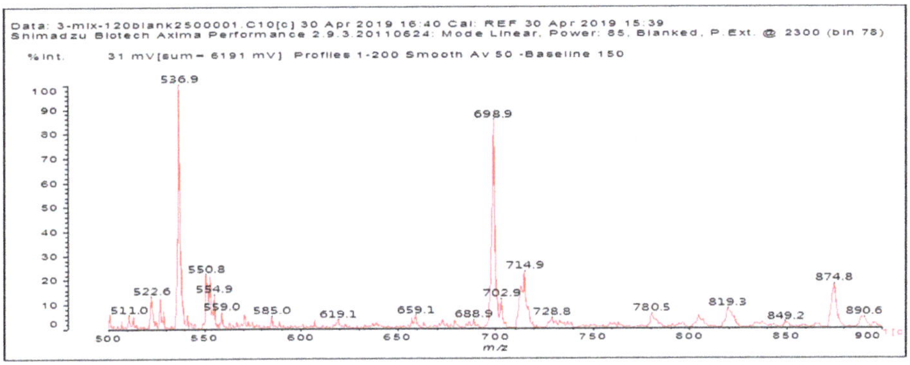

(b)

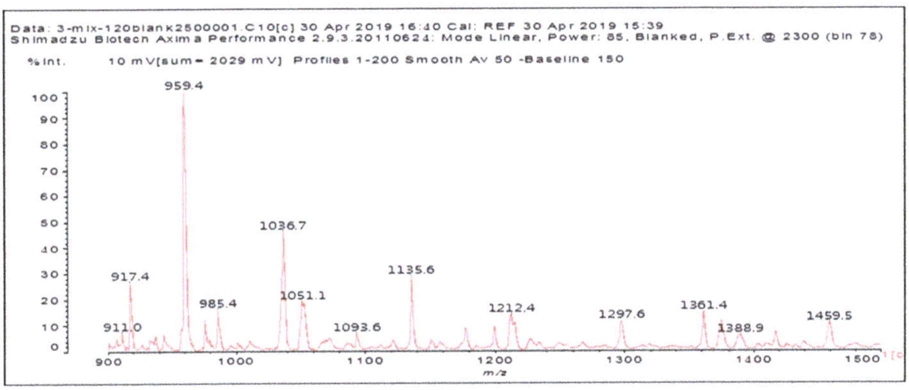

(c)

Figure 3. The MALDI ToF spectra of the reaction of the mix by a weight of 50% soy protein + 25% sucrose + 25% insoluble soy carbohydrates + 15% $NaIO_4$ at 120 °C for 1 h. (**a**) 150 Da to 500 Da range. (**b**) 500 Da to 900 Da range. (**c**) 900 Da to 1500 Da range.

It is of interest to observe the peaks present in the reaction of the reconstituted mix SPI + sucrose + soy insoluble carbohydrates + $NaIO_4$ in Figure 3a–c. Among all the present peaks, the higher molecular weight (hence less reactive) aldehydes generated by sucrose oxidation are at 365 Da, 393 Da, and 523 Da. There are peaks from the protein-generated aldehydes, namely at 377 Da, 409 Da,

453 Da, and 659 Da, and peaks from protein fragments without aldehyde groups, such as at 959 Da. The generated aldehydes by oxidation of the insoluble soy carbohydrates are represented by the peaks at 689 Da, 1036 Da, 1093 Da, 1135 Da, 1212 Da, and 1361 Da and the non-aldehyde fragment at 1297 Da. The DHB matrix enhancer are represented by the peaks at 312 Da, 361 Da, 536 Da, 550 Da and 698 Da. The question is now about which peaks that are present are due to the reaction of an aldehyde, for which it does not matter how they were generated, and which are due to a reaction with amino acids. First of all, it is clear that at a high molecular weight range, many species that are formed will be out of the possible range. Second, the majority of species generated by a reaction with the amino acids of aldehyde-generated species will be found in the middle of the examined molecular weight range. Thus, the co-reacted species are those represented by the peaks that do not belong to any of the single components at 555 Da, 559 Da, 585 Da, 619 Da, 729 Da, 780 Da, 819 Da, 849 Da 890 Da, 1389 Da and 1459 Da. These are due to the reaction of some generated aldehydes with some amino acids, and their assignments are given in Table 3. The more interesting interpretable peak is the one at 585 Da. This is obtained by the reaction of the sucrose-generated tri-aldehyde at 177 Da. While this peak is also present in SPI alone representing just an amino acid, it is also a branched aldehyde generated from sucrose via the aldol condensation of simpler sucrose-generated aldehydes. Thus, the 585 Da peak is due to the reaction of three molecules of the most abundant SPI amino acid, namely glutamic acid, with:

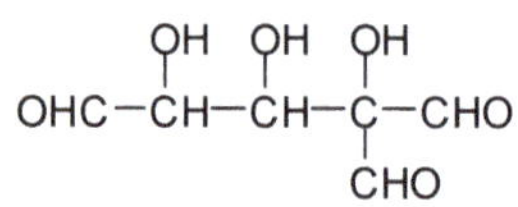

To yield the two alternative structures at 585 Da, one is derived directly by the reaction of three molecules of glutamic acid with the 177 Da tri-aldehyde and the second one is generated via the reaction of three glutamic acid residues with glyoxal and a hydroxyl malonic aldehyde directly generated by the specific oxidation cleavage by the periodate:

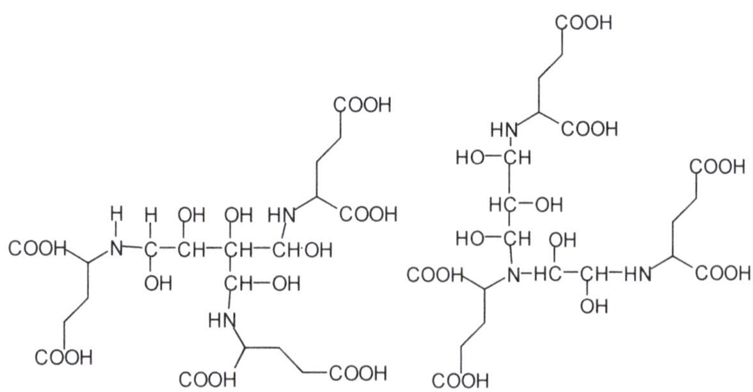

The same peak at 585 Da can also be assigned to an oligomer of:

Lysine-CHOH-CHOH-Lysine-CHOH-CHOH-CHOH-Lysine

As amino acids presenting at least two amino groups can form linear oligomers, the peaks at 554–555 Da, 619 Da, 728 Da, 780 Da, 819 Da, and 849–850 Da can be interpreted as linear oligomers of the reaction of (mainly) lysine and arginine, with glyoxal and hydroxy malonic aldehyde generated by periodate-specific oxidation from simple sugars such as glucose [3] and sucrose, as is indeed the case for one of the assignments for the species at 585 Da (Table 3).

Table 3. Some possible assignments from the MALDI ToF spectra in Figure 3a–c of the reaction of SPI + sucrose + insoluble soy carbohydrates + NaIO4 for peaks not belonging to any of the single periodate-treated constituent components.

554–555 Da, no Na^+
Lysine–CHOH–CHOH–Lysine–CHOH–CHOH–Lysine
585 Da, no Na^+, two possible species from glutamic acid

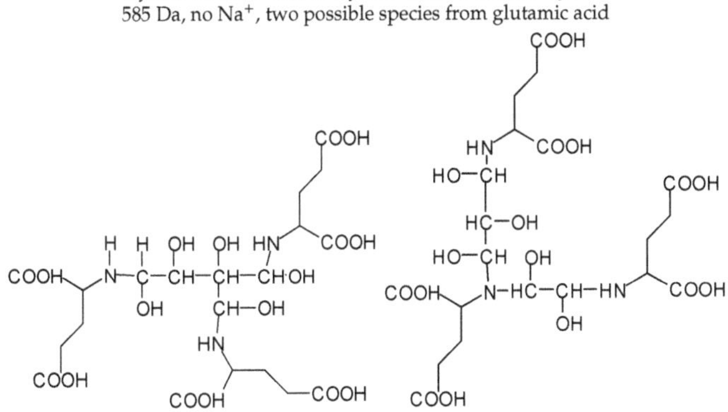

and/or one from Lysine
Lysine–CHOH–CHOH–Lysine–CHOH–CHOH–CHOH–Lysine
619 Da, no Na+
Lysine–CHOH–CHOH–CHOH–Lysine–CHOH–CHOH–CHOH–Lysine
728 Da, no Na+
Lysine–CHOH–CHOH–Lysine–CHOH–CHOH–Lysine–CHOH–CHOH–Proline
780 Da, no Na+
Lysine–CHOH–CHOH–Lysine–CHOH–CHOH–Lysine–CHOH–CHOH–Phenylalanine
819 Da, no Na+
Arginine–CHOH–CHOH–Arginine–CHOH–CHOH–Arginine–CHOH–CHOH–Cysteine
849–850 Da, no Na+
Lysine–CHOH–CHOH–CHOH–Lysine–CHOH–CHOH–CHOH–Lysine–CHOH–CHOH–CHOH–Lysine
and/or
Lysine–CHOH–CHOH–CHOH–Lysine–CHOH–CHOH–CHOH–Lysine–CHOH–CHOH–CHOH–glutamic acid
985 Da, no Na+
Lysine–(–CHOH)$_3$–Lysine–(–CHOH)$_3$–Lysine–(–CHOH)$_3$–Lysine CHOH–CHOH–Glycine
1388–1389 Da
Arginine–(–CHOH)2–Arginine–(–CHOH)2–Arginine–(–CHOH)2–Arginine–(–CHOH)2–Arginine–(–CHOH) 2–Tryptophan
Where the structure is not necessarily linear and can be better expressed as
[Arginine–(–CHOH)$_2$–]$_5$–Tryptophan
1459–1460 Da
[Arginine–(–CHOH)$_2$–]$_5$–Histidine–(–CHOH)$_4$–CHO
or [Arginine–(–CHOH)$_2$–]$_5$–Histidine–(–CHOH)$_2$–Alanine

4. Conclusions

To summarize, in the spectra of the reaction of SPI + sucrose + insoluble carbohydrates + $NaIO_4$, the following peaks ascribed to the action of periodate on SPI and insoluble soy flour carbohydrates are present: 1053 Da, 1092–1093 Da, 1135 Da, 1153 Da, 1198 Da, 1212–1213 Da, 1230 Da, 1254 Da, 1361–1362 Da and 1416 Da coming from the insoluble carbohydrates, and 377 Da, 408–409 Da, 453–455 Da and 659–660 Da coming from the SPI. This shows that some of the carbohydrate aldehyde species are no longer present when the SPI is present; hence, they have reacted with the SPI. Furthermore, the co-reacted species are those represented by the peaks that do not belong to any of the single components, at 555 Da, 559 Da, 585 Da, 619 Da, 729 Da, 780 Da, 819 Da, 849 Da 890 Da, 1389 Da and 1459 Da.

The peak at 231 Da represents just traces (almost non-existent) in the insoluble carbohydrates at 120 °C without periodate, but it is a huge peak in the insoluble carbohydrates with periodate at 120 °C. It does not seem to appear in the SPI + sucrose + insoluble carbohydrates + NaIO4. It could be that with the protein present, the aldehyde groups of the species 231 Da have reacted very readily with SPI. It is one of the best indications that the carbohydrates have not only formed an aldehyde (double here) but also that this reacts with the SPI.

The following peaks coming from the periodate cleavage of the protein are present: 377 Da, 408–409 Da, 455 Da and 659–660 Da. Of these, the 408–409 Da, 455 Da and 659–660 Da could present an aldehyde function due to the cleavage but could also be interpreted otherwise. However, they are most likely carrying aldehyde groups, as without periodate these peaks do not seem to appear. However, the peaks at 408–409 Da and 453–455 Da also appear, although small, in the case of the carbohydrates at 120 °C without periodate, and in the SPI alone at 120 °C without periodate; thus, their interpretation in favour of aldehyde formation is not at all certain. Thus, one cannot be sure at all that they come from the protein.

The final conclusions for the reactions at 120 °C with and without periodate are then as follows:

1. There are clear indications that aldehydes are formed from the insoluble carbohydrates by periodate specific oxidation and cleavage. There are carbohydrate oligomers presenting single, double and multiple aldehyde groups coming from the cleavage of the soy flour insoluble carbohydrates.
2. There is only one, different, low molecular weight aldehyde at 231 Da, which is in considerable proportion when the insoluble carbohydrates are treated with periodate but only in traces when treated at 120 °C without periodate. This peak disappears completely when the SPI is present, indicating that the aldehyde (a dialdehyde) has reacted with the SPI.
3. The peak at 231 Da is a dialdehyde derived by the aldol condensation of fragments issued by the degradation of soluble and insoluble carbohydrates.
4. There are oligomeric peptides coming from the cleavage of the protein by periodate. However, only one species (at 659–660 Da) could (we stress could) be ascribed to an oligomeric aldehyde, as other interpretations, such as peptide oligomers without an aldehyde group, are also possible.
5. Thus, the clear indications are that the carbohydrates do form aldehydes that co-react with the SPI and crosslink it.
6. There is at least one indication that aldehyde groups are formed in the cleavage of the protein, but these are not clear and could also be interpreted otherwise.
7. In the SPI + sucrose + insoluble carbohydrates + $NaIO_4$, several aldehydes are formed, as presented in the Introduction and already reported in the literature [5].
8. In the SPI + sucrose + insoluble carbohydrates + $NaIO_4$, mainly aldehydes generated from sucrose by periodate-specific oxidation react with amino acids, showing that the cross-linking of the protein oligomers by these aldehydes does indeed occur. This entails that at a higher temperature or longer reaction times, the other aldehydes generated from the insoluble carbohydrates and from the protein are also likely to react and contribute to cross-linking.
9. Finally, it must be noted that the peaks at 177 Da, 360–361 Da, 536 Da, 550 Da, 698–699 Da, 717 Da, and 874 Da belong to the DHB enhancer used in the MALDI analysis.

What are the future prospects of this type of approach? The first interest is that an almost classical reaction and cross-linking system based on aldehydes and a protein are generated, but with the generated aldehydes being non-volatile and non-toxic. The second interest is that all parts of soy flour, much cheaper than soy protein isolates, can be directly used for adhesives and can participate in cross-linking. Third, while one can use soy flour, nothing impedes the use of a separate mix of oligomeric and or monomeric carbohydrates to generate aldehydes via periodate specific oxidation and then adding this to a soy protein isolate, another protein isolate (such as gluten for example), or other aldehyde-reacting materials. In this respect, periodate oxidized carbohydrates could be used as hardeners for a variety of different natural materials, such as tannins, lignin, humins, etc., or possibly even mixtures of these.

Supplementary Materials: The following are available in the Supplementary Material online at http://www.mdpi.com/2073-4360/11/9/1478/s1, Figure S1: MALDI Tof Spectra of insoluble carbohydrates treated at 120 °C without periodate: (a) 150 Da to 500 da range, (b) 500 Da to 900 Da range, and (c) 900 Da to 1200 Da range. Figure S2: sucrose + 15% $NaIO_4$ at 120 °C for 1 h: (a) 50 Da to 1000 Da range, (b) 50 Da to 500 Da range, (c) 300 Da to 500 Da range, (d) 500 Da to 800 Da range, and (e) 800 Da to 1600 Da range.

Author Contributions: C.R.F., A.P. and L.F.L. conceived, planned and designed the experiments, X.X. did the reactions and performed the MALDI analysis, A.P. analyzed the data, interpreted the MALDI analysis, coordinated the work and wrote the paper.

Acknowledgments: The LERMAB of the University of Lorraine is supported by a grant overseen by the French National Research Agency (ANR) as part of the Laboratory of Excellence (Labex) ARBRE, which is herewith thanked by the second and third author. The third author thanks the China Scholarship Council for the study bursary granted to him.

Conflicts of Interest: The authors declare no conflict of interest.

References

1. Jones, G.H.; Moffat, J.G. Oxidation of carbohydrates by the sulfoxide-carbodiimide and related methods. In *Methods in Carbohydrate Chemistry, General Carbohydrate Method*; Whistler, R.D.J.N., Ed.; Academic Press: New York, NY, USA, 1972; Volume 6, pp. 315–319.
2. Cimecioglu, A.L.; Thomaides, J.S. Polysaccharide Aldehydes Prepared by Oxidation Method and Used as Strength Additives in Papermaking. U.S. Patent 6,586,588, 1 January 2003.
3. Breton, T.; Bashiardes, G.; Leger, J.-M.; Kokoh, K.B. Selective Oxidation of Unprotected Carbohydrates to Aldehyde Analogues by Using TEMPO Salts. *Eur. J. Organic Chem.* **2007**, *10*, 1567–1570. [CrossRef]
4. Sharma, P.R.; Chattopadhyay, A.; Sharma, S.K.; Hsiao, B.S. Efficient Removal of UO22+ from Water Using Carboxycellulose Nanofibers Prepared by the Nitro-Oxidation Method. *Ind. Eng. Chem. Res.* **2017**, *56*, 13885–13893. [CrossRef]
5. Sharma, P.R.; Varma, A.J. Thermal stability of cellulose and their nanoparticles: Effect of incremental increases in carboxyl and aldehyde groups. *Carbohydr. Polym.* **2014**, *114*, 339–343. [CrossRef] [PubMed]
6. Bobbitt, J.M. Periodate oxidation of carbohydrates. *Adv. Carbohydr. Chem.* **1956**, *11*, 1–41.
7. Hoda, S.; Kakehi, K.; Takiura, K. Periodate oxidation analysis of carbohydrates: Part IV. Simultaneous determination of aldehydes in di-aldehyde fragments as 2,4-dinitrophenylhydrazones by thin-layer or liquid chromatography. *Anal. Chim. Acta* **1975**, *77*, 125–131. [CrossRef]
8. De Wit, D.; Tournois, H.; Veelaert, S. Method for the Oxidation of Carbohydrates. U.S. Patent 5,747,658, 5 May 1998.
9. Niipelo, T.; Amer, H.; Konnerth, J.; Potthast, A.; Rosenau, T. Self-standing nanocellulose Janus-type films with aldehyde and carboxyl functionalities. *Biomacromolecules* **2018**, *19*, 973–979. [CrossRef] [PubMed]
10. Wescott, J.M.; Frihart, C.R.; Traska, A.E. High-soy-containing water-durable adhesives. *J. Adhesion Sci. Technol.* **2006**, *20*, 859–873. [CrossRef]
11. Petruccelli, S.; Anon, M.C. Soy Protein Isolate Components and their Interactions. *J. Agric. Food Chem.* **1995**, *43*, 1762–1767. [CrossRef]
12. Frihart, C.R.; Lorenz, L. Specific oxidants improve the wood bonding strength of soy and other plant flours. *J. Polym. Sci. A Polym. Chem.* **2019**, *57*, 1017–1023. [CrossRef]
13. Xi, X.; Pizzi, A.; Firhart, C.R.; Lorenz, L.; Gerardin, C. Tannin plywood bioadhesives with non-volatile aldehydes generation by specific oxidation of mono-and disaccharides. *Int. J. Adhes. Adhes.* **2019**. submitted.
14. Guigo, N.; Mazeau, K.; Putaux, J.L.; Heux, L. Surface modification of cellulose microfibrils by periodate oxidation and subsequent reductive amination with benzylamine: A topochemical study. *Cellulose* **2014**, *21*, 4119–4133. [CrossRef]
15. Codou, A.; Guigo, N.; Heux, L.; Sbirrazzuoli, N. Partial periodate oxidation and thermal cross-linking for the processing of thermoset all-cellulose composites. *Compos. Sci. Technol.* **2015**, *117*, 54–61. [CrossRef]
16. McMurray, J. *Organic Chemistry*, 3rd ed.; Brooks/Cole Publishing Co.: Pacific Grove, CA, USA, 1992.

17. Despres, A.; Pizzi, A.; Pasch, H.; Kandelbauer, A. Comparative ^{13}C NMR and MALDI-TOF of species variation and structure maintenance during MUF resins preparation. *J. Appl. Polym. Sci.* **2007**, *106*, 1106–1128. [CrossRef]
18. Pizzi, A.; Pasch, H.; Rode, K.; Giovando, S. Polymer structure of commercial hydrolisable tannins by MALDI-TOF mass spectrometry. *J. Appl. Polym. Sci.* **2009**, *113*, 3847–3859. [CrossRef]
19. Petrovic, M.; Schiller, J.; Muller, M.; Suss, R.; Arnold, K.; Arnhold, J. Detection of adducts with matrix clusters in the positive and negative ion mode MALDI-TOF mass spectra of phospholipids. *Z. Für Nat. B* **2009**, *64*, 331–334. [CrossRef]
20. Mohsen, S.M.; Fadel, H.H.M.; Bekhit, M.A.; Edris, A.E.; Ahmed, M.Y.S. Effect of substitution of soy protein isolate on aroma volatiles, chemical composition and sensory quality of wheat cookies. *Int. J. Food Sci. Technol.* **2009**, *44*, 1705–1712. [CrossRef]
21. Cavins, J.F.; Wolek, Y.F.K.; Inglett, G.E.; Cowan, J.C. *Amino Acid Analysis of Soybean Meal: Interlaboratory Study*; USDA Northern Marketing and Nutrition Research Division, Agricultural Research Service, U.S. Department of Agriculture: Peoria, IL, USA, 1972; p. 686. Available online: https://naldc.nal.usda.gov/download/31371/PDF (accessed on 12 July 2019).

polymers

MDPI

Article

Photo-Oxidative and Soil Burial Degradation of Irrigation Tubes Based on Biodegradable Polymer Blends

Marco Rapisarda [1], Francesco Paolo La Mantia [2,*], Manuela Ceraulo [2], Maria Chiara Mistretta [2], Carmelo Giuffrè [3], Roberto Pellegrino [3,4], Graziella Valenti [1] and Paola Rizzarelli [1,*]

1 Istituto per i Polimeri, Compositi e Biomateriali, Consiglio Nazionale delle Ricerche, Via Paolo Gaifami 18, 95126 Catania, Italy; marcorapis7@gmail.com (M.R.); valentigraziella1985@gmail.com (G.V.)
2 Department of Engineering, University of Palermo, Viale delle Scienze, Ed.6, 90128 Palermo, Italy; manuela.ceraulo@unipa.it (M.C.); mc.mistretta@gmail.com (M.C.M.)
3 IRRITEC, Cda Scarpitta, 98071 Capo d'Orlando, Italy; carmelo.giuffre@irritec.com (C.G.); roberto.pellegrino@anton-paar.com (R.P.)
4 Anton Paar, Ankerstraße 6, 8054 Graz, Austria
* Correspondence: francescopaolo.lamantia@unipa.it (F.P.L.M.); paola.rizzarelli@cnr.it (P.R.)

Received: 6 August 2019; Accepted: 10 September 2019; Published: 12 September 2019

Abstract: Irrigation tubes based on biodegradable polymers were prepared via an extrusion-drawing process by Irritec and compared to conventional pipes made of high-density polyethylene (HDPE). A commercial polylactide/poly (butyleneadipate-*co*-butyleneterephthalate) (PLA/PBAT) blend (Bio-Flex®) and Mater-Bi® were used. The polymers were characterized from rheological and mechanical points of view. Irrigation pipes were subjected to photoaging with continued exposure to UV radiation up to 22 days. The degradability in the soil of irrigation tube samples was studied. The influence of temperature and UV irradiation on soil burial degradation was investigated. A soil burial degradation test was carried out at 30 °C and 50 °C for up to 70 days. The degree of degradation was evaluated from the weight loss percentage. The degradation rate of irrigation tube samples based on Mater-Bi® was higher at 30 °C and was stimulated after 14 days of UV irradiation. Higher temperatures or UV aging encouraged the disintegration in soil of Bio-Flex®-based irrigation tubes. Furthermore, tube samples, before and after UV and soil burial degradation, were analyzed by Attenuated Total Reflection-Fourier Transform Infra-Red (ATR-FTIR) spectroscopy.

Keywords: biodegradable polymers; rheological properties; irrigation pipes; soil burial test; polyesters; Bio-Flex®; Mater-Bi®; polymer degradation; photo-oxidation

1. Introduction

Nowadays, polymeric materials are widely used in agricultural field for several types of plastic products (i.e., mulch films, containers, filaments, clips, pots, packaging, drip irrigation tubes, greenhouse covers, etc.). Nevertheless, the widespread and increasing use of plastic materials in agriculture and their persistence in the environment contribute seriously to environmental pollution due to plastic wastes. Plastic items manufactured by traditional polymers, such as mulch films, have to be collected after use and recycled or burned for energy recovery. The removal and disposal can be very expensive and difficult to achieve; in addition, plastic residues can persist in the field with a negative environmental impact. Recently, an interesting overview of agricultural plastic waste generation in Europe was published, highlighting this great concern [1].

The worldwide increase in plastic wastes has involved a great deal of strategies aimed at minimizing the negative impact of the cumulative production and consumption of polymeric materials.

The reduction of fossil resources and waste accumulation problems have stimulated a growing interest toward replacing conventional plastics by biodegradable ones, including in agriculture. By using biodegradable items, in fact, both recovery and final disposal can be avoided, since they are optimized to degrade in situ.

Biodegradable plastic systems can potentially replace commonly used PE ones in agriculture. Biodegradable items must ensure the same performance of traditional ones, with the advantage of being left in soil or disposed of in industrial composting plants as common organic waste. Indeed, several commercially biodegradable polymers show rheological and mechanical properties suitable for the production of irrigation pipes, as they can be easily processed in extrusion and can also be drawn [2–4].

Aliphatic polyesters are among the most employed biodegradable materials in packaging applications as well as in agricultural and sanitary fields [5]. Some studies have been carried out on commercial biodegradable polymers to verify their applicability in mulching or irrigation pipes [6,7]. However, the effect of UV on their performances and degradation rates in soil has not been investigated in depth. In fact, both mulching and irrigation pipes, during their normal use, are exposed to sunlight. Therefore, their photoaging becomes of utmost concern and can affect their performances as well as their biodegradation rates.

The degradation of traditional plastic materials in nature is a very slow process that involves environmental factors and microorganism activities. Polymer degradation implies different processes stimulated by one or more environmental factors, such as heat, light, microorganisms, or chemicals that deteriorate polymers, producing alterations in their properties. The degradation is the result of irreversible structural changes that are usually undesirable or, in some cases, essential, as in biodegradation, or are else induced to support structure determinations, such as in pyrolysis-gas chromatography-mass spectrometry (Py-GC-MS) studies [8,9]. The biodegradation of polymeric materials depends on several chemical-physical features (chemical bonds, composition, crystallinity, morphology, etc.), environmental factors (oxygen, temperature, biological agents, microbial cells, and enzymes), and their combined effect [10].

The degradability in soil of poly(vinyl alcohol) [11], copolyesters [12], and poly(ester amide) [13] film samples has been investigated under controlled soil burial conditions. Despite extensive research on materials, few studies have been focused on polymer degradation due to a combined effect of UV irradiation and soil burial [6,7]. Briassoulis et al. evaluated the degradation behavior of MaterBi®-based films and irrigation tubes under real field conditions [7]. However, laboratory tests under controlled conditions can provide more reproducible data for both degradation mechanisms and kinetics.

Biodegradation can be schematically represented in three stages: in the first one, plastic is depolymerized into monomers and oligomers; in the second step, the monomers and oligomers are taken up as biomass; and in the third, the respiration of biomass consumes O_2 and produces CO_2 and H_2O (under aerobic conditions). The measurement of reactant consumption (i.e., the plastic material) does not allow us to prove whether the process has actually been completed or has stopped, for example, at depolymerization. Therefore, all of the standardized methods for determining biodegradation are based on the measurement of respiration, i.e., the conversion into CO_2 of the carbon initially present in the plastic through the use of the oxidant (O_2). However, most of the papers in the literature concerning polymer and composite biodegradation have been based on weight loss measurements [10,14,15].

The aim of this work is to report the possible application of two different classes of biodegradable polymers in pipes for irrigation. An evaluation of their performance is reported. Irrigation tubes based on biodegradable polymer blends were prepared by IRRITEC through a proprietary process and compared to conventional pipes made of high-density polyethylene (HDPE). A commercial polylactide/poly(butyleneadipate-*co*-butyleneterephthalate) (PLA/PBAT) blend (Bio-Flex®) and a commercial Mater-Bi® blend were used. The polymers were characterized from rheological and mechanical points of view in order to investigate their suitability to be processed in an industrial apparatus to obtain pipes. Irrigation pipes were subjected to photoaging, and their degradability in soil was studied. The influence of temperature and UV irradiation on soil burial degradation was investigated.

2. Experiment

2.1. Materials

Two biodegradable polymer systems were used in this work: a Mater-Bi® CF04P grade (MB) made by Novamont (Novara, Italy) and a Bio-Flex® F2110 grade (BF) made by FKUR (Willich, Germany). The Mater-Bi® sample had a melt flow rate (MFI) of about 7 g/10 min at 160 °C (under a 5-kg load), and the Bio-Flex® sample (BF) had an MFI of approximately 5 g/10 min (190 °C, 2.16 kg). The composition of the Mater-Bi® biodegradable system was proprietary, and its formulation was based on a combination of biodegradable aliphatic polyesters, biodegradable aliphatic–aromatic copolyesters, and starch [2,7]. The Bio-Flex® was a blend based on PLA and PBAT [16] (ratio ~3:1). A high-density polyethylene (HDPE) sample, utilized for similar applications, was used for comparison. Carbon black (2%) was added in all the compositions. For the soil burial test, tube portions of 2.5 × 5 cm (thickness about 290 μm) were cut. Specimens of 10 × 2.5 cm were used instead for the accelerating weathering test.

2.2. Rheological, Mechanical, and Structural Characterization

In order to assess if these two biodegradable polymers could be processed in industrial apparatuses for the production of pipes, a rheological characterization was carried out both in shear and in nonisothermal elongational flow. A rheological characterization was performed in shear flow in a rotational rheometer, a TA (USA) Ares G2, and in a capillary viscometer (a Rheologic 1000 CEAST (Pianezza, TO, Italy)).

The flow curves were measured at T = 170 °C for the two biodegradable polymers and at 230 °C for the HDPE.

The behavior of these polymers in nonisothermal elongational flow in this application was very important because the emerging tube from the die extruder was drawn in a molten state to achieve the final desired dimensions. Measures of the properties in nonisothermal elongational flow, of the melt strength (MS), and of the breaking stretching ratio (BSR) were achieved according to procedures described elsewhere [2–4].

MS is the force in the molten filament at breaking, while the BSR is the ratio between the drawing speed at breaking and the extrusion velocity in runs in which the drawing velocity increases with steady acceleration. The capillary has a length-to-diameter (L/D) ratio = 40. The reproducibility of all of the results was good (±5%).

Mechanical properties, the elastic modulus (E), the tensile strength (TS), and the elongation at break (EB) were measured in tensile tests with an Instron (USA) machine at room temperature.

Samples (10 mm/90 mm/~0.6 mm) for mechanical characterization and samples (D = 25 mm, thickness about 2 mm) for rheological characterization in shear flow of the unprocessed materials were prepared by compression-molded sheets obtained with a Carver (USA) laboratory press at a temperature of ~170 °C for the two biodegradable polymers and of 230 °C for the HDPE. Attenuated Total Reflection-Fourier Transform Infra-Red (ATR-FTIR) spectra were collected by using a Spectrum Two spectrometer (Perkin-Elmer) using Spectrum software. Spectra were measured through 8 scans at a 4 cm^{-1} resolution.

In order to compare the different effects of photo-oxidation and degradation in soil on the structures, the area ratio between the sums of the peaks at 1550 cm^{-1} and 1850 cm^{-1} and at 2750 cm^{-1} and 3100 cm^{-1}, which were related to the stretching of the CH_2 and CH_3 groups, was evaluated for the virgin and degraded samples.

2.3. Production of the Pipes

The pipes were produced by Irritec (Capo d'Orlando, Italy) through an extrusion-drawing proprietary process. The tube merging from the extruder was drawn to reach the desired dimensions (thickness about 290 μm). The die temperature was ~170 °C for the two biodegradable polymers and 230 °C for the HDPE.

Mater Bi®-based pipes were named T1, and the Bio-Flex® ones were called T2.

2.4. Accelerated Weathering Test

Specimen portions were exposed to an accelerated weathering test. A QUV Panel system (UVA lamps: λ_{max} = 340 nm) at 60 °C was used (for up to 22 days). Irrigation pipes photo-oxidized at different UV exposure times were selected for a soil burial test in triplicate.

2.5. Soil Burial Test

Tests were carried out at 30 and 50 ± 0.1 °C under moisture-controlled conditions. Triplicate specimens at 30 and 50 °C of each irrigation pipe were placed in a series of darkened vessels containing a multilayer substrate [13]. At 30 °C, the specimens of the irrigation pipe (initial weight 0.7 ÷ 1.2 g) were sandwiched between two layers of a mixture of milled perlite (50 g) and commercial soil (200 g) moistened with 100 mL of distilled water (150 mL during the test at 50 °C). The bottom and top layers were filled with 60 g (70 g at 50 °C) of perlite moistened with 120 mL (240 mL at 50 °C) of distilled water. Perlite was added to increase the aeration of the soil and the amount of water retained. A flow of moistened air was supplied from the bottom of each vessel every 12 h for 15 min (every 6 h at 50 °C). The specimens were removed after regular intervals, brushed softly, washed with distilled water several times, and dried under vacuum in the presence of P_2O_5 at room temperature to a constant weight [17]. The degree of degradation was evaluated by weight loss (WL) by using the following equation:

$$WL\ (\%) = (W_i - W_t)/W_i \times 100, \quad (1)$$

where W_i is the initial weight of the sample, and W_t is the weight after an established time. Filter paper and polyethylene samples were used, respectively, as a positive and negative control.

3. Results and Discussion

3.1. Rheological and Mechanical Properties

The flow properties of the two biodegradable samples and of the reference HDPE are reported in Figure 1.

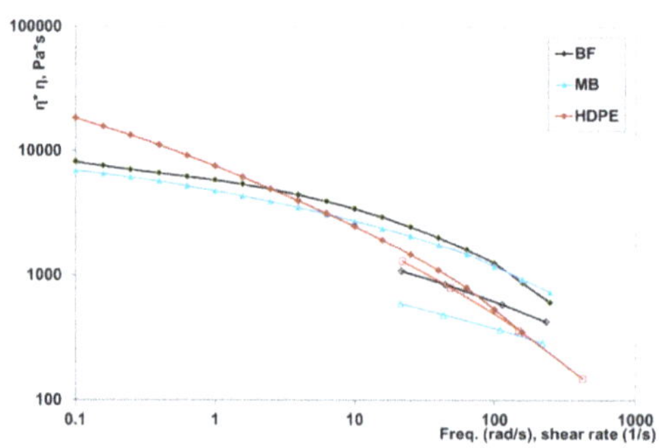

Figure 1. Flow curves of the investigated samples in a rotational rheometer (full symbols) and a capillary viscometer (open symbols).

The flow curves obtained in the rotational rheometer and in the capillary viscometer were superimposed, according to the empirical Cox-Merz rule, only for the HDPE sample, while this was not true for the two biodegradable polymers. This phenomenon has already been reported in the literature for similar systems and was in agreement with the results already found in other studies on heterogeneous multiphases [2,18].

The flow curves, moreover, showed that the biodegradable samples present at these temperatures had a viscosity slightly larger and a non-Newtonian behavior less pronounced than the conventional HDPE in the shear rate range (about 100 s^{-1}) typical of the extrusion process. However, this difference can be considered acceptable for good extrusion in the production of pipes.

The melt strength (MS) and the breaking stretching ratio (BSR) of the same polymers are reported in the Figures 2 and 3 as a function of the apparent shear rate. In the shear rate range typical of the extrusion process, the melt strength of the biodegradable polymers was very similar to that of HDPE. This means that it is possible to draw the melt to obtain films or pipes that are relatively thick.

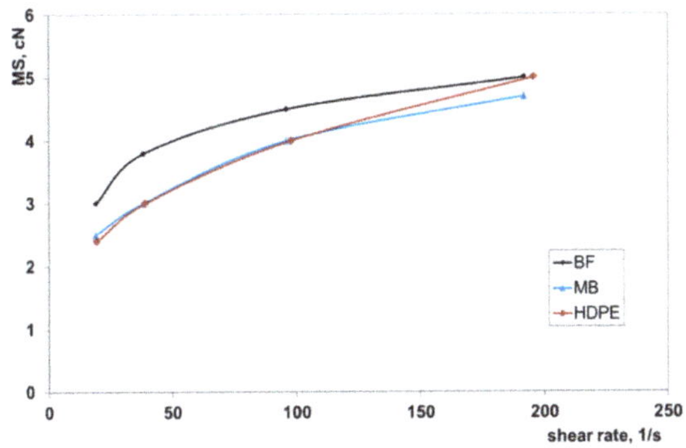

Figure 2. Melt strength (MS) of the investigated samples.

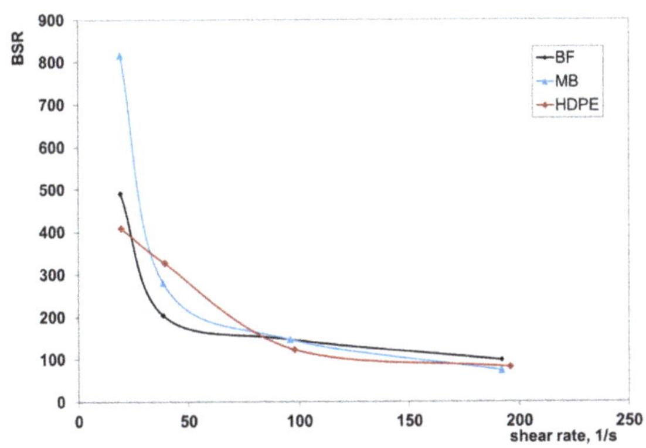

Figure 3. Breaking stretching ratio (BSR) of the investigated samples.

High values of the BSR indicated that the molten polymer can be drawn in nonisothermal elongational flow to produce thin films or tubes. All three polymers showed very similar values of the BSR in the shear rate region of the extrusion. This means that the two investigated biodegradable samples will show rheological behavior in the extrusion-drawing process that is very similar.

Mechanical properties, the elastic modulus (E), the tensile strength (TS), and the elongation at break (EB) of the isotropic samples are reported in Table 1.

Table 1. Mechanical properties: elastic modulus (E), tensile strength (TS), and elongation at break (EB) of the isotropic samples.

Materials	Modulus (MPa)	Tensile Strength (MPa)	Elongation at Break (%)
MB	81	12.5	400
BF	167	12.4	167
HDPE	316	21.3	700

The elastic modulus of HDPE was higher than that of the two biodegradable samples, and the MB sample showed the lowest value. In addition, the tensile strength and the elongation at break of the HDPE were higher than the biodegradable samples. However, while the TS values of the two biodegradable samples were very similar, the EB value of the Mater-Bi® was much larger than that of the Bio-Flex®.

3.2. Soil Burial Test

Soil burial degradation tests were carried out under moisture-controlled conditions at 30 ± 0.1 °C and 50 ± 0.1 °C to evaluate the influence of temperature. Triplicate specimens of each irrigation tube were placed in a series of darkened vessels containing a multilayer substrate [13,17]. The polymer films were sandwiched between two layers of a mixture of milled perlite and commercial soil, which were moistened with distilled water. The bottom and top layers were filled with perlite moistened with distilled water. Perlite was added to increase the amount of water retained and the aeration in the soil. A flow of moistened air was supplied from the bottom of each vessel every 24 h for 15 min.

In Figure 4a,b, the average values of the weight loss percentages as a function of degradation time for the samples T1, T2, and HDPE at temperatures of (a) 30 and (b) 50 °C are compared. The carbon black contribution was evaluated and subtracted. T1 (Mater-Bi®-based tubes) showed the highest weight losses both at 30 and at 50 °C, with an almost linear trend at 50 °C. On the contrary, for the T2 specimens (Bio-Flex®-based tubes), the WL values were remarkably higher at 50 than at 30 °C, reaching 4% at 80 days. By increasing the degradation, the standard deviation grew due to the embrittlement of the samples and the increasing difficulties in the removal of soil from the surface. Unsurprisingly, the WL values for the HDPE specimens were close to zero and remained almost constant throughout the test period. Overall (in agreement with the literature [6]), the higher temperature encouraged the degradation of both biodegradable polymer-based tubes.

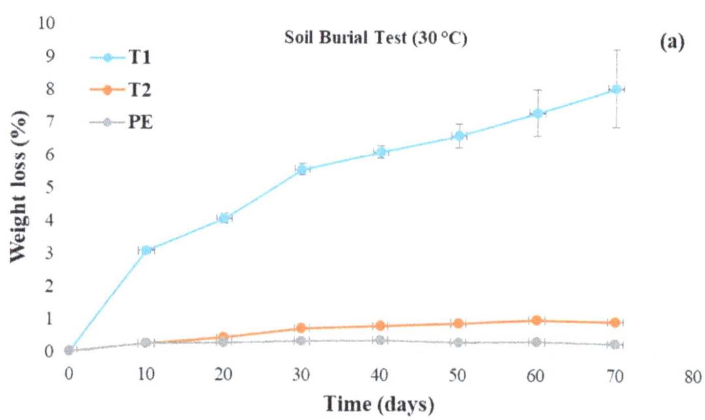

Figure 4. *Cont.*

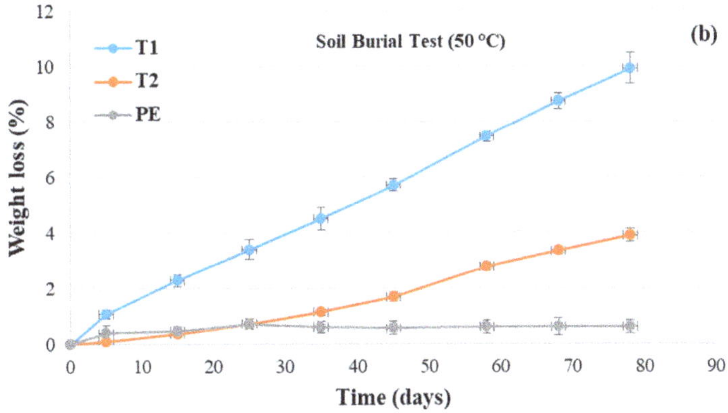

Figure 4. Weight loss (%) versus degradation time for the samples T1, T2, and HDPE at (**a**) 30 °C and (**b**) 50 °C. T1 = Mater Bi®-based pipes, and T2 = Bio-Flex®-based pipes.

Table 2 shows some representative photos of the samples after different soil burial degradation intervals and temperatures. Despite the T1 samples showing the highest WL values, macroscopic degradative effects, such as embrittlement, were observed at 30 °C only after 70 days (Table 2). T2 samples appeared to be stable for a longer time. In the T1 samples, degradation appeared to be increasingly evident after 58 days at 50 °C. At 50 °C, the degradation of the T1 samples produced embrittlement and longitudinal cracks after 68 days (Table 2).

Table 2. Representative photos of the film samples recovered after a soil burial degradation test at different intervals and temperatures.

Temperature Test (°C)	Degradation Time in Soil (Days)	Sample	
		T1	T2
/	0		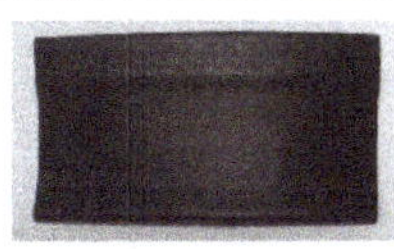
30	60		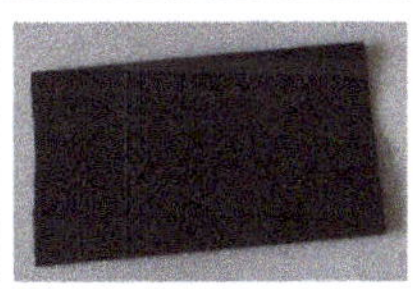
	70	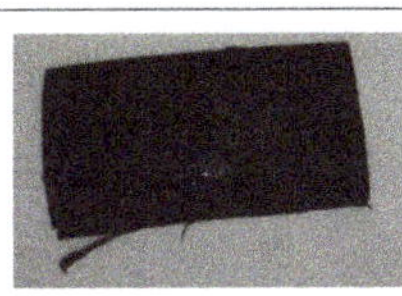	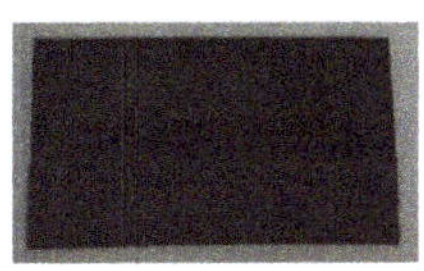

Table 2. *Cont.*

Temperature Test (°C)	Degradation Time in Soil (Days)	Sample	
		T1	T2
50	58		
	68	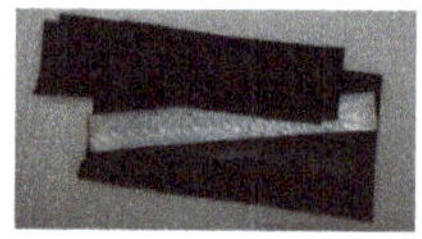	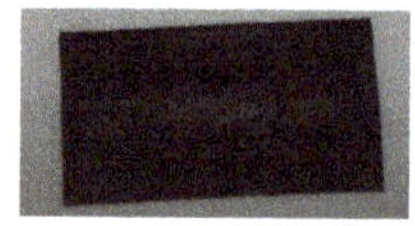
	78		

T1 and T2 samples photo-oxidized for 8 and 14 days were selected for the soil burial test at 30 °C. Their weight losses (%) were evaluated for up to 70 days.

In Figure 5, the average values of the WL values were plotted as a function of degradation time at 30 °C for the samples (a) T1 and (b) T2, which were photo-oxidized for 8 and 14 days. For all the samples, photodegradation favored successive degradation in the soil (but more significantly for the specimens of T1) after 30 days of soil burial degradation. Additionally, the WL values increased with the time of UV exposure. Reasonably, UV exposure yielded a molecular weight (MW) decrease producing oligomeric chains, which are more susceptible to the attacks of microorganisms in the soil. In fact, it must be kept in mind that biodegradation proceeds in three stages. Whenever WL is used to monitor the degradation of polymer samples, just the first step is involved, i.e., macromolecular chain depolymerization into monomers and oligomers that are eroded from the surface. UV irradiation enhances and accelerates the formation of monomers and oligomers, increasing consequently the weight loss rate.

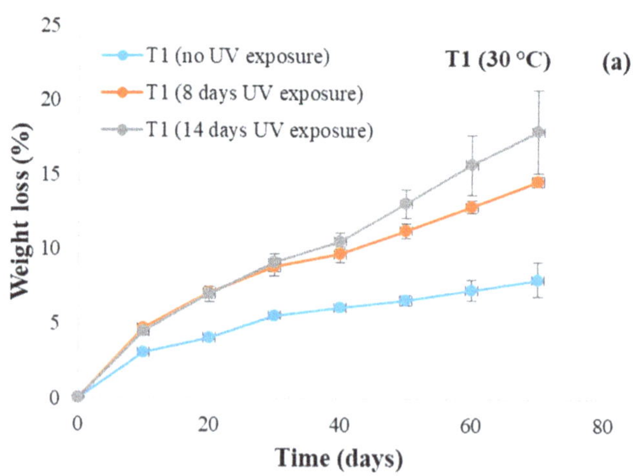

Figure 5. *Cont.*

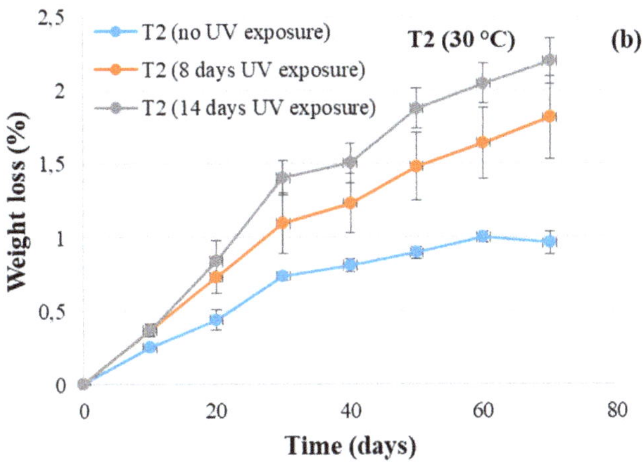

Figure 5. Weight loss (%) versus degradation time for the UV-treated and untreated (**a**) T1 and (**b**) T2 samples. T1 = Mater Bi®-based pipes, and T2 = Bio-Flex®-based pipes.

The weight loss values that were lower for the T2 samples (i.e., the Bio-Flex®-based tubes) than for the T1 specimens (Figures 4b and 5b) could have been related to the presence of PLA, whose biodegradation is MW-dependent. It could be assumed that the higher temperature encouraged the hydrolysis of ester bonds and that UV exposure induced bond-breaking within the macromolecular chains. Consequently, the molecular weight reduction assisted the degradation process in the soil. After the time needed to decrease the molecular weight, the oligomeric products were more easily eroded from the surface of the tubes, and the weight loss values increased. On the other hand, the WL trend at 50 °C, which was close to the T_g value of PLA, could have affected the weight loss data due to the higher macromolecular chain mobility. The trend of weight loss % versus time with increasing UV exposure (Figure 5) could suggest that chain scission overrules crosslinking reactions. In fact, according to the literature, photo-oxidation gives rise to the rearrangement of polymeric chains, leading, in the case of aliphatic-aromatic copolyesters, to polymeric chain crosslinking and the formation of an insoluble polymeric gel fraction. On the contrary, PLA photochemical reactions produce chain scissions instead of crosslinking [19,20].

Overall (and in agreement with the literature [6]), the photo-oxidation accelerated the soil burial degradation. Therefore, in order to investigate the structural changes induced by the different types of degradation, ATR-FTIR was carried out on virgin, photo-oxidized, soil-buried, and photo-oxidized + buried samples (see spectra reported in Figure 6a,b for T1 and T2, respectively). Unsurprisingly, the spectra of the two samples showed that both photo-oxidation and degradation in the soil increased the ratio of the area in the 1550–1850 cm^{-1} region (of carbonyl groups), as well as the area of the functional groups at 2750 cm^{-1} and 3100 cm^{-1}, which were related to the stretching of the CH_2 and CH_3 groups. It has to be underlined that carbon back is a UV absorber, and its photostabilizing action is due to the absorption of ultraviolet energy that then is no longer available for the formation of radicals and for the consequent propagation of oxidation reactions [21]. This means that carbon black does not interfere with the photo-oxidative mechanisms of different matrices, and its effect is just to shift photo-oxidation toward longer times, increasing the induction time of the process. Carbon black slows down molecular weight reduction and consequently the degradation process in soil. The preliminary experimental data on the T2 samples without carbon black (data not shown) confirmed our conclusions. The shift toward longer times to start photo-oxidation is relevant from a commercial point of view above all in biodegradable polymeric systems whose performances could be impaired by photo-oxidation processes more than traditional ones would be [6].

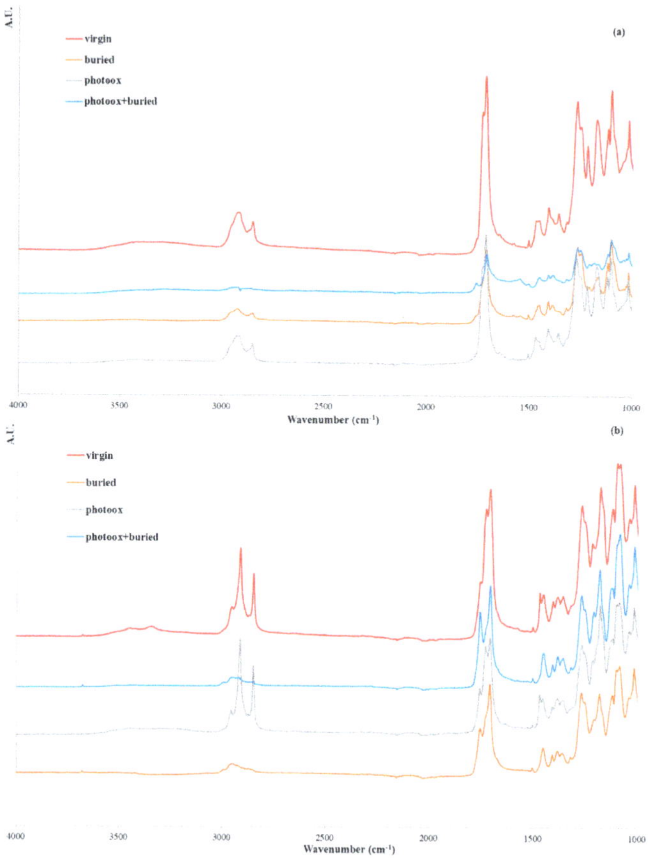

Figure 6. Attenuated Total Reflection-FTIR spectra of virgin, photo-oxidized, soil-buried, and photo-oxidized + buried (**a**) T1 and (**b**) T2 samples. T1 = Mater Bi®-based pipes, and T2 = Bio-Flex®-based pipes.

To better follow this trend, in Figure 7 the dimensionless values of the area ratio are reported for the two samples.

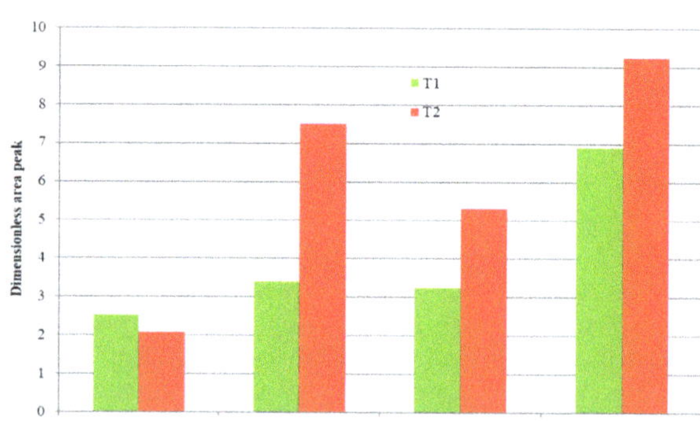

Figure 7. Dimensionless values of area ratio in the 1550–1850 cm^{-1} region for the two biodegradable samples. T1 = Mater Bi®-based pipes, and T2 = Bio-Flex®-based pipes.

According to these data (Figure 7), the Mater-Bi®-based sample (T1) showed similar behavior with both photo-oxidation (14 days) and soil degradation (70 days). A synergistic effect of photo-oxidation and soil degradation clearly occurred for both samples. Indeed, after photo-oxidation, the pipe samples seemed much more prone to soil degradation (Figure 5). Interestingly, the Bio-Flex®-based sample (T2) showed an apparently peculiar behavior. In fact, the CO increase was more evident in the T2 sample whatever the degradation process, which could have been related to the presence of the PLA component [22]. The increase of the carbonyl groups was an indicator of the photo-oxidation progress as well as of hydrolytic reactions occurring during soil degradation. As a result, T2 was clearly more susceptible to photo-oxidation than to hydrolysis. Even though the CO functional groups increased, due to the combined photo-oxidation and soil degradation, there were more in the T2 sample than in the T1 sample, and the UV irradiation did not accelerate the weight loss rate as it did in the T1 sample (Figure 5). These results were in agreement with the biodegradation process and with our choice of monitoring soil degradation through weight loss. Actually, the T1 sample was much more prone than the T2 sample, since monomers and low-molecular-weight oligomers were originated and removed in the erosion step faster than in the T2 sample.

4. Conclusions

The biodegradable polymer systems investigated in this work showed rheological properties both in shear flow and in nonisothermal elongational flow similar to those of a typical polymer used in the production of pipes for irrigation.

As for the mechanical properties, the high-density polyethylene sample showed higher values of elastic modules and of tensile strength and elongation at break than did the two biodegradable samples. However, the values of these mechanical properties were compatible with those required by this application.

Soil burial tests showed that the Mater-Bi®-based irrigation tubes were more susceptible to degradation in the soil than the Bio-Flex®-based ones were. The increase of the temperature in the soil burial tests had a negligible effect on the Mater-Bi®-based tube degradation rate. For the Bio-Flex®-based samples, the degradation in soil occurred faster at 50 °C rather than at 30 °C. Additionally, for all of the samples, soil degradation appeared to be encouraged by UV exposure. By comparing the changes in the CO groups measured by the ATR-FTIR spectra to the data of the soil degradation, a larger increase of the CO groups was highlighted for the Bio-Flex®-based samples, which showed a minor soil degradation kinetic. The combined results of the soil degradation tests and the ATR-FTIR spectra suggest that the carbon atoms of both materials were oxidized but that the kinetics of soil degradation were faster in the T1 pipe samples than in the T2 pipe samples, whose formation and removal of oligomers containing carbonyl groups was slower.

Author Contributions: Conceptualization, F.P.L.M. and P.R.; Investigation, M.R., M.C., M.C.M. and G.V.; Project administration, F.P.L.M. and P.R.; Resources, C.G. and R.P.; Supervision, F.P.L.M. and P.R.; Visualization, P.R.; Writing—Original draft, F.P.L.M., M.C.M., G.V. and P.R.

Funding: This research received no external funding.

Acknowledgments: Thanks are due to POR FSE Sicily 2020 (project "Polymeric systems: innovative aspects and applications in the biomedical and agri-food fields—SPIN OFF of Polymers") and Call 11/2017 ("Strengthening employability in the R&D system and the emergence of research SPIN OFFS in Sicily") for their partial financial support.

Conflicts of Interest: The authors declare no conflicts of interest.

References

1. Briassoulis, D.; Babou, E.; Hiskakis, M.; Scarascia, P.P.; Guarde, D.; Dejean, C. Review, mapping and analysis of the agricultural plastic waste generation and consolidation in Europe. *Waste Manag. Res.* **2013**, *31*, 1262–1278. [CrossRef] [PubMed]
2. La Mantia, F.P.; Arrigo, R.; Morreale, M. Effect of the orientation and rheological behaviour of biodegradable polymer nanocomposites. *Eur. Polym. J.* **2014**, *54*, 11–17. [CrossRef]
3. Morreale, M.; Mistretta, M.C.; Ceraulo, M.; La Mantia, F.P. Rheological behavior under shear and non-isothermal elongational flow of biodegradable polymers for foam extrusion. *J. Polym. Environ.* **2014**, *22*, 112–118. [CrossRef]
4. La Mantia, F.P.; Ceraulo, M.; Mistretta, M.C.; Morreale, M. Rheological behaviour, mechanical properties and processability of biodegradable polymer systems for film blowing. *J. Polym. Environ.* **2018**, *26*, 749–756. [CrossRef]
5. Avérous, L.; Pollet, E. Biodegradable polymers. In *Environmental Silicate Nano-biocomposites. Green Energy and Technology*; Avérous, L., Pollet, E., Eds.; Springer: London, UK, 2012; pp. 13–39.
6. Brodhagen, M.; Peyron, M.; Miles, C.; Inglis, D.A. Biodegradable plastic agricultural mulches and key features of microbial degradation. *Appl. Microbiol. Biotechnol.* **2015**, *99*, 1039–1056. [CrossRef] [PubMed]
7. Briassoulis, D.; Babou, E.; Hiskakis, M. Degradation behaviour and field performance of experimental biodegradable drip irrigation systems. *J. Polym. Environ.* **2011**, *19*, 341–361. [CrossRef]
8. Wypych, G. *Handbook of UV Degradation and Stabilization*, 2nd ed.; ChemTec Publishing: Toronto, ON, Canada, 2015; pp. 1–420.
9. Rizzarelli, P.; Rapisarda, M.; Perna, S.; Mirabella, E.F.; La Carta, S.; Puglisi, C.; Valenti, G. Determination of polyethylene in biodegradable polymer blends and in compostable carrier bags by Py-GC/MS and TGA. *J. Anal. Appl. Pyrol.* **2016**, *117*, 72–81. [CrossRef]
10. Emadian, S.M.; Onay, T.T.; Demirel, B. Biodegradation of bioplastics in natural environments. *Waste Manag.* **2017**, *59*, 526–536. [CrossRef] [PubMed]
11. Solaro, R.; Corti, A.; Chiellini, E. A new respirometric test simulating soil burial conditions for the evaluation of polymer biodegradation. *J. Environ. Polym. Degr.* **1998**, *6*, 203–208. [CrossRef]
12. Kijchavengkul, T.; Auras, R.; Rubino, M.; Alvarado, E.; Montero, J.R.C.; Rosales, J.M. Atmospheric and soil degradation of aliphatic-aromatic polyester films. *Polym. Degrad. Stab.* **2010**, *95*, 99–107. [CrossRef]
13. Rizzarelli, P.; Cirica, M.; Pastorelli, G.; Puglisi, C.; Valenti, G. Aliphatic poly(ester amide)s from sebacic acid and aminoalcohols of different chain length: Synthesis, characterization and soil burial degradation. *Polym. Degrad. Stab.* **2015**, *121*, 90–99. [CrossRef]
14. Eubeler, J.P.; Zok, S.; Bernhard, M.; Knepper, T.P. Environmental biodegradation of synthetic polymers I. Test methodologies and procedures. *Trends Anal. Chem.* **2009**, *28*, 1057–1072. [CrossRef]
15. Rizzarelli, P.; Degli Innocenti, F.; Valenti, G.; Rapisarda, M. Biodegradation of green polymer composites: Laboratory procedures and standard test methods. In *Green Composites: Materials and Applications*, 1st ed.; Inamuddin, M.P., Ed.; Springer: London, UK, 2019.
16. Gregorova, A.; Riedl, E.; Sedlarik, V.; Stelzer, F. Effect of 4,4′-methylenediphenyl diisocyanate on thermal and mechanical properties of Bioflex/lactic acid polycondensate blends. *Asia-Pac. J. Chem. Eng.* **2012**, *7*, 317–323. [CrossRef]
17. Rizzarelli, P.; Puglisi, C.; Montaudo, G. Soil burial and enzymatic degradation in solution of aliphatic co-polyesters. *Polym. Degrad. Stab.* **2004**, *85*, 855–863. [CrossRef]
18. Ren, J.; Krishnamoorti, R. Nonlinear viscoelastic properties of layered-silicate-based intercalated nanocomposites. *Macromolecules* **2003**, *36*, 4443–4451. [CrossRef]
19. Stloukal, P.; Verney, V.; Commereuc, S.; Rychly, J.; Matisova-Rychla, L.; Pis, V.; Koutny, M. Assessment of the interrelation between photooxidation and biodegradation of selected polyesters after artificial weathering. *Chemospere* **2012**, *88*, 1214–1219. [CrossRef] [PubMed]
20. Yasuda, N.; Wang, Y.; Tsukegi, T.; Shirai, Y.; Nishida, H. Quantitative evaluation of photodegradation and racemization of poly(L-lactic acid) under UV-C irradiation. *Polym. Degrad. Stab.* **2010**, *95*, 1238–1243. [CrossRef]

21. Mistretta, M.C.; Botta, L.; Vinci, A.D.; Ceraulo, M.; La Mantia, F.P. Photooxidation of poypropylene/graphene nanoplatelets nanocomposites. *Polym. Deg. Stab.* **2019**, *160*, 35–43. [CrossRef]
22. Copinet, A.; Bertrand, C.; Govindin, S.; Coma, V.; Couturier, Y. Effects of ultraviolet light (315 nm), temperature and relative humidity on the degradation of polylactic acid plastic films. *Chemosphere* **2004**, *55*, 763–773. [CrossRef] [PubMed]

polymers

MDPI

Article

Safely Dissolvable and Healable Active Packaging Films Based on Alginate and Pectin

Maziyar Makaremi [1], Hosnieh Yousefi [2], Giuseppe Cavallaro [3], Giuseppe Lazzara [3], Calvin Bok Sun Goh [4], Sui Mae Lee [4], Atefeh Solouk [2] and Pooria Pasbakhsh [1,*]

1 Advanced Engineering Platform, Mechanical Engineering Discipline, School of Engineering, Monash University Malaysia, Selangor 47500, Malaysia; maziyar.makaremi@gmail.com
2 Biomedical Engineering faculty, Amirkabir University of Technology (Tehran Polytechnic), Tehran 15875-4413, Iran; hosniehyousefi@yahoo.com (H.Y.); atefeh.solouk@aut.ac.ir (A.S.)
3 Department of Physics and Chemistry, University of Palermo, Viale delle Scienze, pad. 17, 90128 Palermo, Italy; giuseppe.cavallaro@unipa.it (G.C.); giuseppe.lazzara@unipa.it (G.L.)
4 School of Science, Monash University Malaysia, Selangor 47500, Malaysia; calvin.goh@monash.edu (C.B.S.G.); lee.sui.mae@monash.edu (S.M.L.)
* Correspondence: pooria.pasbakhsh@monash.edu

Received: 30 August 2019; Accepted: 24 September 2019; Published: 29 September 2019

Abstract: Extensive usage of long-lasting petroleum based plastics for short-lived application such as packaging has raised concerns regarding their role in environmental pollution. In this research, we have developed active, healable, and safely dissolvable alginate-pectin based biocomposites that have potential applications in food packaging. The morphological study revealed the rough surface of these biocomposite films. Tensile properties indicated that the fabricated samples have mechanical properties in the range of commercially available packaging films while possessing excellent healing efficiency. Biocomposite films exhibited higher hydrophobicity properties compared to neat alginate films. Thermal analysis indicated that crosslinked biocomposite samples possess higher thermal stability in temperatures below 120 °C, while antibacterial analysis against *E. coli* and *S. aureus* revealed the antibacterial properties of the prepared samples against different bacteria. The fabricated biodegradable multi-functional biocomposite films possess various imperative properties, making them ideal for utilization as packaging material.

Keywords: pectin; alginate; biocomposites; food packaging; degradable films

1. Introduction

Packaging has rapidly developed in different industries, especially in the food industry, because many foodstuffs are being supplied in packages in developed countries. During the past decades, polymers were used for food packaging extensively because they have advantages in comparison to traditional material (e.g., glass, tin plate) [1]. Plastics have many different compositions which allow designing appropriate packaging for each product specifically [2]. They are also inexpensive, thermo sealable, and low weight materials that are able to print and fit in different shapes [2,3].

In the last few decades, the population growth has increased the use of petroleum based non-biodegradable polymers enormously and this led to an increase in the amount of non-biodegradable waste materials in our environment [4]. A big portion of this waste plastic will end up in the marine and land ecosystem where it suffers degradation and fragmentation [5]. Microplastic debris (<5 mm), such as microcapsules or microbeads proliferate, migrate and accumulate in natural habitats [6,7]. The debris can be dangerous for marine animals in the ocean and also immigrate from the ocean and deposited on beaches [6].

In the past few decades, many studies have been done about biodegradable materials for various applications such as biomedical, packaging, agronomical, and textile industry. This is due to their excellent advantages over non-degradable materials including biodegradation and biocompatibility. They are also processable and eco-friendly [8]. Poly glycolic acid, poly lactic acid, and polydioxanone are the poly(α-hydroxy acid) and most usual synthetic biodegradable polymers [9]. Furthermore, other types of biopolymer that have attracted researchers include cellulose, chitosan, collagen, pectin, and alginate [10].

Pectin is a hetero-polysaccharide composed of (1,4)-α-linked galacturonic acid and (1,2)-linked rhamnose. There are also galactosyl and L-arabinosyl side chains in the pectin structure [11]. It is found in plants which cause strength and pliability and is common in many fruits and vegetables. This polysaccharide has several functions. For example, cell signaling, adhesion, proliferation, and differentiation [12]. Pectin is applicable in foodstuffs production as a stabilizer and gelling ingredient because of its gel-forming ability. Moreover, its good gelling properties, biocompatibility, and biodegradability make it an interesting novel biomaterial for biomedical applications (e.g., pharmaceutic and cosmetic) [13,14]. Pectins can be low-methoxyl or high-methoxyl based on their degree of methylation [15]. Properties of pectin such as gelation, solvability, and film formation are dependent on the degree of methylation [16]. The origin and processing manners determine the degree of methyl-esterification, and the main source of commercial pectins is apple pomace (low degree of methylation = 24%) and citrus peels (high degree of methylation = 74%).

Alginate as a natural polysaccharide is isolated from brown algae, and its monomers are α-L-guluronic acid and (1,4)-linked β-D-mannuronic acid [17]. This natural biopolymer is widely used for producing films, gels, suspensions, and emulsions [18]. About 30,000 metric tons of alginate is produced annually, originating from brown algae (genera Laminaria and Macrocystis) [19]. Alginate is renowned for its biocompatible and biodegradable properties as well as its low price. An important characteristic of alginate is the ability of functional groups (carboxyl and hydroxyl) of the G blocks to react with polyvalent cations (e.g., Ca^{2+}, Al^{3+}, and Fe^{3+}) [20]. Among the divalent ions, calcium ions are commonly reacted with alginate to form a low solubility polymer. In general, G block longitude characterizes the ability and selectivity of the alginate in forming these interactions, whereas M and MG blocks are almost without selectivity. M and G blocks associate through ions to form a 3D structure ("egg-box") [21,22]. This triggers an anion exchange process in which the water-soluble alginate exchanges its counter ions with Ca^{2+}. This ionic crosslinking leads to cold-setting and heat-stable films. Due to the hydrophilicity of alginate films, they are crosslinked for improvement of their resistance in water, mechanical properties, and coherence [23].

Selecting convenient materials and packaging techniques are necessary for the maintenance of the product freshness and modality. Therefore, packaging materials which react with the foodstuffs and preserve them are of interest [2]. Active packaging is described as the packaging with components in the material or packaging empty space which improves the performance of the package [24]. Antimicrobial packaging has a great influence on the food packaging industry due to controlling the bacteria growth, lengthening the shelf-life, and ensuring the health of foodstuffs [25]. In this system, antimicrobial agents are embedded in the polymeric matrix. Biopolymers can be used for the construction of a biodegradable antimicrobial packaging. Essential oils [26]), plant extract [27], bacteriocin [28], enzymes [29], chitosan [30], organic acids [31], metallic nanoparticles [32], and chelating agents [33] are the antimicrobial agents which have been used in biodegradable polymers.

Various antimicrobial agents have been incorporated in the pectin films to create antimicrobial active packaging which prolongs the shelf-life and reduces the bacteria growth on the product surface [34]. For instance, Grau et al. [35] prepared pectin and apple puree based films with oregano, cinnamon, or lemongrass essential oil incorporated in them. Developed samples possessed antimicrobial properties for *E. coli* bacteria, and the oregano edible film had most efficiency against assayed bacteria. In the other study, Ravishankar et al. [36] incorporated cinnamaldehyde and carvacrol in apple puree and pectin films at 0.5%, 1.5%, and 3% (*w*/*w*) and evaluated their antibacterial activity.

Results revealed that films including carvacrol have stronger antibacterial activity against *E. coli* and *S. enterica* in comparison to samples containing cinnamaldehyde.

On the ground of wide use of biopolymers and biocomposite in recent technologies, researchers are interested in developing mechanisms which are capable of healing and repairing the bio-based material after fracture, in order to restore the mechanical strength of primary material. The accustomed repairing methods of bio-based materials are quite complicated and can be expensive and time-consuming [37,38]. Hence, a novel healing method which employs a simple, safe, and environmentally friendly approach is highly desirable in order to bridge the damaged zones in bio-based materials by the formation of new covalent bonds.

In this study, healable and safely dissolvable alginate–pectin based biocomposites have been prepared. These films are applicable in the food packaging industry. Physico-chemical properties of all biocomposites were extensively investigated with mechanical and thermal analysis and wettability test. The morphological study was used to determine the structural properties. Tensile properties indicated that the fabricated samples have mechanical properties in the range of commercially available packaging films while possessing excellent healing efficiency. Biocomposite films exhibited highly hydrophobic properties, while thermal analysis showed that crosslinked biocomposite films possess higher thermal stability in temperatures below 120 °C. Biocomposite films exhibited antibacterial activity against various bacteria. Therefore, this study deals with the development of biodegradable films for foodstuff packaging application.

2. Materials and Methods

2.1. Materials

Pectin from the citrus peel (LM pectin, galacturonic acid, >74.0%) and lactic acid (ACS reagent, ≥85%) were obtained from Aldrich (Saint Louis, MO, USA). Medium range molecular mass sodium alginate (Manugel GHB, FMC Biopolymer, Ayrshire, UK, 37%(M)–63%(G)) was used as received. Ascorbic acid (99%) and calcium chloride dehydrate (99+%) were obtained from ACROS Organics (Geel, Belgium), while tri-sodium citrate dehydrate (294.10 g/mol, 99+%) was purchased from Merck (Darmstadt, Germany).

2.2. Preparation of Biocomposite Films

Aqueous alginate and pectin solutions containing 1 wt% of each biopolymer with mixing ratio of 1:1 was prepared by dissolving 0.7 g of pectin and 0.7 g of alginate together in 70 mL of deionized water and stirring for 4 h (500 rpm) using a magnetic stirrer at room temperature. Then, 5 wt% AA and 5 wt% LA, with reference to the total polymer weight, were added to the solution and stirred for another 4 h. The resulting solution was poured into a glass petri dish (diameter, 15 cm) and was dried in the oven for 24 h at 35 °C to form a thin film (Figure 1).

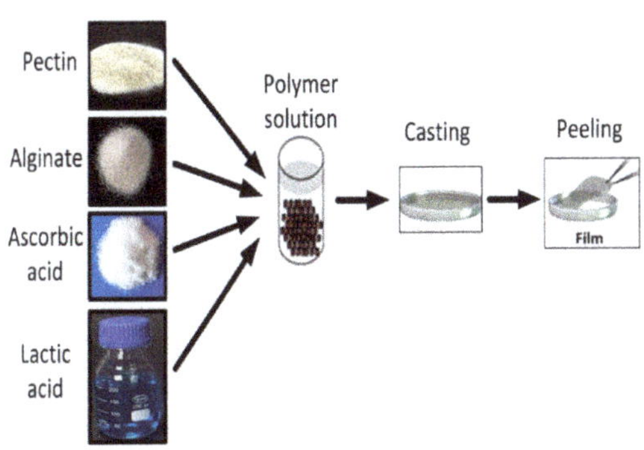

Figure 1. Schematic diagram explaining preparation of functionalized pectin–alginate films by solution casting method.

2.3. Crosslinking and Decrosslinking of Biocomposite Films

Samples were crosslinked in 2 wt% calcium chloride solution and then decrosslinked in 5 wt% sodium citrate solution. The crosslinking time of 2 min was observed to be sufficient for complete crosslinking of the biocomposite films. However, the decrosslinking time varied for different compositions and has been reported.

2.4. Healing of Biocomposite Films

Separated strips of pectin-alginate based biocomposite films were healed and stuck together after the tensile test. Briefly, strips were decrosslinked in a solution of 5 wt% sodium citrate and then, after drying, a thin line of water applied on the edge of the films by a syringe. Then, the edge of the films were pressed together (Figure 2). Samples were crosslinked again in 2 wt% calcium chloride solution after drying.

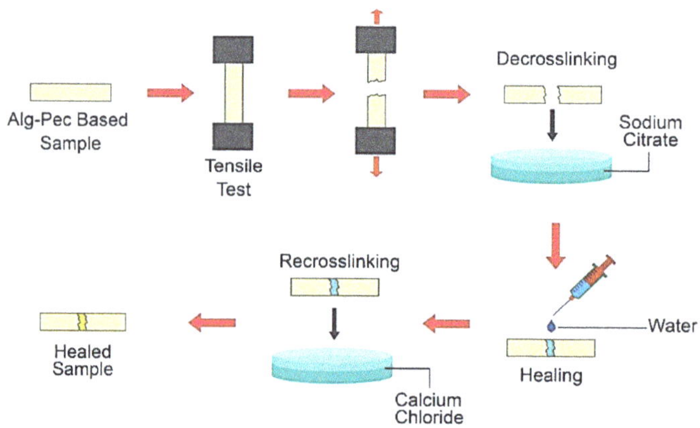

Figure 2. Healing process of alginate–pectin (Alg–Pec) based samples.

2.5. FE-SEM Observation

Morphological information of biocomposite films were revealed with ultra-high resolution field emission scanning electron microscope (Model SU8010, Hitachi, Tokyo, Japan). Before imaging, the samples were coated with a thin layer of platinum (4–7 nm) using a Q150R S rotary-pumped sputter coating system (Quorum Technologies, Lewes, UK) at 2.5 KV before being observed to prevent electrostatic charging.

2.6. FT-IR Spectrometry

FT-IR spectra were recorded using a spectrometer (Model IS10, Thermo scientific, Waltham, MA, USA) to investigate the chemical composition and interactions of the biocomposite films. All spectra were recorded using a 0.4 cm^{-1} resolution with 32 scans per sample in the 4000 to 600 cm^{-1} wavelength region.

2.7. Hydrophilicity Analysis

The Water contact angle of the samples was measured with an optical device (Model 250, rame-hart instrument, Succasunna, NJ, USA). After fixing the samples on a support surface, the water contact angle was determined with the sessile drop method for 80 s. At least five measurements were performed on each sample and the volume of the water droplet was 2 ± 0.2 μL.

2.8. Thermogravimetric Analysis

The thermogravimetric analysis (TGA) is a thermoanalytical technique that provides a clear picture of the thermal stability and the interaction of polymer blend components. TGA was performed using a thermogravimeter (Model Q50, TA Instruments, New Castle, DE, USA). The weighed samples (6 mg) were heated with a temperature range of 25–700 °C at a rate of 10/min under the nitrogen flow 60 cm^3 min^{-1} for the specimen and 40 cm^3 min^{-1} for the balance.

2.9. Tensile Property

Mechanical characterization of alginate, pectin, and their biocomposite samples was carried out by tensile analyzer (Model 5982, Instron, Norwood, MA, USA) through stress-strain curves. Rectangular samples (10.00 × 2.00 mm^2) were prepared and the tensile test was performed at rate of 2 mm/min. Tensile stress at the fracture point and elongation at rupture were determined. In addition, the tensile properties of healed samples were obtained. The tensile measurements were repeated 3 times and average values with standard deviation are provided. The healing efficiency (η) is defined as the ratio of the tensile strength of the healed sample (σ healed) to that of pristine ones (σ pristine):

$$\eta = \frac{\sigma \text{ healed}}{\sigma \text{ pristine}} \quad (1)$$

2.10. Antimicrobial Evaluation

Gram-negative *Escherichia coli* ATCC 25922, *Pseudomonas aeruginosa* ATCC 10145, and Gram-positive *Staphylococcus aureus* ATCC 29213 were employed to test biocomposite films antibacterial activity by disk diffusion method. All bacteria were cultured in nutrient broth (Merck, Darmstadt, Germany) overnight at 37 °C. After absorption of each culture on the nutrient agar (Merck, Darmstadt, Germany) for 15 min, prepared discs from biocomposite films (diameter = 6 mm) were placed on the agar plate. The Diameter of inhibition zone was measured after an overnight incubation at 37 °C for 18 h to determine the inhibitory activity of the biocomposite films.

2.11. Calcium Concentration

The measurements of the calcium concentration in the crosslinked samples were performed by inductive coupled plasma-optical emission spectrophotometer (Model Optima 8000, Perkin Elmer, Waltham, MA, USA). Composite films (about 10 mg) were dissolved in 20 mL of a 2 wt% sodium citrate solution and three measurements were carried out on each sample.

3. Results and Discussion

3.1. Dissolvability and Reusability Analysis

Dissolvability and reusability of casted films of pectin, alginate, and their biocomposites were analyzed by a series of crosslinking and decrosslinking processes (Table 1). Samples were crosslinked in 2 wt% calcium chloride solution and then decrosslinked in 5 wt% sodium citrate solution. It can be seen that alginate samples can go through the crosslinking–decrosslinking process multiple times without deterioration (>5 times), while they can fully dissolve in water after decrosslinking in less than 20 min. Similarly, alginate–pectin (Alg–Pec) and Alg–Pec–AA–LA samples have the ability to be crosslinked and then decrosslinked for several times while having the ability to completely dissolve in water in 10 min. In contrast, pectin samples did not possess the ability to be crosslinked after the first decrosslinking. Interestingly, these samples were able to be dissolved in the decrosslinking solution, while this was not the case for other samples and they had to be taken to a beaker containing water to start dissolving.

Table 1. Dissolvability and reusability analysis of pectin, alginate, and their biocomposite films.

Sample	Thickness (mm)	Crosslinking Time (min)	Decrosslinking Time (min)	Crosslinking Repeatability (times)	Decrosslinking Repeatability (times)	Complete Dissolving after Decrosslinking (min)
Alginate	0.043 ± 0.008	2	7	>5	>5	15
Pectin	0.038 ± 0.15	2	1	-	-	2
Alg–Pec	0.043 ± 0.013	2	6	>5	>5	10
Alg–Pec–AA–LA	0.056 ± 0.01	2	6	>5	>5	9

Potential utilization and safe disposal of the fabricated alginate–pectin based films are illustrated in Figure 3.

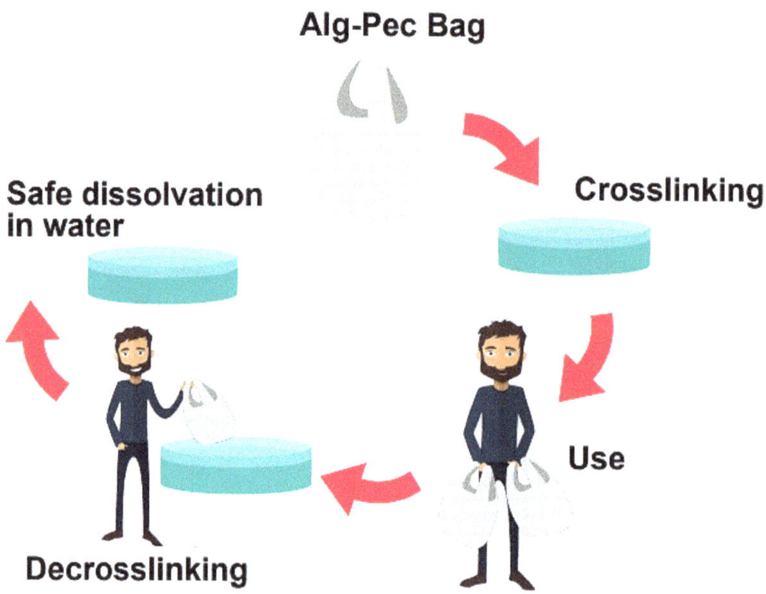

Figure 3. Potential utilization and disposal of Alg–Pec based films.

FE-SEM micrographs of crosslinked alginate, pectin, and their biocomposites are reported in Figure 4. As shown in Figure 4A, alginate films have a semi-rough surface with obvious patches of smooth surfaces, while high content of roughness can be observed all over the surface of the pectin films (Figure 4B). In the case of Alg–Pec biocomposite films (Figure 4C) and Alg–Pec–AA–LA (Figure 4D) it can be seen that the surface roughness in these samples are similar to the pectin samples. It has been reported that surface roughness has a direct effect on the hydrophilicity of material [39]. Hence, it is expected that an Alg–Pec biocomposite film possesses different degrees of hydrophilicity than pristine alginate film.

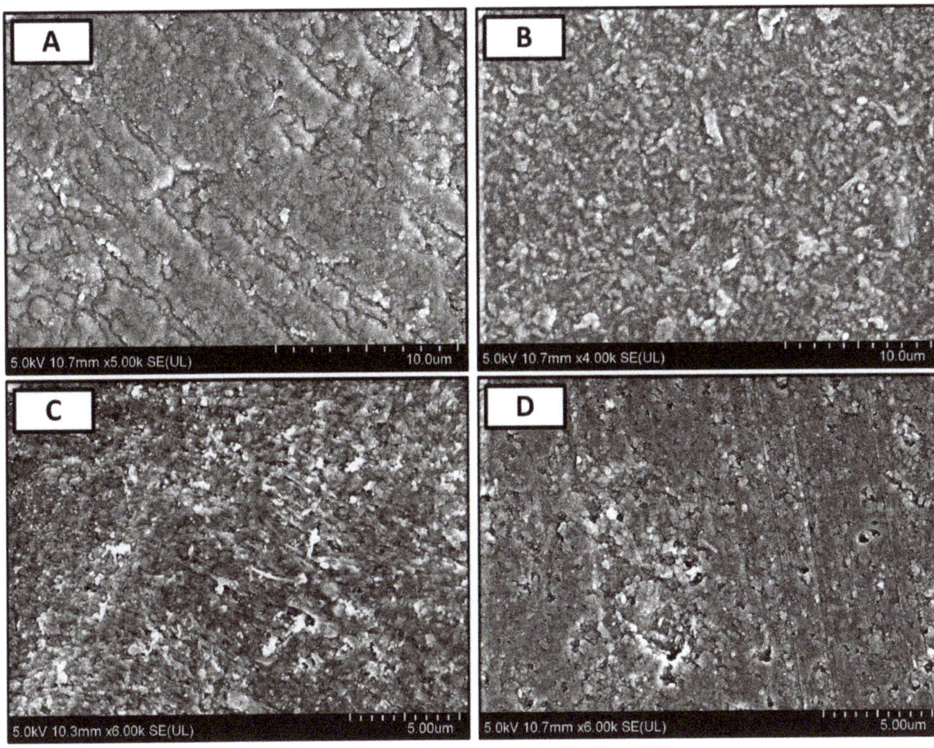

Figure 4. FE-SEM micrographs of (**A**) crosslinked alginate, (**B**) pectin, (**C**) Alg–Pec, and (**D**) Alg–Pec–AA–LA biocomposite films.

3.2. FT-IR Spectrometry

The FT-IR spectra of crosslinked films of alginate, pectin, and their biocomposites are reported in Table 2 and Figure 5. For a pure pectin film, a wide peak between 3600 and 3000 cm^{-1} is attributed to the stretching of O–H because of hydrogen bonding interactions in the galacturonic acid [40]. Due to bending vibrations of CH, CH_2, and CH_3, a mildly sharp peak between 3000 and 2500 cm^{-1} is observed. Strong peaks at 1732 and 1605 cm^{-1} are assigned to the –CO of the methyl ester group (–$COOCH_3$) and asymmetric stretching of the carbonyl group of the carboxylate ion (COO^-), respectively, and a weaker peak can be seen at 1435 cm^{-1} due to the symmetric stretching of the carboxylate ion [40]. Stretching vibrations of C–O–C and C–C of the glucose ring appeared between 1360–800 cm^{-1} which is the "finger print" region and is specific for each compound [41,42].

Table 2. FTIR-ATR main vibrational frequencies of pectin–alginate based films.

	Band Frequency (cm^{-1})			
Sample	**–OH**	**COO^- (asym)**	**COO^- (sym)**	**$COOCH_3$**
Pectin	3378	1605	1435	1732
Alginate	3264	1591	1416	-
Alg–Pec	3300	1601	1417	1738
Alg–Pec–AA–LA	3282	1594	1417	1736

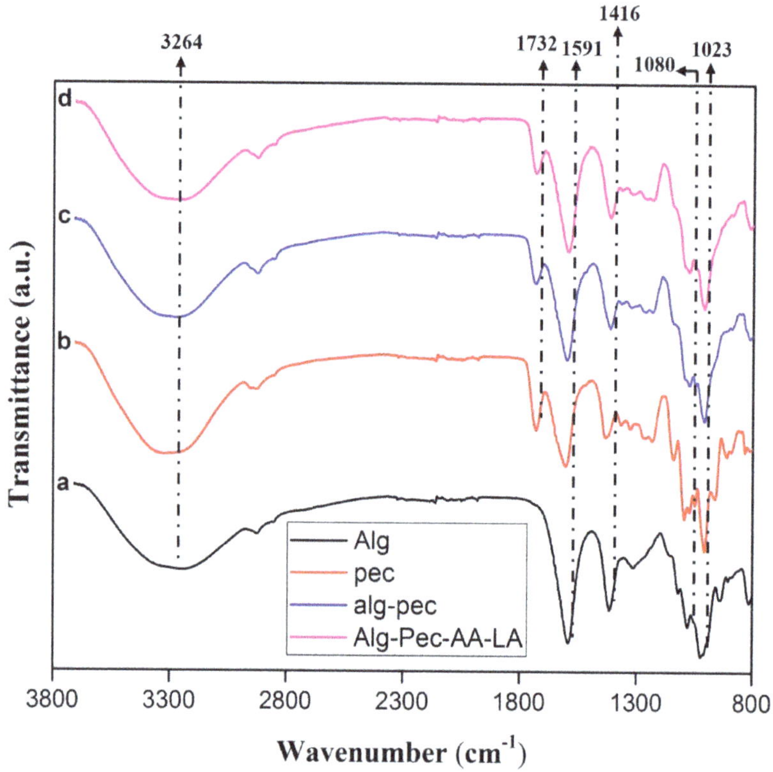

Figure 5. FT-IR spectra of crosslinked (**a**) alginate, (**b**) pectin, (**c**) Alg–Pec, and (**d**) Alg–Pec–AA–LA films.

In the Alg–Pec samples, the shifting of –OH and asymmetric stretching vibrations of COO^- peak from 3378 to 3300 cm^{-1} and from 1605 to 1601 cm^{-1}, respectively, is due to the hydrogen bonding interactions between the polar groups of both polymers and polyglycerol. Actually, the shifting of the vibrational peaks to the lower wavenumbers is due to the mixing of the miscible Alg–Pec blend which leads to a balance between the enthalpy and entropy and consequently decreasing the free energy of the system. This result revealed that hydrogen bonding interaction between polymers in the blend system is more powerful in comparison to hydrogen bonding in the pure polymers [41,42].

3.3. Hydrophilicity Analysis

The water contact angle of biocomposite samples is shown in Figure 6. It was observed that the water contact angle values of Alg–Pec films decreased in the presence of alginate which is attributed to the hydrophilicity of Alg–Pec samples due to the hydrophilic functional groups such as hydroxyl, carboxylate, and ether in the alginate structure. In addition, as described in the morphology analysis section, pectin and Alg–Pec films possess higher surface roughness in comparison to alginate films; hence, they have higher values of contact angle. It has been reported that surface roughness has a direct effect on the contact angle of surfaces, and an increase in surface roughness will lead to an increase in hydrophobicity of surfaces [39]. Moreover, the water contact angle decreased during measurements due to the droplet absorption and extension. Similar results were reported for pectin based biocomposites [43,44] and other biopolymers [45]. Alginate–pectin samples functionalized by AA and LA had an insignificant increase in their surface hydrophilicity (50°) in comparison to pristine Alg–Pec samples (52°), indicating that an incorporation of AA and LA did not have an effect on surface properties of these samples.

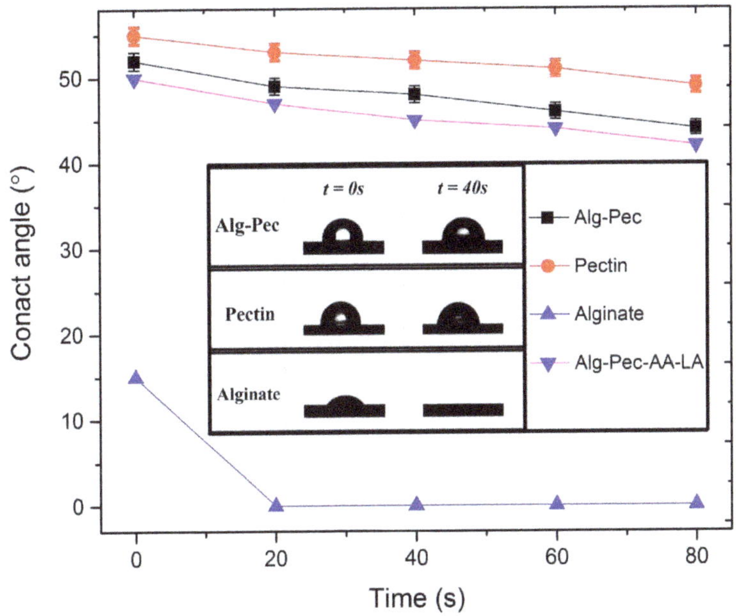

Figure 6. Water contact angle evaluation of alginate–pectin based films over time.

3.4. Tensile Property

Mechanical properties of the alginate, pectin, and their biocomposite films are presented in Table 3. It can be seen that the alginate film exhibit higher tensile strength in comparison to the pectin and Alg–Pec films, attributed to the stronger molecular structure and lower thickness. Oakefull et al. [46] investigated the interaction between alginate and pectin through polyguluronate blocks and methyl ester region that caused a rigid packed structure.

Table 3. Tensile properties of the alginate, pectin, and their biocomposite films before and after crosslinking [a].

Sample	Crosslinking Time (min)	Thickness (mm)	Tensile (MPa)	Elongation (%)
Alginate	0	0.024 ± 0.01	29.1 ± 3.1	8.16 ± 0.9
Alginate	2	0.043 ± 0.008	26.05 ± 4.7	7.3 ± 1.8
Pectin	0	0.035 ± 0.012	18.7 ± 3.2	3.9 ± 0.5
Pectin	2	0.038 ± 0.15	16.8 ± 2.4	3.5 ± 1.2
Alg–Pec	0	0.033 ± 0.014	20.1 ± 1.7	11.8 ± 1.5
Alg–Pec	2	0.043 ± 0.013	23.4 ± 0.9	9.7 ± 2.6
Alg–Pec–AA–LA	0	0.039 ± 0.14	19.3 ± 2.7	10.9 ± 1.3
Alg–Pec–AA–LA	2	0.048 ± 0.01	22.7 ± 1.8	9.1 ± 0.8

[a] Errors represent standard deviation for three independent measurements.

Tensile strength and elongation at break of alginate and pectin samples were decreased after crosslinking perhaps due to the fact that a crosslinking reaction occurred in an aqueous solution containing a crosslinking agent. Therefore, some water stayed in the interchain space in the network. Indeed, crosslinked dry samples are thicker in comparison to uncrosslinked ones. Russo et al. [47] reported a study about the crosslinking of alginate–polyglycerol with calcium ion that confirms the mentioned hypothesis. The crosslinked samples lost their flexibility due to the reduction of chain mobility resulting from crosslinking reaction. It should be noted that differences are significant considering the experimental errors. Conversely, a small increase was observed in the tensile value of crosslinked Alg–Pec based films compared to the same uncrosslinked film. This could be attributed to the fact that chain entanglement and connection parts in the blended film reduced the swelling. Incorporation of AA and LA led to a small decrease in value of tensile strength and elongation of Alg–Pec films.

3.5. Healing Property

In order to evaluate the heal-ability of these films, rectangular shaped strips of biocomposite films of Alg–Pec with a size of 10.00 × 2.00 mm^2 were pulled by the tensile machine until failure and then un-crosslinked by being dipped in a 2% sodium citrate solution for 5 min. After drying, both pieces of the film were patched together by applying a thing line of water, which acted like glue. A FE-SEM micrograph of these healed films is illustrated in Figure 7. A smooth surface can be seen at the patched line and intersection of both films, which indicates excellent adhesion between the surfaces of both films.

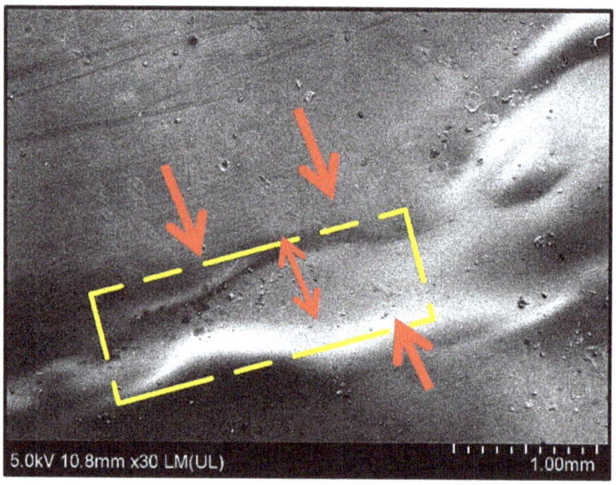

Figure 7. FE-SEM micrographs of healed Alg–Pec films. Arrows are pointing at healed regions and boundaries.

Healing efficiency of alginate and alginate–pectin based films is tabulated in Table 4. It can be seen that the healing process is considered to be highly effective in providing the strong bonding between two ruptured strips. This is due to the fact that the line of water applied to the decrosslinked area of the sample dissolves the surface in contact on both samples and this surface provides new bonding between samples after drying, which leads to impressive healing efficiency of the sample. A pristine alginate film with 92.5% healing efficiency pretty much sustains its mechanical strength, even after a process of crosslink–decrosslink. However, the pristine pectin film lost its rigidity after the decrosslinking process. Hence, the healing efficiency of this sample was not obtained. Healing efficiency of alginate–pectin based samples with (87.6%) or without AA and LA (84.1%) revealed high healing capacity of these samples, indicating the high potential of these samples to be utilized in packaging applications.

Table 4. Healing efficiency of crosslinked alginate and alginate–pectin based films.

Sample	Crosslinking Time (min)	Thickness (mm)	Tensile (1st healing) (MPa)	Elongation (1st healing) (%)	Healing Efficiency (1st healing) (%)	Tensile (2nd healing) (MPa)	Elongation (2nd healing) (%)	Tensile (3rd healing) (MPa)	Elongation (3rd healing) (%)
Alginate	2	0.043 ± 0.008	24.1 ± 2.4	6.5 ± 1.1	92.5	20.5 ± 1.8	5.1 ± 0.3	16.4 ± 1.3	3.8 ± 0.4
Alg–Pec	2	0.043 ± 0.013	20.5 ± 1.7	9.1 ± 1.5	87.6	15.8 ± 1.5	7.4 ± 1.2	12.1 ± 0.9	4.9 ± 0.5
Alg–Pec–AA–LA	2	0.048 ± 0.01	19.1 ± 1.8	8.5 ± 0.8	84.1	14.5 ± 0.7	6.1 ± 0.5	10.7 ± 1.1	4.2 ± 0.7

In order to examine the repeatability of the healing process on alginate–pectin based films, tensile strength was obtained from samples after every healing. As shown in Table 4, the value of tensile strength gradually reduces after every healing. For instance, in pristine alginate film, the value of the tensile strength decreased from 26 MPa to 16.4 after three healings. The noteworthy and interesting observation was that these samples never break from the healed line, indicating that the healing process completely bonds the separated pieces of the film after every breakage (Figure 8). Hence, the reduction in tensile strength of the films was associated with elongation of the polymer chains which reduces the overall strength of the samples. Even though the mechanical properties of these films reduced after every healing, the heal-ability of these films could be beneficial for various applications, since it is favorable for the environment and eliminates the costs associated with the replacement.

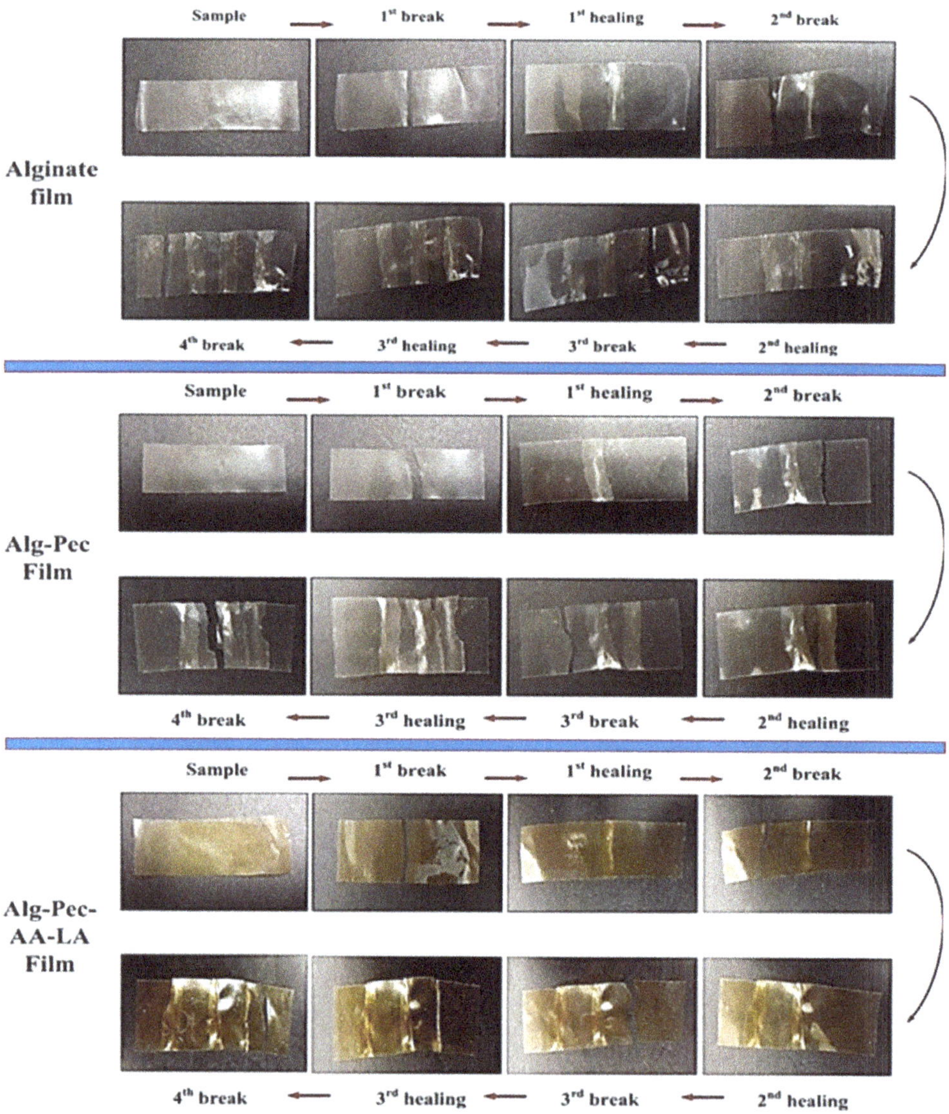

Figure 8. Repeated healing of alginate and alginate–pectin based biocomposites.

Polymeric packaging can be damaged by the mechanical, chemical, and thermal stimulus [37]. Each year an overwhelming amount of foodstuff with damaged packaging is described due to deterioration caused while transportation. Healable polymeric packaging material will effectively prolong the usability and reusability of the packages, which directly reduces the food wastage and plastic pollution (Figure 9). Since progress in healable materials is important for developed societies, several researchers proposed healing mechanisms for packaging applications. For instance, Andersson et al. [48] investigated specific microcapsules as self-healing agents in paperboard coatings. The microcapsules were coated on the surface of paperboard and the healing agent was released while external forces were applied during creasing and folding operations. Results indicated that plasticization of the coating hindered the crack propagation while releasing the hydrophobic healing agent from the microcapsules increased the hydrophobicity and consequently coating properties were preserved.

Figure 9. Features and functions of safely healable packaging materials.

3.6. *Thermogravimetric Analysis*

Figure 10 represents TGA curves of alginate, pectin, and their biocomposites before and after crosslinking. The weight of the samples decreased in the range of 25–120 °C due to the removal of moisture, followed by sharp weight loss at ca. 220 °C because of the biopolymer degradation. Lower rate of weight loss can be seen in crosslinked samples from 25 to 120 °C, indicating that crosslinking leads to an increase in thermal stability at temperatures below 120 °C and delays the degradation temperature. The removal of water has caused the weight loss of crosslinked samples. Since the weight loss below 120 °C is associated with the elimination of retained water, the crosslinked outer layer of the films provides a barrier that reduces the amount of evaporation of these water molecules and therefore diminishes the rate of mass loss.

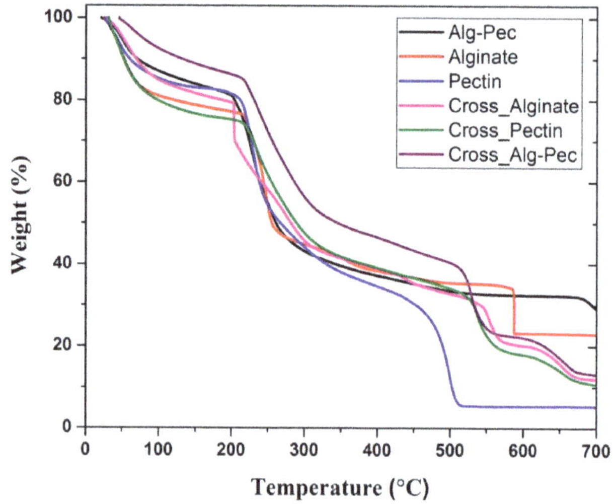

Figure 10. Thermogravimetric analysis (TGA) curves of alginate, pectin, and their biocomposites before and after crosslinking.

Thermal properties of biocomposite samples are reported in Table 5. Crosslinked films of Alg–Pec show the lowest value of weight loss at 100 °C which is important for packaging applications. This indicates the stability of these samples and they can be used in the packaging of hot food and beverages. In addition, the value of weight loss at 100 °C reduced for both alginate and Alg–Pec samples after crosslinking, while the value of temperature at 5% weight loss increased. This indicates the notable effect crosslinking on improving the thermal stability of these samples in temperatures below 100 °C. In contrast, crosslinking did not lead to an improvement of thermal properties of pectin samples.

Table 5. Thermal stability parameters of alginate, pectin, and their biocomposite films before and after crosslinking.

Sample	Temperature at 5% Weight Loss (°C)	Weight loss at 100 °C (%)	Residual Matter at 700 °C (%)	Temperature at Maximum Weight Loss Rate (°C)
Alginate	41.05	18.96	22.95	249, 587
Pectin	42.51	14.72	5.33	228, 498
Alg–Pec	49.10	12.98	29.10	229, 690
Crosslinked Alginate	51.18	15.13	12.07	203, 553
Crosslinked Pectin	40.97	20.17	10.67	232, 537
Crosslinked Alg–Pec	78.11	7.48	13.21	240, 527

Interestingly, studying the percentage of residual matter at 700 °C revealed that crosslinked samples of alginate and Alg–Pec had lower values of residue char in comparison to their pristine samples, while the opposite was true for pectin samples. This indicates that the existence of Ca ions between the bonds of alginate leads to further degradation and reduces the thermal stability at high temperatures in comparison to the pristine alginate sample. However, it was observed that temperature values at maximum weight loss rate of crosslinked samples of pectin and Alg–Pec were lower in comparison to their pristine samples, indicating that crosslinking improved the thermal stability of samples containing pectin at high temperatures. It must be mentioned that the TGA result of Alg–Pec–AA–LA samples was not included in the discussion since no major variation in the result was observed in this sample in comparison to the pristine Alg–Pec sample.

3.7. Antimicrobial Evaluation

Different bacteria were employed to test the antimicrobial activities of the alginate–pectin biocomposite films containing AA and LA by disk diffusion method. As shown in Table 6, biocomposite films were more effective against *E. coli* and *S. aureus*, while no antibacterial activity was observed in the pristine Alg–Pec sample or Alg–Pec samples containing either AA or LA. It has been reported that AA has effective antibacterial activity when it is used in combination with LA. In a study conducted by Tajkarimi et al. [49], it was reported that the incorporation of 0.4% AA and 0.2% LA inhibited the growth of *E. coli* O157:H7 in carrot juice, indicating their potential to be utilized as preservative. AA absorbs the oxygen and prevents reaching of oxygen to *E. coli* O157:H7, while the synergistic effect of AA and LA can be same as the synergetic effect of air, oxidizing agents, transition metal ions against enteroviruses [49].

Table 6. Antimicrobial activity of functional biocomposite films.

Bacteria	Diameter of Zone of Inhibition (mm)
E. coli	9.0 ± 1.5
P. aeruginosa	6.0 ± 2.0
S. aureus	8.0 ± 1.0

Varying antimicrobial effects of the biocomposite films were attributed to the different types of test bacterial pathogens used for antimicrobial evaluation, particularly in terms of the pathogen microbial structures. *P. aeruginosa* is known to be a heavy producer of extracellular matrices. Such matrixes blocks and resists the movement of the antimicrobial component of the film towards the cells themselves [50]. Conversely, as *S. aureus* is the only gram-positive bacteria tested, it has a peptidoglycan cell wall that enables it to avoid lysis upon damage of its sensitive membrane [51]. Being gram-negative, *E. coli* and *P. aeruginosa* both posess outer membranes that coat their cells and enable them to further repel antibacterial components away from the cells. These findings point towards the antibacterial potential of the biocomposites rather than of its mode of action.

Non-polar components of the film, however, may also be antibacterial and hence only be visually active upon direct cell contact with the film itself. This is because the disk diffusion assay employed here depends largely on the polarity, size, and chemical structure of the tested compounds as the agar medium comprises largely of water. *S. aureus* and *E. coli* are both inhibited (zones) by the film away from the edges of the disk due to diffusion of antibacterial compounds of which at varying concentrations, have a stronger effect against *E. coli* than *S. aureus* at a lower concentration (further away from the disk). However, *P. aeruginosa* is resistant against the film.

3.8. Calcium Concentration

The calcium content of alginate, pectin, and their biocomposite films is shown in Table 7. Alginate samples have more calcium content followed by composite and pectin samples. It should be noted that differences are far above the statistical error (1%). These results are consistent with previous results about the swelling and mechanical properties of the samples and augment the hypothesis that alginate film was crosslinked more effectively in comparison to pectin sample. Same results were observed in studies conducted by Sriamornsak et al. [52] and Da Silva et al. [53] in which alginate films had more calcium in comparison to pectin films.

Table 7. Calcium content of pectin, alginate, and their biocomposite films.

Sample	Calcium Content (mg/g sample)
Alg	32.18
Pec	22.7
Alg–Pec	27.8
Alg–Pec–AA–LA	30.78

4. Conclusions

In this study, we have prepared active, healable, and safely dissolvable alginate–pectin based biocomposites which have potential to be used in the food packaging industry. The morphological study revealed the rough surface of these biocomposite films, while antibacterial analysis indicated effective activity of functionalized films against various bacteria. Mechanical properties indicated that the fabricated biocomposite films have mechanical properties in the range of commercially available packaging films while possessing excellent healing efficiency. On the grounds of extensive use of biopolymers and biocomposites in various applications, the ability to recover the mechanical and physical properties of pristine materials after damage is an attractive feature. The fabricated multifunctional biocomposites with simple, low-cost, and environmentally friendly heal-ability properties proposed in this study have tremendous potential to be utilized in various applications such as packaging material.

Author Contributions: Conceptualization, G.L. and P.P.; data curation, M.M., H.Y., G.C., C.B.S.G., S.M.L., and A.S.; funding acquisition, P.P.; writing—original draft, M.M.

Funding: This research received no external funding.

Conflicts of Interest: The authors declare no conflict of interest.

References

1. de Souza, A.C.; Ditchfield, C.; Tadini, C. Biodegradable Films Based on Biopolymers for Food Industries. In *Innovation in Food Engineering: New Techniques and Products*; CRC Press: Boca Raton, FL, USA, 2009; pp. 511–537. ISBN 0309-1740.
2. López-Rubio, A.; Almenar, E.; Hernandez-Muñoz, P.; Lagarón, J.M.; Catalá, R.; Gavara, R. Overview of active polymer-based packaging technologies for food applications. *Food Rev. Int.* **2004**, *20*, 357–387. [CrossRef]
3. Camilloto, G.P.; Pires, A.C.S.; Soares Nde, F.; Araújo, E.A.; Andrade, N.J.; Ferreira, S.O. Effect of active packaging incorporated with triclosan on bacteria adhesion. *J. Food Sci.* **2010**, *75*, E557–E564. [CrossRef] [PubMed]
4. Srikanth, P. *Handbook of Bioplastics and Biocomposites Engineering Applications*; John Wiley & Sons, Inc.: Hoboken, NJ, USA, 2011; ISBN 9781118203699.
5. Andrady, A.L. Microplastics in the marine environment. *Mar. Pollut. Bull.* **2011**, *62*, 1596–1605. [CrossRef] [PubMed]
6. do Sul, J.A.I.; Costa, M.F. The present and future of microplastic pollution in the marine environment. *Environ. Pollut.* **2014**, *185*, 352–364. [CrossRef] [PubMed]
7. Barnes, D.K.A.; Galgani, F.; Thompson, R.C.; Barlaz, M. Accumulation and fragmentation of plastic debris in global environments. *Philos. Trans. R. Soc. B Biol. Sci.* **2009**, *364*, 1985–1998. [CrossRef] [PubMed]
8. Lasprilla, A.J.R.; Martinez, G.A.R.; Lunelli, B.H.; Jardini, A.L.; Filho, R.M. Poly-lactic acid synthesis for application in biomedical devices—A review. *Biotechnol. Adv.* **2012**, *30*, 321–328. [CrossRef] [PubMed]
9. Middleton, J.C.; Tipton, A.J. Synthetic Biodegradable Polymers as Medical Devices. *Med. Device Diagn. Ind. News Prod. Suppliers* **1998**, *21*, 1–8.
10. Porta, R.; Mariniello, L.; di Pierro, P.; Sorrentino, A.; Giosafatto, C.V.L. Transglutaminase crosslinked pectinand chitosan-based edible films: A review. *Crit. Rev. Food Sci. Nutr.* **2011**, *51*, 223–238. [CrossRef]
11. Chen, C.H.; Sheu, M.T.; Chen, T.F.; Wang, Y.C.; Hou, W.C.; Liu, D.Z.; Chung, T.C.; Liang, Y.C. Suppression of endotoxin-induced proinflammatory responses by citrus pectin through blocking LPS signaling pathways. *Biochem. Pharmacol.* **2006**, *72*, 1001–1009. [CrossRef]

12. Ciriminna, R.; Chavarría-Hernández, N.; Inés Rodríguez Hernández, A.; Pagliaro, M. Pectin: A new perspective from the biorefinery standpoint. *Biofuels Bioprod. Biorefin.* **2015**, *9*, 368–377. [CrossRef]
13. Rao, M.A. Pectins: Structure, Functionality, and Uses. *Food Polysacch. Appl.* **2006**, *160*, 353–411.
14. Chen, J.; Liu, W.; Liu, C.M.; Li, T.; Liang, R.H.; Luo, S.J. Pectin Modifications: A Review. *Crit. Rev. Food Sci. Nutr.* **2015**, *55*, 1684–1698. [CrossRef]
15. Saha, N.R.; Sarkar, G.; Roy, I.; Rana, D.; Bhattacharyya, A.; Adhikari, A.; Mukhopadhyay, A.; Chattopadhyay, D. Studies on methylcellulose/pectin/montmorillonite nanocomposite films and their application possibilities. *Carbohydr. Polym.* **2016**, *136*, 1218–1227. [CrossRef]
16. Gorrasi, G. Dispersion of halloysite loaded with natural antimicrobials into pectins: Characterization and controlled release analysis. *Carbohydr. Polym.* **2015**, *127*, 47–53. [CrossRef]
17. Benavides, S.; Villalobos-Carvajal, R.; Reyes, J.E. Physical, mechanical and antibacterial properties of alginate film: Effect of the crosslinking degree and oregano essential oil concentration. *J. Food Eng.* **2012**, *110*, 232–239. [CrossRef]
18. Rhim, J.-W. Physical and mechanical properties of water resistant sodium alginate films. *LWT Food Sci. Technol.* **2004**, *37*, 323–330. [CrossRef]
19. Hay, I.D.; Rehman, Z.U.; Moradali, M.F.; Wang, Y.; Rehm, B.H.A. Microbial alginate production, modification and its applications. *Microb. Biotechnol.* **2013**, *6*, 637–650. [CrossRef] [PubMed]
20. Bedian, L.; Villalba-rodríguez, A.M.; Hernández-vargas, G.; Parra-saldivar, R.; Iqbal, H.M.N. International Journal of Biological Macromolecules Bio-based materials with novel characteristics for tissue engineering applications—A review. *Int. J. Biol. Macromol.* **2017**, *98*, 837–846. [CrossRef]
21. Abdul-Khalil, H.P.S.; Tye, Y.Y.; Saurabh, C.K.; Leh, C.P.; Lai, T.K.; Chong, E.W.N.; Fazita, M.R.N.; Hafiidz, J.M.; Banerjee, A.; Syakira, M.I. Seaweed biodegradable polymer films: A review on cellulose as reinforcement materials. *Express Polym. Lett.* **2016**, *11*, 1–51. [CrossRef]
22. Grant, G.T.; Morris, E.R.; Rees, D.A.; Smith, P.J.C.; Thom, D. Biological interactions between polysaccharides and divalent cations: The egg-box model. *FEBS Lett.* **1973**, *32*, 195–198. [CrossRef]
23. Rhim, J.-W.; Kim, J.-H.; Kim, D.-H. Modification of Na-Alginate Films by CaCl2 Treatment. *Korean J. Food Sci. Technol.* **2003**, *35*, 217–221.
24. Parker, S. Principles and Practice. *IFLA J.* **2006**, *32*, 179–180. [CrossRef]
25. García Ibarra, V.; Sendón, R.; Rodríguez-Bernaldo De Quirós, A. Antimicrobial Food Packaging Based on Biodegradable Materials. In *Antimicrobial Food Packaging*; Academic Press: San Diego, CA, USA, 2016; pp. 363–384. ISBN 9780128007235.
26. Perdones, Á.; Vargas, M.; Atarés, L.; Chiralt, A. Physical, antioxidant and antimicrobial properties of chitosan-cinnamon leaf oil films as affected by oleic acid. *Food Hydrocoll.* **2014**, *36*, 256–264. [CrossRef]
27. Bof, M.J.; Jiménez, A.; Locaso, D.E.; García, M.A.; Chiralt, A. Grapefruit Seed Extract and Lemon Essential Oil as Active Agents in Corn Starch–Chitosan Blend Films. *Food Bioprocess Technol.* **2016**, *9*, 2033–2045. [CrossRef]
28. Imran, M.; Klouj, A.; Revol-Junelles, A.M.; Desobry, S. Controlled release of nisin from HPMC, sodium caseinate, poly-lactic acid and chitosan for active packaging applications. *J. Food Eng.* **2014**, *143*, 178–185. [CrossRef]
29. Fabra, M.J.; Sánchez-González, L.; Chiralt, A. Lysozyme release from isolate pea protein and starch based films and their antimicrobial properties. *LWT Food Sci. Technol.* **2014**, *55*, 22–26. [CrossRef]
30. Bonilla, J.; Fortunati, E.; Atarés, L.; Chiralt, A.; Kenny, J.M. Physical, structural and antimicrobial properties of poly vinyl alcohol-chitosan biodegradable films. *Food Hydrocoll.* **2014**, *35*, 463–470. [CrossRef]
31. Ma, Q.; Zhang, Y.; Zhong, Q. Physical and antimicrobial properties of chitosan films incorporated with lauric arginate, cinnamon oil, and ethylenediaminetetraacetate. *LWT Food Sci. Technol.* **2016**, *65*, 173–179. [CrossRef]
32. Rhim, J.W.; Wang, L.F. Preparation and characterization of carrageenan-based nanocomposite films reinforced with clay mineral and silver nanoparticles. *Appl. Clay Sci.* **2014**, *97–98*, 174–181. [CrossRef]
33. Liu, L.S.; Jin, T.; Coffin, D.R.; Liu, C.K.; Hicks, K.B. Poly(lactic acid) membranes containing bacteriocins and edta for inhibition of the surface growth of gram-negative bacteria. *J. Appl. Polym. Sci.* **2010**, *117*, 486–492. [CrossRef]
34. Espitia, P.J.P.; Du, W.X.; Avena-Bustillos, R.d.J.; Soares, N.d.F.F.; McHugh, T.H. Edible films from pectin: Physical-mechanical and antimicrobial properties—A review. *Food Hydrocoll.* **2014**, *35*, 287–296. [CrossRef]

35. Rojas-Graü, M.A.; Avena-Bustillos, R.J.; Friedman, M.; Henika, P.R.; Martín-Belloso, O.; Mchugh, T.H. Mechanical, barrier, and antimicrobial properties of apple puree edible films containing plant essential oils. *J. Agric. Food Chem.* **2006**, *54*, 9262–9267. [CrossRef] [PubMed]
36. Ravishankar, S.; Zhu, L.; Olsen, C.W.; McHugh, T.H.; Friedman, M. Edible apple film wraps containing plant antimicrobials inactivate foodborne pathogens on meat and poultry products. *J. Food Sci.* **2009**, *74*, M440–M445. [CrossRef] [PubMed]
37. Wu, D.Y.; Meure, S.; Solomon, D. Self-healing polymeric materials: A review of recent developments. *Prog. Polym. Sci.* **2008**, *33*, 479–522. [CrossRef]
38. Ferrari, F.; Greco, A. Thermal analysis of self-healing thermoplastic matrix nanocomposite from cyclic butylene terephthalate. *J. Therm. Anal. Calorim.* **2018**, *134*, 567–574. [CrossRef]
39. Lee, W.; Jin, M.; Yoo, W.; Lee, J. Nanostructuring of a Poymeric Substrate with Wll-defined Nanometer-Scale Topography and Tailored Surface Wettability. *Langmuir* **2004**, *20*, 7665–7669. [CrossRef]
40. Cerna, M.; Barros, A.S.; Nunes, A.; Rocha, S.M.; Delgadillo, I.; Copikova, J.; Coimbra, M.A. Use of FT-IR spectroscopy as a tool for the analysis of polysaccharide food additives. *Carbohydr. Polym.* **2003**, *51*, 383–389. [CrossRef]
41. Nešić, A.; Onjia, A.; Davidović, S.; Dimitrijević, S.; Errico, M.E.; Santagata, G.; Malinconico, M. Design of pectin-sodium alginate based films for potential healthcare application: Study of chemico-physical interactions between the components of films and assessment of their antimicrobial activity. *Carbohydr. Polym.* **2017**, *157*, 981–990. [CrossRef]
42. Coleman, M.M.; Moskala, E.J. FTi.r. studies of polymer blends containing the poly(hydroxy ether of bisphenol A) and poly(ε-caprolactone). *Polymer (Guildf.)* **1983**, *24*, 251–257. [CrossRef]
43. Cavallaro, G.; Donato, D.I.; Lazzara, G.; Milioto, S. Films of halloysite nanotubes sandwiched between two layers of biopolymer: From the morphology to the dielectric, thermal, transparency, and wettability properties. *J. Phys. Chem. C* **2011**, *115*, 20491–20498. [CrossRef]
44. Cavallaro, G.; Lazzara, G.; Milioto, S. Sustainable nanocomposites based on halloysite nanotubes and pectin/polyethylene glycol blend. *Polym. Degrad. Stab.* **2013**, *98*, 2529–2536. [CrossRef]
45. Farris, S.; Introzzi, L.; Biagioni, P.; Holz, T.; Schiraldi, A.; Piergiovanni, L. Wetting of biopolymer coatings: Contact angle kinetics and image analysis investigation. *Langmuir* **2011**, *27*, 7563–7574. [CrossRef] [PubMed]
46. Oakenfull, D.; Scott, A.; Chai, E. The mechanism of formation of mixed gels by high methoxyl pectins and alginates. *Gums Stabilisers Food Ind.* **1990**, *5*, 243–264.
47. Russo, R.; Malinconico, M.; Santagata, G. Effect of cross-linking with calcium ions on the physical properties of alginate films. *Biomacromolecules* **2007**, *8*, 3193–3197. [CrossRef]
48. Andersson, C.; Järnström, L.; Fogden, A.; Mira, I.; Voit, W.; Zywicki, S.; Bartkowiak, A. Preparation and incorporation of microcapsules in functional coatings for self-healing of packaging board. *Packag. Technol. Sci.* **2009**, *22*, 275–291. [CrossRef]
49. Gyawali, R.; Ibrahim, S.A.; Abu Hasfa, S.H.; Smqadri, S.Q.; Haik, Y. Antimicrobial Activity of Copper Alone and in Combination with Lactic Acid against *Escherichia coli* O157:H7 in Laboratory Medium and on the Surface of Lettuce and Tomatoes. *J. Pathog.* **2011**, *2011*, 650968. [CrossRef] [PubMed]
50. Davies, D. Understanding biofilm resistance to antibacterial agents. *Nat. Rev. Drug Discov.* **2003**, *2*, 114–122. [CrossRef]
51. Derouaux, A.; Sauvage, E.; Terrak, M. Peptidoglycan Glycosyltransferase Substrate Mimics as Templates for the Design of New Antibacterial Drugs. *Front. Immunol.* **2013**, *4*, 78. [CrossRef]
52. Sriamornsak, P.; Kennedy, R.A. Swelling and diffusion studies of calcium polysaccharide gels intended for film coating. *Int. J. Pharm.* **2008**, *358*, 205–213. [CrossRef]
53. da Silva, M.A.; Bierhalz, A.C.K.; Kieckbusch, T.G. Alginate and pectin composite films crosslinked with Ca^{2+} ions: Effect of the plasticizer concentration. *Carbohydr. Polym.* **2009**, *77*, 736–742. [CrossRef]

polymers

MDPI

Article

Study on the Printability through Digital Light Processing Technique of Ionic Liquids for CO_2 Capture

Matteo Gillono [1,2], Annalisa Chiappone [1], Lorenzo Mendola [1], Manuel Gomez Gomez [1], Luciano Scaltrito [1], Candido Fabrizio Pirri [1,2] and Ignazio Roppolo [1,*]

1 Department of Applied Science and Technology, Politecnico di Torino, Corso Duca degli Abruzzi 24, 10129 Torino, Italy; matteo.gillono@iit.it (M.G.); annalisa.chiappone@polito.it (A.C.); lollomen92@live.it (L.M.); manuel.gomezgomez@polito.it (M.G.G.); luciano.scaltrito@polito.it (L.S.); fabrizio.pirri@polito.it (C.F.P.)

2 Center for Sustainable Future Technologies @Polito, Istituto Italiano di Tecnologia, Via Livorno 60, 10144 Torino, Italy

* Correspondence: ignazio.roppolo@polito.it; Tel.: +39-0110907412

Received: 4 November 2019; Accepted: 20 November 2019; Published: 23 November 2019

Abstract: Here we present new 3D printable materials based on the introduction of different commercially available ionic liquids (ILs) in the starting formulations. We evaluate the influence of these additives on the printability of such formulations through light-induced 3D printing (digital light processing—DLP), investigating as well the effect of ionic liquids with polymerizable groups. The physical chemical properties of such materials are compared, focusing on the permeability towards CO_2 of the different ILs present in the formulations. At last, we show the possibility of 3D printing high complexity structures, which could be the base of new high complexity filters for a more efficient CO_2 capture.

Keywords: ionic liquids; 3D printing; digital light processing; CO_2 capture

1. Introduction

Public and political opinion is currently always more interested in global warming phenomenon. Data collected through the last 150 years demonstrate that the greenhouse effect is the main reason of the global temperature increase and carbon dioxide is one of the main gases involved in this effect. The emission of a huge quantity of this gas (nearly 40 Gtonn/year in 2018 only from fossil fuels) [1] is causing an abrupt increase of its concentration from 300 ppm in the 50s to more than 400 ppm currently [2,3]. So, the capture of the gases present in the atmosphere is an urgent demand in order to decrease the greenhouse gases emissions, in particular for carbon dioxide. In this context, different technologies are available for CO_2 capture (CC), such as pre-combustion, post-combustion and oxyfuel capture [4–8]. These technologies can employ liquids, both in pre-combustion and post-combustion processes [9,10], but also solid materials in form of beads and powders packed in columns, used as gas contactors [11]. Such materials can undergo two main capture processes: physical adsorption and chemical absorption. In the physical adsorption process, the capture is mainly due to Van der Waals forces generated by the chemical potential and the affinity between the CO_2 molecules and the surface of the sorbents. The weak bond created guarantees a minimum amount of energy to be used in the regeneration phase. The most used solid CO_2 adsorbents are zeolites, activated carbons and metal organic framework (MOF) [12–15]. The chemical absorption process is more employed in post combustion processes, and involves the creation of strong irreversible chemical bonds that bring to the creation of new chemical species. The main problem related to this process is the high amount of energy required to break the chemical bonds to regenerate the CO_2 absorbed molecules. Among the most used chemical absorbents, there are the amine and lithium based materials [16–19].

In the last years, ionic liquids (ILs) have gained a lot of attention in the application of CC technologies [5,20–22]. They are salts with low melting point, usually under 100 °C and below room temperature for RTILs (room temperature ionic liquids); they possess very low vapor pressure with a consequent almost zero volatility. Furthermore, their cations and anions can be tuned and engineered in such a way that their CO_2 solubility can be increased. ILs have been largely used in adsorbing liquid blends, like aqueous alkanolamines. The main differences of these systems consist in a decreased degradation in oxidizing environments, minor corrosion rate and lower regeneration energy [23]; nevertheless, the ILs present problems caused by their high viscosity and lower CO_2 solubility compared with standard capturing liquid systems [24,25]. Supported liquid membranes (SLM), solid membranes containing ILs, have been also proposed [26–29]. Among those, some are formed of ionic liquids that were functionalized with reactive groups and used to create polymeric chains, the so-called poly ionic liquids (PILs) [30–33]. In 2007 Noble et al. proposed for the first time the use of PILs based membranes for CO_2 capture, obtaining better results than the ones of the relative neat ILs [34]. Merging the intriguing CO_2 solubility ability of ILs with the handiness of solid materials results extremely attractive for the development of CO_2 capturing devices [35]; furthermore, the possibility to shape the material with complex geometries could enable the production of smart filters. In this perspective 3D printing can be a forwarding solution [36].

3D printing is a fabrication method that enables the direct fabrication of three dimensional objects starting from a digital model [37]. Compared to traditional methods, 3D printing enables a higher level of complexity, gathered with saving of raw materials and lower energy consumption [38]. ILs were already used in 3D printing both for direct Ink Writing (DIW) [39–41], Fused Filament Fabrication (FFF) technology [42] and in photopolymerization-based technologies (i.e., stereolithography—SL—and digital light processing—DLP) [43]. The union between ILs and 3D printing was explored in many applications, such as sensors [39,44], optical data storage systems [45], chemical [46] and biomedical devices [47].

Among the different 3D printing technologies, DLP results particularly interesting for the production of functional components, in which materials and design properties are gathered. DLP is a vat polymerization 3D printing method that employs photosensitive liquid resins able to be cure upon light irradiation. In a DLP equipment, a digital micromirror device (DMD) shines the light on a photocurable formulation following a two-dimensional pixel pattern, which represents a single layer of a digital model (CAD file), inducing the layer solidification. Repeating the same procedure along the z axis, the object is built [48]. If compared to the other 3D printing technologies, DLP presents one of the highest resolution together with high output [49]. The polymeric materials typically employed in DLP are acrylic- and methacrylic- resins, which are cured by means of free-radical photopolymerization [50]. Considering the wide range of suitable resins, objects produced by DLP can now cover a wide range of applications, from biomedical to mechanical, electronic, energy and many more [51].

In this paper, we present the development of polymeric formulations printable with the DLP technique containing commercial available ILs. Those ILs are particularly suitable for DLP technology due to their intrinsic characteristic: first, in DLP liquid photocurable formulations are processed, so perfectly compatible with RTILs; second, cross-linked polymers are produced by this technology, trapping ILs and possibly avoiding phase separation or leakage; third, DLP allows a higher shape freedom and superior precision, which can be straightforward in order to produce complex devices containing ILs [52]. Here, we selected six commercially available ILs (Bmim[BF_4], Bmim[Tf_2N], Bvim[Tf_2N], Amim[Tf_2N], Bmim[ac] and $C_4N_{MA,11}$[Tf_2N], Table 1) and we fabricated 3D printed structures based on these compounds, both trapped and co-polymerized in the polymeric network. First, a polymerization study showing the photo-rheological and infrared spectroscopy kinetics is presented to assess the reactivity of the formulations containing different ILs. Second, the formulations are loaded in the DLP system and 3D printed structures. The thermo-mechanical properties of the printed structures were then investigated. Afterwards, CO_2 permeability on photo-polymerized membranes is done to study their gas transport properties. At last, very complex geometries (gyroids) are 3D printed.

2. Materials and Methods

2.1. Materials

Poly(ethyleneglycol)diacrylate (PEGDA, Mn. 575), (Bis(2,4,6-trimethylbenzoyl)-phenylph osphineoxide (BAPO) and Reactive Orange 16 were purchased from Merck (Milan, Italy) and used as received. Regarding the ionic liquids employed, 1-Butyl-3-Methylimidazolium tetrafluoroborate, 1-Butyl-3-Methylimidazolium bis (trifluorometilsulfonyl)imide, 1-Allyl-3-Methylimidazolium bis (trifluoromethylsulfonyl)imide and 1-Butyl-3-Methylimidazolium acetate were purchased from Merck (Milan, Italy); *N,N,N,N*-ButyldimethylMethacryloyloxyethylammonium bis(trifluoromethyl sulfonyl)imide and 1,4-Butandiyl-3,3′-bis-1-vinylimidazoliumbis (trifluoromethylsulfonyl)imide were purchased from Solvionic (Toulouse, France). The ionic liquids used are summarized in Table 1.

Table 1. Ionic liquids tested in this study.

Butyl-3-Methylimidazolium tetrafluoroborate Bmim[BF_4] 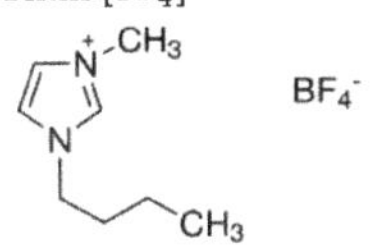	*N,N,N,N*-ButyldimethylMethacryloyloxyethyl ammonium bis(trifluoromethylsulfonyl)imide$C_4N_{MA,11}$[Tf_2N] 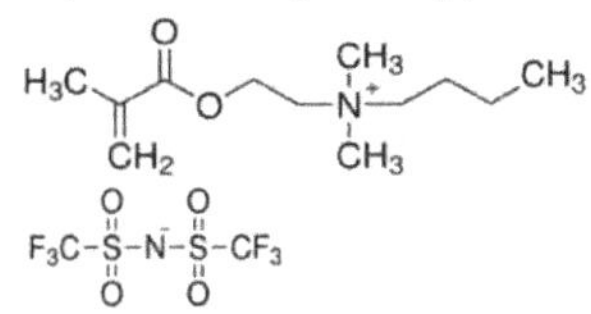
1-Butyl-3-Methylimidazolium bis (trifluorometilsulfonyl)imide Bmim[Tf_2N] 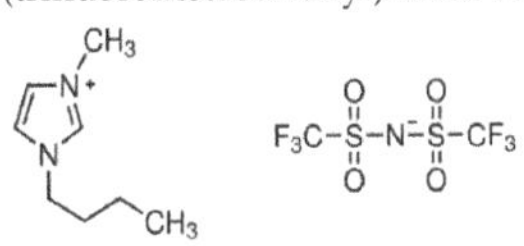	1-Allyl-3-Methylimidazolium bis (trifluoromethylsulfonyl)imide Amim[Tf_2N] 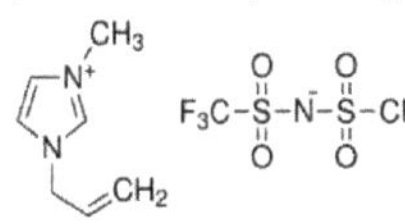
1-Butyl-3-Methylimidazolium acetate Bmim[ac] 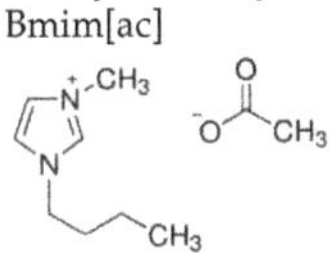	1,4-Butandiyl-3,3′-bis-1-vinylimidazoliumbis (trifluoromethylsulfonyl)imide Bvim[Tf_2N] 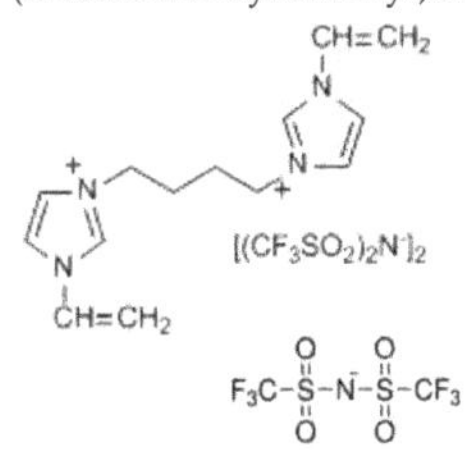

2.2. Formulation Preparation and Printing

A reference formulation (named PEGDA) was obtained by mixing Poly(ethyleneglycol)diacrylate with Reactive Orange 16 at 0.2 phr (per hundred resin), as a dye, and BAPO at 2 phr as a photoinitiator.

The formulations containing the different ILs were produced by adding 0.4 mmol/g of each IL to the reference formulation. These were named as P_Bmim[BF_4], P_Bmim[Tf_2N], P_Bmim[ac], P_$C_4N_{MA,11}$[Tf_2N], P_Amim[Tf_2N] and P_Bvim[Tf_2N] accordingly to the IL used. The influence of the IL concentration was tested for Bmim[BF_4] and for $C_4N_{MA,11}$[Tf_2N], in this case multiples of 0.4 mmol/g were added to the reference formulation. The samples were named as P_(ILname)_X, where X is the multiple of 0.4 used.

The different formulations were then printed using a DLP printer HD 2.0 (Robot Factory S.r.l., Venice, Itay), equipped with a projector as light source (X–Y resolution 50 μm corresponding to 1920 pixels × 1080 pixels, intensity 10 mW/cm^2). The layer thickness was set to 50 μm and the layer exposure time was set for each sample according to preliminary tests (values reported in Table 2). The obtained samples were first washed, gently rinsing in ethanol, and then underwent a post curing process (5 min), performed with a medium pressure mercury lamp also provided by Robotfactory equipped with a rotating platform (intensity 10 mW/cm^2).

2.3. Characterization

FTIR-ATR spectra were collected using a Tensor 27 FTIR Spectrometer (Bruker, Billerica, MA, USA) equipped with ATR tool, 32 scans were collected with a resolution of 4 cm^{-1} from 4000 to 400 cm^{-1}. Real-time photo-rheology measurements were performed using an Anton Paar rheometer Physica MCR 302 (Graz, Austria) in parallel plate mode equipped with a bottom glass plate. For the measurement a light source Hamamatsu Photonic LC8 lamp (Hamamatsu City, Japan) with visible bulb and a cut-off filter below 400 nm equipped with 8 mm light guide) was placed below the glass plate. The upper aluminum plate was placed at 0.1 mm from the glass one and the sample was kept at a constant temperature (25 °C) and under constant shear frequency of 5 rad s^{-1} at 1% of amplitude accordingly to preliminary amplitude sweep tests; light was turned on after 1 min in order to stabilize the system. Dynamic mechanical thermal analyses (DMTAs) were performed with a Triton Technology TTDMA (Mansfield, MA, USA). All of the experiments were conducted with a temperature ramp of 3 °C/min, applying a force with a frequency of 1 Hz and a displacement of 20 μm with a temperature range between −80 and 30 °C. TGA analysis were carried out with a Themo Netzsch TG 209 F1 Libra (Selb, Germany) instrument with a temperature ramp of 10 °C/min in air. For the permeability measurements, a single chamber Extrasolution Multiperm permeometer was used (Extrasolution, Pieve Fosciana, Italy). For the CO_2 uptake measurements, FT-IR was used to check the interaction between the samples and the gas. The apparatus for CO_2 atmosphere is composed by a pressurizing test chamber connected with a CO_2 cylinder and a vacuum pump used to clean the chamber and regenerate the samples. A thin layer of the printable formulations (25 μm) was deposited on the silicon wafer and photocured in nitrogen atmosphere. The samples were then degassed in vacuum (5 mbar) for 24 h to desorb the gases adsorbed. The CO_2 uptake phase was carried out by inserting the samples in the CO_2 unit for 4 h at a pressure of 4 bar. Afterwards, the samples were again subjected to a vacuum cycle (24 h at 5 mbar) and a further heating (80 °C). After every step, the samples were tested by FT-IR in order to check the CO_2 peaks in the 2400–2200 cm^{-1} range.

3. Results and Discussion

In order to determine the printability of the different formulations, investigations on the polymerization of the prepared mixtures containing the different ionic liquids were performed. The photo-rheological measurements show how the liquid formulations behave during light irradiation; in particular, the variation of the elastic modulus G′ vs. the irradiation time was investigated. The tests, reported in Figure 1, showed that three of the chosen ILs (i.e., Bmim[BF_4], Bmim[Tf_2N] and Bvim[Tf_2N]) do not influence the kinetic of polymerization of the monomer, indicating that neither the imidazole cation nor the corresponding anions inhibit the polymerization process. Instead, $C_4N_{MA,11}$[Tf_2N] showed slightly slower kinetics, which led to a delayed plateau value of G′, probably due to the ammonium-based cation, which could induce radical quenching [53]. Moreover, the IL Amim[Tf_2N], which bears an allyl group, slows the polymerization: this could be expected considering the poor reactivity of allyl monomers [54,55]. At last, it was not possible to measure the polymerization of P_Bmim[ac]: in this case no variation of the G′ modulus was observed, indicating that the acetate-based IL completely hinders the reaction. The data obtained gave an indication of the required layer exposure time during printing that was then experimentally adjusted (Table 2).

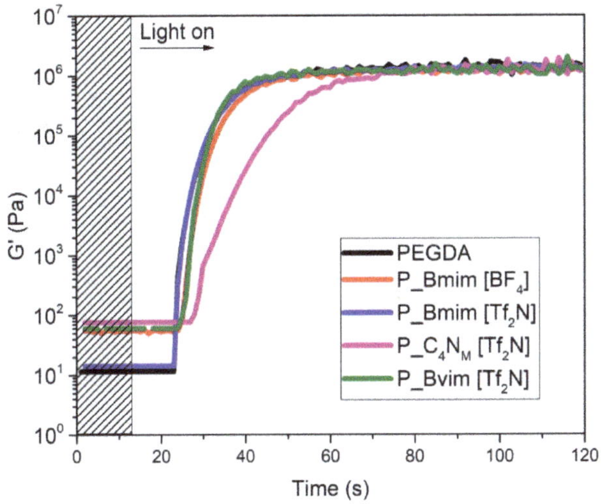

Figure 1. Photorheology tests of PEGDA/ionic liquid (IL) formulations containing different ILs and neat PEGDA as reference. Light irradiation starts after 12 s.

After this preliminary study, 3D printing tests were performed to create both planar films and three-dimensional filter-like structures with cubical structures (Figure 2). The cubical hollow structure was chosen both to prove the possibility to manufacture complex features with high resolution and to create a structure with a high surface area to improve the gas contact with the material during the CO_2 uptake analysis.

Figure 2. 3D printed complex structure (hollow cube).

The key parameters to control in the printing process are the layer thickness and the layer exposure time. In fact, to adjust both parameters is essential to obtain the best printing performance, resulting in increased-X, Y and Z resolution. The exposure times (Table 2) were empirically obtained during the production of membranes and 3D printed parts, considering a good trade-off between ease of handling and precision of printing (i.e., no over-polymerization); those values are in good agreement with the trends obtained from photorheology tests, showing that all the samples containing IL required layer exposure times similar to neat PEGDA monomer. The only exception was the sample P_Amim[Tf_2N], which needed longer exposure times for the polymerization in the form of membrane, and it demonstrated unsuitability to obtain more complex shapes. Sample Bmim[ac] could not be polymerized thus it was not further considered for the next experiments.

Table 2. Layer exposure times needed for the samples.

Sample	Layer Exposure Time (s)	
	Membrane	Cubic Filter-Like
PEGDA	1	2
P_Bmim[BF4]	1	1.7
P_Bmim[Tf2N]	1	2
P_$C_4N_{MA,11}$[Tf2N]	1	1.8
P_Amim[Tf2N]	2	Not obtained
P_Bvim[Tf2N]	1	1.7

FTIR-ATR analyses of the 3D printed ionic liquids blended with PEGDA were carried out in order to further investigate the interaction of the different ILs with PEGDA matrix, in particular to observe weather if the ILs containing double bonds ($C_4N_{MA,11}[Tf_2N]$, Bvim[Tf_2N] and Amim[Tf_2N]) co-polymerize. First, the FTIR-ATR spectra of all neat ionic liquids and PEGDA as reference were collected (Figure S1 in Supplementary Materials), then the polymerized samples of each formulation were analyzed before and after being washed in acetone for 24 h (Figure S2 in Supplementary Materials). The spectra confirmed that $C_4N_{MA,11}[Tf_2N]$ and Bvim[Tf_2N] copolymerized with PEGDA acrylic bonds, proving that these ionic liquids remained linked to the polymer chains during the polymerization reaction, being a part of polymeric network. So, the structures with these ILs could be used even in liquid environment without leakage of the components. Instead, in the spectra of the samples containing the remaining ILs (Bmim[BF_4], Bmim[Tf_2N] and Amim[Tf_2N]), we observed that the peaks related to the ionic liquids disappeared after solvent rinsing, demonstrating that they were only trapped in the polymer matrix and no reaction occurred during polymerization. This confirms also that the allyl group in Amim[Tf_2N] was not reactive enough to react with photoactivated process. Moreover, the weight of the samples was measured before and after the solvent washing. Table 3 reports the values of the concentration of the ionic liquids in the samples and the loss in weight after solvent washing. The results confirmed the considerations reported for FT-IR measurements.

The evaluation of the effect of the presence of the ILs on the thermo-mechanical properties of the 3D printed structures was carried out by DSC and DMTA. The data are reported in Table 4 while the curves are reported in Figures S3 and S4 (in Supplementary Materials) in the supporting information file. Starting from neat PEGDA as reference materials, it was possible to distinguish two distinct behavior: while the ILs that do not react with PEGDA (Bmim[BF_4], Bmim[Tf_2N] and Amim[Tf_2N]) induced a decrease of the T_g, for the sample in samples in which copolymerization occurred, an increase of T_g was evidenced. This could be explained considering that not polymerizable ILs remained as liquid in the formulation, swelling the matrix and acting as plasticizers. On the other hand, when the ILs entered in the polymeric network, the chain mobility decreased, leading to an increase of T_g. The results followed a similar trend both on DSC and DMA, confirming the assumptions. For DMTA, unfortunately, it was not possible to perform measurements on P_Amim[Tf_2N] sample due to brittleness and poor mechanical properties of this membrane.

According to literature [56], DMTA measurements were used also to calculate the cross-linking density according to the Equation

$$E' = \upsilon \cdot R \cdot T \quad (1)$$

where E' (MPa) is the conservative modulus in the rubbery plateau, υ (mol/cm^3) is the cross-linking density, R (cm^3 MPa/mol K) the gas constant and T the temperature in K at which the measure is performed (in our case, 293 K for all the samples). The values (reported in Table 3) confirmed that the ILs without reactive groups drastically reduced the cross-linking density, acting as swelling agent in the polymeric matrix. A different behavior was measured for the IL with a single cross-linkable moiety ($C_4N_{MA,11}[Tf_2N]$): even in this case the cross-linking density decreased as expected, since a monofunctional monomer was introduced in the network. On the other hand, the T_g slightly increased, related to the fact that the IL was not anymore liquid but integrated in the network. This was measured even in the presence of Bvim[Tf_2N]. Moreover, the sample P_Bvim[Tf_2N] showed also an increase of cross-linking density, related to the fact that this IL is bifunctional, thus introducing additional cross-linking points.

Table 3. Ionic liquid content in the samples in relation to the weight variation after solvent treatment.

Sample	Ionic Liquid Concentration (wt %)	Weight Variation (wt %)	T_g (DSC; °C)	T_g (DMTA; °C)	υ (mmol/cm^3)
PEGDA	0	0	−19.9	−17	10.8
P_Bmim[BF_4]	9	8.5	−22.6	−21.3	1.3
P_Bmim[Tf_2N]	15.1	13.9	−22.5	−21.6	1.4
P_$C_4N_{MA,11}$[Tf_2N]	17.5	0	−15.1	−12.8	1.6
P_Amim[Tf_2N]	14.7	14.2	−44.8	Not calculated	-
P_Bvim[Tf_2N]	14.7	0	−11.2	−10	12.2

Table 4. Calculated diffusivity (D), solubility (S) and permeability (P) values for 3D printed membranes containing different ionic liquids.

Sample	D ($\frac{cm^2}{s}$)	S	P ($\frac{cm^2}{s \times bar}$)
PEGDA	6.73	4.71	3.17
P_Bmim[BF_4]	11.33	2.99	3.39
P_Bmim[Tf_2N]	8.65	5.26	4.55
P_$C_4N_{MA,11}$[Tf_2N]	4.46	4.22	1.88
P_Amim[Tf_2N]	9.53	14.47	13.79
P_Bvim[Tf_2N]	8.45	3.20	2.70

The first approach on the study of the interaction between CO_2 and the polymerized structures containing dispersed and linked ionic liquid species was to carry out gas barrier measurements on the membranes. These experiments are fundamental for the evaluation of the gas transport properties such as permeability, diffusivity and solubility. Furthermore, they are a first step towards the understanding of the usage of these materials as supported liquid membranes (SLM) for CO_2 separation. The membranes were tested with the same IL concentrations (0.4 mmol/g) in order to evaluate the real effect of each IL. The values of diffusivity, solubility and permeability, calculated from the CO_2 transmission rate curves of the membranes as described in a precedent work [57], are shown in Table 4. With the addition of ionic liquids in the formulations, the diffusivity increases in all cases, with the exception of $C_4N_{MA,11}$[Tf_2N]. The best result is the one with Bmim[BF_4] that presents a diffusivity value almost double with respect the neat PEGDA. For what regards solubility, the only sample with a significant variation from neat PEGDA is Amim[Tf_2N] that unfortunately has got issues in the 3D printing process. The characteristic that allows to evaluate the effectiveness of a SLM system is the permeability and, among the analyzed samples, the most promising are the Bmim[BF_4] and Bmim[Tf_2N].

Afterwards, further permeability tests were conducted by increasing the amount of Bmim[BF_4] and $C_4N_{MA,11}$[Tf_2N] in the printable formulations. The change in the amount of the two ILs did not affect the printability of the resins and the same irradiation times were used for producing the membranes to be tested. Those two ILs were selected among the others because they seemed to give the best (Bmim[BF_4]) and the worst $C_4N_{MA,11}$[Tf_2N] performance when introduced in PEGDA. In Figure S5 (in Supplementary Materials) of the supporting information the values of diffusivity, solubility and permeability are displayed with the increase of the ILS concentration. Unfortunately, it is not present a clear trend between the amount of the ILs and the gas permeation properties. Surprisingly, increasing the amount of $C_4N_{MA,11}$[Tf_2N] a great increase of both solubility and diffusivity was measured, indicating that an optimization work could be performed individuating the most effective amount of IL. This is part of ongoing studies.

The interaction between polymers containing ILs and carbon dioxide was monitored also by FTIR. As an example, in Figure 3, the absorption spectra of P_Bmim[BF_4] in the range between 2300 and 2400 cm^{-1} after different steps was reported. While after the production step the two typical peaks of this gas, related to the IR active v_3 vibrating mode centered at 2330 and 2360 cm^{-1} [58,59], were not present in the sample, as expected, they appeared after the treatment in CO_2 atmosphere. The following treatment in vacuum produced a reduction of the absorbance in the range, indicating desorbing of CO_2. However, CO_2 was still present in the film due to the interaction with the embedded IL. A following thermal step led to a further but incomplete desorption of CO_2, indicating that IL-gas interaction is strong and could be used for gas trapping.

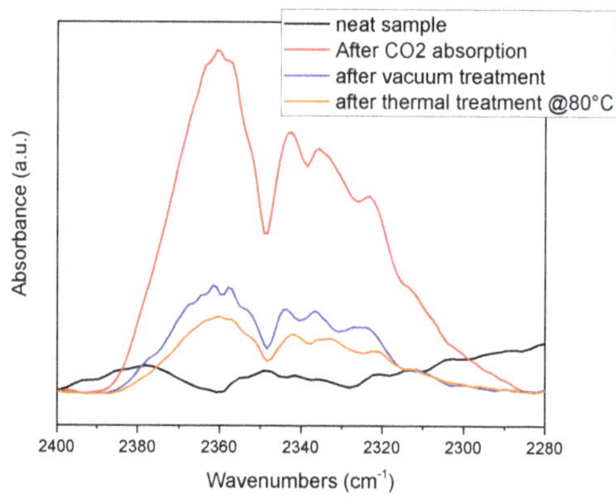

Figure 3. FT-IR of P_Bmim[BF4] in the CO_2 absorption area after different absorption/desorption steps.

At last, taking advantage of the high printability of the developed formulations, we printed components of very complex geometry like gyroids. In Figure 4 we reported gyroids of different sizes printed with P_Bmim[BF_4] formulation, but the same structures were obtained with the all the other tested ILs except P_Amim[Tf_2N]. This highly complex geometry applied to 3D printing offers improved mechanical properties compared to other structures [60] and a high surface to volume ratio [61] that can be exploited to create active filter for CO_2 capture with improved gas/surface interaction. In particular, these shapes present interconnected continuous channels that allow the passage of the gas without a pressure drop but, at the same time, higher interaction with the structure for enabling gas adsorption. Those complex shape components are at the moment under investigation as well and will be presented in future works. In fact, different parameters should be taken into account in order to have an optimized device including material properties, channel dimensions, tortuosity and available surface. In this context 3D printing could be fully exploited integrating CAD design with formulation development.

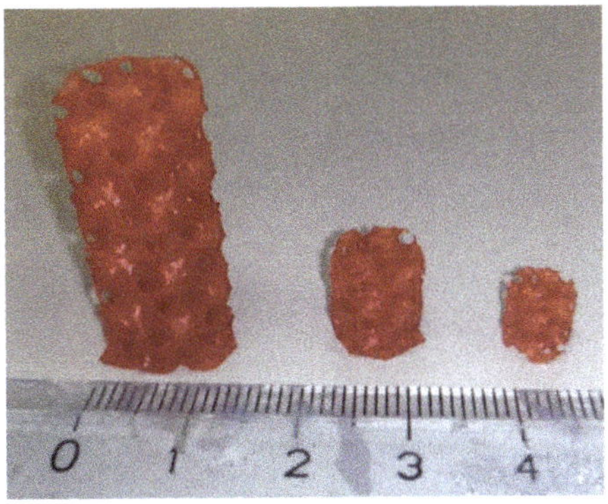

Figure 4. Gyroid-like structures printed with P_Bmim[BF_4].

4. Conclusions

In this work, we tested different commercially available ionic liquids in order to develop 3D printable formulations processable with DLP technology, with potential application in CO_2 capture technology. Six ionic liquids were tested (Bmim[BF_4], Bmim[Tf_2N], Bvim[Tf_2N], Amim[Tf_2N], Bmim[ac] and $C_4N_{MA,11}$[Tf_2N]) mixed with a bi-functional acrylate monomer (PEGDA). The photo-rheological analyses carried out brought us to exclude Bmim[ac] since the acetate anion inhibits the photopolymerization. Regarding the other ILs, we evidenced that imidazole based cations as well as tetrafluoroborate and bis (trifluorometilsulfonyl)imide anions did not affect the photopolymerization, while the other cations (containing ammonium salt and allyl group) slowed down the process. Afterwards, a 3D printing test was conducted on the formulations to build simple (membranes) and complex (hollow cubic) structures and to find the optimal printing parameters. All the ILs samples allowed the fabrication of complex 3D structures, with the exception of the samples containing Amim[Tf_2N]. Moreover we demonstrated that the ILs bearing a reactive group ($C_4N_{MA,11}$[Tf_2N] and Bvim[Tf_2N]) participated to the polymerization reaction entering in the polymeric network, while it was possible to remove the other ILs by solvent extraction. This resulted even in thermomechanical properties, with a plasticizing effect of the dispersed ILs. The samples were then tested for the CO_2 capture capacity by means of FTIR. The spectra showed the characteristic peaks related to the vibrating modes of the CO_2 molecule, demonstrating the physical adsorption of the gas in the samples. Moreover an absorption/desorption experiment was conducted to check the regeneration of the samples. Regarding the possible application of these materials as supported liquid membranes (SLM) for CO_2 separation, gas transport analyses were carried out to determine the transport properties of the membranes, unfortunately resulting in not univocal trends. As a final experiment, we 3D printed part with high complexity as gyroids in order to demonstrate the possibility to produce complex geometries that could be employed as active filters for CO_2 capture. The present work successfully demonstrated the possibility to employ 3D printing for CO_2 capture technologies, opening the way to new investigations in which CAD design and materials properties could lead to more efficient devices.

Supplementary Materials: The following are available online at http://www.mdpi.com/2073-4360/11/12/1932/s1, Figure S1: FTIR-ATR spectra of neat ILs and neat PEGDA; Figure S2: FTIR-ATR spectra of polymerized formulations before (red) and after (blue) solvent washing; Figure S3: DSC analysis of the studied formulations; Figure S4: DMTA analysis on the polymerized samples; Figure S5: Diffusivity, solubility and permeability values of membranes containing increasing concentrations Bmim[BF4] (red) e C4NMA, 11[Tf2N] (green). The black line indicates the value calculated for neat PEGDA.

Author Contributions: M.G. and L.M. conducted the experiments, M.G.G. methodology, A.C. and I.R. writing the paper and data analysis, L.S. and C.F.P. discussion and supervision.

Funding: This research received no external funding.

Acknowledgments: The authors would like to acknowledge Federica Piccatto for the support in CAD design.

Conflicts of Interest: The authors declare no conflict of interest.

References

1. Figueres, C.; Le Quéré, C.; Mahindra, A.; Bäte, O.; Whiteman, G.; Peters, G.; Guan, D. Emissions are still rising: Ramp up the cuts. *Nature* **2018**, *564*, 27–30. [CrossRef]
2. Jacobson, M.Z. *Air Pollution and Global Warming: History, Science, and Solutions*, 2nd ed.; Cambridge University Press: Cambridge, UK, 2012.
3. Hoffmann, M.J. Global climate change. In *The Handbook of Global Climate and Environment Policy*; Wiley-Blackwell: Oxford, UK, 2013; pp. 3–18.
4. Araújo, O.D.Q.F.; de Medeiros, J.L. Carbon capture and storage technologies: Present scenario and drivers of innovation. *Curr. Opin. Chem. Eng.* **2017**, *17*, 22–34. [CrossRef]
5. Theo, W.L.; Lim, J.S.; Hashim, H.; Mustaffa, A.A.; Ho, W.S. Review of pre-combustion capture and ionic liquid in carbon capture and storage. *Appl. Energy* **2016**, *183*, 1633–1663. [CrossRef]
6. Jansen, D.; Gazzani, M.; Manzolini, G.; Dijk, E.V.; Carbo, M. Pre-combustion CO2 capture. *Int. J. Greenh. Gas Control* **2015**, *40*, 167–187. [CrossRef]
7. Hadjipaschalis, I.; Kourtis, G.; Poullikkas, A. Assessment of oxyfuel power generation technologies. *Renew. Sustain. Energy Rev.* **2009**, *13*, 2637–2644. [CrossRef]
8. Wang, C.A.; Zhang, X.; Liu, Y.; Che, D. Pyrolysis and combustion characteristics of coals in oxyfuel combustion. *Appl. Energy* **2012**, *97*, 264–273. [CrossRef]
9. Machida, H.; Ando, R.; Esaki, T.; Yamaguchi, T.; Norinaga, K. Modelling of CO2 solubility in phase separation solvent composed of amine/ether/water system for CO2 capture. *J. Mol. Liq.* **2019**, *292*, 111411. [CrossRef]
10. Kussainova, D.; Shah, D. Monoethanolamine based DESs for CO2 absorption: Insights from molecular dynamics simulations. *Sep. Purif. Technol.* **2020**, *231*, 115931. [CrossRef]
11. Meisen, A.; Shuai, X. Research and development issues in CO2 capture. *Energy Convers. Manag.* **1997**, *38*, S37–S42. [CrossRef]
12. Andirova, D.; Cogswell, C.F.; Lei, Y.; Choi, S. Effect of the structural constituents of metal organic frameworks on carbon dioxide capture. *Microporous Mesoporous Mater.* **2016**, *219*, 276–305. [CrossRef]
13. Nandasiri, M.I.; Jambovane, S.R.; McGrail, B.P.; Schaef, H.T.; Nune, S.K. Adsorption, separation, and catalytic properties of densified metal-organic frameworks. *Coord. Chem. Rev.* **2016**, *311*, 38–52. [CrossRef]
14. Venna, S.R.; Carreon, M.A. Metal organic framework membranes for carbon dioxide separation. *Chem. Eng. Sci.* **2015**, *124*, 3–19. [CrossRef]
15. Li, W.; Zhang, Y.; Li, Q.; Zhang, G. Metal−organic framework composite membranes: Synthesis and separation applications. *Chem. Eng. Sci.* **2015**, *135*, 232–257. [CrossRef]
16. Heo, Y.J.; Seong, D.B.; Park, S.J. Synthesis of polyethylenimine-impregnated titanate nanotubes for CO2 capture: Influence of porosity and nitrogen content on amine-modified adsorbents. *J. CO2 Util.* **2019**, *34*, 472–478. [CrossRef]
17. Zhao, P.; Zhang, G.; Xu, Y.; Lv, Y. Amine functionalized hierarchical bimodal mesoporous silicas as a promising nanocomposite for highly efficient CO2 capture. *J. CO2 Util.* **2019**, *34*, 543–557. [CrossRef]
18. Stefanelli, E.; Puccini, M.; Vitolo, S.; Seggiani, M. CO2 sorption kinetic study and modeling on doped-Li4SiO4 under different temperatures and CO2 partial pressures. *Chem. Eng. J.* **2020**, *379*, 122307. [CrossRef]
19. Yang, Y.; Liu, W.; Hu, Y.; Sun, J.; Tong, X.; Li, Q.; Zhou, Z. Novel low cost Li4SiO4-based sorbent with naturally occurring wollastonite as Si-source for cyclic CO2 capture. *Chem. Eng. J.* **2019**, *374*, 328–337. [CrossRef]
20. Halder, K.; Khan, M.M.; Grünauer, J.; Shishatskiy, S.; Abetz, C.; Filiz, V.; Abetz, V. Blend membranes of ionic liquid and polymers of intrinsic microporosity with improved gas separation characteristics. *J. Membr. Sci.* **2017**, *539*, 368–382. [CrossRef]
21. Marsh, K.N.; Boxall, J.A.; Lichtenthaler, R. Room temperature ionic liquids and their mixtures—A review. *Fluid Phase Equilibria* **2004**, *219*, 93–98. [CrossRef]
22. Zeng, S.; Zhang, X.; Bai, L.; Wang, H.; Wang, J.; Bao, D.; Li, M.; Liu, X.; Zhang, S. Ionic-Liquid-Based CO2 Capture Systems: Structure, Interaction and Process. *Chem. Rev.* **2017**, *117*, 9625–9673. [CrossRef]

23. Aroonwilas, A.; Veawab, A. Characterization and comparison of the CO2 absorption performance into single and blended alkanolamines in a packed column. *Ind. Eng. Chem. Res.* **2004**, *43*, 2228–2237. [CrossRef]
24. Sistla, Y.S.; Khanna, A. CO2 absorption studies in amino acid-anion based ionic liquids. *Chem. Eng. J.* **2015**, *273*, 268–276. [CrossRef]
25. Makino, T.; Kanakubo, M. CO2 absorption property of ionic liquid and CO2 permselectivity for ionic liquid membrane. *J. Jpn. Pet. Inst.* **2016**, *59*, 109–117. [CrossRef]
26. Lozano, L.J.; Godínez, C.; de los Ríos, A.P.; Hernández-Fernández, F.J.; Sánchez-Segado, S.; Alguacil, F.J. Recent advances in supported ionic liquid membrane technology. *J. Membr. Sci.* **2011**, *376*, 1–14. [CrossRef]
27. Malik, M.A.; Hashim, M.A.; Nabi, F. Ionic liquids in supported liquid membrane technology. *Chem. Eng. J.* **2011**, *171*, 242–254. [CrossRef]
28. Parhi, P.K. Supported liquid membrane principle and its practices: A short review. *J. Chem.* **2013**. [CrossRef]
29. Dai, Z.; Noble, R.D.; Gin, D.L.; Zhang, X.; Deng, L. Combination of ionic liquids with membrane technology: A new approach for CO2 separation. *J. Membr. Sci.* **2016**, *497*, 1–20. [CrossRef]
30. Zulfiqar, S.; Sarwar, M.I.; Mecerreyes, D. Polymeric ionic liquids for CO2 capture and separation: Potential, progress and challenges. *Polym. Chem.* **2015**, *6*, 6435–6451. [CrossRef]
31. Tang, J.; Sun, W.; Tang, H.; Radosz, M.; Shen, Y. Enhanced CO2 Absorption of Poly(ionic liquid)s. *Macromolecules* **2005**, *38*, 2037–2039. [CrossRef]
32. Tang, J.; Shen, Y.; Radosz, M.; Sun, W. Isothermal Carbon Dioxide Sorption in Poly(ionic liquid)s. *Ind. Eng. Chem. Res.* **2009**, *48*, 9113–9118. [CrossRef]
33. Tomé, L.C.; Gouveia, A.S.L.; Freire, C.S.R.; Mecerreyes, D.; Marrucho, I.M. Polymeric ionic liquid-based membranes: Influence of polycation variation on gas transport and CO2 selectivity properties. *J. Membr. Sci.* **2015**, *486*, 40–48. [CrossRef]
34. Bara, J.E.; Lessmann, S.; Gabriel, C.J.; Hatakeyama, E.S.; Noble, R.D.; Gin, D.L. Synthesis and Performance of Polymerizable Room-Temperature Ionic Liquids as Gas Separation Membranes. *Ind. Eng. Chem. Res.* **2007**, *46*, 5397–5404. [CrossRef]
35. Zhou, X.; Weber, J.; Yuan, J. Poly(ionic liquid)s: Platform for CO2 capture and catalysis. *Curr. Opin. Green Sustain. Chem.* **2019**, *16*, 39–46. [CrossRef]
36. Steldinger, H.; Esposito, A.; Brunnengräber, K.; Gläsel, J.; Etzold, B.J.M. Activated Carbon in the Third Dimension—3D Printing of a Tuned Porous Carbon. *Adv. Sci.* **2019**, *6*, 1901340. [CrossRef]
37. Chiappone, A.; Fantino, E.; Roppolo, I.; Lorusso, M.; Manfredi, D.; Fino, P.; Pirri, C.F.; Calignano, F. 3D Printed PEG-Based Hybrid Nanocomposites Obtained by Sol–Gel Technique. *ACS Appl. Mater. Interfaces* **2016**, *8*, 5627–5633. [CrossRef]
38. Farahani, R.D.; Dubé, M.; Therriault, D. Three-Dimensional Printing of Multifunctional Nanocomposites: Manufacturing Techniques and Applications. *Adv. Mater.* **2016**, *28*, 5794–5821. [CrossRef]
39. Emon, M.O.F.; Alkadi, F.; Philip, D.G.; Kim, D.-H.; Lee, K.-Ċ.; Choi, J.-W. Multi-material 3D printing of a soft pressure sensor. *Addit. Manuf.* **2019**, *28*, 629–638. [CrossRef]
40. Crump, M.R.; Gong, A.T.; Chai, D.; Bidinger, S.L.; Pavinatto, F.J.; Reihsen, T.E.; Sweet, R.M.; Mackenzie, J.D. Monolithic 3D printing of embeddable and highly stretchable strain sensors using conductive ionogels. *Nanotechnology* **2019**, *30*, 364002. [CrossRef]
41. Wong, J.; Gong, A.T.; Defnet, P.A.; Meabe, L.; Beauchamp, B.; Sweet, R.M.; Sardon, H.; Cobb, C.L.; Nelson, A. 3D Printing Ionogel Auxetic Frameworks for Stretchable Sensors. *Adv. Mater. Technol.* **2019**, *4*, 1900452. [CrossRef]
42. Chen, G.; Chen, N.; Wang, Q. Preparation of poly (vinyl alcohol)/ionic liquid composites with improved processability and electrical conductivity for fused deposition modeling. *Mater. Des.* **2018**, *157*, 273–283. [CrossRef]
43. Schultz, A.R.; Lambert, P.M.; Chartrain, N.A.; Ruohoniemi, D.M.; Zhang, Z.; Jangu, C.; Zhang, M.; Williams, C.B.; Long, T.E. 3D Printing Phosphonium Ionic Liquid Networks with Mask Projection Microstereolithography. *ACS Macro Lett.* **2014**, *3*, 1205–1209. [CrossRef]
44. Lee, J.; Faruk Emon, M.O.; Vatani, M.; Choi, J.W. Effect of degree of crosslinking and polymerization of 3D printable polymer/ionic liquid composites on performance of stretchable piezoresistive sensors. *Smart Mater. Struct.* **2017**, *26*, 035043. [CrossRef]
45. Wales, D.J.; Cao, Q.; Kastner, K.; Karjalainen, E.; Newton, G.N.; Sans, V. 3D-Printable Photochromic Molecular Materials for Reversible Information Storage. *Adv. Mater.* **2018**, *30*, 1800159. [CrossRef] [PubMed]

46. Wu, J.; Yan, Y.; Zhang, L.; Qin, Z.; Tao, S. Enhanced Mass Transfer and Improved Catalyst Recovery in a Stirred Reactor by Polymeric Ionic Liquids Modified 3D Printed Devices. *Adv. Mater. Technol.* **2019**, *4*, 1800515. [CrossRef]
47. Ding, H.; Chang, R.C. Printability study of bioprinted tubular structures using liquid hydrogel precursors in a support bath. *Appl. Sci. (Switzerland)* **2018**, *8*, 403. [CrossRef]
48. Melchels, F.P.W.; Feijen, J.; Grijpma, D.W. A review on stereolithography and its applications in biomedical engineering. *Biomaterials* **2010**, *31*, 6121–6130. [CrossRef]
49. Frascella, F.; González, G.; Bosch, P.; Angelini, A.; Chiappone, A.; Sangermano, M.; Pirri, C.F.; Roppolo, I. Three-Dimensional Printed Photoluminescent Polymeric Waveguides. *ACS Appl. Mater. Interfaces* **2018**, *10*, 39319–39326. [CrossRef]
50. Roppolo, I.; Chiappone, A.; Angelini, A.; Stassi, S.; Frascella, F.; Pirri, C.; Ricciardi, C.; Descrovi, E. 3D printable light-responsive polymers. *Mater. Horiz.* **2017**, *4*, 396–401. [CrossRef]
51. Ligon, S.C.; Liska, R.; Stampfl, J.; Gurr, M.; Mülhaupt, R. Polymers for 3D Printing and Customized Additive Manufacturing. *Chem. Rev.* **2017**, *117*, 10212–10290. [CrossRef]
52. Nulwala, H.; Mirjafari, A.; Zhou, X. Ionic liquids and poly(ionic liquid)s for 3D printing—A focused mini-review. *Eur. Polym. J.* **2018**, *108*, 390–398. [CrossRef]
53. Jakubiak, J.; Allonas, X.; Fouassier, J.; Sionkowska, A.; Andrzejewska, E.; Linden, L.; Rabek, J. Camphorquinone–amines photoinitating systems for the initiation of free radical polymerization. *Polymer* **2003**, *44*, 5219–5226. [CrossRef]
54. Litt, M.; Eirich, F.R. Polymerization of allyl acetate. *J. Polym. Sci.* **1960**, *45*, 379–396. [CrossRef]
55. Matsumoto, A.; Kumagai, T.; Aota, H.; Kawasaki, H.; Arakawa, R. Reassessment of Free-Radical Polymerization Mechanism of Allyl Acetate Based on End-Group Determination of Resulting Oligomers by MALDI-TOF-MS Spectrometry. *Polym. J.* **2009**, *41*, 26–33. [CrossRef]
56. Chiappone, A.; Jeremias, S.; Bongiovanni, R.; Schönhoff, M. NMR study of photo-crosslinked solid polymer electrolytes: The influence of monofunctional oligoethers. *J. Polym. Sci. Part B Polym. Phys.* **2013**, *51*, 1571–1580. [CrossRef]
57. Gillono, M.; Roppolo, I.; Frascella, F.; Scaltrito, L.; Pirri, C.F.; Chiappone, A. CO2 permeability control in 3D printed light responsive structures. *Appl. Mater. Today* **2019**, 100470. [CrossRef]
58. Yamazaki, T.; Katoh, M.; Ozawa, S.; Ogino, Y. Adsorption of CO2 over univalent cation-exchanged ZSM-5 zeolites. *Mol. Phys.* **1993**, *80*, 313–324. [CrossRef]
59. Veyrié, D.; Lellouchi, D.; Roux, J.; Pressecq, F.; Tetelin, A.; Pellet, C. FTIR spectroscopy for the hermeticity assessment of micro-cavities. *Microelectron. Reliab.* **2005**, *45*, 1764–1769. [CrossRef]
60. Al-Ketan, O.; Rezgui, R.; Rowshan, R.; Du, H.; Fang, N.X.; Abu Al-Rub, R.K. Microarchitected stretching-dominated mechanical metamaterials with minimal surface topologies. *Adv. Eng. Mater.* **2018**, *20*, 1800029. [CrossRef]
61. Yoo, D.-J. Advanced porous scaffold design using multi-void triply periodic minimal surface models with high surface area to volume ratios. *Int. J. Precis. Eng. Manuf.* **2014**, *15*, 1657–1666. [CrossRef]

polymers

MDPI

Article

Understanding the Polymerization of Polyfurfuryl Alcohol: Ring Opening and Diels-Alder Reactions

Gianluca Tondi [1,2,3,*], Nicola Cefarin [4], Thomas Sepperer [1,2], Francesco D'Amico [4], Raphael J. F. Berger [5], Maurizio Musso [5], Giovanni Birarda [4], Andreas Reyer [5], Thomas Schnabel [1] and Lisa Vaccari [4]

1 Forest Products Technology & Timber Constructions Department, Salzburg University of Applied Sciences, Marktstrasse 136a, 5431 Kuchl, Austria; thomas.sepperer@fh-salzburg.ac.at (T.S.); thomas.schnabel@fh-salzburg.ac.at (T.S.)
2 Salzburg Center for Smart Materials, Jakob-Haringer-strasse 2, 5020 Salzburg, Austria
3 Land, Environment, Agriculture and Forestry Department, University of Padua, Via dell'Università 16, 35020 Legnaro (PD), Italy
4 Elettra-Sincrotrone Trieste S.C.p.A., Strada Statale 14—km 163,5 in AREA Science Park, 34149 Basovizza, Trieste, Italy; nicola.cefarin@elettra.eu (N.C.); francesco.damico@elettra.eu (F.D.); giovanni.birarda@elettra.eu (G.B.); lisa.vaccari@elettra.eu (L.V.)
5 Department of Chemistry and Physics of Materials, Paris Lodron University Salzburg, Jakob-Haringer-Strasse 2a, 5020 Salzburg, Austria; raphael.berger@sbg.ac.at (R.J.F.B.); maurizio.musso@sbg.ac.at (M.M.); andreas.reyer@sbg.ac.at (A.R.)
* Correspondence: gianluca.tondi@unipd.it; Tel.: +39-049-827-27-76

Received: 2 December 2019; Accepted: 9 December 2019; Published: 17 December 2019

Abstract: Polyfurfuryl alcohol (PFA) is one of the most intriguing polymers because, despite its easy polymerization in acid environment, its molecular structure is definitely not obvious. Many studies have been performed in recent decades, and every time, surprising aspects came out. With the present study, we aim to take advantage of all of the findings of previous investigations and exploit them for the interpretation of the completely cured PFA spectra registered with three of the most powerful techniques for the characterization of solid, insoluble polymers: Solid-State ^{13}C-NMR, Attenuated Total Reflectance (ATR), Fourier Transform Infrared (FTIR) spectroscopy, and UV-resonant Raman spectroscopy at different excitation wavelengths, using both an UV laser source and UV synchrotron radiation. In addition, the foreseen structures were modeled and the corresponding ^{13}C-NMR and FTIR spectra were simulated with first-principles and semi-empiric methods to evaluate their matching with experimental ones. Thanks to this multi-technique approach, based on complementary analytical tools and computational support, it was possible to conclude that, in addition to the major linear unconjugated polymerization, the PFA structure consists of Diels-Alder rearrangements occurring after the opening of some furanic units, while the terminal moieties of the chain involves γ-lactone arrangements. The occurrence of head-head methylene ether bridges and free hydroxyl groups (from unreacted furfuryl alcohol, FA, or terminal chains) could be excluded, while the conjugated systems could be considered rather limited.

Keywords: furanic polymer; spectroscopy; solid-state ^{13}C-NMR; FTIR; Raman; linear structure; ring opening; γ-lactone; Diels-Alder

1. Introduction

The polymerization of furfuryl alcohol (FA) in acid environment to polyfurfuryl alcohol (PFA) is a subject that has attracted the interest of many scientists between the 1950s and today. This dark thermosetting polymer has been investigated in consideration of various synthetic and analytic perspectives, and new information about its structure has always jumped out. In Table 1, some of the most important studies on PFA are summarized by presenting the molecular arrangements proposed by the authors.

Table 1. Possible structural arrangements of PFA, their chemical formula and the related scientific papers that first proposed them. Repetitive PFA core moieties are drawn in black, while furanic rings in grey represent the terminal units.

Structure	Arrangement	Chemical Formula	Reference(s)
1	Linear		Dunlop & Peters—1953 [1]
2	Ring opening		Conley & Metil—1963 [2]
3	α,β-unsaturated γ-lactons	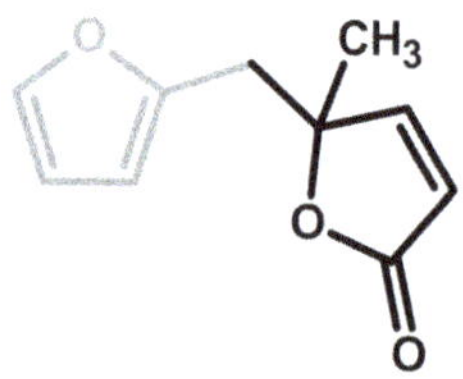	Wewerka—1971 [3]
4	Methylene bridge	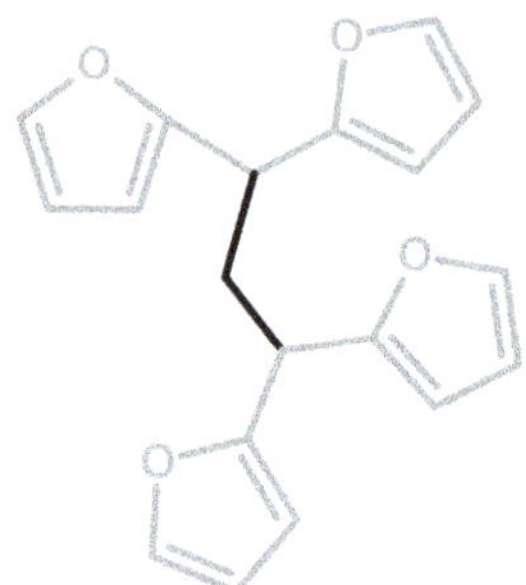	Chuang et al.—1984 [4]
5	Conjugated	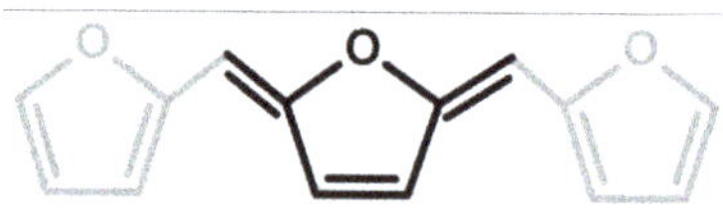	Buchwalter—1985 [5]
6	Diels-Alder	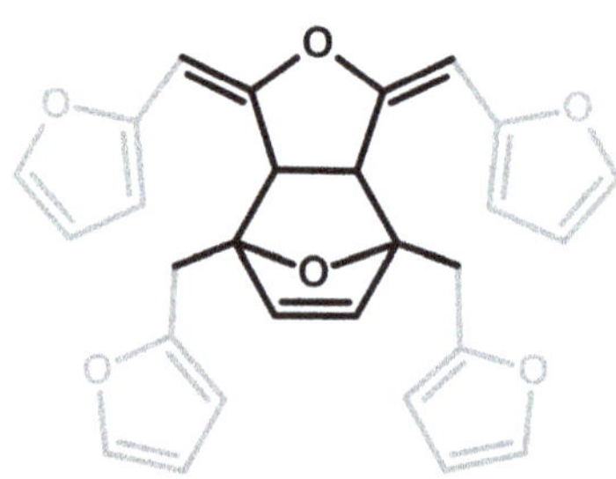	Choura et al.—1996 [6]

Table 1. *Cont.*

Structure	Arrangement	Chemical Formula	Reference(s)
7	Ring-Opening + Diels-Alder	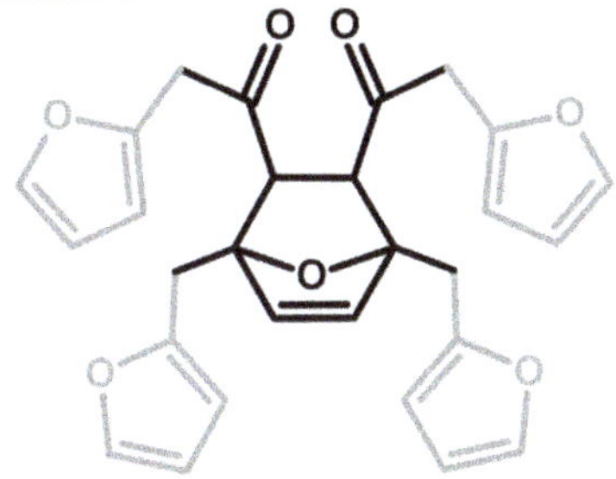	Present paper

The first investigation of this intriguing polymer was published by Dunlop and Peters in 1953 [1], presenting the linear furanic polymerization through methylene (**1**) and di-methylene ether bridges. This model explained the linear structures and attributed the crosslinking to other methylene bridges obtained from the formaldehyde (coming from the head-head condensation of two molecules of FA). During the early 1960s, Conley and Metil [2] performed their studies by means of FTIR spectroscopy, where they identified the ring opening due to the formation of γ-diketones (**2**). A decade later, Wewerka [3] proposed a polymerization pathway for the termination of the chains through formation of a γ-lactone (**3**), which occurred for its alumina-catalyzed PFAs. Specifically, in the aforementioned study, a large amount of low polymerized units were produced and an increased amount of γ-lactone was observed. In 1984 Chuang et al. [4] first published ^{13}C-NMR results on PFA, stating that the presence of dimethylene ether bridges was very limited, as well as the substitution of C3 and C4 of the furanic ring. The Chuang group observed a peak above 200 ppm, assigned to the presence of γ-diketones due to ring opening and confirming the observation of Conley and Metil [2], and also a small signal at 177 ppm that they attributed to levulinic acid. This study supported crosslinking through formaldehyde-derived methylene bridges (**4**) first proposed by Dunlop and Peters [1]. One year later, Buchwalter [5] published the ^{1}H and ^{13}C-NMR analysis on furfuryl acetate, and it was the first study to propose a conjugated structure (**5**). Hence, in 1996, Choura et al. [6,7] published a systematic study on the polymerization of FA using several model compounds [6]. In this study, which exploited FTIR spectroscopy and ^{1}H-NMR as analytic tools for the characterization of the polymers, the research group concluded that the polymerization of FA occurs in two steps: the first was linear polymerization though methylene bridges (as already proposed by Dunlop and Peters [1]), while the second was far more complex than classically expected. In particular, the chromophore effect of the polymer was attributed to the conjugation of the linear furanic chain, while the branching was expected to involve Diels-Alder reactions between conjugated and unconjugated chains (**6**) [7]. These solid studies did not involve ^{13}C-NMR investigations, and therefore did not completely consider the possibility of ring opening. In the 2000s, other scientists came across this polymer: Bertarione et al. [8] focused on the polymerization of FA in confined spaces and studied this macromolecule with several vibrational techniques, suggesting the polymerization mechanism to occur through carbocations. Guigo et al. [9,10] studied the chemo-rheological kinetics of this polymerization, confirming again the two reaction steps and attributing the curing step to Diels-Alder reactions. Barsberg and Thygesen [11] studied PFA formation using Attenuated Total Reflectance (ATR) FTIR spectroscopy, supporting their spectral attributions with theoretical predictions using density functional theory of the vibrational bands of the most probable PFA model structures. The aforementioned study confirmed the dominance of the methylene bridge formation in the initial polymerization phase and the formation of conjugated systems in the second one, supporting the model of Buchwalter. PFA curing was also studied by Raman spectroscopy, and in this field Kim et al. [12,13] analyzed the acid polymerization of FA focusing on the thermodynamics of the system and suggesting that, based on the comparison of measured and calculated Raman spectra, the formation of a conjugated diene structures (the conjugation of the furanic group) was preferred to the diketonic structures.

Recently, Falco et al. [14] presented a study in which major importance was given to the opening of the furanic ring, confirming the presence of Diels-Alder crosslinking. Both the study of the group of Gandini [6,7] and that of Falco agreed on the presence of Diels-Alder arrangements, but Falco et al. [14] gave major importance to the ring opening and also confirmed the inter-chained methylene bridge which was "almost" excluded by Gandini.

Overall, we can conclude that there is no general consensus on the possible structure(s) of PFA, and more comprehensive studies are needed to further clarify the structure of the reaction products. To shed more light on the PFA molecular structure after polymerization completion, we performed an analysis of PFA synthesized through simple acid catalysis at room temperature by complementing spectroscopic techniques, such as solid-state ^{13}C-NMR, ATR FTIR and Raman spectroscopy with UV laser excitation and using UV synchrotron light, with theoretical modeling.

2. Materials and Methods

2.1. Sample Preparation

PFA was prepared at room temperature by adding 0.1 mL of sulfuric acid (Merck) 32% in an open test tube containing 2 mL FA (Transfuran chemicals, Geel, Belgium). The reaction started with a darkening of the original yellow solution, which developed into an orange, then brown solution that turned into a black solid after around 60–80 s through exothermal reaction. The solid polymer was kept in an oven at 103 °C for 1 h to ensure the complete curing and then it was grinded into fine powder. This powder was leached with water to remove the catalyst and the unreacted FA, and then dried again at 103 °C for 4 h until constant weight. The material resulted almost completely insoluble (only 2% was leached out) and the resulting dried powder was the subject of this study.

2.2. Solid-State ^{13}C-NMR

The ^{13}C-NMR spectrum of PFA powder was obtained with a Bruker Avance NEO 500 wide bore system (Bruker BioSpin, Rheinstetten, Germany) at the NMR centre of the Faculty of Chemistry at the University of Vienna. A 4 mm triple resonance magic angle spinning (MAS) probe was used with a resonance frequency for ^{13}C of 125.78 MHz, and the MAS rotor spinning was set to 14 kHz. Cross polarization (CP) was achieved by a ramped contact pulse with a contact time of 2 ms. During acquisition, ^{1}H was high-power decoupled using SPINAL with 64 phase permutations. The chemical shifts for ^{13}C are reported in ppm and are externally referenced to adamantane by setting the low field signal to 38.48 ppm.

The data elaboration was done with the software Top-spin 4.0.6 (Bruker) and OriginPro (OriginLab, Northampton, MA, USA), while the calculations of the theoretical chemical shifts were done with the software NMR-Predict developed by the University of Lausanne (L. Patiny) and the University of del Valle (J. Wist) [15–17].

2.3. ATR-FTIR Analysis

The FTIR experiments were carried out at the Chemical and Life Sciences branch of the Synchrotron Infrared Source for Spectroscopy and Imaging beamline (SISSI-Bio) at the Elettra Sincrotrone Trieste, Italy [18].

PFA powder was measured by ATR-FTIR spectroscopy using the Platinum™ Single Reflection ATR box (Bruker Optik GmbH, Ettlingen, Germany) with diamond internal reflective element (IRE). Spectra were acquired using the Vertex 70 v in-vacuum interferometer (Bruker Optik GmbH, Ettlingen, Germany) equipped with a wide band deuterated triglycine sulfate detector and silicon FIR-MIR beamsplitter. ATR-FTIR spectra were collected from 6000 to 60 cm^{-1} in double side, forward/backward acquisition mode with a scanner velocity of 5 kHz. For each spectrum, 256 and 128 scans were averaged with a spectral resolution of 2 cm^{-1} for the background and the sample respectively. Fourier transform was carried out with Mertz Blackman-Harris 3-terms apodization function. Each sample was measured 5 times and the results averaged before comparison and further analyses. Background was collected on clean diamond IRE. The spectra were cut in the 4000 to 400 cm^{-1} region, then baseline corrected with the concave rubberband function (5 iterations) and finally vector normalized (4000 to 400 cm^{-1} region) and corrected for the offset with OPUS 7.5 (Bruker Optik GmbH, Ettlingen, Germany) software. Peak assignment procedure was guided by second derivative analysis (Savitzky-Golay algorithm, 19-smoothing points) in order to enhance the separation of overlapping peaks.

2.4. *Raman Investigation*

UV Resonant Raman (UVRR) measurements were carried out at the IUVS beamline of the Elettra Sincrotrone Trieste, Italy [19]. The synchrotron-based radiation source was employed to get the excitation wavelengths of 226 nm and 249 nm, with a beam power reaching the sample of approximately 10 μW. A UV laser source has been adopted for the excitation wavelength at 266 nm with 100 μW beam power. The Raman signal was collected in backscattering configuration. During the measurements, the samples were continuously moved with oscillations of 1 Hz and 1 mm to avoid photodegradation phenomena. A Czery-Turner spectrometer with focal length of 750 mm (Acton SP-2750, Acton, MA, USA), coupled with a holographic reflection grating of 1800 g/mm and with a Peltier-cooled back-thinned CCD (Princeton Instruments 7510, Acton, MA, USA), was employed to collect the Raman signal. Spectral resolution (FWHM of the apparatus function) was set to 25 cm^{-1} for the synchrotron-based Raman measurements and to 7 cm^{-1} for the UV laser-based Raman measurements. Raman wavenumbers were calibrated by using cyclohexane spectra [20].

2.5. *Computational Details*

The starting geometries for the model structures (**1**), (**2**), (**3**) and (**7**) (see Table 1) were set up using Marvin suite [21], and the structures were pre-optimized with Marvin's molecular dynamics module using standard settings. The resulting lowest energy conformer was used in all four cases for the further steps. The pre-optimized structures were optimized with Turbomole [22] at the ri-DFT level of theory [23] using the BP86 functional [24,25] together with the RI and MARIJ [23] approximations, the def-SV(P) basis set [26,27] and standard settings otherwise. After geometry convergence, a single point DFT calculation with tightened numerical convergence criteria ($denconv 0.1d–07 and $scfconf 8) prior to the NMR shielding calculation with Turbomoles mpshift module was performed [28]. Relative isotropic chemical shifts with respect to tetramethylsilane (TMS) were calculated (the computation of TMS and its properties was performed in a fully analogous way) for all carbon atoms for the selected (**1**), (**2**), (**3**) and (**7**) molecules. Intensities of 0.6 for (**1**), 0.1 for (**2**), 0.1 for (**3**), and 0.2 for (**7**), respectively, were assigned to the signals (irrespective of the chemical connectivity of the carbon atoms) to obtain rough ^{13}C-NMR model spectra. All four model spectra were merged, and the merged spectrum was folded with a Lorentzian line shape of 5 ppm half-width. Finally, the chemical shielding values were scaled by a factor of 108/120 in order to obtain a match of the respective peak at 108 ppm in the experimental ^{13}C-NMR spectrum.

For the calculation of the IR model spectrum, the lowest energy conformer search for all four structures from the DFT optimization has been performed with Grimme's xtb method and the "confscript" script using all standard settings [29]. For the obtained minimum conformers, a hessian calculation using the hessian module and standard settings of xtb was performed to arrive at IR frequencies and intensities. To obtain the model spectrum, as in the case of the NMR spectra, the four individually calculated IR spectra were scaled and merged (0.6 (**1**) + 0.1 (**2**) + 0.1 (**3**) + 0.2 (**7**)), and folded in this case with a Gaussian line shape of 0.05% of the full width of the spectrum (20 cm^{-1}) as half width.

3. Results and Discussion

The numerous studies performed on PFA are fully justified, because this polymer cannot be explained by simple chemistry. Almost all authors agreed that this polymerization does not follow a single pattern, but many arrangements may occur simultaneously, and they also depend on the procedure adopted. The approach we consider in the present paper consists of evaluating the proposed arrangements summarized in Table 1, discriminating between those structures that are possible and those that are not on the basis of the results obtained by ^{13}C-NMR, FTIR and Raman spectroscopies on our thermoset PFA sample, and finally comparing the registered spectra with the computational data to strengthen the spectral interpretation.

3.1. *Solid-State ^{13}C-NMR*

The solid-state ^{13}C-NMR spectrum of PFA is reported in Figure 1.

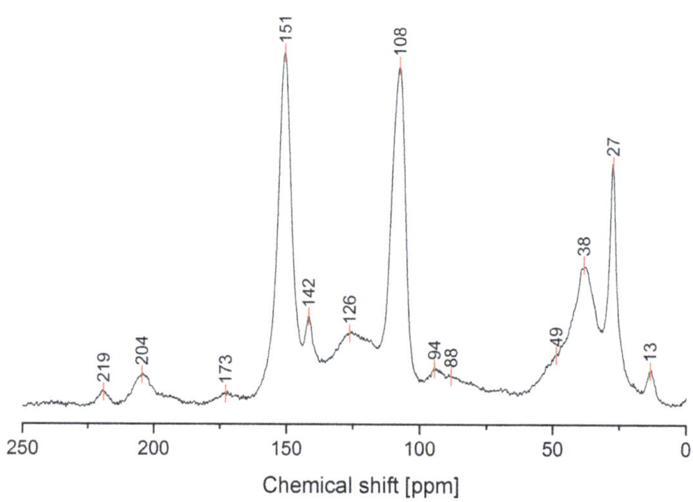

Figure 1. Solid-state ^{13}C-NMR spectrum of hardened PFA.

In Table 2, the interpretation of the ^{13}C-NMR signals is summarized, and the presence of the identified chemical moieties in the possible PFA structures of Table 1 is highlighted.

The ^{13}C-NMR results, summarized in Table 2, suggest that all the presented structures contribute to explaining the signals. However, the major contribution to the PFA structure is certainly given by the linear structure (**1**). Indeed, the high intensities of the signals at 151, 108 and 27 ppm suggest that this structure represents a major proportion of the whole macromolecule, even if it alone cannot justify the presence of the other signals. The broad peak at 38 ppm can be attributed to Diels-Alder arrangements (**6**,**7**) and to the methylene bridge structure (**4**) only, while the presence of a shoulder at around 49 ppm can be attributed to the methylene bridge between furans and carbonyl groups (**2**,**7**). The latter are certainly present due to the signals of the keto-carbonyl at 219 and 204 ppm. The signal at 142 ppm is due to CH=CH inside the unsaturated ring, while the signal at around 88 ppm is due to quaternary carbons. Both assignments can be attributed to Diels-Alder arrangements (**6**,**7**) and also to γ-lactone formation (**3**). The presence of the bands at 13 and 172 ppm can be attributed exclusively to the methyl and carbonyl groups of the lactone (**3**), which possibly occurs in the terminal units of the chain.

Table 2. Summary of the attribution of the chemical shifts of PFA obtained by solid-state ^{13}C-NMR.

Chem. Shift	Chemical Structures from Table 1							Attributions
	1	2	3	4	5	6	7	
219	No	**Yes**	No	No	No	No	**Yes**	C=O diketones vicinal dienophyle
204	No	**Yes**	No	No	No	No	**Yes**	C=O diketones
173	No	No	**Yes**	No	No	No	No	γ-lactones or Levulinic acid/ester
151	**Yes**	No	No	No	**Yes**	**Yes**	No	C2, C5 Furan
142	No	No	**Yes**	No	No	**Yes**	**Yes**	C=C in Diels-Alder bicycle or lactones
126	No	**Yes**	**Yes**	No	**Yes**	No	No	C=C in β between ketones or ester, C=C in the conjugated furan ring.
108	**Yes**	No	No	No	No	No	No	C3, C4 Furan
94	No	No	No	No	**Yes**	**Yes**	No	-CH= Bridge in conjugated systems
88	No	No	**Yes**	No	No	**Yes**	**Yes**	Quaternary C in Diels-Alder or lactones
49	No	**Yes**	No	No	No	No	**Yes**	$-CH_2-$ Bridge between furans & C=O
38	No	No	No	**Yes**	No	**Yes**	**Yes**	Tertiary C in Diels-Alder, -CH- bridge in methylene bridge (4)
27	**Yes**	No	**Yes**	**Yes**	No	**Yes**	**Yes**	$-CH_2-$ bridge in between furans, $-CH_2-$ bridge (4)
13	No	No	**Yes**	No	No	No	No	$-CH_3$ in lactone structures

The relative intensities of these signals suggest that ring opening (**2**,**7**) and Diels-Alder (**6**,**7**) are secondary structural arrangements of completely cured PFA, while γ-lactone (**3**) occurs in a lower percentage. In addition, the assignments given in Table 2 also make it possible to postulate the presence of conjugated structures (**5**, **6**), since the region at around 126 and 94 ppm, presenting some broad peaks, cannot be completely justified by the presence of γ-lactone structures (**3**) and ring opening (**2**). Nevertheless, the more probable Diels-Alder occurrence pathway should be the one presented in structure (**7**), with the γ-diketonic structure (**2**) being more abundant and more dienophilic than conjugated furanic (**5**). We propose a mechanism that encompasses the ring opening of some furanic rings, producing a double bond which promptly combines with the linear PFA to produce structure (**7**) through Diels-Alder crosslinking (See Figure 2).

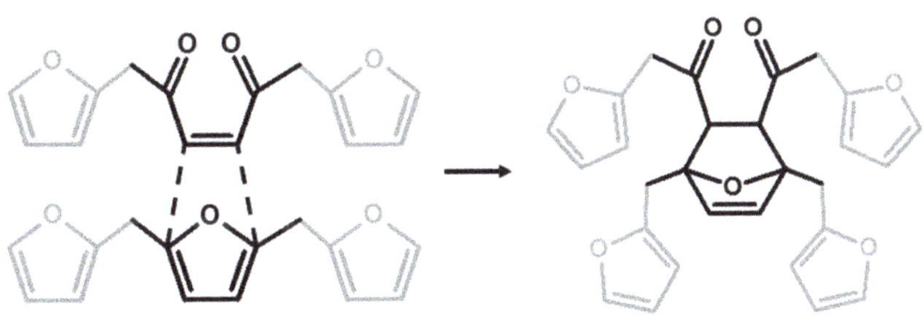

Figure 2. Diels-Alder crosslinking reaction between linear (**1**) and ring opened (**2**) PFA structures.

Moreover, the presence of structure (**4**) could not be excluded, since the broad signal at 38 ppm might be due to both -CH- methyne bridged furanic rings in structure (**4**) and tertiary carbons in Diels-Alder products (**6**,**7**), while, at the same time, the signal at 27 ppm of $-CH_2-$ methylene bridge overlaps with the methylene bridge of structure (**1**). In this investigation, the presence of levulinic derivatives, proposed by Chuang et al. [4], is less probable because simple acid and esters will be leached out.

In summary, on the basis of the observations of the solid-state ^{13}C-NMR spectrum of hardened PFA, it is possible to assert that the dominant PFA structure is the linear one (**1**), while structures (**2**) and (**7**) significantly contribute to the final composition of the polymer, and that structure (**3**) may occur in the terminal part of the chains. Conjugated systems (**5**,**6**) and the methylene bridged structure (**4**) could not be clearly identified and anyway their contribution to the final structure should be negligible with respect to the other proposed arrangements. At the same time, solid-state ^{13}C-NMR excludes the presence of: (i) dimethylene-ether bridges (head-head combination of FA molecules [1]); (ii) unreacted FA; and (iii) opening of two consecutive furanic rings, since β-diketones signals were not detected.

3.2. ATR-FTIR Spectroscopy

The ATR-FTIR spectrum of PFA registered with conventional Mid-IR sources is reported in Figure 3, while the attributions of most of the diagnostic peaks identified here are summarized in Table 3, following the same organization scheme proposed in Table 2.

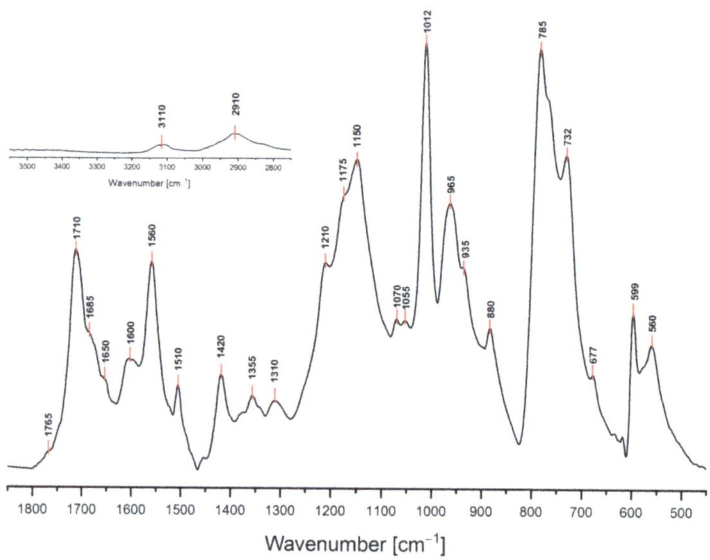

Figure 3. FTIR spectrum of hardened PFA in the spectral region 1850–450 cm^{-1}. In the inset, the spectral region 3550–2750 cm^{-1} is plotted, keeping the original intensity scale.

As a preliminary observation from the analysis of Figure 3, it is evident that the broad band that characterizes the FA in the 3600–3000 cm^{-1} spectral region (see Supplementary data for FA spectrum), associated with hydroxyl group stretching, is not present in the hardened PFA spectrum, confirming that the percentage of unreacted FA is below the detection limit of the technique. At the same time, for the same reason, it is possible to exclude the presence of levulinic acid.

Analyzing the proposed structures, we can observe that the signals expected for structure (**1**) are all present and give major absorptions at 1560, 1420 and 780 cm^{-1}, which is in agreement with the ^{13}C-NMR findings. Structure (**2**) is also certainly present, as proven by the signals of the α,β-unsaturated ketone at around 1690 and 1670 cm^{-1}, assigned to the carbonyl group conjugated with the vicinal C=C, and at 1615–1590 cm^{-1}, related to the carbon-carbon double bond. The presence of structure (**3**) is also possible. Indeed, the α,β-unsaturated γ-lactones generate two intense and characteristic peaks in the range between 1790–1740 cm^{-1}, due to the stretching of the carbonyl group, and two more components in the 1345–1290 cm^{-1} spectral window, usually assigned to the C-O stretching of the lactone ring [30]. Although the aforementioned signals are present, their intensities are quite low. Therefore, structure (**3**) could be present only in small amounts, confirming this pattern as a possible end-chain arrangement. Structure (**4**) with methylene bridges can also not be excluded, because the aliphatic stretching region, between 3000–2800 cm^{-1}, is composed of several subcomponents, which may be explained by C-H stretching modes of different moieties, including the methyne and methylene bridges. The signals possibly generated by the conjugated structure (**5**) have already been discussed in previous papers [11,12,31], but an unambiguous assignment of the vibrational modes to the peaks in the infrared spectra has still not yet been established, even if vibrational analysis has been supported by theoretical simulations. As a matter of fact, the peaks at 1650 cm^{-1} and 1600 cm^{-1} could be assigned either to diene structures [31], or as reported in Table 3. Therefore, on the basis of the literature, FTIR alone cannot exclude structure (**5**), and hence (**6**). At the same time, structures (**2**) and (**7**) explain all the remaining peaks in the 1710–1520 cm^{-1} region. Specifically, the presence of the Diels-Alder rearrangement product (**7**) seems to be very reasonable, due to the presence of two distinctive signals: the isolated C=O stretching at 1710 cm^{-1} and the isolated C=C stretching at 1650 cm^{-1}; formed after the reaction between the furan ring (diene) and the carbon-carbon double bond of the structure (**2**) (dienophile), accordingly with the reactions scheme proposed in Figure 2.

Table 3. Assignment summary of the hardened PFA ATR-FTIR absorption bands.

Wavenumber (cm^{-1})	Chemical Structures from Table 1							Attributions [11,30,31]
	1	2	3	4	5	6	7	
3150–3050	**Yes**	**Yes**	**Yes**	No	**Yes**	**Yes**	**Yes**	C-H stretching aromatic and vinyl
2950–2850	**Yes**	**Yes**	**Yes**	**Yes**	No	**Yes**	**Yes**	C-H stretching aliphatic
1790–1740	No	No	**Yes**	No	No	No	No	C=O stretching α,β-unsat γ -lactone
1720–1700	No	No	No	No	No	No	**Yes**	C=O stretching (isolated)
1690–1670	No	**Yes**	No	No	No	No	No	C=O vicinal to C=C (α,β-unsat. ketone)
1650	No	No	No	No	No	**Yes**	**Yes**	C=C stretching in D.A. (isolated) or conjugated diene
1615–1590	No	**Yes**	No	No	No	No	No	C=C vicinal to C=O (α,β-unsat. ketone) or conjugated diene
1560	**Yes**	No	No	No	**Yes**	No	No	C=C stretching (ring vibr. 2,5-disubstituted furans)
1510	**Yes**	No	No	No	**Yes**	No	No	C=C stretching (ring vibr. 2,5-disubstituted furans)
1450–1345	**Yes**	**Yes**	**Yes**	**Yes**	No	**Yes**	**Yes**	-CH_2 scissoring and wagging
1345–1290	No	No	**Yes**	No	No	No	No	C-O stretching γ-lactone
1230–1100	No	No	No	**Yes**	No	**Yes**	**Yes**	Complex network of several vibrational modes associated with C-O ring stretching, C-C furan stretching,-CH_2 in plane wagging.The peak at 1175 might be due to the C-O-C stretching of the D.A. (difficult to assign)
1100–1040	**Yes**	No	No	No	**Yes**	No	No	=C-O-C= ring vibration (associated with another peak in the range 1200–1120)
1012	**Yes**	No	No	No	No	No	No	-CH in plane wagging 2,5-disubstituted furan (Barsberg simulation)
980–900	No	**Yes**	No	No	No	No	**Yes**	-CH out of plane deformation vibration of alkenes –CH=CH– (usually 2 peaks and they are both present)
880–860	**Yes**	No	No	No	**Yes**	No	No	Furan ring C-H out-of-plane deform. vibration
810–745	**Yes**	No	No	No	**Yes**	No	No	Wagging/twisting -CH-ring structure
745–700	**Yes**	No	No	No	**Yes**	No	No	Furan ring -CH out of plane bend
677	No	**Yes**	**Yes**	No	No	**Yes**	**Yes**	-CH out of plane bending, cis -CH=CH-
599	**Yes**	No	No	No	**Yes**	No	No	Ring deformation vibration
550	No	Yes	**Yes**	No	No	**Yes**	**Yes**	-CH out of plane bending, cis -CH=CH-

Finally, it is also possible to point out that all the signals present in the 1710–1520 cm^{-1} spectral window, mostly involving C=O and C=C vibrational modes, are directly correlated with the polymerization reaction. Indeed, no signals are present for the FA spectrum in the aforementioned region (see Figure S1), since they are due to the new chain formations and rearrangements. This highlights that the ring opening (**2**) certainly occurs and that lactone arrangement (**3**) is reasonable.

3.3. UV-Raman Spectroscopy

To verify the presence of conjugation inside the PFA sample, we performed Raman measurements employing excitation wavelengths of 266 nm, 259 nm and 226 nm. The collected spectra are plotted in Figure 4. At the chosen excitation wavelengths, resonance conditions with the aromatic compounds and the conjugated part of the polymer are achieved. As a matter of fact, as reported by Asher et al. [32,33], UV-Raman spectroscopy uses selective excitation in the UV absorption bands of molecules to produce spectra of their chromophoric segments, enhancing the selectivity of the technique. Specifically, conjugated π-bond systems act as chromophores, exhibiting an absorption wavelength the longer the higher the conjugation degree. In addition, UV-Raman measurements of condensed-phase samples excited below 270 nm are negligibly plagued by fluorescent interferences, further improving the spectral quality.

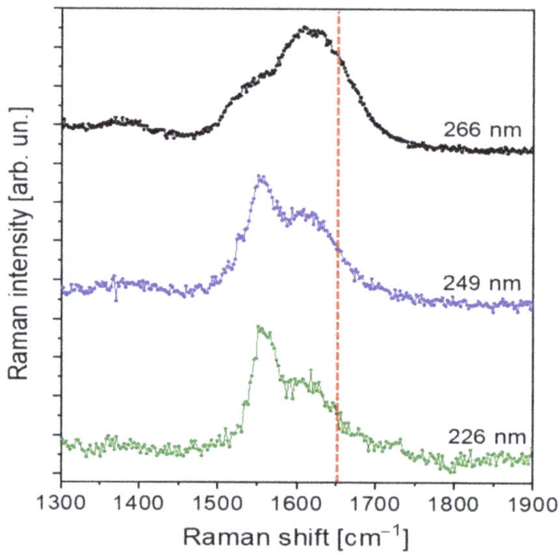

Figure 4. UV-Raman spectra of hardened PFA collected with UV laser excitation at 266 nm (black curve), and with synchrotron radiation excitation at 249 nm and 226 nm (blue and green curves respectively). Red dotted line pinpoints the 1650 cm^{-1} wavenumber.

The careful comparison of our spectra with those obtained by Kim et al. [12,31] evidences important differences, which can be addressed by a different polymer composition, possibly due to the curing process. Specifically, the spectra in Kim et al. are characterized by well-defined vibrational peaks with spectral widths lower than 50 cm^{-1} in the 1400–1700 cm^{-1} spectral region. In particular, the Kim spectra are dominated by the vibrational mode centered at 1650 cm^{-1}, which has been assigned to conjugated C=C, and therefore to a conjugated-type PFA polymer. On the contrary, our spectra show more broadened peaks, not so well defined as those of Kim et al. [12,31], which can therefore be associated with a complex mixture of different PFA structures. Particularly relevant is the absence of a well-defined, isolated intense peak at 1650 cm^{-1} at all the UV excitation wavelengths employed. Even if, due to the peak broadening, the presence of C=C conjugated vibrational components could not be excluded, at the same time, we can safely assert that the contribution of conjugated structure to the final hardened PFA composition is negligible.

3.4. Computational Results

To strengthen the conclusions drawn out complementing the spectroscopic results, a computational effort was made in order to simulate both ^{13}C-NMR and FTIR spectra of a mixture of the most probable FPA arrangements, identifiable in the structures (**1**), (**2**), (**3**) and (**7**) of Table 1. It has to be clearly stated that, due to the chemical complexity of the hardened PFA, as is certainly deducible from the mosaic information obtained with ^{13}C-NMR, FTIR and UV-Raman analyses, it is almost impossible to precisely model the chemical arrangements of this macromolecule and their relative proportions. For complex mixtures of partially known constituents that result in very broad multi-component FTIR and ^{13}C-NMR peaks, only speculative evaluations can be made. On these premises, the final aim of the performed simulations is to demonstrate that our proposed chemical model does not contrast the experimental results. Therefore, for the calculation of the simulated spectra, we tentatively considered a model of the polymer constituted as follows: 60% linear (**1**), 20% Diels-Alder (**7**), 10% ring opening (**2**) and 10% γ-lactone terminal moieties (**3**). In Figure 5, the ^{13}C-NMR spectrum calculated on the base of the proposed model (0.6 (**1**) + 0.1 (**2**) + 0.1 (**3**) + 0.2 (**7**)) is compared with the experimental ^{13}C-NMR one.

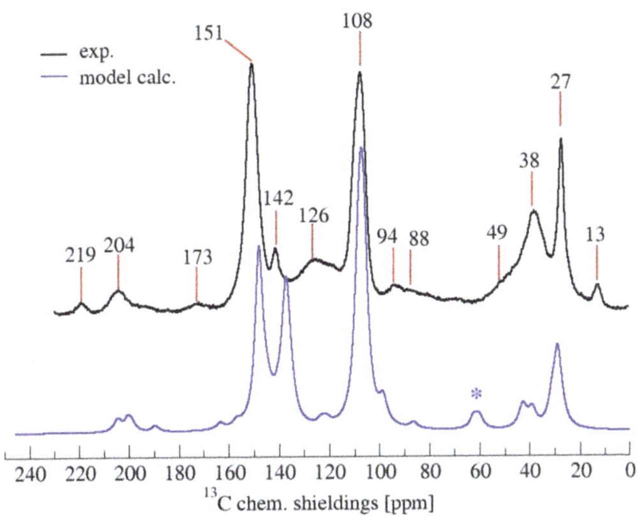

Figure 5. Comparison of experimental and calculated ^{13}C-NMR spectrum of PFA. With * the bands due to the artifact of the model.

It can be appreciated that the simulated ^{13}C-NMR spectrum of the model PFA structure is in good agreement with the experimental one. In particular, the prominent furan-carbon signals at 108 and 151 ppm, as well as the methylene carbon signal at 27 ppm and the carbonyl carbon signals at 204 and 219 ppm, find their close counterparts in the simulated spectra. The greatest gap between the calculated and experimental ^{13}C-NMR is due to the signal at 60 ppm (marked by an asterisk) in the simulated spectrum, which has no counterpart in the experimental spectrum. This signal most likely originates from the two terminal furan-units of structure (**7**), the contribution of which is an artifact of modeling the polymer by considering only a short moiety.

Similarly to the ^{13}C-NMR spectrum, the simulated IR spectrum also does not contradict the experimental FTIR one (See Figure 6).

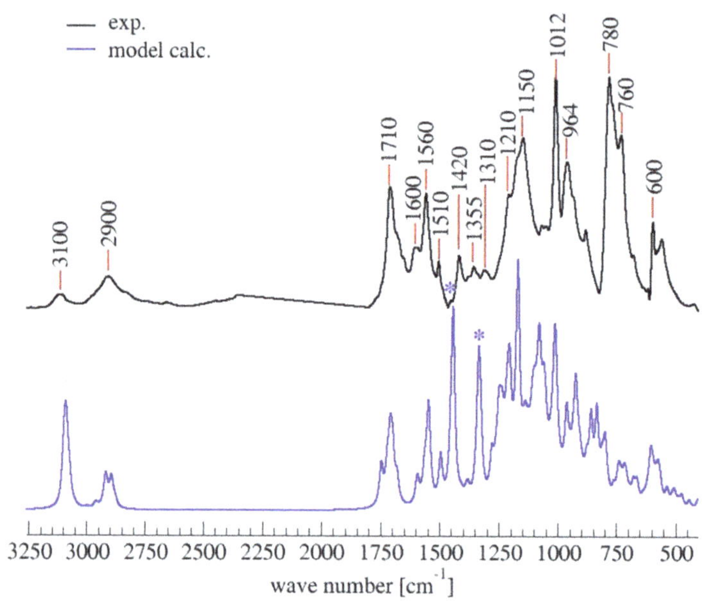

Figure 6. Comparison of experimental and calculated FTIR spectrum of PFA. With * the bands due to the artifact of the model.

In brief, all peaks in the experimental FTIR spectrum also appear in the simulated one. The largest discrepancy between the spectra lies in the two peaks of the simulated spectrum centered at 1350 and 1450 cm^{-1} (marked by asterisks in Figure 6), which do not have a counterpart in the experimental spectrum. The two signals can be assigned to the C=C stretching mode of the two terminal furan moieties in structure (**1**), and their prominent intensity is due to their overweight with respect to the repetitive core, which is also an artifact of the modeling.

To summarize the outcome from the spectroscopic techniques exploited in this work, it is possible to state that structure (**1**), as reported in literature, is the predominant species in this complex matrix. This statement is confirmed by both the ^{13}C-NMR and FTIR data. At the same time, the linear aliphatic structure (**1**) might undergo ring opening, and therefore form the γ-diketonic structure (**2**). Both ^{13}C-NMR and FTIR suggest that structure (**2**) can still be present in the final PFA product, even if at much lower concentrations with respect to (**1**). Indeed, structure (**2**) likely reacts with structure (**1**) through the Diels-Alder rearrangement, leading to the formation of structure (**7**). Also, this species is highly probable, due to the presence of its characteristic peaks in both ^{13}C-NMR and FTIR spectra. The possibility to find conjugated structure (**5**) in this polymer is matter of controversial interpretations. Both the FTIR and UV-Raman were unable to exclude it, since some signals due to this monomeric unit are present. At the same time, these signals might be easily assigned to other species (see Table 3) characterized by the presence of different C=C bonds, as in the case of structures (**2**), (**3**) and (**7**). From the ^{13}C-NMR analysis, one broad signal (94 ppm) with low intensity seems to be due to the conjugated moiety. In any case, even if present, the concentration of the conjugated arrangement is negligible with respect to structures (**1**), (**2**) and (**7**), and its eventual Diels-Alder product, (**6**), is less probable than (**7**). Structure (**3**) has some characteristic peaks in the FTIR region due to the α,β-unsaturated γ-lactones, as well as one peak (173 ppm) in the ^{13}C-NMR spectrum. The intensities of the aforementioned signals are low, suggesting only a minor presence of this species, possibly as an end-chain of the polymeric mixture. Structure (**4**), characterized by the methylene bridge, is an intriguing polymerization product but its presence is very difficult both to confirm and to deny. Indeed, this structure does not generate any peculiar signal in any of the spectroscopic techniques used in this work so far. The aliphatic moieties (-CH and -CH_2) are very common and highly repeated functional groups all along the polymer, preventing saying anything about the presence or not of this chemical unit as a product.

Moreover, it is possible to assert that no presence of unreacted FA could be detected, while the possibility of having carboxylic acids in our hardened PFA polymer is also excluded, since there are no signals in the 3600–3300 cm^{-1} region of the FTIR spectra.

Further confirmation of the interpretation reported by the authors comes from the simulated ^{13}C-NMR and FTIR spectra. The good matching between the theoretical outcome and the experimental data supports the chemical and structural interpretation of the polymeric PFA mixture just summarized.

4. Conclusions

Although this paper focuses its attention on a fairly simple polymerization reaction of a monomeric unit (FA) in acid environment, the final product is not as straightforward as expected from both a chemical and a structural point of view. As highlighted by other authors [9], the polymerization conditions (such as temperature and acid amount) lead to different curing degrees and this affects the relative proportions of the structures proposed in this article. The possible rearrangements and reactions between FA monomers that might occur lead to a wide range of different products, making it possible to postulate that even minimal variations in the synthesis parameters can affect the final PFA composition. The complexity of the final product requires a multi-technique approach that encompasses both experimental and theoretical methods.

By complementing ^{13}C-NMR, FTIR and UV-Raman spectroscopies with spectra modelling, it was possible to confirm the complexity of the heterogeneous final structure of the PFA polymer. The interpretation of the data so far collected helped the authors to clarify the presence or not of some structures and to propose a new one as a possible product. In particular, the linear aliphatic polymer (**1**) is the dominant specie in our hardened PFA. At the same time, due to the acidic environment, this structure might evolve into the γ-diketonic structure (**2**) after ring opening. The mutual presence of these two products (**1**,**2**) suggests their Diels-Alder rearrangement in structure (**7**), which has never been proposed before. The presence of structure (**7**) is supported and confirmed by both ^{13}C-NMR and FTIR techniques. At the same time, the presence in low concentration of the α,β-unsaturated γ-lactone structure (**3**) seems to be probable as the terminal unit of the polymeric chains.

Despite that in this work three of the most powerful techniques (^{13}C-NMR, FTIR, UV-Raman) for investigating solid organic materials were exploited, it was still not possible to achieve an unambiguous overall picture of the FA polymerization reaction. In particular, with respect to the presence of structure (**4**) with methylene bridge, this was almost impossible to assert due to the lack of specific signals. Likewise, from the ^{13}C-NMR data, the conjugated structure (**5**) and the Diels-Alder outcome (**6**) seems not to be so favorable, and the same holds true for FTIR and UV-Raman, although they could not exclude their presence. This final remark further strengthens the need for multi-technique approaches for the characterization of macromolecules, even in the case of well-known ones, such as PFA, the complexity of which should never be underestimated.

Supplementary Materials: The following are available online at http://www.mdpi.com/2073-4360/11/12/2126/s1, Figure S1: FT-IR spectra of furfuryl alcohol and PFA.

Author Contributions: Conceptualization, G.T., L.V., M.M., R.J.F.B.; methodology, G.T, L.V., G.B.; software, G.B., R.J.F.B.; validation, R.J.F.B.; formal analysis, G.T., M.M., L.V.; investigation, T.S. (Thomas Sepperer), G.T.,G.B., N.C., A.R.; resources, G.T., L.V., M.M.; data curation, T.S. (Thomas Sepperer), F.D., N.C.; writing—original draft preparation, G.T., N.C., F.D.; writing—review and editing, G.T., M.M., N.C., L.V.; visualization, T.S. (Thomas Sepperer), N.C., F.D.; supervision, G.T., L.V.; project administration, G.T., L.V., T.S. (Thomas Schnabel); funding acquisition, G.T., L.V., T.S. (Thomas Schnabel).

Funding: This research was funded by the European Regional Development Fund and Interreg V-A Italy Austria 2014–2020 through the Interreg Italy Austria project ITAT 1023 InCIMa "Smart Characterization of Smart Materials".

Acknowledgments: The authors would like to acknowledge the financial support provided by the European Regional Development Fund and Interreg V-A Italy Austria 2014–2020 through the Interreg Italy Austria project

ITAT 1023 InCIMa "Smart Characterization of Intelligent Materials". R.J.F.B. gratefully acknowledges. Dage Sundholm and the University of Helsinki for the generous provision of computational resources and the DFG SPP 1807 "Control of London dispersion interactions in molecular chemistry" (BE 4632/2-1) for funding.

Conflicts of Interest: The authors declare no conflict of interest.

References

1. Dunlop, A.P.; Peters, F.N. *The Furans*; Reinhold Pub. Corp: New York, NY, USA, 1953; Volume 1.
2. Conley, R.T.; Metil, I. An investigation of the structure of furfuryl alcohol polycondensates with infrared spectroscopy. *J. Appl. Polym. Sci.* **1963**, *7*, 37–52. [CrossRef]
3. Wewerka, E.M. Study of the γ-alumina polymerization of furfuryl alcohol. *J. Polym. Sci. Part A-1 Polym. Chem.* **1971**, *9*, 2703–2715. [CrossRef]
4. Chuang, I.S.; Maciel, G.E.; Myers, G.E. Carbon-13 NMR study of curing in furfuryl alcohol resins. *Macromolecules* **1984**, *17*, 1087–1090. [CrossRef]
5. Buchwalter, S.L. The polymerization of furfuryl acetate in acetonitrile. *J. Polym. Sci. Polym. Chem. Ed.* **1985**, *23*, 2897–2911. [CrossRef]
6. Choura, M.; Belgacem, N.M.; Gandini, A. Acid-catalyzed polycondensation of furfuryl alcohol: Mechanisms of chromophore formation and cross-linking. *Macromolecules* **1996**, *29*, 3839–3850. [CrossRef]
7. Choura, M.; Belgacem, N.M.; Gandini, A. The acid-catalyzed polycondensation of furfuryl alcohol: Old puzzles unravelled. In *Macromolecular Symposia*; Hüthig & Wepf Verlag: Basel, Switzerland, 1997; Volume 122, pp. 263–268.
8. Bertarione, S.; Bonino, F.; Cesano, F.; Damin, A.; Scarano, D.; Zecchina, A. Furfuryl Alcohol Polymerization in H−Y Confined Spaces: Reaction Mechanism and Structure of Carbocationic Intermediates. *J. Phys. Chem. B* **2008**, *112*, 2580–2589. [CrossRef]
9. Guigo, N.; Mija, A.; Vincent, L.; Sbirrazzuoli, N. Chemorheological analysis and model-free kinetics of acid catalysed furfuryl alcohol polymerization. *Phys. Chem. Chem. Phys.* **2007**, *9*, 5359–5366. [CrossRef]
10. Guigo, N.; Mija, A.; Zavaglia, R.; Vincent, L.; Sbirrazzuoli, N. New insights on the thermal degradation pathways of neat poly (furfuryl alcohol) and poly (furfuryl alcohol)/SiO_2 hybrid materials. *Polym. Degrad. Stab.* **2009**, *94*, 908–913. [CrossRef]
11. Barsberg, S.; Thygesen, L.G. Poly (furfuryl alcohol) formation in neat furfuryl alcohol and in cymene studied by ATR-IR spectroscopy and density functional theory (B3LYP) prediction of vibrational bands. *Vib. Spectrosc.* **2009**, *49*, 52–63. [CrossRef]
12. Kim, T.; Assary, R.S.; Marshall, C.L.; Gosztola, D.J.; Curtiss, L.A.; Stair, P.C. Acid Catalyzed Furfuryl Alcohol Polymerization: Characterizations of Molecular Structure and Thermodynamic Properties. *ChemCatChem* **2011**, *3*, 1451–1458. [CrossRef]
13. Kim, T.; Assary, R.S.; Kim, H.; Marshall, C.L.; Gosztola, D.J.; Curtiss, L.A.; Stair, P.C. Effects of solvent on the furfuryl alcohol polymerization reaction: UV Raman spectroscopy study. *Catal. Today* **2013**, *205*, 60–66. [CrossRef]
14. Falco, G.; Guigo, N.; Vincent, L.; Sbirrazzuoli, N. Opening Furan for Tailoring Properties of Bio-based Poly (Furfuryl Alcohol) Thermoset. *ChemSusChem* **2018**, *11*, 1805–1812. [CrossRef] [PubMed]
15. Banfi, D.; Patiny, L. www.nmrdb.org: Resurrecting and Processing NMR Spectra On-line. *Chim. Int. J. Chem.* **2008**, *62*, 280–281. [CrossRef]
16. Steinbeck, C.; Krause, S.; Kuhn, S. NMRShiftDB constructing a Free Chemical Information System with Open-Source Components. *J. Chem. Inf. Comput. Sci.* **2003**, *43*, 1733–1739. [CrossRef] [PubMed]
17. Castillo, A.M.; Patiny, L.; Wist, J. Fast and accurate algorithm for the simulation of NMR spectra of large spin systems. *J. Magn. Reson.* **2011**, *209*, 123–130. [CrossRef] [PubMed]
18. Lupi, S.; Nucara, A.; Perucchi, A.; Calvani, P.; Ortolani, M.; Quaroni, L.; Kiskinova, M. Performance of SISSI, the infrared beamline of the ELETTRA storage ring. *J. Opt. Soc. Am. B* **2007**, *24*, 4. [CrossRef]
19. D'Amico, F.; Saito, M.; Bencivenga, F.; Marsi, M.; Gessini, A.; Camisasca, G.; Principi, E.; Cucini, R.; Di Fonzo, S.; Battistoni, A.; et al. UV resonant Raman scattering facility at Elettra. *Nucl. Instrum. Methods Phys. Res. Sect. A Accel. Spectrom. Detect. Assoc. Equip.* **2013**, *703*, 33–37. [CrossRef]
20. Richard, L. McCreery, Photometric Standards for Raman Spectroscopy. In *Handbook of Vibrational Spectroscopy*; Chalmers, J.M., Griffiths, P.R., Eds.; Wiley: Hoboken, NJ, USA, 2006; Volume 1.

21. ChemAxon. MarvinSketch, Version 15.2.16.0. 2014. Available online: http://www.chemaxon.com (accessed on 1 June 2019).
22. Turbomole V7.0 2015, a Development of University of Karlsruhe and Forschungszentrum Karlsruhe GmbH, 1989-2007, TURBOMOLE GmbH, since 2007. Available online: http://www.turbomole.com (accessed on 1 June 2019).
23. Sierka, M.; Hogekamp, A.; Ahlrichs, R. Fast evaluation of the Coulomb potential for electron densities using multipole accelerated resolution of identity approximation. *J. Chem. Phys.* **2003**, *118*, 9136–9148. [CrossRef]
24. Becke, A.D. Density-functional exchange-energy approximation with correct asymptotic behavior. *Phys. Rev. A* **1988**, *38*, 3098. [CrossRef]
25. Perdew, J.P. Density-functional approximation for the correlation energy of the inhomogeneous electron gas. *Phys. Rev. B* **1986**, *33*, 8822. [CrossRef]
26. Weigend, F.; Häser, M.; Patzelt, H.; Ahlrichs, R. RI-MP2: Optimized auxiliary basis sets and demonstration of efficiency. *Chem. Phys. Lett.* **1998**, *294*, 143–152. [CrossRef]
27. Weigend, F.; Ahlrichs, R. Balanced basis sets of split valence, triple zeta valence and quadruple zeta valence quality for H to Rn: Design and assessment of accuracy. *Phys. Chem. Chem. Phys.* **2005**, *7*, 3297–3305. [CrossRef] [PubMed]
28. Kollwitz, M.; Gauss, J. A direct implementation of the GIAO-MBPT (2) method for calculating NMR chemical shifts. Application to the naphthalenium and anthracenium ions. *Chem. Phys. Lett.* **1996**, *260*, 639–646. [CrossRef]
29. Grimme, S.; Bannwarth, C.; Shushkov, P. A robust and accurate tight-binding quantum chemical method for structures, vibrational frequencies, and noncovalent interactions of large molecular systems parametrized for all spd-block elements (Z = 1–86). *J. Chem. Theory Comput.* **2017**, *13*, 1989–2009. [CrossRef] [PubMed]
30. Socrates, G. *Infrared and Raman Characteristic Group Frequencies: Tables and Charts*; Wiley: Hoboken, NJ, USA, 2004.
31. Kim, T.; Jeong, J.; Rahman, M.; Zhu, E.; Mahajan, D. Characterizations of furfuryl alcohol oligomer/polymerization catalyzed by homogeneous and heterogeneous acid catalysts. *Korean J. Chem. Eng.* **2014**, *31*, 2124–2129. [CrossRef]
32. Asher, S.A. Ultraviolet resonance Raman spectrometry for detection and speciation of trace polycyclic aromatic hydrocarbons. *Anal. Chem.* **1984**, *56*, 720–724. [CrossRef]
33. Asher, S.A.; Johnson, C.R. Raman Spectroscopy of a Coal Liquid Shows That Fluorescence Interference Is Minimized with Ultraviolet Excitation. *Science* **1984**, *225*, 311–313. [CrossRef]

polymers

MDPI

Article

Eudragit: A Novel Carrier for Controlled Drug Delivery in Supercritical Antisolvent Coprecipitation

Paola Franco and Iolanda De Marco *

Department of Industrial Engineering, University of Salerno, Via Giovanni Paolo II, 132, 84084 Fisciano, Italy; pfranco@unisa.it

* Correspondence: idemarco@unisa.it; Tel.: +39-089-964066

Received: 13 December 2019; Accepted: 16 January 2020; Published: 18 January 2020

Abstract: In this work, the supercritical antisolvent (SAS) process was used to coprecipitate Eudragit L100-55 (EUD) with diclofenac (DICLO) and theophylline (THEOP), with the aim of obtaining composite microparticles with a prolonged drug release for oral delivery. Working at the optimized conditions in terms of pressure and overall concentration in the liquid solution (10.0 MPa and 50 mg/mL), microparticles of EUD/DICLO 20/1 and 10/1 *w/w* were produced with a mean size of 2.92 μm and 1.53 μm, respectively. For the system EUD/THEOP, well-defined spherical microspheres with a mean diameter ranging from 3.75 μm and 5.93 μm were produced at 12.0 MPa. The produced composite systems were characterized by various techniques, such as scanning electron microscopy, differential scanning calorimetry, X-ray microanalysis, FT-IR and UV–vis spectroscopy. Dissolution studies showed the potential of EUD to prolong the drug release, significantly, up to a few days.

Keywords: SAS coprecipitation; controlled drug delivery systems; microparticles; eudragit; supercritical carbon dioxide

1. Introduction

The inflammatory process is a defense response of body tissues against different harmful agents, irritants or pathogens. Inflammation can be acute when a rapid resolution is possible, or chronic when it is characterized by longer duration. Musculoskeletal disorders, such as rheumatoid arthritis and osteoarthritis, are chronic inflammations for which long-term therapies are required [1]. Non-Steroidal Anti-Inflammatory Drugs (NSAIDs) are generally used to reduce this kind of inflammatory conditions, as well as for their painkilling and antipyretic effects. Among the different drug delivery routes, oral administration is the preferred one because of its high patient compliance. Diclofenac (DICLO) is one of the most frequently prescribed NSAIDs; however, due to its short half-life, high dosage and frequent administration of diclofenac are necessary, resulting in many side effects, especially gastrointestinal ones. Another example of chronic inflammatory diseases is asthma, which affects the respiratory tract, due to an obstruction of the bronchi. Nowadays, the therapy goal is to achieve control of chronic bronchial asthma with fewer drug dosages. Therefore, drugs to be taken daily are recently prescribed in order to ensure long-term control of the symptoms and the maintenance of normal lung function. Theophylline (THEOP) is included in this category of drugs. However, high drug concentrations increase the risk of side effects (even toxicity effects), since the therapeutic range of this drug is very narrow [2]; to overcome the consequent limitations, prolonged-release formulations are preferred.

A valid solution to reduce the frequency of administration and unwanted effects is represented by controlled drug delivery systems. Generally speaking, the production of polymer/drug composite particles has a relevant importance in the pharmaceutical industry. By selecting a specific polymer as the carrier, it is possible to avoid the oxidation and/or the deactivation of the drug, to mask its organoleptic properties (color, taste, and odor) or modify its release kinetics. As a consequence, the bioavailability of a poorly water-soluble drug can be enhanced, or a prolonged drug release can be reached [3,4].

Spray-drying [5], freeze-drying [6], emulsification/solvent evaporation [7], centrifugal extrusion [8], jet-milling [9], and coacervation [10] are only some of the traditional techniques employed to produce polymer/drug coprecipitated powders. However, their application often involves the production of irregular particles with a wide particle size distribution (PSD), low encapsulation efficiencies, in addition to the possible degradation of the materials caused by mechanical or thermal stresses and/or organic solvents residues in the product. On the contrary, supercritical fluids (SCFs) based techniques seem to be a successful alternative to overcome these limits, especially when supercritical carbon dioxide ($scCO_2$) is used, being non-toxic, non-polluting and cheap with moderate critical parameters (T_c = 31.1 °C, P_c = 7.38 MPa). Supercritical AntiSolvent (SAS) precipitation is one of the most used supercritical fluids based techniques, which has been successfully employed to obtain drug or polymer/drug nanoparticles, microparticles and expanded microparticles [11–16].

According to the SAS literature, in order to achieve a massive polymer/drug coprecipitation by the SAS process, microdroplets have to be formed, and after the subsequent removal of the solvent by $scCO_2$, composite microspheres are produced.

However, the coprecipitation implies the formation of a quaternary system solvent + $scCO_2$ + polymer + drug with a complex phase behavior, which can lead to the failure of the coprecipitation [16]. Up to now, a few polymers allow reaching a successful coprecipitation in the form of microspheres by the SAS process, namely, polyvinylpyrrolidone (PVP), poly (L-lactic acid) (PLLA), and recently, zein [11,12,17–21].

Commonly used for oral dosage forms, Eudragits are methacrylic acid copolymers, a registered trademark of Rohm Pharmaceuticals (Darmstadt, Germany). The various Eudragit polymers differ in the ratios of dimethylaminoethyl methacrylates, methacrylic acid and methacrylic esters. Eudragit E, RL, RS, and NE are polycations, because of the presence of dimethylamino groups or quaternary amino groups; whereas, Eudragit L and S are polyanions, due to the presence of carboxylate groups [22,23].

To our knowledge, very few literature studies were focused on the use of Eudragit polymers using the SAS process, although they are really interesting from a pharmaceutical point of view [13,14,24,25]. Indeed, they offer protection from moisture and masking of smells/flavors; moreover, they are versatile, since the choice of a specific Eudragit can be made based on the type of desired drug release (immediate, delayed or sustained). On the market, there are various kinds of Eudragits defined as pH-sensitive polymers, because they are soluble at different pH values [26,27]. Among them, Eudragit E100 and Eudragit EPO are soluble in the gastric juice up to pH 5.0, and therefore, are generally used for immediate release delivery systems [28]. Eudragit L100-55 (EUD) promotes a controlled drug release at pH higher than 5.5, starting from the first intestinal tract (duodenum), thus, protecting the drug against the acid gastric fluid [29,30]. Eudragit L100 differs from Eudragit L100-55 only for the presence of a methyl group instead of an ethyl group [31], which influences the slightly different dissolution pH threshold for the two polymers. Eudragit S100 is soluble at pH above 7.0; it can be used, in combination with other Eudragits, for colon-targeted oral drug delivery [32].

Montes et al. tried to coprecipitate Eudragit L100, which guarantees the drug release at pH values above 6, starting from the central section of the small intestine (jejunum), with ibuprofen [13], and naproxen [14] by SAS process. Eudragit L100/ibuprofen and Eudragit L100/naproxen particles with mean sizes in the range 0.08–0.51 μm and 0.08–0.31 μm, respectively were precipitated. Low loading efficiencies were achieved (0.94–7.88% for ibuprofen and 4.45–25.55% for naproxen), which can be attributed to the nanometric dimension of the powders; indeed, it is presumable that the particles at nanodimensions tend to precipitate separately, considering that it is difficult to obtain heterogeneous nucleation. Similar results were obtained by the same authors in another study [33], coprecipitating Eudragit L100 and ellagic acid at different polymer/drug ratios. From the shown FESEM images, it is possible to observe that the precipitation of crystals occurred using Eudragit L100/ellagic acid ratios equal to 1/1 and 2/1, whereas, coalescent sub-microparticles precipitated using a polymer/drug ratio of 4/1. Garay et al. processed Eudragit EPO using the gas antisolvent process [25]; using EtOH/H_2O mixtures as the solvent, they obtained micrometric particles with a wide particle size distribution. Duarte et al. coprecipitated acetazolamide with Eudragit L100, Eudragit S100, and their mixtures [24]; the SEM images reported in the paper show the co-presence of crystals and microparticles; this suggests that the polymer and the drug precipitated separately and that an effective coprecipitation did not occur.

Up to now, it would appear that SAS coprecipitation using Eudragit as the polymeric carrier is not successful. Therefore, with the aim of obtaining prolonged drug delivery devices for oral administration, Eudragit L100-55 was selected as the polymeric carrier for SAS coprecipitation. Firstly, the effect of some process conditions on the micronization of Eudragit L100-55 (without the drug) was studied to obtain spherical microparticles. Then, several SAS coprecipitation experiments were performed using an NSAID (diclofenac) and a bronchodilator drug (theophylline) as model compounds.

2. Materials, Methods and Procedures

2.1. Materials

Eudragit L100-55 (EUD) was generously donated by Degussa (Darmstadt, Germany). Diclofenac sodium salt (DICLO, average molecular weight 318.13 g/mol), anhydrous theophylline (THEOP, average molecular weight 180.16 g/mol) and Dimethylsulfoxide (DMSO, purity 99.5%) were purchased by Sigma-Aldrich (Milan, Italy). CO_2 (purity 99%) was supplied from Morlando Group s.r.l. (Sant'Antimo - NA, Italy). All materials were used as received. The solubility in DMSO at room temperature is approximately 55 mg/mL for EUD, 150 mg/mL for DICLO and 25 mg/mL for THEOP.

2.2. SAS Apparatus and Procedure

The homemade laboratory plant used to perform SAS experiments is sketched in Figure 1.

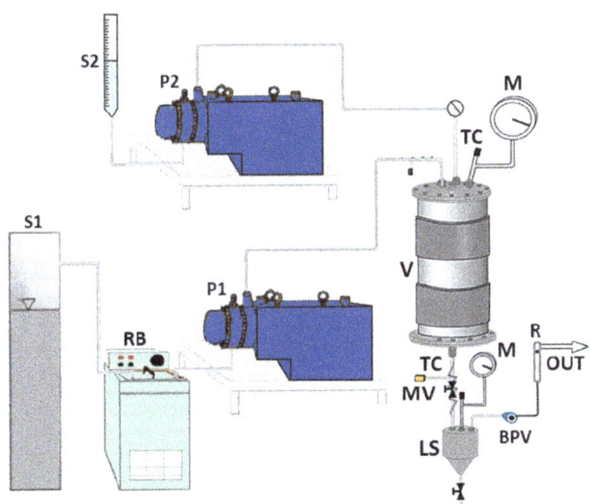

Figure 1. A sketch of the supercritical antisolvent (SAS) laboratory plant. S1, tank for the CO_2; S2, organic solution; RB, refrigerating bath; P1, P2, pumps; V, vessel; M, manometer; TC, thermocouple; MV, micrometric valve; LS, liquid separator; BPV, back-pressure valve; R, rotameter.

The precipitation vessel (V) is a cylindrical chamber with an internal volume of 500 cm^3. The CO_2 is stored in a tank (S1) and feed to the vessel through a high-pressure pump (P1). The liquid solution contained in a burette (S2) is co-currently injected into the vessel by another high-pressure pump (P2), passing through a 100 μm internal diameter stainless steel nozzle. The temperature control

(TC) is assured by a proportional integral derivative (PID) controller connected with electrically thin bands and the pressure in the vessel is measured using a test gauge manometer (M) and regulated by a micrometric valve (MV). At the bottom of the precipitator, the precipitated powder is collected on a stainless steel filter (pores size of 0.1 μm), which also allows the passage of CO_2—solvent solution. Downstream the precipitation chamber, the liquid solvent is recovered in a second collection vessel (liquid separator, LS) at a lower pressure (1.8–2.0 MPa) regulated by a back-pressure valve (BPV). The flow rate of delivered CO_2 is measured at the exit of the second vessel by a rotameter ®.

A SAS test starts pumping CO_2 at a constant flow rate in the precipitation vessel until the desired pressure is reached; then, the antisolvent steady flow is established. After that, pure solvent (DMSO) is injected through the nozzle in co-current mode with CO_2, since the steady-state solvent/antisolvent composition is reached. Then, the solvent flow rate is stopped, and the liquid solution (DMSO + solutes) is injected into the vessel at the same flow rate as the pure DMSO with consequent precipitation of the solute/solutes. At the end of the solution injection, the liquid pump is stopped, and only the $scCO_2$ is sent to the precipitation chamber to eliminate the solvent residues. The CO_2 flow is stopped when the washing step is completed, and the precipitator is depressurized up to the atmospheric pressure. Finally, the precipitated powder can be collected and characterized.

All SAS experiments were performed using an operating temperature of 40 °C, DMSO as the liquid solvent, a CO_2 flow rate and a solution flow rate equal to 30 g/min and 1 mL/min, respectively. At the selected temperature, the flow rates were chosen to obtain molar fractions on the right of the mixture critical point (MCP) of the binary system solvent/antisolvent; i.e., at CO_2 molar fractions approximately equal to 0.98 that ensures the supercritical mixture conditions [34,35]. In the case of EUD micronization, the effect of the operating pressure (P) and of the overall concentration in the liquid solution (C_{tot}) on particles' morphology, particle mean diameter (m.d.) and particle size distribution (PSD) was studied; in the case of EUD/DICLO and EUD/THEOP coprecipitation, the effect of the polymer/drug ratio, P and C_{tot} on morphology, m.d., and PSD was investigated.

2.3. Characterization Methods

The morphology of the samples was evaluated by a Field Emission Scanning Electron Microscope (FESEM, mod. LEO 1525, Carl Zeiss SMT AG, Oberkochen, Germany). The powder was dispersed on a carbon tab previously stuck to an aluminum stub (Agar Scientific, Stansted, United Kingdom) and coated with gold-palladium (layer thickness 250 Å) using a sputter coater (mod. 108 A, Agar Scientific, Stansted, United Kingdom). Mean dimensions and standard deviations of particles were measured from FESEM photomicrographs by using an image analysis software (SigmaScan Pro, Jandel Scientific, Bangalore, India), considering about 1000 particles for each sample. Particle size distributions (PSDs) were determined by Microcal Origin Software (release 8.0, Microcal Software, Inc., Northampton, MA, USA).

Fourier transform infrared (FT-IR) analyses were performed by M2000 FT-IR (MIDAC Co, Costa Mesa, CA, USA). The scan wavenumber range was 4000–450 cm^{-1}, at a resolution of 0.5 cm^{-1} and 16 scan signals were averaged to reduce the noise. Approximately 1 mg of the powder samples was well-mixed with 100 mg of potassium bromide (KBr), which was used as an infrared transparent matrix. The discs to be analyzed were prepared by compressing the powders in a hydraulic press.

Calorimetric analyses were obtained by means of a Differential Scanning Calorimeter (DSC, mod. TC11, Mettler-Toledo, Inc., Columbus, OH, USA) using the Mettler STARe system. The accurately weighed powders (5 ± 0.5 mg) were heated from 25 °C up to 300 °C, using a heating rate of 5 °C/min and a nitrogen gas flow equal to 50 mL/min. Each analysis was performed twice.

An X-ray diffractometer (Bruker D8-Advance, Bruker AXS GmbH, Karlsruhe, Germany) with a Cu sealed tube source was used to record the X-ray diffractograms (XRD) of samples. The measuring conditions were: Ni-filtered CuK_α radiation, $\lambda = 1.54$ Å, 2θ angle ranging from 10° to 60° with a scan rate of 0.5 s/step and a step size of 0.08°.

In vitro drug dissolution was monitored using a UV/vis spectrophotometer (model Cary 50, Varian, Palo Alto, CA, USA) at a wavelength of 276 nm and 271 nm for DICLO and THEOP, respectively. An equivalent amount of drug equal to 5 mg was considered to compare the dissolution rate of unprocessed drug and that of coprecipitated powders. The samples were suspended in 3 mL of phosphate-buffered saline solution (PBS) at pH 7.4 and placed into a 12,000 Da molecular weight cutoff dialysis sack, which was incubated in 300 mL of PBS, continuously stirred at 200 rpm and heated at 37 ± 0.1 °C. The value of the absorbance was measured by the instrument every minute from time zero up to 5 h, then every hour until the maximum value was reached. Each analysis was performed in triplicate: Considering that the difference between the triplicates was less than 1%, the mean release profiles were reported in this paper.

3. Results and Discussion

A summary of several experiments performed at different process conditions is reported in Table 1, indicating the obtained morphology (M), the mean diameter (m.d.) and the standard deviation (s.d.) on a volumetric basis. In particular, some results reported in Table 1 are related to the micronization of EUD alone, followed by coprecipitation studies on the systems EUD/DICLO and EUD/THEOP.

Table 1. A list of SAS tests for Eudragit L100-55 (EUD) micronization, EUD/ diclofenac (DICLO) coprecipitation and EUD/ theophylline (THEOP) coprecipitation (C, crystals; NP, nanoparticles; MP, microparticles; cMP, coalescing microparticles; EMP, expanded microparticles); when in a single test more than one morphology is obtained, * indicates the one whose diameter is calculated.

#	Polymer/Drug [*w/w*]	P, [Mpa]	C_{tot} [mg/mL]	M	m.d. ± s.d. [μm]
			EUD		
1	1/0	9	20	MP	1.99 ± 0.49
2	1/0	10	20	MP	1.69 ± 0.51
3	1/0	12	20	cMP	1.64 ± 0.72
4	1/0	10	40	MP	1.95 ± 0.54
			EUD/DICLO		
5	0/1	9	20	NP	0.14 ± 0.05
6	20/1	9	40	MP * + cMP	* 2.16 ± 0.69
7	20/1	10	40	MP	2.47 ± 0.71
8	20/1	10	20	cMP	-
9	20/1	10	50	MP	2.92 ± 0.81
10	10/1	10	50	MP	1.53 ± 0.45
			EUD/THEOP		
11	0/1	9	20	C	-
12	20/1	10	40	C	-
13	20/1	10	20	C + MP *	* 6.79 ± 1.84
14	20/1	12	20	MP	5.93 ± 1.62
15	20/1	15	20	cMP	1.64 ± 0.32
16	20/1	12	40	MP	5.65 ± 1.66
17	10/1	12	40	MP * + EMP	* 3.75 ± 1.08

3.1. *Micronization of Eudragit L100-55*

The first set of experiments (runs #1–4 in Table 1) was performed by micronizing EUD alone, at different P and C_{tot}, in order to optimize the process conditions for the attainment of polymer microparticles.

Effect of the Operating Pressure

The effect of the operating pressure was investigated setting a polymer concentration in DMSO equal to 20 mg/mL and gradually increasing the pressure from 9 to 12 Mpa (runs #1–3 in Table 1).

FESEM analyses revealed that well-separated spherical microparticles were obtained working at 9 and 10 Mpa (Figure 2a,b, respectively), whereas, coalescing and more irregular in shape and dimensions microparticles were formed at 12 Mpa (Figure 2c). Moreover, a slight increase of the particle dimensions was observed by decreasing the pressure, as shown in the last column of Table 1. Considering that it has been previously demonstrated that SAS coprecipitation is favored at pressures in correspondence of which well-defined micrometric particles are produced, for the subsequent experimentation the pressure of 12 Mpa was not considered [16].

(c)

Figure 2. FESEM images of Eudragit particles precipitated from DMSO at 40 °C and 20 mg/mL. Effect of the operating pressure. (**a**) 9 MPa; (**b**) 10 MPa; (**c**) 12 MPa.3.1.2 Effect of polymer concentration in DMSO.

Then, fixing the pressure at 10 MPa, the polymer concentration in DMSO was increased at 40 mg/mL (run #4 in Table 1). Spherical microparticles were produced; the comparison of the volumetric cumulative PSDs in Figure 3 (related to runs #2 and #4) demonstrated that the mean particle size increased and the particle size distribution widened by increasing the EUD concentration in DMSO. Indeed, the higher the concentration of the polymer in the liquid solution, the more the phenomenon of growth prevails over that of nucleation.

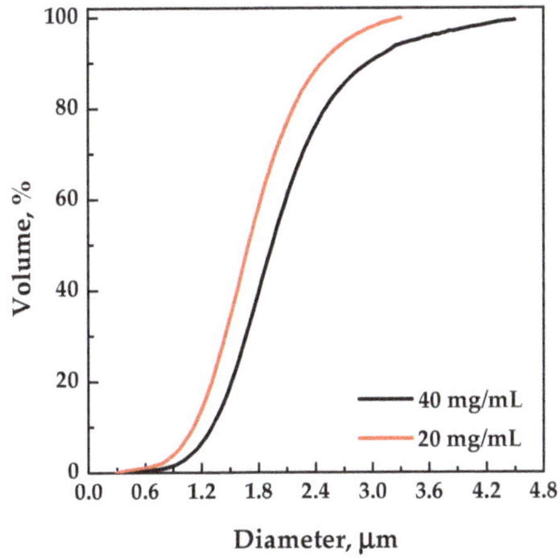

Figure 3. Volumetric cumulative particle size distributions (PSDs) of EUD precipitated from DMSO at 40 °C and 10 MPa; effect of the polymer concentration in DMSO.

3.2. Coprecipitation Using Eudragit as the Carrier

Once verified the attainment of EUD microparticles using the SAS process, the polymer coprecipitation was attempted using diclofenac and theophylline as model drugs.

Preliminary experiments were performed by processing each drug separately at 40 °C, 9 MPa using a solute concentration in DMSO equal to 20 mg/mL (runs #5 and #11 in Table 1). From FESEM analyses, it is possible to note that SAS processed diclofenac (Figure 4a) precipitated in the form of nanoparticles (mean diameter of about 0.13 μm). On the contrary, after processing theophylline alone, no powder was found in the precipitation chamber, because it was probably extracted by the mixture formed by the organic solvent and supercritical CO_2. Long crystals (Figure 4b) were observed by analyzing the very few granules that were recovered.

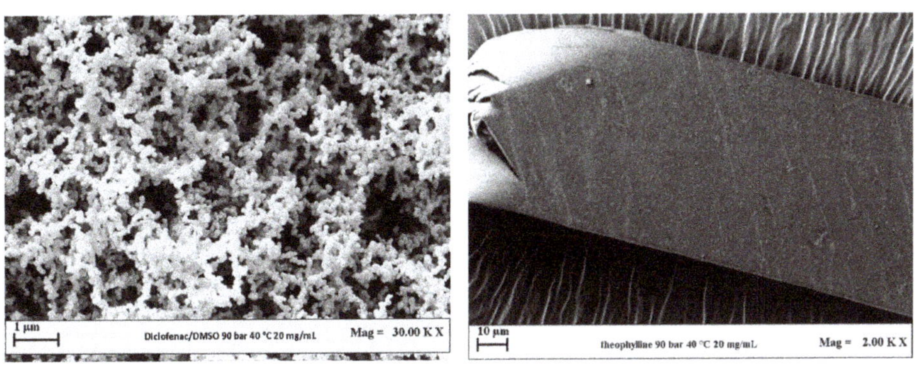

(**a**) (**b**)

Figure 4. FESEM images of the drugs precipitated from DMSO at 9 MPa, 40 °C and 20 mg/mL. (**a**) DICLO; (**b**) THEOP.

In order to successfully coprecipitate the two model drugs with EUD, the effect of pressure, of polymer + active principle concentration in DMSO and of polymer/drug ratio *w*/*w* on the morphology and on the mean size of the produced composite particles was investigated.

3.2.1. Effect of the Operating Pressure on Coprecipitated Particles

Firstly, the effect of the pressure was investigated for the system EUD/DICLO by selecting 9 and 10 MPa as pressures, according to the previous results obtained by studying EUD micronization. Moreover, an overall concentration of solutes in DMSO of 40 mg/mL and a polymer/drug ratio equal to 20/1 *w/w* were fixed. A less degree of coalescence of the particles was observed operating at 10 MPa (run #7 in Table 1) with respect to 9 MPa (run #6 in Table 1), as it is possible to observe from the FESEM images reported in Figure 5.

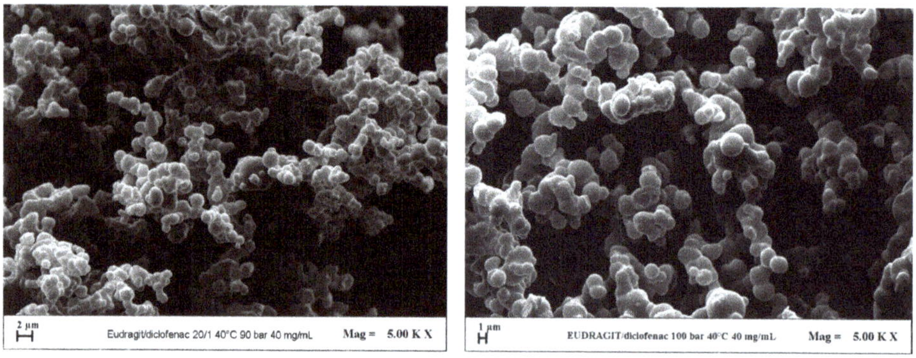

(a) (b)

Figure 5. FESEM images of EUD/DICLO 20/1 particles precipitated from DMSO at 40 °C and 40 mg/mL. Effect of the operating pressure. (**a**) 9 MPa (run #6); (**b**) 10 MPa (run #7).

Considering that the pressure of 10 MPa gave the best results in terms of sphericity and mean diameter of the powders in the case of EUD micronization and for its coprecipitation with DICLO, the preliminary tests with the system EUD/THEOP 20/1 *w/w* at 40 mg/mL (run #12) and 20 mg/mL (run #13) were performed in correspondence of this pressure. Using this system, crystals and crystals precipitated with coalescing microparticles (as shown in Figure 6a,b for run #13) were obtained at 40 mg/mL and 20 mg/mL, respectively. This outcome may be explained, taking into account the thermodynamic aspects. Considering the temperature of 40 °C, the mixture critical point (MCP) of the binary system DMSO/CO_2 is located at 8.61 MPa [34]; it means that the operating point at 10 MPa is above the MCP. However, the presence of the solutes can modify the high-pressure vapor-liquid equilibria (VLEs) of the system, and the MCP of the quaternary system EUD/THEOP/DMSO/scCO_2 could shift towards higher pressures with respect to the one of the binary system formed by the solvent and the antisolvent. This shift of the MCP towards higher pressures is generally more evident by increasing the concentration of the solutes in the liquid solution [36,37]. In this specific case, the presence of THEOP probably modified the high-pressure VLEs; thus, the operating point at 10 MPa could be below the MCP and lie in the biphasic region, from which a split of the precipitated solute in two morphologies may occur; i.e., crystals and microparticles as in case of run #13.

In the following set of experiments, with the aim of shifting the operating point above the MCP, the total concentration in DMSO was fixed at 20 mg/mL, and the effect of pressure in the range 10–15 MPa (runs #13–15) was evaluated. Well-defined spherical microparticles (Figure 6c) were obtained in correspondence of a pressure of 12 MPa, whereas, slightly coalescing microparticles precipitated at 15 MPa (Figure 6d). Moreover, it was observed that the particles mean diameter decreased by increasing the pressure, as shown by the comparison of the volumetric cumulative PSDs reported in Figure 7.

Figure 6. FESEM images of EUD/THEOP 20/1 powders precipitated from DMSO at 40 °C and 20 mg/mL. Effect of the operating pressure. (**a**) 10 MPa (run #13) filter; (**b**) 10 MPa (run #13) precipitating chamber; (**c**) 12 MPa (run #14); (**d**) 15 MPa (run #15).

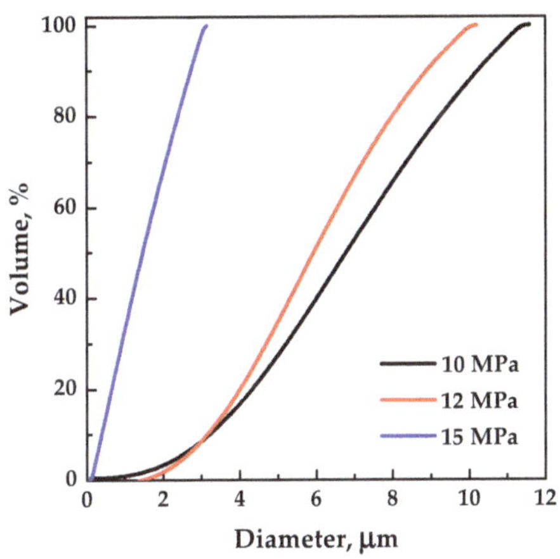

Figure 7. Volumetric cumulative PSDs of EUD/THEOP 20/1 particles precipitated from DMSO at 40 °C and 20 mg/mL; effect of the operating pressure.

3.2.2. Effect of Total Concentration on Coprecipitated Particles

Since the pressure of 10 MPa allows the attainment of defined EUD/DICLO particles, the influence of the overall concentration in DMSO at a polymer/drug ratio of 20/1 *w/w* was evaluated in correspondence of that pressure. Coalescing particles (Figure 8a) precipitated at 20 mg/mL (run #8), whereas, less coalescing particles were obtained by increasing the total concentration up to 50 mg/mL (run #9), as shown in Figure 8b.

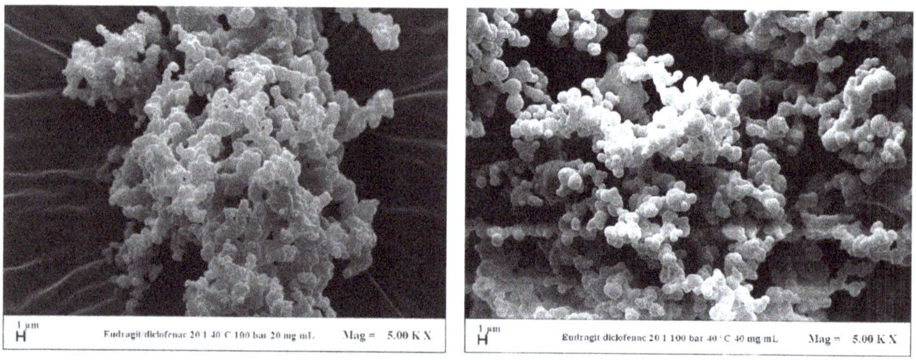

(**a**) (**b**)

Figure 8. FESEM images of EUD/DICLO 20/1 particles precipitated from DMSO at 40 °C, 10 MPa. (**a**) 20 mg/mL (run #8); (**b**) 50 mg/mL (run #9).

For the system EUD/THEOP 20/1 *w/w*, the effect of the overall concentration in DMSO was evaluated at 12 MPa in the range 20 mg/mL (run #14)–40 mg/mL (run #16). Well-defined spherical microparticles were obtained at 40 mg/mL, as well as at 20 mg/mL; no significant changes in terms of particle diameters occurred by increasing the concentration.

3.2.3. Effect of Polymer/Drug Ratio on Coprecipitated Particles

For the system EUD/DICLO, the pressure was fixed at 10 MPa and the total concentration at 50 mg/mL; in correspondence of these conditions, the polymer/drug ratio was decreased from 20/1 to 10/1 *w/w*. Slightly irregular and coalescing microparticles were produced working at 10/1 *w/w*, and a decrease of the mean diameter was observed by decreasing the polymer/drug ratio.

EUD/THEOP ratio *w/w* was instead decreased from 20/1 to 10/1 *w/w*, fixing the pressure at 12 MPa and the total concentration at 40 mg/mL. Well-defined microparticles (Figure 9a) were obtained even at 10/1 *w/w* (run #17 in Table 1), with a lower average diameter than the particles produced at 20/1, as shown from the comparison of the volumetric cumulative PSDs reported in Figure 10. The presence of few larger particles (Figure 9b) with an internal structure characterized by holes was also noted working at 10/1 *w/w*. This outcome can be ascribed to the stronger influence of theophylline on the VLEs, since a higher concentration in the liquid solution can cause the MCP shift at higher pressures.

3.3. Characterization of Samples

Fourier transform infrared (FT-IR) analyses were performed in order to identify the presence of the polymer and the drug in the composite powders and possible interactions between the two compounds. FT-IR spectra of unprocessed drug, unprocessed and processed EUD, physical mixture EUD/drug 10/1 *w/w* and SAS processed EUD/drug 20/1, and 10/1 are reported in Figure 11a,b for DICLO and THEOP, respectively.

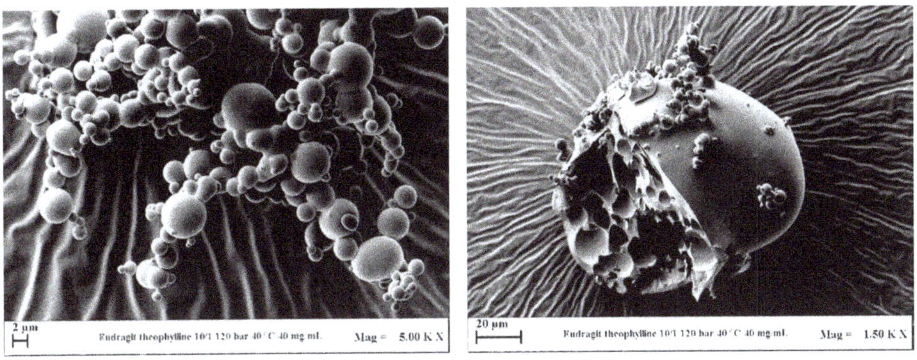

(**a**) (**b**)

Figure 9. FESEM images of EUD/THEOP 10/1 *w/w* precipitated from DMSO at 40 °C, 12 MPa and 40 mg/mL (run #17). (**a**) Microparticles and (**b**) expanded microparticles.

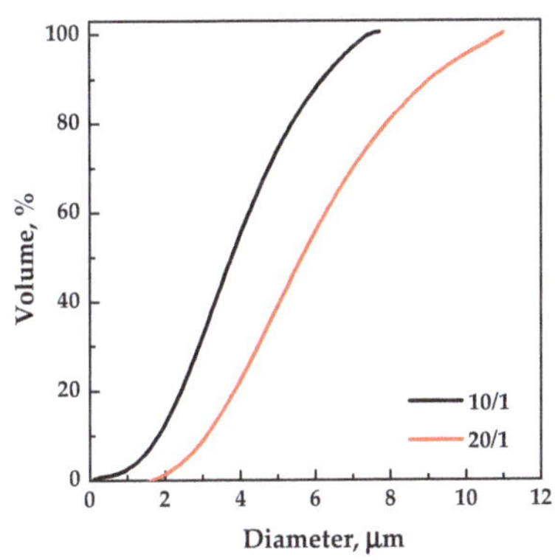

Figure 10. Volumetric cumulative PSDs of EUD/THEOP particles precipitated from DMSO at 40 °C, 12 MPa and 40 mg/mL; effect of the polymer/drug ratio.

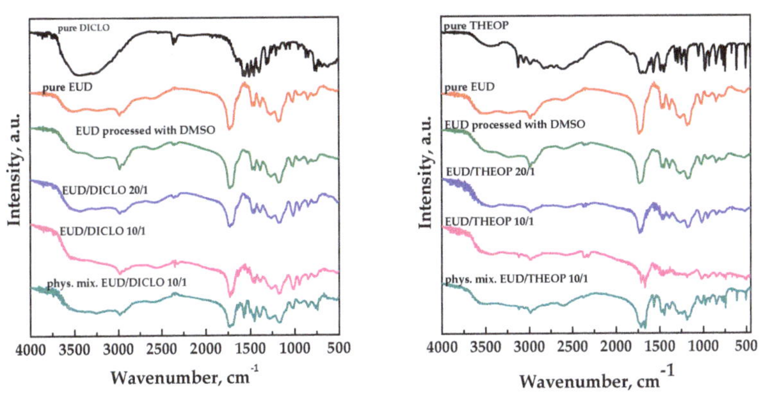

(a) (b)

Figure 11. FT-IR spectra for unprocessed and SAS processed Eudragit L100-55, unprocessed drugs, physical mixture polymer/drug and SAS processed Eudragit/drug powders. (**a**) DICLO; (**b**) THEOP.

FT-IR spectra of pure and unprocessed EUD is characterized by absorption bands at around 1157 cm^{-1}, 1184 cm^{-1} e 1261 cm^{-1}, corresponding to the ester vibrations, peaks at about 1701 cm^{-1} and at 1736 cm^{-1} attributed to the C = O stretching of the carboxylic acid and the vibrations of the esterified carboxyl groups, respectively; moreover, the presence of peaks at 1387 cm^{-1}, 1479 cm^{-1} e 2979 cm^{-1} corresponded to CHX vibrations [29]. No changes were observed in the polymer at the level of the functional group after SAS processing. FT-IR spectra of the physical mixture and SAS processed EUD/DICLO powders clearly showed the characteristic bands of the polymer, since it is present in more quantity, and few peaks related to DICLO as the C–Cl stretching at about 720 cm^{-1} [38]. In addition to EUD characteristic peaks, FT-IR spectra of physical mixture and SAS processed EUD/THEOP powders exhibited several absorption bands attributed to the theophylline, such as at 3120 cm^{-1} assigned to the N–H stretching, at 3060 and 2989 cm^{-1} related to the C–H stretching, at about 1718 and 1667 cm^{-1} assigned to the carbonyl stretching and at 1307 cm^{-1} assigned to the C–O stretching [39–41].

DSC thermograms of the unprocessed drug, unprocessed and processed EUD and SAS processed EUD/drug 20/1 and 10/1 *w/w* are reported in Figure 12a,b for DICLO and THEOP, respectively.

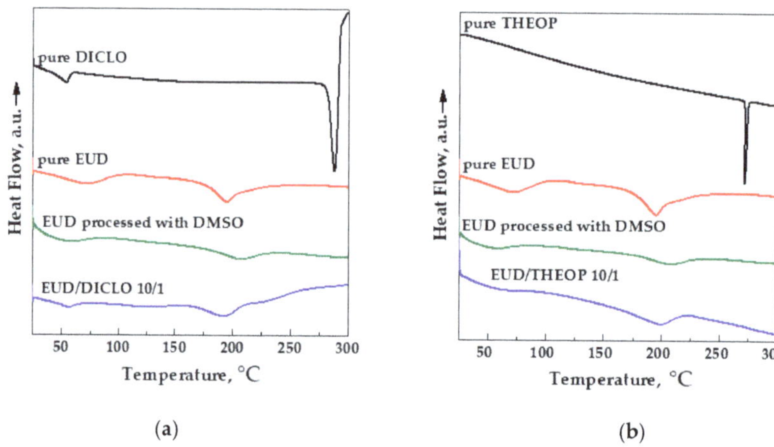

Figure 12. DSC thermograms of unprocessed and SAS processed EUD, unprocessed drugs, and SAS processed EUD/drug coprecipitated powders. (**a**) DICLO; (**b**) THEOP.

The DSC curve of pure DICLO showed two endothermic peaks, the first one related to the dehydration and the second one at about 288 °C corresponding to the melting point [42]. The DSC thermogram of pure THEOP showed an endothermic peak at around 272 °C, which is its melting point [43]. The thermogram of unprocessed EUD had two endothermic peaks, the first one, ascribable to the loss of water and the other one at about 200 °C related to the melting of the crystalline portion of the polymer [44]. Moreover, in agreement with the literature [45], unprocessed EUD is characterized by a glass transition temperature (T_g) at around 119 °C, slightly shifted at a lower temperature (around 112 °C) for SAS processed EUD, because of the $scCO_2$ plasticizing effect in the presence of semicrystalline polymers [46,47]. The thermal behavior of SAS coprecipitated powders is similar to the polymer one. The absence of the melting point of the drug in the thermograms of the coprecipitated powders can be ascribed to the amorphization. Moreover, because of the particle size reduction, a slightly higher degree of amorphism is detected both in SAS processed polymer and SAS coprecipitated powders with respect to unprocessed EUD, as observable from the less pronounced melting peak linked to the crystalline portion of the polymer.

XRD patterns of the unprocessed drug, unprocessed EUD and SAS processed EUD/drug 10/1 *w/w* are reported in Figure 13a,b for DICLO and THEOP, respectively.

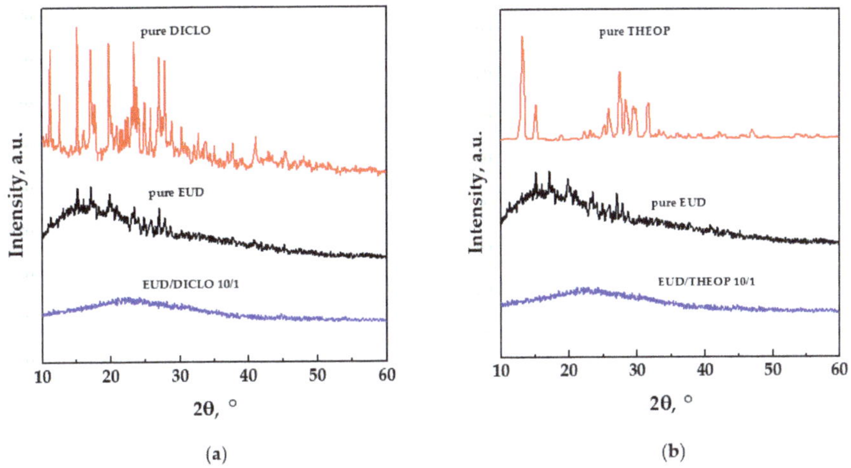

Figure 13. XRD patterns of unprocessed EUD, unprocessed drugs, and SAS processed EUD/drug coprecipitated powders. (**a**) DICLO; (**b**) THEOP.

XRD analyses confirmed the outcomes deduced by DSC thermograms; i.e., the pure drugs were in a crystalline state, the pure EUD showed a semi-crystalline pattern, whereas, the coprecipitated powders were characterized by an amorphous behavior.

Dissolution tests were performed using UV-vis spectroscopy to compare the dissolution rate of each unprocessed drug with the drug coprecipitated with EUD. The dissolution profiles in PBS of the unprocessed drug, physical mixture EUD/drug 10/1 *w/w* and SAS processed EUD/drug 20/1, and 10/1 are reported in Figure 14a,b for DICLO and THEOP, respectively.

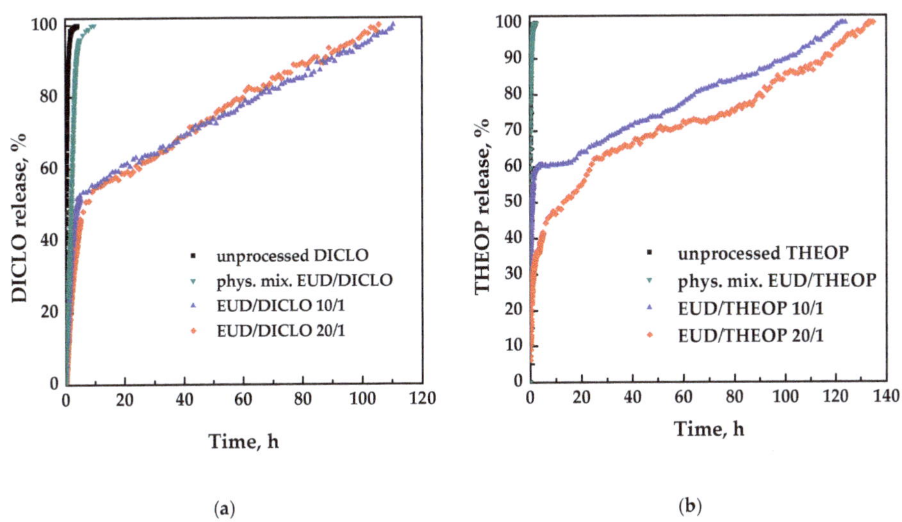

Figure 14. Dissolution profiles in PBS at 37 °C and pH 7.4. (**a**) DICLO; (**b**) THEOP.

Pure DICLO and DICLO present in the physical mixture were completely released in about 4 h and 10 h, respectively. The delayed effect of the drug release, due to the coprecipitation with the polymer is evident since the complete dissolution is achieved for both SAS coprecipitated powders 20/1 and 10/1 in about 100 h with a similar burst effect (about 45%) ascribable to the drug portion located near/on the surface of the particles. In summary, the dissolution of DICLO coprecipitated with EUD was about 28 times slower with respect to the unprocessed NSAID.

As for pure theophylline and the corresponding physical mixture, the time taken for the complete release is practically the same (about 2.3 e 2.6 h, respectively). Differently, SAS coprecipitated EUD/theophylline 10/1, and 20/1 *w/w* powders require for the complete drug dissolution about 118 h and 130 h, respectively; therefore, also, in this case, the release was considerably prolonged using EUD, up to about 57 times. Moreover, by increasing the polymer/drug ratio from 10/1 to 20/1 *w/w*, the drug release burst decreased from 60% to 20%. The difference in the release of EUD/THEOP 10/1 and 20/1 *w/w* can be explained, according to the literature [48], to the size of the microparticles. Indeed, increasing the powders' diameter, the portion of the drug finely dispersed in the polymer microparticle increased, and correspondingly, the portion of the drug disposed on the polymer surface decreased. This corresponds to a reduction of the initial drug release burst observed in the case of EUD/THEOP 20/1 *w/w*.

Considering the obtained results, the coprecipitation was achieved only in part in the case of DICLO, for which smaller and less-defined microparticles were generally obtained with respect to the other studied drug; moreover, a burst-like effect of about 45% was observed. Well-defined spherical microparticles EUD/THEOP were instead produced, assuring a massive coprecipitation with a burst effect that reduced up to a value of 20% for EUD/THEOP 20/1 *w/w*. As reported in the literature [16], in this work it was also observed that the polymeric carrier has a great influence on the precipitate morphology, and by selecting the proper one, it is possible to coprecipitate even an active principle which revealed to be a bad candidate for SAS micronization, as occurred with THEOP. Moreover, different results were obtained with different drugs even using the same polymer, proving once again that the interactions between polymer, active principle, $scCO_2$, and liquid solvent can influence the high-pressure vapor-liquid equilibria and modify the processability of materials.

4. Conclusions

In this work, it was demonstrated that Eudragit L100-55 is an effective carrier for SAS coprecipitation as composite microspheres EUD/diclofenac and EUD/theophylline were produced; the operating conditions have to be accurately selected depending on the drug used because the drug can influence the high-pressure vapor-liquid equilibria. Dissolution tests showed that the release of the drugs was significantly delayed, up to 28 and 57 times for DICLO and THEOP, respectively; therefore, Eudragit L100-55 is suitable for applications focused on controlled-release to reduce the side effects, due to drug overuse.

Author Contributions: I.D.M. provided the concept and the experimental design; P.F. performed the experiments; P.F. wrote the draft of the manuscript; I.D.M. reviewed and edited the manuscript. All authors have read and agreed to the published version of the manuscript.

Funding: This research received no external funding.

Acknowledgments: The authors acknowledge Antonietta Mancuso for her help in performing the experiments during her Master Thesis at the Department of Industrial Engineering of the University of Salerno.

Conflicts of Interest: The authors declare no conflict of interest.

References

1. Todd, P.A.; Sorkin, E.M. Diclofenac sodium. *Drugs* **1988**, *35*, 244–285. [CrossRef] [PubMed]
2. Mengozzi, G.; Intorre, L.; Bertini, S.; Giorgi, M.; Soldani, G. Comparative bioavailability of two sustained-release theophylline formulations in the dog. *Pharm. Res.* **1998**, *38*, 481–485. [CrossRef] [PubMed]
3. Huang, Q.; Yu, H.; Ru, Q. Bioavailability and delivery of nutraceuticals using nanotechnology. *J. Food Sci.* **2010**, *75*, R50–R57. [CrossRef] [PubMed]
4. Tønnesen, H.H.; Karlsen, J. Alginate in drug delivery systems. *Drug Dev. Ind. Pharm.* **2002**, *28*, 621–630. [CrossRef] [PubMed]

5. Vehring, R. Pharmaceutical particle engineering via spray drying. *Pharm. Res.* **2008**, *25*, 999–1022. [CrossRef] [PubMed]
6. Heidebach, T.; Först, P.; Kulozik, U. Influence of casein-based microencapsulation on freeze-drying and storage of probiotic cells. *J. Food Eng.* **2010**, *98*, 309–316. [CrossRef]
7. O'Donnell, P.B.; McGinity, J.W. Preparation of microspheres by the solvent evaporation technique. *Adv. Drug Del. Rev.* **1997**, *28*, 25–42. [CrossRef]
8. Oxley, J. Coextrusion for Food Ingredients and Nutraceutical Encapsulation: Principles and Technology. In *Encapsulation Technologies and Delivery Systems for Food Ingredients and Nutraceuticals*; Woodhead Publishing: Cambridge, UK, 2012; pp. 131–150.
9. Nykamp, G.; Carstensen, U.; Müller, B. Jet milling—A new technique for microparticle preparation. *Int. J. Pharm.* **2002**, *242*, 79–86. [CrossRef]
10. Nihant, N.; Grandfils, C.; Jérôme, R.; Teyssié, P. Microencapsulation by coacervation of poly (lactide-co-glycolide) IV. Effect of the processing parameters on coacervation and encapsulation. *J. Control. Release* **1995**, *35*, 117–125. [CrossRef]
11. Franco, P.; Reverchon, E.; De Marco, I. Zein/diclofenac sodium coprecipitation at micrometric and nanometric range by supercritical antisolvent processing. *J. CO2 Util.* **2018**, *27*, 366–373. [CrossRef]
12. Franco, P.; Reverchon, E.; De Marco, I. Production of zein/antibiotic microparticles by supercritical antisolvent coprecipitation. *J. Supercrit. Fluids* **2019**, *145*, 31–38. [CrossRef]
13. Montes, A.; Gordillo, M.D.; Pereyra, C.; De los Santos, D.M.; Martínez de la Ossa, E.J. Ibuprofen–polymer precipitation using supercritical CO_2 at low temperature. *J. Supercrit. Fluids* **2014**, *94*, 91–101. [CrossRef]
14. Montes, A.; Kin, N.; Gordillo, M.D.; Pereyra, C.; Martínez de la Ossa, E.J. Polymer–naproxen precipitation by supercritical antisolvent (SAS) process. *J. Supercrit. Fluids* **2014**, *89*, 58–67. [CrossRef]
15. Montes, A.; Wehner, L.; Pereyra, C.; Martínez de la Ossa, E.J. Mangiferin nanoparticles precipitation by supercritical antisolvent process. *J. Supercrit. Fluids* **2016**, *112*, 44–50. [CrossRef]
16. Prosapio, V.; De Marco, I.; Reverchon, E. Supercritical antisolvent coprecipitation mechanisms. *J. Supercrit. Fluids* **2018**, *138*, 247–258. [CrossRef]
17. Matos, R.L.; Lu, T.; Prosapio, V.; McConville, C.; Leeke, G.; Ingram, A. Coprecipitation of curcumin/PVP with enhanced dissolution properties by the supercritical antisolvent process. *J. CO2 Util.* **2019**, *30*, 48–62. [CrossRef]
18. Zahran, F.; Cabañas, A.; Cheda, J.A.R.; Renuncio, J.A.R.; Pando, C. Dissolution rate enhancement of the anti-inflammatory drug diflunisal by coprecipitation with a biocompatible polymer using carbon dioxide as a supercritical fluid antisolvent. *J. Supercrit. Fluids* **2014**, *88*, 56–65. [CrossRef]
19. Lim, R.T.Y.; Ng, W.K.; Tan, R.B. Amorphization of pharmaceutical compound by co-precipitation using supercritical anti-solvent (SAS) process (Part I). *J. Supercrit. Fluids* **2010**, *53*, 179–184. [CrossRef]
20. Wang, W.; Liu, G.; Wu, J.; Jiang, Y. Co-precipitation of 10-hydroxycamptothecin and poly (l-lactic acid) by supercritical CO_2 anti-solvent process using dichloromethane/ethanol co-solvent. *J. Supercrit. Fluids* **2013**, *74*, 137–144. [CrossRef]
21. Lee, S.; Kim, M.; Kim, J.; Park, H.; Woo, J.; Lee, B.; Hwang, S.J. Controlled delivery of a hydrophilic drug from a biodegradable microsphere system by supercritical anti-solvent precipitation technique. *J. Microencapsul.* **2006**, *23*, 741–749. [CrossRef]
22. Kibbe, A. *Handbook of Pharmaceutical Excipients*; Americal Pharmaceutical Association and Pharmaceutical Press: Washington, DC, USA, 2000; pp. 401–406.
23. Moustafine, R.; Kabanova, T.; Kemenova, V.; Van den Mooter, G. Characteristics of interpolyelectrolyte complexes of eudragit E100 with eudragit L100. *J. Control. Release* **2005**, *103*, 191–198. [CrossRef]
24. Duarte, A.R.C.; Roy, C.; Vega-González, A.; Duarte, C.M.; Subra-Paternault, P. Preparation of acetazolamide composite microparticles by supercritical anti-solvent techniques. *Int. J. Pharm.* **2007**, *332*, 132–139. [CrossRef]
25. Garay, I.; Pocheville, A.; Madariaga, L. Polymeric microparticles prepared by supercritical antisolvent precipitation. *Powder Technol.* **2010**, *197*, 211–217. [CrossRef]
26. Singh, S.; Neelam, S.A.; Singla, Y.P. An overview of multifaceted significance of eudragit polymers in drug delivery systems. *Asian J. Pharm. Clin. Res.* **2015**, *8*, 1–6.
27. Sonje, A.; Chandra, A. Comprehensive review on eudragit polymers. *Int. Res. J. Pharm* **2013**, *4*, 2230–8407. [CrossRef]

28. Shidhaye, S.; Malke, S.; Kadam, V. Taste masked, orally disintegrating tablet containing microspheres for immediate release. *J. Pharm. Res.* **2008**, *1*, 225–229.
29. Majeed, A.; Ranjha, N.M.; Hanif, M.; Abbas, G.; Khan, M.A. Development and evaluation of ivabradine HCl-loaded polymeric microspheres prepared with eudragit L100-55 (methacrylic acid-ethyl acrylate copolymer) and ethyl cellulose for controlled drug release. *Acta Pol. Pharm.* **2017**, *74*, 565–578.
30. Moustafine, R.I.; Zaharov, I.M.; Kemenova, V.A. Physicochemical characterization and drug release properties of eudragit® E PO/eudragit® L 100-55 interpolyelectrolyte complexes. *Eur. J. Pharm. Biopharm.* **2006**, *63*, 26–36. [CrossRef]
31. Hamman, J.H. Chitosan based polyelectrolyte complexes as potential carrier materials in drug delivery systems. *Mar. Drugs* **2010**, *8*, 1305–1322. [CrossRef]
32. Khan, M.Z.I.; Prebeg, Ž.; Kurjaković, N. A pH-dependent colon targeted oral drug delivery system using methacrylic acid copolymers: I. Manipulation of drug release using Eudragit® L100-55 and Eudragit® S100 combinations. *J. Control. Release* **1999**, *58*, 215–222. [CrossRef]
33. Montes, A.; Wehner, L.; Pereyra, C.; Martínez de la Ossa, E.J. Generation of microparticles of ellagic acid by supercritical antisolvent process. *J. Supercrit. Fluids* **2016**, *116*, 101–110. [CrossRef]
34. Andreatta, A.E.; Florusse, L.J.; Bottini, S.B.; Peters, C.J. Phase equilibria of dimethyl sulfoxide (DMSO) + carbon dioxide, and DMSO + carbon dioxide + water mixtures. *J. Supercrit. Fluids* **2007**, *42*, 60–68. [CrossRef]
35. Reverchon, E.; De Marco, I. Mechanisms controlling supercritical antisolvent precipitate morphology. *Chem. Eng. J.* **2011**, *169*, 358–370. [CrossRef]
36. Campardelli, R.; Reverchon, E.; De Marco, I. Dependence of SAS particle morphologies on the ternary phase equilibria. *J. Supercrit. Fluids* **2017**, *130*, 273–281. [CrossRef]
37. Campardelli, R.; Reverchon, E.; De Marco, I. PVP microparticles precipitation from acetone-ethanol mixtures using SAS process: Effect of phase behavior. *J. Supercrit. Fluids* **2019**, *143*, 321–329. [CrossRef]
38. Edavalath, S.; Shivanand, K.; Prakasam, K.; Rao, B.P.; Divakar, G. Formulation development and optimization of controlled porosity osmotic pump tablets of diclofenac sodium. *Int. J. Pharm. Pharm. Sci.* **2011**, *3*, 80–87.
39. Lin, H.-L.; Hsu, P.-C.; Lin, S.-Y. Theophylline–citric acid co-crystals easily induced by DSC–FTIR microspectroscopy or different storage conditions. *Asian J. Pharm. Sci.* **2013**, *8*, 19–27. [CrossRef]
40. Kesavan, J.G.; Peck, G.E. Solid-state stability of theophylline anhydrous in theophylline anhydrous-polyvinylpyrrolidone physical mixtures. *Drug Dev. Ind. Pharm.* **1996**, *22*, 189–199. [CrossRef]
41. Kumar Singh Yadav, H.; Shivakumar, H. In vitro and in vivo evaluation of pH-sensitive hydrogels of carboxymethyl chitosan for intestinal delivery of theophylline. *ISRN Pharm* **2012**, *2012*, 1–9. [CrossRef]
42. Tiţa, D.; Fuliaş, A.; Tiţa, B. Thermal stability of ketoprofen—active substance and tablets: Part 1. Kinetic study of the active substance under non-isothermal conditions. *J. Therm. Anal. Calorim.* **2011**, *105*, 501–508.
43. Ochoa, L.; Igartua, M.; Hernandez, R.; Gascon, A.; Pedraz, J.L. Preparation of sustained release hydrophilic matrices by melt granulation in a high-shear mixer. *J. Pharm. Pharm. Sci. A Publ. Can. Soc. Pharm. Sci. Soc. Can. Sci. Pharm.* **2005**, *8*, 132–140.
44. Ceballos, A.; Cirri, M.; Maestrelli, F.; Corti, G.; Mura, P. Influence of formulation and process variables on in vitro release of theophylline from directly-compressed Eudragit matrix tablets. *Farmaco* **2005**, *60*, 913–918. [CrossRef]
45. Jadhav, N.R.; Gaikwad, V.L.; Nair, K.J.; Kadam, H.M. Glass transition temperature: Basics and application in pharmaceutical sector. *Asian J. Pharm. Free Full Text Artic. Asian J. Pharm* **2014**, *3*. [CrossRef]
46. Lian, Z.; Epstein, S.A.; Blenk, C.W.; Shine, A.D. Carbon dioxide-induced melting point depression of biodegradable semicrystalline polymers. *J. Supercrit. Fluids* **2006**, *39*, 107–117. [CrossRef]
47. Campardelli, R.; Franco, P.; Reverchon, E.; De Marco, I. Polycaprolactone/nimesulide patches obtained by a one-step supercritical foaming+ impregnation process. *J. Supercrit. Fluids* **2019**, *146*, 47–54. [CrossRef]
48. Brzeziński, M.; Kost, B.; Wedepohl, S.; Socka, M.; Biela, T.; Calderón, M. Stereocomplexed PLA microspheres: Control over morphology, drug encapsulation and anticancer activity. *Coll. Surf. B Biointerfaces* **2019**, *184*, 110544. [CrossRef]

polymers

MDPI

Article

Multifunctional Bioactive Resin for Dental Restorative Materials

Loredana Tammaro [1,*,†], Anna Di Salle [2,†], Anna Calarco [2,*], Ilenia De Luca [3], Francesco Riccitiello [4], Gianfranco Peluso [2], Vittoria Vittoria [5] and Andrea Sorrentino [6]

1 Nanomaterials and Devices Laboratory (SSPT-PROMAS-NANO), Italian National Agency for New Technologies, Energy and Sustainable Economic Development, ENEA, P.le E. Fermi 1, 80055 Portici (Na), Italy

2 Research Institute on Terrestrial Ecosystems (IRET)—CNR, via P. Castellino 111, 80131 Napoli, Italy; anna.disalle@cnr.it (A.D.S.); gianfranco.peluso@cnr.it (G.P.)

3 Elleva Pharma s.r.l., via P. Castellino 111, 80131 Napoli, Italy; ilenia.deluca@ellevapharma.com

4 Department of Restorative Dentistry, University of Naples Federico II, via S. Pansini 5, 80131 Napoli, Italy; francesco.riccitiello@unina.it

5 NICE FILLER s.r.l., via Loggia dei Pisani 25, 80133 Napoli, Italy; vvittoria@nicefiller.it

6 Institute for Polymer, Composites and Biomaterials (IPCB)—CNR, P.le E. Fermi 1, 80055 Portici (Na), Italy; andrea.sorrentino@cnr.it

* Correspondence: loredana.tammaro@enea.it (L.T.); anna.calarco@cnr.it (A.C.)

† These authors contributed equally to this work.

Received: 17 December 2019; Accepted: 25 January 2020; Published: 5 February 2020

Abstract: Resin-based composites are widely used as dental restorative materials due to their excellent properties. They must have high modulus, high hardness, and be chemically inert while minimizing moisture uptake. To fulfill these higher standard prerequisites and properties, continuous improvements in each of their components are required. This study develops novel composites with multiple biofunctions. Light-cured Bis-GMA/TEGDMA dental resin (RK)/layered double hydroxide intercalated with fluoride ions (LDH-F)/calcium bentonite (Bt) hybrid composites were prepared. The loading ratio of LDH-F to Bt was varied, ranging from 2.5/2.5 to 10/10 parts per hundred RK and structural, mechanical, and biological properties were studied. The incorporation of even small mass fractions (e.g., 2.5 wt% of LDH-F and 2.5 wt% of Bt) in RK dental resin significantly improved the mechanical properties of the pristine resin. The synthetized materials showed antibacterial and antibiofilm effects against three bacterial strains isolated from healthy volunteers' saliva (*Streptococcus* spp., *Bacteroides fragilis*, and *Staphylococcus epidermidis*) without affecting its ability to induce dental pulp stem cells differentiation into odontoblast-like cells. The capability to balance between the antibiofilm activity and dental pulp stem cells differentiation in addition with improved mechanical properties make these materials a promising strategy in preventive and restorative dentistry.

Keywords: dental materials; composite resin; layered double hydroxide; calcium bentonite; antibiofilm activity

1. Introduction

In recent years, numerous synthetic materials have been developed in the field of dentistry with the aim of reconstructing and maintaining the oral function, health, and aesthetics of the patients [1,2]. Among these materials, resin-based composites are probably the most important from the aesthetics and performance point of view [3,4]. Hard and chemically stable filler particles are generally dispersed in these composites to provide the structural reinforcement necessary for dental applications [5]. Common fillers are glass particles or fibers, ceramic materials, and natural minerals. Among the natural minerals, bentonite clays are aluminum phyllosilicates with interesting properties [6,7]. They are named based on dominant elements, such as potassium (K), sodium (Na), calcium (Ca), and aluminum (Al). In the presence of water, bentonite easily swells and can be reduced to nanometric dimension providing self-sealing and barrier properties [8,9]. Several researchers have demonstrated that the incorporation of fillers such as montmorillonite into dental materials results in the increase of their mechanical properties [10,11]. Atai and colleagues, indeed, reported the incorporation of nanoclay into the dental adhesives as a solution to overcome the lower mechanical properties of the adhesive layer and the problem of rapid sedimentation of particles into adhesives [12]. In addition, Nikolaidis et al. reported the reinforcement of dental nanocomposite resins with different organo-modified montmorillonite nanofillers [13]. Dental materials still suffer several significant problems, such as polymerization shrinkage at the tooth–composite interface, which can lead to micro-leakage, and secondary caries [14–18]. In 2007, Bernardo et al. [19] evaluated 1748 restorations and revealed that the use of composites increased the risk of secondary caries by about 3.5 times as compared with the use of amalgams. Different approaches have been adopted to improve the quality and performance of these composites [20–22]. One of these is the development of fillers that can release substances with antibacterial effects or buffer the acids produced by bacteria [23–26]. Fluorine-containing filler particles have been introduced in order to eliminate the cariostatic demineralization effect produced by the filling restoration. Soluble fluorine-based salts have been introduced in the formulation of these composites in order to have a release over time. However, these salts present an initial burst release and, when removed, tend to leave micropores in the bulk material with a consequent loss of mechanical properties. These problems can be reduced by introducing fluorine, aluminum, and silicate based fillers with less solubility. Fused modified fillers such as silica whiskers and dicalcium phosphate anhydrous or tetracalcium phosphate have been proposed to release Ca and PO4 ions [27].

Despite the large amounts of ions released by these systems, their ability to prevent secondary caries is questionable. Technological improvements are needed to increase and prolong the release of active components during the working life of these dental composites. In our previous work, we report the synthesis and characterization of fluoride-intercalated layered double hydroxide (LDH-F) showing a long-term controlled delivery of micromolar amounts of fluoride. This material is able to elicit beneficial effects on proliferation and mineralization of human dental pulp stem cells (hDPSC) [28], as well as on the migration and differentiation of STRO-1+ cells, a DPSC subpopulation able to differentiate into dental hard tissue-forming cells [29].

This study presents the synthesis of a light-cured Bis-GMA/TEGDMA dental resin (RK)/layered double hydroxide intercalated with fluoride ions (LDH-F)/calcium bentonite (Bt), named RK-FBtx (where x is the total mass fraction of LDH-F and Bt), as a new hybrid composite material for use in dentistry. Here, we demonstrate that the addition of bentonite improves mechanical properties, without influencing the bioactive properties of the composite. This result is of great importance in dentistry by defining the ability of dental material to act effectively and safely over extended periods of time. Moreover, we demonstrate the ability of RK-FBtx with the highest F/Bt rate (RK-FBt20) to elicit antibacterial and antibiofilm effects, making it useful as dental material to prevent the formation of secondary caries [30]. Indeed, recent studies have highlighted that secondary caries are increasingly related to the development of bacterial biofilm [31,32]. The oral cavity represents the second most complex microbiota in the body after the colon hosting around 1000 species [33]. These commensal microbes in the oral cavity represent the principal source of the formation of biofilm in the root canal, and colonize the numerous surfaces in the mouth, hydrating and feeding on the nutrients provided by either saliva or gingival crevicular fluid, or dietary intake. The new synthesized material RK-FBt20 was able to release fluoride even in acidic conditions, inhibiting the formation of *Streptococcus* spp., *Bacteroides fragilis*, and *Staphylococcus epidermidis* (three bacterial strains isolated from saliva of healthy volunteers) biofilm.

2. Materials and Methods

2.1. Sample Preparation

Two clays were used for the preparation of the hybrid composites, a layered double hydroxide intercalated with fluoride ions and a calcium bentonite. A fluoride-intercalated layered double hydroxide (LDH-F) was prepared by intercalation of fluoride anions in LDH in nitrate form, using an aqueous solution of fluoride sodium salt, NaF, (F/NO_3 molar ratio = 1.3). The resulting precipitate was collected, washed three times with deionized water and, finally, dried at rt over a saturated NaCl solution. The material was characterized by XRD in order to verify the sample. The resulting LDH-F has the following composition: $[Mg_{0.65}Al_{0.35}(OH)_2](F)_{0.35}\cdot 0.8H_2O$ [28]. Calcium bentonite (Bt) with the following chemical composition: Na_2O (0.2%), MgO (3.5%), Al_2O_3 (14.5%), SiO_2 (71.9%), P_2O_5 (0.01%), K_2O (1.9%), CaO (1.6%), TiO_2 (0.03%), MnO (0.01%), Fe_2O_3 (0.85%), and LOI (5.5%) was kindly supplied by Laviosa Chimica Mineraria SpA (Livorno LI, Italy). Commercial light-activated restorative material (RK), provided by Kerr s.r.l. (Scafati, Italy), consists of bisphenol-A glycidyl methacrylate (Bis-GMA), tri-ethylene glycol dimethacrylate (TEGDMA), camphorquinone (CQ), ethoxylated bisphenol A dimethacrylate (EBPADMA), and glass fillers. The inorganic solids LDH-F and Bt were used as fillers and mixed into the dental resin system (RK) to obtain hybrid composites with three different total filler concentration (5, 10, and 20 wt%). The composites containing only LDH-F or bentonite solids are coded as RK-Fx and RK-Bty, where x and y are the percentages by weight of the LDH-F and bentonite, respectively, in the neat resin RK. Composites containing both clays are named RK-FBtz, where z is the percentage by weight of the total amount of the same content of the inorganic solids (LDH-F and Bt) presents in the resin RK (e.g., RK-FBt5 means RK 95%, LDH-F 2.5%, and Bt 2.5%). The neat RK resin was used as control material.

The neat resin and all the hybrid composites obtained were cured by photo-polymerization using a visible light-curing unit with a light intensity of 550 mW/cm^2 and an irradiation time of 60 s. Specimen disks 20 mm in diameter and 1 mm thick were fabricated using steel molds.

2.2. Characterization and Evaluation

2.2.1. Wide-Angle X-ray Diffraction (WAXD)

Wide-angle X-ray diffraction (WAXD) patterns with nickel-filtered Cu Kα radiation (λ = 1.54050 Å) were obtained at room temperature, in reflection, by an automatic Bruker diffractometer operating at 40 kV and 40 mA. The patterns were recorded from 2 to 40° of 2θ. The counting time was 3 s per 0.05° of 2θ step scan.

2.2.2. Gravimetric Measurements

To evaluate the effect of the moisture uptake on the composite properties, tests were carried out according to ASTM 570. The test specimens, in the form of a disk 30 ± 3 mm in diameter and 2 ± 0.2 mm in thickness, were firstly dried in a vacuum oven for 24 h at 30 °C and, then, weighed to the nearest 0.001 g. Specimens were then completely immersed in distilled water at 37 °C for about 1 year and periodically taken out, wiped with a dry cloth and carefully weighed. The percentage weight gain of the samples was then measured by using the following Equation (1):

$$\text{Weight gain (\%)} = (P1 - P2)/P2 \times 100 \quad (1)$$

where P1 and P2 are the weight of the wet and dry samples, respectively. For each condition, at least three different test specimens were prepared and measured in independent runs to confirm the reproducibility of the data. Test results were accepted only if the results are within ± 0.05%.

2.2.3. Dynamic Mechanical Analysis

The mechanical properties of the samples were tested using a dynamic mechanical analyzer (DMA 2980, TA Instruments, New Castle, DE, USA). The samples were tested by applying a variable flexural deformation in single cantilever mode. The displacement amplitude was set to 0.1%, whereas the measurements were performed at the frequency of 1 Hz. Rectangular sample with dimensions $18 \times 10 \times 1$ mm^3 were cut from the specimen disks. The range of temperature analyzed was −50 to 150 °C with scanning rate of 3 °C/min.

2.2.4. Fluoride Release Study

Weighed disks of RK-FBt5, RK-FBt10, and RK-FBt20 were placed in a mineral medium with composition similar to saliva (SAGF, 15 mL) at 37 °C under magnetic stirring. After specific time intervals, free fluoride ion concentration (ppm) was determined using an ion chromatograph (DX 100; Dionex, Camberley, UK) with suppressed conductivity as reported by Calarco et al. [34]. The analysis was made in triplicate and averaged the values.

2.3. Biological Test Methods

2.3.1. Primary Cell Culture

The human dental pulp cell (hDPSC) were enzymatically isolated from impacted third molars obtained from 10 adults (18 to 22 years of age), as previously described [28,35], in compliance with Italian legislation (including informed consent and institutional review board approval of the protocol number 7413). The cells were cultured with α-minimum essential medium (α-MEM) supplemented with 15% fetal bovine serum (FBS), 2 mM L-glutamine, 100 mM L-ascorbic acid-2-phosphate, 100 U/mL penicillin-G, 100 mg/mL streptomycin, and 0.25 mg/mL fungizone (HyClone, Milan, Italy) and maintained in 5% CO_2 at 37 °C. Proliferation, clonogenic potential, and stem cell markers were analyzed. STRO-1+ stem cells were directly sorted from pulp cell at passage 3 with mouse anti-human STRO-1 IgM (Life Technologies, Milan, Italy) with immune magnetic beads according to the manufacturer's protocol (Dynabeads; Life Technologies, Milan, Italy). After cell sorting, each of the following experiments was performed in triplicate on pooled STRO-1 sorted cells (STRO-1+ cells).

2.3.2. Cytotoxicity Assay

The effect of synthetized materials on STRO-1+ cell viability was determined using Cell Counting Kit-8. Briefly, STRO-1+ cells were plated at 1 × 104 cells/well on RK, RK-FBt5, RK-FBt10, and RK-FBt20 materials in 24-well flat-bottomed plates in culture medium. After 1, 3, and 7 days, CCK-8 solution was added to each well followed by 3 h incubation at 37 °C. Absorbance was measured at 450 nm using a microplate reader (Cytation 3; AHSI, Milan, Italy). Cells cultured on tissue culture polystyrene were used as the control. The experiment was repeated 3 times and the mean value calculated.

2.3.3. Alkaline Phosphatase Activity

Alkaline phosphatase activity is a typical marker for early odontoblastic differentiation. Alkaline phosphatase (ALP) activity was evaluated using ALP Assay Kit according to Tammaro et al. [28]. Briefly, STRO-1+ cells were cultured on RK-FBtx disks (RK, RK-FBt5, RK-FBt10, and RK-FBt20, 14 mm diameter) for 1, 7, 14. and 28 days. Then, the cells were scraped into cold PBS, sonicated in an ice bath and centrifuged at 1500× *g* for 15 min. ALP activity was measured in the supernatant using p-nitrophenyl phosphate as a phosphatase substrate and alkaline phosphatase supplied by the kit as a standard. The absorbance was measured at 405 nm and the amount of ALP in the cells was normalized against total protein content.

2.4. Microbiological Procedures

2.4.1. Bacteria

The bacteria used in this study, *Streptococcus* spp., *Bacteroides fragilis*, and *Staphylococcus epidermidis* isolated from 3 healthy volunteers' saliva as reported in Di Salle et al. [35], were grown at 37 °C in nonselective Nutrient Broth (NB, Oxoid, Basingstoke, Hants, UK).

2.4.2. Direct Contact Test (DCT)

For this test, 10 µL of each bacterial suspension containing a microbial concentration of approximately 2.0×10^8 cells/mL was placed on each disk surface and allowed to evaporate at 37 °C for 60 min to ensure close direct contact between the bacteria and the disk surface. Then, the disks were placed in a 12-well plate, covered with 500 µL of NB, and placed in a microplate reader (Cytation 3; AHSI, Milan, Italy). Bacterial growth curves were obtained recoding the absorbance at 600 nm (OD 600 nm) every 30 min during 48 h at 37 °C under orbital agitation.

2.4.3. Biofilm Development under Dynamic Conditions

The drip-flow reactor (M-DFR) employed in the present study was constructed as described in the Webworks Laboratory of the Montana State University [36]. Briefly, a plastic storage box was modified inserting rubber stoppers for the feed tubes, the air vent, and the waste tube (Scheme 1).

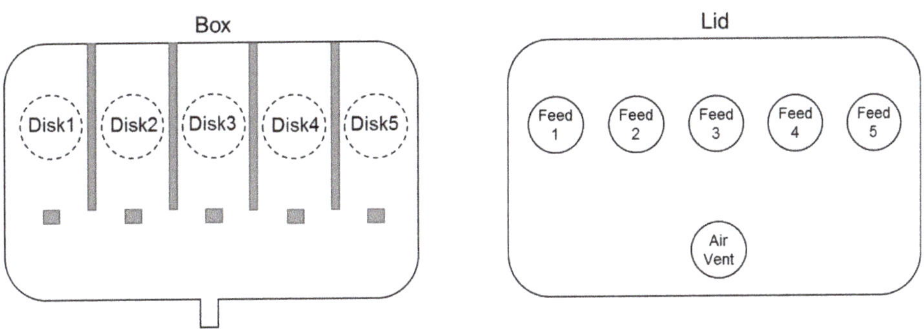

Scheme 1. Drip-flow reactor scheme.

The modified plastic box allows placement of five specimen disks 20 mm in diameter on the bottom of the flow cells, ensuring complete immersion of disks' surfaces in the flow medium. Before the experiments, in order to minimize the risk of microbial contamination, the DFR was UV sterilized. For each bacterial strain, biofilm formation was studied for 48 and 96 h simulating or not the formation of salivary pellicle for 24 h. For simulation of salivary pellicle formation, before inoculation the disks' surfaces were completely covered with thawed sterile saliva and incubated at 37 °C for 24 h. Successively, the supernatant was carefully removed. Biofilm was developed inoculating 10 mL of each bacterial suspension in early log phase in the DFR flow cells to allow bacterial adhesion. After 4 h, a constant flow (9.0 mL/h) of NB was provided using a peristaltic pump and the temperature maintained at 37 °C for 48 or 96 h.

2.4.4. MTT Assay

The metabolic activity of biofilms was determined by 3-(4,5-Dimethylthiazol-2-yl)-2,5-diphenyl tetrazolium bromide (MTT) assay [37]. After 48 or 96 h of incubation ion DFR, the feed flows to the selected flow cells were discontinued. The disks were removed, immediately washed with sterile PBS to remove nonadherent bacteria and placed in a 12-well plate. Then, 500 µL of 0.3 mg/mL MTT was added to each well and incubated at 37 °C in dark. After 3 h, the MTT solution was removed and the formed formazan crystals were dissolved by adding 500 μL dimethyl sulfoxide (DMSO). OD510 nm was recorded using a microplate reader.

2.5. Statistical Analysis

All quantitative data are presented as the mean ± SD. Each experiment was performed at least 3 times. For DCT, data were statistically analyzed using Kruskal–Wallis test followed by Dunn's post hoc analysis. All the data were analyzed with the GraphPad Prism version 6.01 statistical software package (GraphPad, San Diego, CA, USA).

3. Results and Discussion

3.1. Structural Investigation

The incorporation and dispersion of the inorganic solids into the dental resin was investigated by X-ray analysis. The diffractograms of the LDH-F and Bt are displayed in Figures 1a and 1b, respectively. The X-ray reflection of LDH-F at 2θ = 11.7° corresponds to an interlayer distance of 0.757 nm. The second and third patterns at 2θ = 23.6° and 34.9°, respectively, correspond to the higher harmonics of the interlayer distance. All the peaks are sharp, and this indicates an ordered accommodation of the inorganic fluoride anion within the interlayer regions.

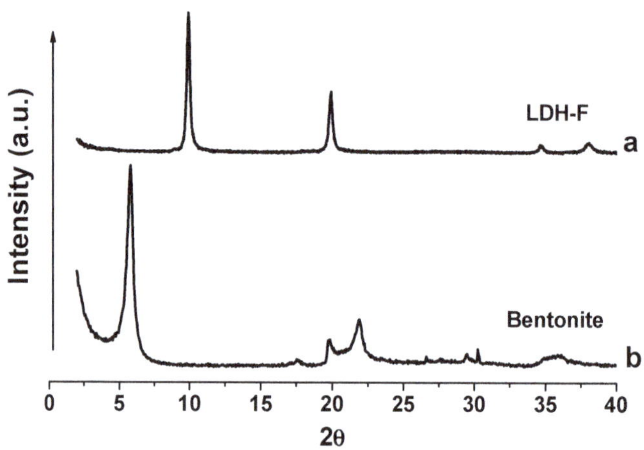

Figure 1. XRD patterns of (**a**) fluoride-intercalated layered double hydroxide (LDH-F) and (**b**) calcium bentonite clay (Bt) powders. The curves are vertically translated for clarity.

Characteristic for bentonites is that they are mainly composed of smectites, a group of expandable clay minerals with a wide range of chemical compositions. The X-ray spectra of the calcium bentonite shows a main diffraction pattern at 2θ = 5.8° and other peaks at 17.5, 19.8, 21.9, 26.6, 30.2, and 36.1 degrees of 2-theta typical of a mixture of several minerals, such as kaolinite, quartz, montmorillonite, and cristobalite. X-ray patterns of the pure resin RK and the hybrid composites are shown in Figure 2. The broad pattern in Figure 2a is attributed to the reflection of amorphous RK, while the diffraction spectra of RK/LDH-FBt composites (Figure 2b–d) show characteristic reflections of LDH-F powders at 2θ = 11.68°, 23.6°, and 34.9° and Bt clay at 2θ = 5.8°, (more evident for samples with higher concentration of LDH-F and Bt), in addition to the broad reflection of pristine RK. The X-ray data indicated that the obtained hybrid composites are a mixture of the resin RK with micro domains of LDH-F and Bt.

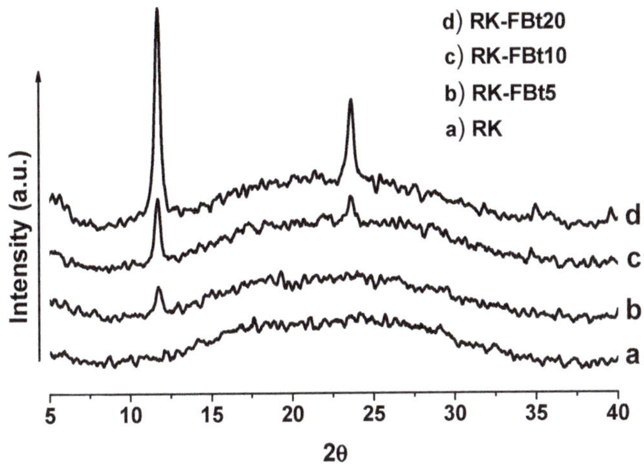

Figure 2. XRD patterns of the neat resin RK (**a**) and the hybrid composites samples at different LDH-F and Bt content, RK-FBt5 (**b**), RK-FBt10 (**c**), and RK-FBt20 (**d**).

3.2. Water Absorption

One of the most important aspects to be improved in the dental resins is represented by the water permeability. In fact, epoxy resins are vulnerable to absorbing water, thus, leading to a strong performance loss especially when used in heat and humid ambient. Figure 3 shows the weight gain (Equation (1)) for all composite samples after immersion in water at 37 °C for a selected time. The neat resin shows a gradual water uptake until about 0.7%. The equilibrium water uptake increases almost linearly with the addition of bentonite clay. The increase in water uptake could be due to both the hydrophilic nature of the bentonite and the interface volume formed around the bentonite particles. The hydrophilic clay lamellae are less embedded in the resin and more available to the water molecules. This generally induces an increase in the sample diffusivity. In addition, nanoparticles could produce a lack of crosslinking, which makes easy the water absorption. An opposite behavior is observed for the LDH-F based samples. In this case, the filler addition produces a gradual reduction in the quantity of water absorbed. As mentioned above, this is an interesting result as the quantity of water absorption is closely related to the composites performances. The absorbing curves show an initial time lag of about 10 days in which the sample weights change little with time. Probably, the introduction of fluorine atoms into polymer systems has the effect of making the matrix more resistant to water molecules, which is a desirable property for dental restorative materials. The hybrid composites containing both clays (LDH-F and Bt) show an intermediate behavior. In addition, at higher filler concentrations, the equilibrium water absorbed is only slightly larger than that of the neat resin.

An in deep analysis of the water sorption mechanism is crucial to understand the interactions between the filler and the resin matrix [38]. The water molecules can be present in two states, as free molecules evenly distributed between the polymer chains or linked to the molecular structure by means of hydrogen bonding. When they are free to move, they can act as diluents, and thus lead to a loss in the composite properties. The sorption process of liquids in resin matrices is usually assumed to be a concentration independent Fickian diffusion process [39]. The following expression is generally used to describe the diffusion process:

$$\frac{M_t}{M_\infty} = 1 - \frac{8}{\pi^2}\sum_{j=0}^{\infty}\frac{1}{(2j+1)^2} exp\left(-\frac{(2j+1)^2\pi^2 D}{h^2}t\right) \quad (2)$$

where M_t is moisture uptake at time t and M_∞ is the equilibrium moisture gain in the specimen at saturation; h is the sample thickness and D is the sample diffusivity. Since the Fickian model considers that the sorbed resides in the free volume and that no interaction is present (i.e., relaxation or degradation effects on the polymer are not considered), it may not be completely accurate as a predictive tool, especially when a strong bonding between water and epoxy groups is observed [40]. However, for the purpose of determining the diffusion coefficient of the epoxy it seems to yield acceptable results. In Figure 4, the water uptake curves for all samples are in reasonable agreement with the Fickian model (Equation (2)). However, the model overestimates the mass uptake at short time. This suggests that the initial uptake behavior is contrasted by the ions release.

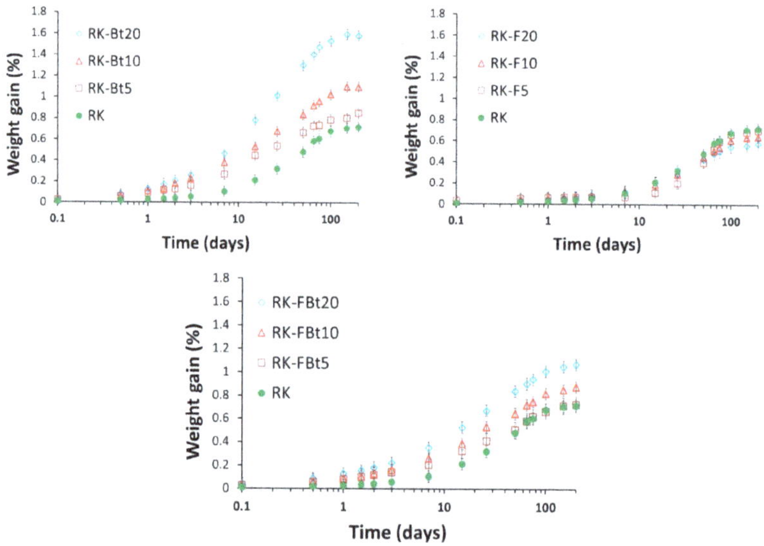

Figure 3. Weight gain as function of conditioning time in water at 37 °C for neat RK resin and all hybrid composites.

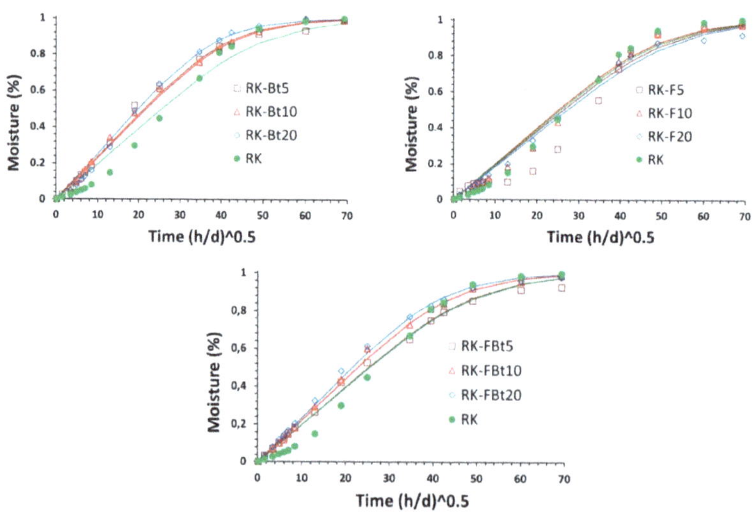

Figure 4. Simulation of the absorption curves by Fickian model.

In Figure 5 the equilibrium moisture concentrations, as well as the diffusion coefficients derived according to Equation (2), are shown. These parameters strongly depend by the filler nature and concentration. In samples containing bentonite clay, the diffusivity decreases with the filler addition. As expected, the tortuous diffusion pathway produced by the impermeable filler reduces the diffusivity of water molecules [38]. This hindrance effect, however, does not have an effect on the equilibrium water uptake, which increases with the filler addition. Probably, the hydrophilic nature of the clay bentonite favors the water molecules accumulation at the equilibrium. An opposite effect is observed with the LDH-F samples, i.e., water uptake decreases and diffusivity increases with the filler concentration. In this case, the presence of fluorine atoms provides water repellence. However, the presence on LDH-F filler probably induces a lack in crosslinking density, and thus an increase in chain mobility with a consequent increase in water diffusivity. The combined effect of the two fillers produces intermediate values. In this case, both water uptake and diffusivity increase with the filler concentration increase.

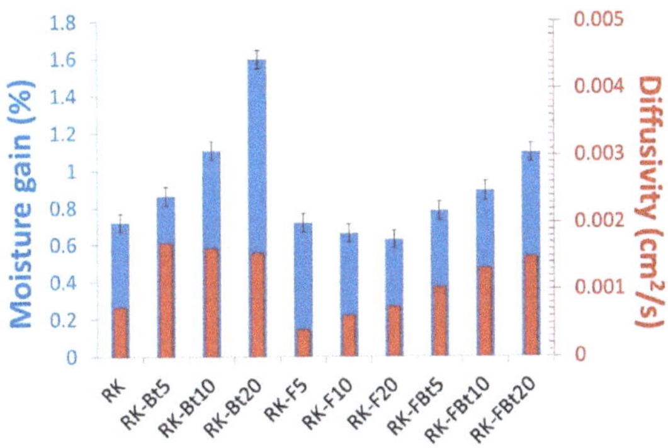

Figure 5. Equilibrium moisture uptake and Fickian diffusivity for all tested composites.

3.3. Mechanical Properties

The mechanical properties were investigated in a wide range of temperatures by performing dynamic mechanical analysis. Figure 6 shows the elastic modulus and the loss modulus for the pristine resin and the resin containing 20 wt% of LDH-F/Bt mixture, as representative sample. As expected, the addition of inorganic filler increases the elastic modulus of the resins in all the temperature range investigated. This effect was observed in all dry samples and at almost all concentrations (Figure 7).

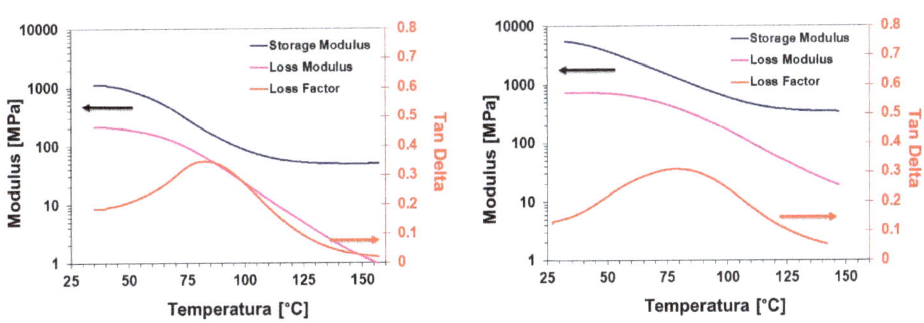

Figure 6. Dynamic mechanical analysis (DMA) results of RK (**left**) and RK-FBt20 (**right**).

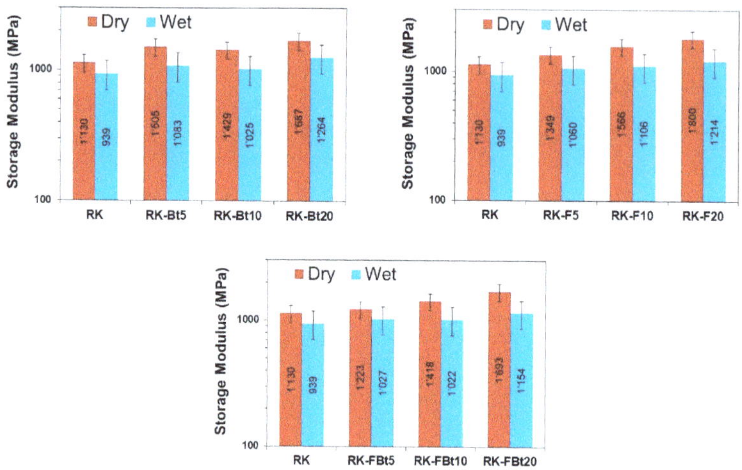

Figure 7. Storage modulus at 37 °C of all samples investigated, in wet and dry conditions.

The loss factor (also known as tanδ) of the resin and RK-FBt20 composite is also reported in Figure 6. The height and the shape of the curve are strictly related to the molecular relaxation and the crosslinking distribution of the composites. The filler addition produces a wide distribution in the molecular relaxation with an apparent increase of the crosslinking density. In literature, the glass transition temperature is generally identified as the maximum in the loss factor curve. This increase is evident after the glass transition temperature.

The values of the glass transition temperatures are reported in Figure 8. We observed that the T_g of the composite resins are consistently higher than the pristine resin. The observed reinforcement increases when the filler concentration increases. As expected, as shown in many composite systems, the deformation at breaking of the composite resin was found slightly lower than the pristine resin. However, since the stress is increasingly higher in the composites, the toughness remained almost unchanged.

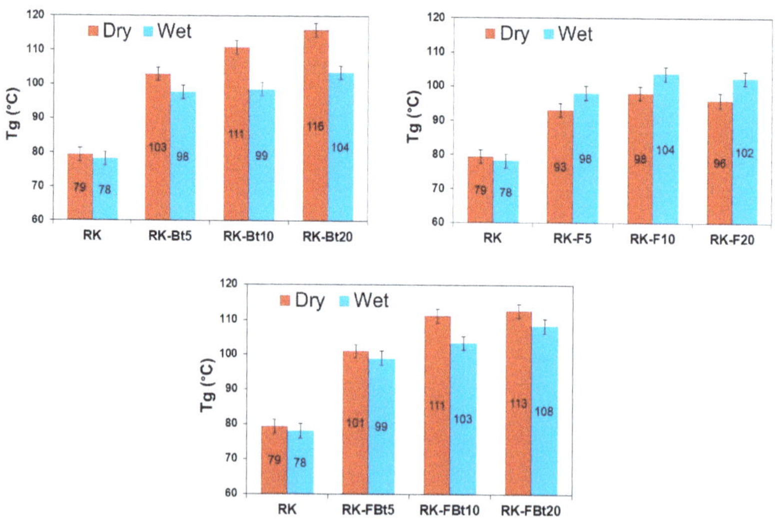

Figure 8. Glass transition temperature of all samples investigated, in wet and dry conditions.

The effect of the moisture on thermomechanical properties of the sample can be observed in Figure 7 (wet samples). The elastic modulus, for almost all wet samples, shows an important decrease in all the temperature ranges. This reduction is probably due to the presence of water molecules in the interface between the polymer matrix and the filler that modify the constraints of the polymer chain mobility near the reinforcements. In addition, the glass transition is affected by the environmental humidity which can be due to a number of reasons, such as changes in reaction chemistry, lower crosslink density, or a plasticizing effect of the water molecules entrapped in the resin-filler interface regions. In particular, the rupture of the hydrogen bonding between polymer chains by water molecules would produce an increment of the chain mobility during the glass transition region.

Composites samples containing fluoride-intercalated layered double hydroxide and bentonite (RK-FBtx) show an interesting increase in the glass transition (Figure 8) because of the synergic effect of the water repellence of LDH-F and the molecular mobility reduction due to the bentonite clay (Figure 5).

3.4. Resins Biocompatibility and Differentiation

The success of dentin/pulp regeneration depends on the development of suitable scaffolding materials as carriers for DPSCs. Stem cells derived from dental pulp (dental pulp stem cells, DPSCs) represent an easy and accessible source of undifferentiated adult stem cells due to their noninvasive collection procedures obtained from both human permanent and supernumerary teeth [41]. Several studies have reported that DPSCs expressing STRO-1, a cell surface marker, represent a subgroup able to differentiate into dental hard tissue-forming cells [42]. In fact, STRO-1+ sorted dental stem cells exhibited a superior predisposition to undergo odontogenesis than unsorted cells establishing reproducible and defined in vitro culture protocol for regenerative purposes. In a previous work, we demonstrated that fluoride-containing resin elicit beneficial effects on STRO-1+ cells inducing differentiation into functional odontoblast-like cells [34]. The new synthesized materials exhibited a time-dependent fluoride release over 28 days of incubation in artificial saliva (SAGF). A slow fluoride release that gradually increases until 21 days was observed for all tested materials (Figure 9) reaching a concentration of fluoride ranging from 0.976 ± 0.116 ppm (RK-FBt5) to 1.826 ± 0.154 ppm (RK-FBt20).

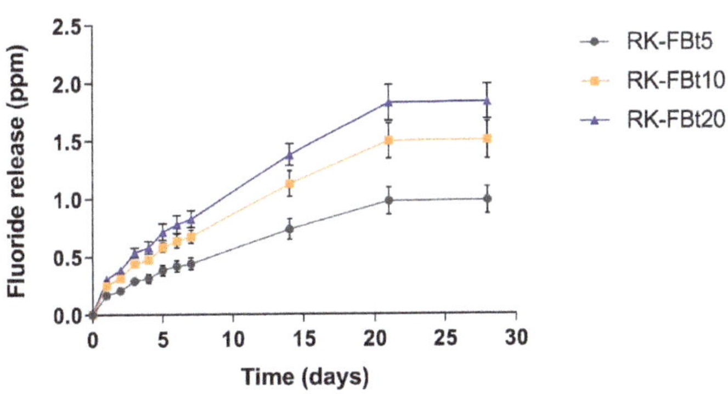

Figure 9. Cumulative release profiles of fluoride ions from RK-FBt5, RK-FBt10, and RK-FBt20 measured by ion chromatography method. Samples were put in mineral medium with composition similar to saliva (SAGF) for 28 days. The bars represent means ± SD (n = 3).

As shown in Figure 10, no significant effect on proliferation was observed between resins (RK, RK-FBt5, RK-FBt10, and RK-FBt20) and control group after one, three, and six days of culture.

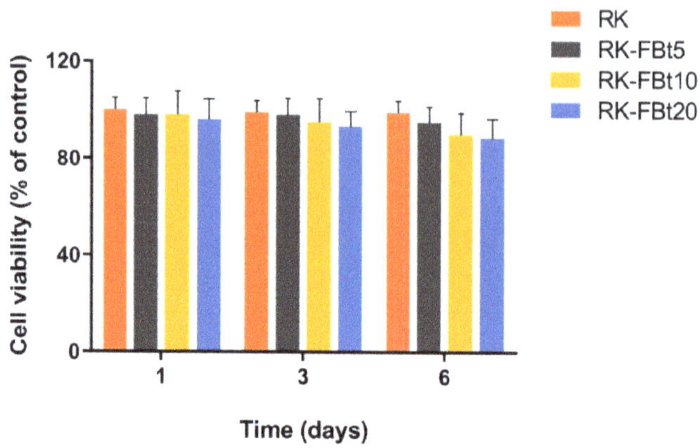

Figure 10. STRO-1+ cell viability assessed by CCK-8 assay on RK, RK-FBt5, RK-FBt10, and RK-FBt20 resins after 1, 3, and 6 days. The bars represent means ± SD ($n = 3$).

In order to determine the effect on STRO-1+ cells, the activity of alkaline phosphatase (ALP) an early marker of immature osteoblast/odontoblast, was evaluated. Figure 11 shows that the ALP activity gradually increased for 28 days in cells grown on RK-FBt5, RK-FBt10, and RK-FBt20 as compared with cells cultured onto tissue culture polystyrene and RK. These results suggest that the resins exhibit equivalent biologic activity.

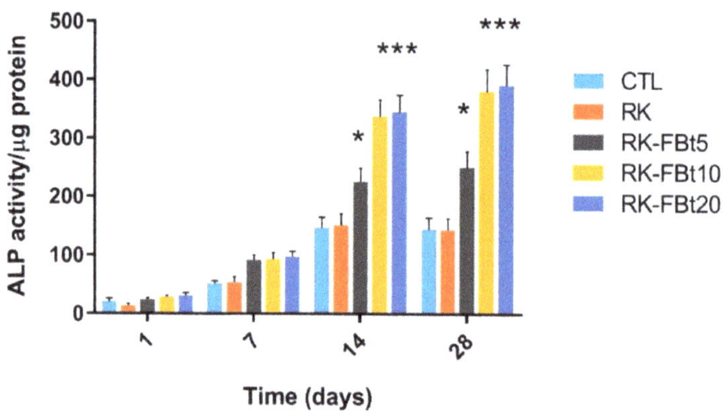

Figure 11. Alkaline phosphatase (ALP) activity of STRO-1+ cells cultured for 28 days on RK-FBt resins. Cell cultured onto tissue culture polystyrene (CTL) or on commercial light-activated restorative materials (RK) were used as controls. * $P < 0.05$ versus CTL and RK; *** $P < 0.001$ versus CTL, RK, and RK-FBt5; The bars represent means ± SD ($n = 3$).

3.5. Direct Contact Test (DCT)

The DCT, introduced in 1996 by Weiss et al. [43] is a quantitative and reproducible method that simulates the contact of the test microorganism with composite resins inside the root canal and evaluates the kinetics of bacterial growth. In particular, this method measures the bacterial growth kinetic and detects the bacteriostatic (prevention of growth) and bactericidal effect of synthetized materials.

The results of the DCT for the time period of 48 h (Figure 12) demonstrated that only RK-FBt20 was able to inhibit the growth of all three bacterial strains, while a slight effect was observed for RK-FBt10. Indeed, the intergroup comparisons between groups analyzed using the Kruskal–Wallis test followed by Dunn's post hoc analysis demonstrated significant difference in overall bacterial kinetics only for RK-FBt10 and RK-FBt20 (Table 1).

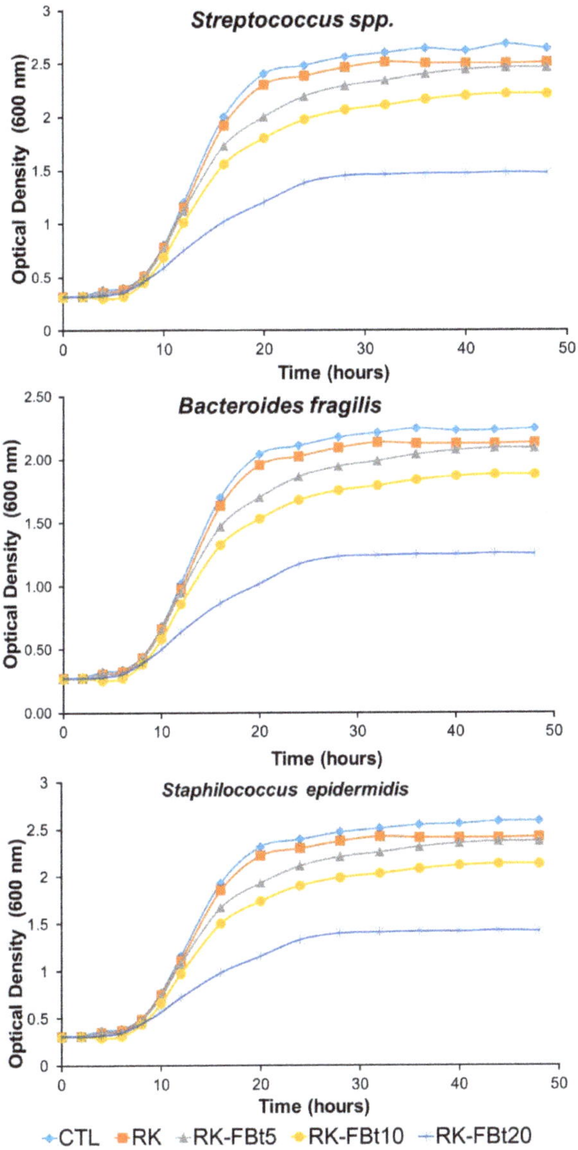

Figure 12. Antibacterial activity evaluated by DCT against *Streptococcus* spp., *Bacteroides fragilis*, and *Staphylococcus epidermidis* of RK, RK-FBt5, RK-FBt10, and RK-FBt20. Single bacterial strain growth in absence of composite resins was used as control (CTL). Each point on the growth curves is the average of optical density measurements in 6 wells at any given time point.

Table 1. Results of inter-group statistical significance of direct contact test (DCT) using Kruskal–Wallis test followed by Dunn's post hoc analysis.

	Streptococcus spp.	*B. fragilis*	*S. epidermidis*
Test details	*P*	*P*	*P*
CTL vs. RK	n.s.	n.s.	n.s.
CTL vs. RK-FBt5	n.s.	n.s.	n.s.
CTL vs. RK-FBt10	0.0041	0.0044	0.0031
CTL vs. RK-FBt20	<0.0001	<0.0001	<0.0001
RK vs. RK-FBt5	n.s.	n.s.	n.s.
RK vs. RK-FBt10	n.s.	n.s.	n.s.
RK vs. RK-FBt20	0.0025	0.0033	0.0040
RK-FBt5 vs. RK-FBt10	n.s.	n.s.	n.s.
RK-FBt5 vs. RK-FBt20	n.s.	n.s.	n.s.
RK-FBt10 vs. RK-FBt20	n.s.	n.s.	n.s.

3.6. Antibiofilm Analyses under Dynamic Conditions

Biofilm architecture, as well as antimicrobial tolerance of biofilms, is dependent on growth conditions, such as hydrodynamics, nutrients, and cell density [44,45]. For these reasons, different microbiological models that recreate a diverse oral environment can lead to different results. The static biofilm formation is the most widely used experimental model to determine the antibiofilm properties of materials, although it is not very representative of the oral cavity environment, where adherence of bacteria is subjected to a salivary flux with the simultaneously formation of salivary pellicle [46]. In particular, the presence of the salivary pellicle, an acellular organic film that covers any type of surface exposed to saliva, can have prominent effects on the biofilm formation and composition. Pellicle proteins, such as lysozyme, histatins, α-amylase, cystatins, lactoferrin, and large salivary mucin, that in vivo provide an array of potential receptors for the attachment of the early colonizer [47,48] in vitro could inactivate the materials antibacterial activity acting as separation layer between bacteria and material surface [46]. Furthermore, the composition of biofilms can be modified by the specificity of the host or site, for example in presence of variations in the intensity of the fluid flow. Indeed, several studies demonstrate that shear conditions enhanced the adhesion of both *Escherichia coli* and *Streptococcus gordonii* in periodontitis [49].

Therefore, in this study a dip-flow reactor was used as a model system to generate a biofilm under a laminar flow of nutrients close to the air–liquid interface, recreating in lab the dynamic conditions of the oral cavity.

As shown in Figure 13, we observed a significant reduction in biofilm formation, regardless of bacterial strain used and the presence or absence of salivary pellicle, only in RK-FBt10 and RK-FBt20 materials. In particular, the greatest effect was observed on *Streptococcus* spp. at 96 h for RK-FBt20 in absence of salivary pellicle, inducing a 70% of biofilm inhibition with respect to a control ($P < 0.001$). Furthermore, the presence of salivary pellicle did not significantly affect the antibiofilm properties of composite resins for every bacterial strain.

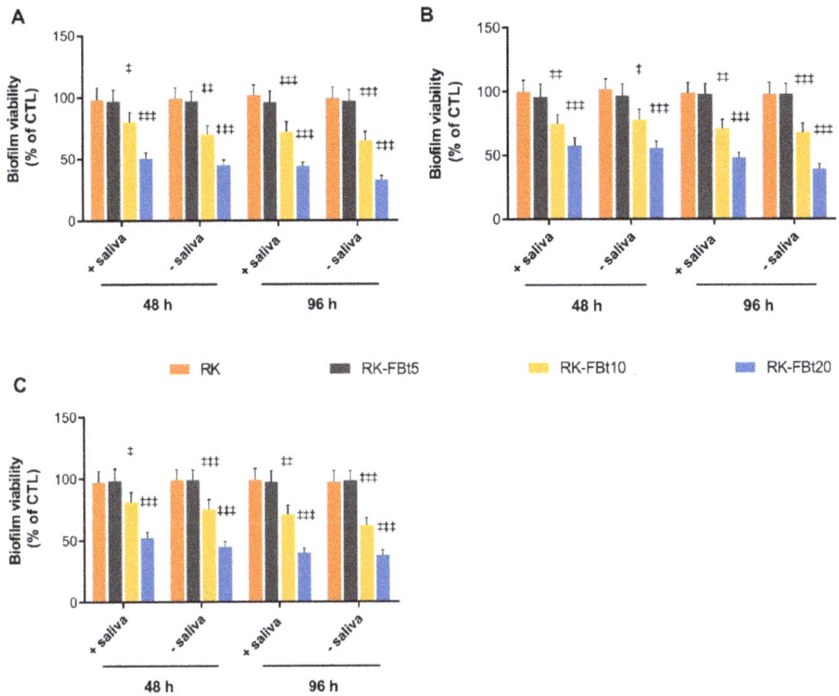

Figure 13. Antibiofilm activity of composite resins. Biofilm viability was evaluated by MTT assay, simulating or not the formation of salivary pellicle, after 48 and 96 h of incubation at 37 °C in the presence of *Streptococcus* spp. (**A**), *B. fragilis* (**B**), and *S. epidermidis* (**C**) as described in the Material and Methods section. The absorbance of each well was read at 510 nm. Biofilm formation was reported as a percentage in comparison with the maximum amount of biofilm produced by *Streptococcus* spp., *B. fragilis*, and *S. epidermidis* grown (positive controls). For each sample, six different experiments were conducted, and the results expressed as the mean of the values obtained (mean ± SD). Statistically significant variations: # $P < 0.05$, ## $P < 0.01$, and ### $P < 0.001$ versus positive control RK and RK-FBt5.

The fluoride concentration released from both RK-FBt10, and RK-FBt20 materials, after 48 h, reaches already a micromolar amount capable of producing a significant antimicrobial effect (Figure 9). The well-known antibacterial and antibiofilm activity [50,51] of fluoride is elicited via the following three principal mechanisms: (i) the ability to form metal-fluoride complexes (i.e., with aluminum and beryllium cations) that can interact with host F-ATPase and nitrogenase enzymes [52]; (ii) the formation of hydrofluoric acid that interfere with the bacterial proton chain [53]; and (iii) the direct inhibition of bacterial specific enzymes such as enolase, acid phosphatase, pyrophosphatase, peroxidase, and catalase [54]. Indeed, fluoride concentrations lower than micromolar are able to inhibit the glycolytic enzyme enolase interfering with a sustained bacterial growth [55].

4. Conclusions

In this paper, the influence of temperature, moisture, and their combination on thermal and mechanical properties was studied for a dental resin filled with various filler. The water absorption has an important effect on the mechanical properties and the T_g of the material. Probably, water absorption causes plasticization in the material, which is associated with a decrease in resistance and T_g. The addition of filler does not affect the trends of the properties, although all the observed decreases are lower for the composites. The new synthesized materials (RK-FBt10 and RK-FBt20) are able to modulate the differentiation of DPSCs expressing STRO-1 into odontoblast-like phenotype leading to an initial mineralization process. Moreover, the released fluoride could elicit antibacterial and antibiofilm activity against three bacteria isolated from human saliva (*Streptococcus* spp., *B. fragilis*, and *S. epidermidis*) and be involved in the formation of dental plaque and caries.

Taking together, these results make these materials a promising strategy in preventive and restorative dentistry, preventing the initial bacterial adhesion and biofilm formation, the first two steps crucial to the formation of dental caries.

Author Contributions: Conceptualization, L.T. and A.C.; data curation A.S.; investigation A.D.S. and I.D.L.; methodology, L.T., A.C., and A.S; writing—original draft L.T. and A.C.; writing—review and editing A.S., L.T., and A.C.; and supervision F.R., G.P., and V.V. All authors have read and agreed to the published version of the manuscript.

Funding: This research received no external funding.

Acknowledgments: The authors acknowledge the support given by "Sviluppo di nanotecnologie Orientate alla Rigenerazione e Ricostruzione tissutale, Implantologia e Sensoristica in Odontoiatria/oculistica—SORRISO" and POR Campania FESR 2014_2020 and "Tecnologie abilitanti per la sintesi eco-sostenibile di nuovi materiali per la restaurativa dentale"—ABILTECH.

Conflicts of Interest: The authors declare no conflict of interest.

References

1. Marghalani, H.Y. Resin-Based Dental Composite Materials. In *Handbook of Bioceramics and Biocomposites*; Springer International Publishing: Berlin, Germany, 2016; pp. 357–405.
2. Cramer, N.B.; Stansbury, J.W.; Bowman, C.N. Recent Advances and Developments in Composite Dental Restorative Materials. *J. Dent. Res.* **2011**, *90*, 402–416. [CrossRef] [PubMed]
3. Bowen, R.L. Use of epoxy resins in restorative materials. *J. Dent. Res.* **1956**, *35*, 360–369. [CrossRef]
4. Zagho, M.; Hussein, E.; Elzatahry, A. Recent Overviews in Functional Polymer Composites for Biomedical Applications. *Polymers* **2018**, *10*, 739. [CrossRef] [PubMed]
5. Zhang, H.; Darvell, B.W. Mechanical properties of hydroxyapatite whisker-reinforced bis-GMA-based resin composites. *Dent. Mater.* **2012**, *28*, 824–830. [CrossRef] [PubMed]
6. Patterson, J.W.H.S.H. Bentonite and fuller's earth resources of the United States. Available online: https://pubs.er.usgs.gov/publication/pp1522 (accessed on 17 December 2019).
7. Sorrentino, A.; Gorrasi, G.; Vittoria, V. Permeability in clay/polyesters nano-biocomposites. In *Green Energy and Technology*; Springer: London, UK, 2012; Volume 50, pp. 237–264. ISBN 9781447141013.
8. Moosavi, M. Bentonite Clay as a Natural Remedy: A Brief Review. *Iran. J. Public Health* **2017**, *46*, 1176–1183.
9. Gorrasi, G.; Milone, C.; Piperopoulos, E.; Lanza, M.; Sorrentino, A. Hybrid clay mineral-carbon nanotube-PLA nanocomposite films. Preparation and photodegradation effect on their mechanical, thermal and electrical properties. *Appl. Clay Sci.* **2013**, *71*, 49–54. [CrossRef]
10. Miyazaki, M.; Ando, S.; Hinoura, K.; Onose, H.; Moore, B.K. Influence of filler addition to bonding agents on shear bond strength to bovine dentin. *Dent. Mater.* **1995**, *11*, 234–238. [CrossRef]
11. Nunes, M.F.; Swift, E.J.; Perdigão, J. Effects of adhesive composition on microtensile bond strength to human dentin. *Am. J. Dent.* **2001**, *14*, 340–343.
12. Atai, M.; Solhi, L.; Nodehi, A.; Mirabedini, S.M.; Kasraei, S.; Akbari, K.; Babanzadeh, S. PMMA-grafted nanoclay as novel filler for dental adhesives. *Dent. Mater.* **2009**, *25*, 339–347. [CrossRef]
13. Nikolaidis, A.K.; Koulaouzidou, E.A.; Gogos, C.; Achilias, D.S. Synthesis and Characterization of Dental Nanocomposite Resins Filled with Different Clay Nanoparticles. *Polymers* **2019**, *11*, 730. [CrossRef]
14. Soares, C.J.; Faria-E-Silva, A.L.; De Paula Rodrigues, M.; Vilela, A.B.F.; Pfeifer, C.S.; Tantbirojn, D.; Versluis, A. Polymerization shrinkage stress of composite resins and resin cements—What do we need to know? *Braz. Oral Res.* **2017**, *31*, 62. [CrossRef] [PubMed]
15. Ersen, K.A.; Gürbüz, Ö.; Özcan, M. Evaluation of polymerization shrinkage of bulk-fill resin composites using microcomputed tomography. *Clin. Oral Investig.* **2019**. [CrossRef] [PubMed]
16. Fronza, B.M.; Lewis, S.; Shah, P.K.; Barros, M.D.; Giannini, M.; Stansbury, J.W. Modification of filler surface treatment of composite resins using alternative silanes and functional nanogels. *Dent. Mater.* **2019**, *35*, 928–936. [CrossRef] [PubMed]

17. Tanner, J.; Tolvanen, M.; Garoushi, S.; Säilynoja, E. Clinical Evaluation of Fiber-Reinforced Composite Restorations in Posterior Teeth—Results of 2.5 Year Follow-up. *Open Dent. J.* **2018**, *12*, 476–485. [CrossRef]
18. Nedeljkovic, I.; Teughels, W.; De Munck, J.; Van Meerbeek, B.; Van Landuyt, K.L. Is secondary caries with composites a material-based problem? *Dent. Mater.* **2015**, *31*, e247–e277. [CrossRef]
19. Bernardo, M.; Luis, H.; Martin, M.D.; Leroux, B.G.; Rue, T.; Leitão, J.; DeRouen, T.A. Survival and reasons for failure of amalgam versus composite posterior restorations placed in a randomized clinical trial. *J. Am. Dent. Assoc.* **2007**, *138*, 775–783. [CrossRef]
20. Imazato, S.; McCabe, J.F. Influence of incorporation of antibacterial monomer on curing behavior of a dental composite. *J. Dent. Res.* **1994**, *73*, 1641–1645. [CrossRef]
21. Mirsayar, M.M. On fracture analysis of dental restorative materials under combined tensile-shear loading. *Theor. Appl. Fract. Mech.* **2018**, *93*, 170–176. [CrossRef]
22. Mirsayar, M.M.; Park, P. Modified maximum tangential stress criterion for fracture behavior of zirconia/veneer interfaces. *J. Mech. Behav. Biomed. Mater.* **2016**, *59*, 236–240. [CrossRef]
23. Tavassoli Hojati, S.; Alaghemand, H.; Hamze, F.; Ahmadian Babaki, F.; Rajab-Nia, R.; Rezvani, M.B.; Kaviani, M.; Atai, M. Antibacterial, physical and mechanical properties of flowable resin composites containing zinc oxide nanoparticles. *Dent. Mater.* **2013**, *29*, 495–505. [CrossRef]
24. Costantino, U.; Nocchetti, M.; Gorrasi, G.; Tammaro, L. Hydrotalcites in nanobiocomposites. In *Multifunctional and Nanoreinforced Polymers for Food Packaging*; Elsevier: Amsterdam, The Netherlands, 2011; pp. 43–85.
25. Bugatti, V.; Esposito, L.; Franzetti, L.; Tammaro, L.; Vittoria, V. Influence of the powder dimensions on the antimicrobial properties of modified layered double hydroxide. *Appl. Clay Sci.* **2013**, *75–76*, 46–51. [CrossRef]
26. Stencel, R.; Kasperski, J.; Pakieła, W.; Mertas, A.; Bobela, E.; Barszczewska-Rybarek, I.; Chladek, G. Properties of Experimental Dental Composites Containing Antibacterial Silver-Releasing Filler. *Materials* **2018**, *11*, 1031. [CrossRef] [PubMed]
27. Xu, H.H.K.; Sun, L.; Weir, M.D.; Antonucci, J.M.; Takagi, S.; Chow, L.C.; Peltz, M. Nano DCPA-whisker composites with high strength and Ca and PO(4) release. *J. Dent. Res.* **2006**, *85*, 722–727. [CrossRef]
28. Tammaro, L.; Vittoria, V.; Calarco, A.; Petillo, O.; Riccitiello, F.; Peluso, G. Effect of layered double hydroxide intercalated with fluoride ions on the physical, biological and release properties of a dental composite resin. *J. Dent.* **2014**, *42*, 60–67. [CrossRef] [PubMed]
29. Conte, R.; De Luise, A.; Valentino, A.; Di Cristo, F.; Petillo, O.; Riccitiello, F.; Di Salle, A.; Calarco, A.; Peluso, G. Hydrogel Nanocomposite Systems. In *Nanocarriers for Drug Delivery*; Elsevier: Amsterdam, The Netherlands, 2019; pp. 319–349.
30. Chenicheri, S.; Usha, R.; Ramachandran, R.; Thomas, V.; Wood, A. Insight into Oral Biofilm: Primary, Secondary and Residual Caries and Phyto-Challenged Solutions. *Open Dent. J.* **2017**, *11*, 312–333. [CrossRef] [PubMed]
31. Zitzmann, N.U.; Berglundh, T. Definition and prevalence of peri-implant diseases. *J. Clin. Periodontol.* **2008**, *35*, 286–291. [CrossRef]
32. Albrektsson, T.; Buser, D.; Chen, S.T.; Cochran, D.; DeBruyn, H.; Jemt, T.; Koka, S.; Nevins, M.; Sennerby, L.; Simion, M.; et al. Statements from the Estepona Consensus Meeting on Peri-implantitis, February 2-4, 2012. *Clin. Implant Dent. Relat. Res.* **2012**, *14*, 781–782. [CrossRef]
33. Wade, W.G. The oral microbiome in health and disease. *Pharmacol. Res.* **2013**, *69*, 137–143. [CrossRef]
34. Calarco, A.; Di Salle, A.; Tammaro, L.; De Luca, I.; Mucerino, S.; Petillo, O.; Riccitiello, F.; Vittoria, V.; Peluso, G. Long-Term Fluoride Release from Dental Resins Affects STRO-1 + Cell Behavior. *J. Dent. Res.* **2015**, *94*, 1099–1105. [CrossRef]
35. Di Salle, A.; Spagnuolo, G.; Conte, R.; Procino, A.; Peluso, G.; Rengo, C. Effects of various prophylactic procedures on titanium surfaces and biofilm formation. *J. Periodontal Implant Sci.* **2018**, *48*, 373. [CrossRef]
36. Constructing and Using a Drip Flow Reactor. Available online: https://www.cs.montana.edu/webworks/projects/stevesbook/contents/chapters/chapter011/section021/blue/page001.html (accessed on 1 September 2019).
37. He, Z.; Liang, J.; Zhou, W.; Xie, Q.; Tang, Z.; Ma, R.; Huang, Z. Effect of the quorum-sensing luxS gene on biofilm formation by Enterococcus faecalis. *Eur. J. Oral Sci.* **2016**, *124*, 234–240. [CrossRef] [PubMed]
38. Sorrentino, A.; Tortora, M.; Vittoria, V. Diffusion behavior in polymer-clay nanocomposites. *J. Polym. Sci. Part B Polym. Phys.* **2006**, *44*, 265–274. [CrossRef]

39. Sorrentino, A.; Gorrasi, G.; Tortora, M.; Vittoria, V. Barrier properties of polymer/clay nanocomposites. In *Polymer Nanocomposites*; Mai, Y.-W., Yu, Z.-Z., Eds.; Elsevier: Cambridge, UK, 2006; pp. 273–296. ISBN 978-1-85573-969-7.
40. Sorrentino, A. Nanocoatings and ultra-thin films for packaging applications. In *Nanocoatings and Ultra-Thin Films*; Makhlouf, A.S.H., Tiginyanu, I., Eds.; Elsevier: Oxford, UK, 2011; pp. 203–234. ISBN 9781845698126.
41. La Noce, M.; Paino, F.; Spina, A.; Naddeo, P.; Montella, R.; Desiderio, V.; De Rosa, A.; Papaccio, G.; Tirino, V.; Laino, L. Dental pulp stem cells: State of the art and suggestions for a true translation of research into therapy. *J. Dent.* **2014**, *42*, 761–768. [CrossRef] [PubMed]
42. Yang, X.; van der Kraan, P.M.; Bian, Z.; Fan, M.; Walboomers, X.F.; Jansen, J.A. Mineralized Tissue Formation by BMP2-transfected Pulp Stem Cells. *J. Dent. Res.* **2009**, *88*, 1020–1025. [CrossRef] [PubMed]
43. Weiss, E.I.; Shalhav, M.; Fuss, Z. Assessment of antibacterial activity of endodontic sealers by a direct contact test. *Dent. Traumatol.* **1996**, *12*, 179–184. [CrossRef]
44. Stewart, P.S. Antimicrobial Tolerance in Biofilms. *Microbiol. Spectr.* **2015**, *3*. [CrossRef]
45. Stoodley, P.; Dodds, I.; Boyle, J.D.; Lappin-Scott, H.M. Influence of hydrodynamics and nutrients on biofilm structure. *J. Appl. Microbiol.* **1998**, *85*, 19S–28S. [CrossRef]
46. Zaltsman, N.; Ionescu, A.C.; Weiss, E.I.; Brambilla, E.; Beyth, S.; Beyth, N. Surface-modified nanoparticles as anti-biofilm filler for dental polymers. *PLoS ONE* **2017**, *12*, e0189397. [CrossRef]
47. Yao, Y.; Berg, E.A.; Costello, C.E.; Troxler, R.F.; Oppenheim, F.G. Identification of protein components in human acquired enamel pellicle and whole saliva using novel proteomics approaches. *J. Biol. Chem.* **2003**, *278*, 5300–5308. [CrossRef]
48. Größner-Schreiber, B.; Teichmann, J.; Hannig, M.; Dörfer, C.; Wenderoth, D.F.; Ott, S.J. Modified implant surfaces show different biofilm compositions under in vivo conditions. *Clin. Oral Implants Res.* **2009**, *20*, 817–826. [CrossRef]
49. Thomas, W.E.; Trintchina, E.; Forero, M.; Vogel, V.; Sokurenko, E.V. Bacterial Adhesion to Target Cells Enhanced by Shear Force. *Cell* **2002**, *109*, 913–923. [CrossRef]
50. Marquis, R.E. Antimicrobial actions of fluoride for oral bacteria. *Can. J. Microbiol.* **1995**, *41*, 955–964. [CrossRef] [PubMed]
51. Marquis, R.E.; Clock, S.A.; Mota-Meira, M. Fluoride and organic weak acids as modulators of microbial physiology. *FEMS Microbiol. Rev.* **2003**, *26*, 493–510. [CrossRef] [PubMed]
52. Sturr, M.G.; Marquis, R.E. Inhibition of proton-translocating ATPases of Streptococcus mutans and Lactobacillus casei by fluoride and aluminum. *Arch. Microbiol.* **1990**, *155*, 22–27. [CrossRef] [PubMed]
53. Guha-Chowdhury, N.; Iwami, Y.; Yamada, T. Effect of Low Levels of Fluoride on Proton Excretion and Intracellular pH in Glycolysing Streptococcal Cells under Strictly Anaerobic Conditions. *Caries Res.* **1997**, *31*, 373–378. [CrossRef] [PubMed]
54. Guha-Chowdhury, N.; Clark, A.G.; Sissons, C.H. Inhibition of purified enolases from oral bacteria by fluoride. *Oral Microbiol. Immunol.* **1997**, *12*, 91–97. [CrossRef]
55. Matsui, R.; Cvitkovitch, D. Acid tolerance mechanisms utilized by Streptococcus mutans. *Future Microbiol.* **2010**, *5*, 403–417. [CrossRef]

polymers

MDPI

Article

Effect of Rapid Mold Heating on the Structure and Performance of Injection-Molded Polypropylene

Sara Liparoti [1], Vito Speranza [1,*], Giuseppe Titomanlio [1,2] and Roberto Pantani [1]

1 Department of Industrial Engineering, University of Salerno, via Giovanni Paolo II, 132-84084 Fisciano (SA), Italy; sliparoti@unisa.it (S.L.); gtitomanlio@unisa.it (G.T.); rpantani@unisa.it (R.P.)

2 Institute of Polymers, Composites and Biomaterials (IPCB), The National Research Council (Cnr), Via Previati 1/C, 23900 Lecco (LC), Italy

* Correspondence: vsperanza@unisa.it

Received: 17 December 2019; Accepted: 23 January 2020; Published: 5 February 2020

Abstract: The tailoring by the process of the properties developed in the plastic objects is the more effective way to improve the sustainability of the plastic objects. The possibility to tailor to the final use the properties developed within the molded object requires further understanding of the relationship between the properties of the plastic objects and the process conduction. One of the main process parameters that allow adjusting the properties of molded objects is the mold temperature. In this work, a thin electrical heater was located below the cavity surface in order to obtain rapid and localized surface heating/cooling cycles during the injection molding process. An isotactic polypropylene was adopted for the molding tests, during which surface temperature was modulated in terms of values and heating times. The modulation of the cavity temperature was found able to control the distribution of relevant morphological characteristics, thus, properties along the sample thickness. In particular, lamellar thickness, crystallinity distribution, and orientation were analyzed by synchrotron X-ray experiments, and the morphology and elastic modulus were characterized by atomic force microscopy acquisitions carried out with a tool for the simultaneous nanomechanical characterization. The crystalline degree slightly increased with the cavity temperature, and this induced an increase in the elastic modulus when high temperatures were adopted for the cavity surface. The cavity temperature strongly influenced the orientation distribution that, on its turn, determined the highest values of the elastic modulus found in the shear layer. Furthermore, although the sample core, not experiencing a strong flow field, was not characterized by high levels of orientation, it might show high values of the elastic modulus if temperature and time during crystallization were sufficient. In particular, if the macromolecules spent adequate time at temperatures close to the crystallization temperature, they could achieve high levels of structuring and, thus, high values of elastic modulus.

Keywords: injection molding; structure/properties relationship; cavity temperature fast modulation

1. Introduction

The injection molding process is one of the most commonly used plastic formation techniques across myriad of industrial sectors, subject to the absolute ease and convenience provided by this methodology. Statistics testimony to the aforementioned fact: global injection molded plastic market size stood at 100 million tons [1]. The high automation, the short processing time, and the possibility to obtain objects with high dimensional and geometrical accuracy are the major advantages of this process [2]. Nowadays, growing interest is devoted to the increase of the sustainability of both the process and the plastic products due to the problems related to the pollution due to the high amount of plastic waste (microplastics) [3].

Generally, in order to improve the performance (mechanical, optical, etc.) of plastic products, fillers are added in the polymeric matrix. For instance, colorants, plasticizers, lubricants, antioxidants are commonly used in plastic packaging [4–6], and flame retardants are commonly used in plastic for electronics [7,8]. With the introduction of fillers during the process, the plastic part fulfills the requirements for the use in specific applications, but, at the same time, it reduces the sustainability of the plastic products [9]. In order to improve the sustainability of plastic products, it is important to have the possibility to tailor the properties without introducing fillers. In other words, the appropriate selection of the operating conditions must assure to tailor the properties developed in the final objects. This would increase the sustainability of the final object, making it suitable for common recycle processes [10,11].

Some attempts to tailor the properties of plastic part by the process was carried out by Turng and coworkers [12] in the production of isotactic polypropylene (iPP) films by extrusion-based processes. They found that the morphology, thus the properties, developed within the film strongly depended on the elongation degree undergone by the polymer chains during drawing. Lamberti [13] found similar behavior in the films made of iPP. Other authors proposed modification of exiting processes to tailor plastic part properties. Wang et al. [14] proposed the water-assisted injection molding for inducing the formation of fibrillar structures, which would improve the mechanical performance of the iPP parts. Also, the foaming injection molding was introduced and adopted for making lighter plastic parts adopted in the automotive field [15,16].

In the injection molding process, one of the main operating parameters that affect the properties of the final object is the mold temperature [17,18]. Surface finishing, morphology distribution, and possibility to produce micro and nano-structures on the molded surface strongly depend on the temperature adopted in the mold and, in particular, on the cavity surface [19–23]. The mechanical properties have been found to be dependent on the mold temperature; in particular, the tensile strength of moldings made of polyether-ether-ketone increases with the mold temperature [24]. Similar results have also been obtained in the case of polylactic acid bio-composites [25]. The impact energy and the flexural modulus have been found to depend on mold temperature for blends made of polyethylene terephthalate and polypropylene [26]. Also, the structure developed, in terms of crystallinity and amounts of crystalline phase fractions, depends on the mold temperature [27], in particular, the crystalline degree increases with mold temperature for moldings made of polypropylene.

Several methodologies can be adopted for controlling cavity surface temperature during the injection molding process, such as proximity heating [28,29], infrared heating [30], and electrical heating [31,32]. Another possibility to control the mold temperature was proposed by Minh and coworkers [33], who developed an external gas-assisted mold temperature control for improving the performance of thin rib parts.

In this work, a thin (220 μm thickness) electrical heating device was adopted to modulate the temperature in the cavity during the injection molding process. The heating device was located just below the cavity surface in order to maximize the heating/cooling rates and to realize a localized control of the cavity temperature. Several temperature conditions, in terms of cavity surface temperature and heating times, were realized by the aforementioned devices, with the aim of tailoring plastic part structure and performance by the cavity temperature modulation. The structure developed, in terms of morphology and crystallinity distributions, and the performance, in terms of mechanical properties, within the molded objects were extensively investigated by advanced analytical techniques with the aim of correlating structure and performance to the operative conditions. In particular, the adopted techniques allowed characterizing the sample from the micrometric to nanometric level, allowing an almost direct correlation among the structures (whose dimensions extend up to tens of microns) and the mechanical performance. The thermomechanical history experienced by the polymer chains was accounted for also by means of an advanced simulation software specifically developed for the injection molding process.

2. Materials and Methods

2.1. Injection Molding Test

The injection-molded samples were obtained adopting a commercial isotactic polypropylene, T30G, supplied by Basell (Ferrara, Italy). An accurate characterization of rheology and crystallization kinetics of the resin, in both quiescent and flow conditions, is reported elsewhere [34–36].

The injection molding tests were performed by a Negri Bossi 70 ton machine (Negri Bossi S.p.A., Cologno Monzese, Milano, Italy) with 220 °C melt injection temperature, 2.9 $cm^3 s^{-1}$ average volumetric flow rate (the cavity filling time was about 0.7 s), and 25 °C mold temperature. The holding stage was performed with a pressure of 720 bar held for 6 s. A thin heating device, whose description is reported elsewhere [37], was located below the cavity surface, covering an area 12.7 mm wide and 70 mm long, realizing a localized control of the cavity surface temperature during the process. A thin steel layer (100 μm thickness) covered the heating device, protecting it from the incoming melt; thus, it was possible to state that the heating device was just below the cavity surface. The heating device adopted was able to increase the cavity surface temperature with a rate of 100 °C/s, and, being very thin (220 μm thick), it allowed a fast decrease in the temperature after deactivation. Passive tests were performed adopting the heating device as an insulant; in this case, the whole mold was left at 25 °C during the process. At least three samples for each condition were analyzed. In Table 1, the conditions of cavity surface temperatures, T, and heating times, t_h adopted during the injection molding experiments are reported.

Table 1. Operating conditions adopted during the injection molding experiments Passive, A and B (T = temperature at the cavity surface; t_h = cavity surface heating time).

Test Name	T Cavity Surface (°C)	t_h (s)
Passive	25	-
A	150	0.7
B	150	6

Five pressure transducers (type 6171BB and 6190 BA, Kistler, Milan, Italy) were located along the flow path (namely P0, P1, P2, P3, and P4). In the position P2, a thermocouple (type T, Omega Engineering Ltd, Manchester, UK) was also located for recording the cavity surface temperature evolution. Figure 1 shows the cavity adopted for the injection molding experiments with the position of each pressure transducer along the flow path. The darker area represents the area where the heating device is located.

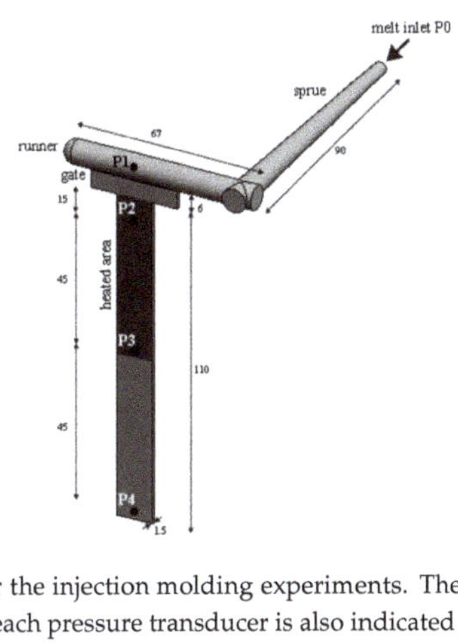

Figure 1. Cavity adopted for the injection molding experiments. The dimensions are referred to in millimeters. The position of each pressure transducer is also indicated along the flow direction.

2.2. Optical Microscopy

Thin slices (100 μm) were cut along the flow-thickness plane by Leica slit microtome (mod. 625, Leica Biosystem, Buccinasco, Milan, Italy) and observed by optical microscopy in polarized light by an Olympus BX51 microscope (Olympus Italia S.R.L., Segrate, Italy) with crossed polarizer-analyzer. Optical micrographs were taken with sample slices oriented at 45° with respect to the analyzer.

2.3. SAXS and WAXS

The slices observed by optical microscopy were also analyzed by small and wide-angle scattering and diffraction apparatus (SAXS and WAXS, Diamond Light Source, Harwell Science and Innovation Campus, Didcot, Oxfordshire, UK) for the analyses of the crystallinity and orientation distribution along the sample thickness.

Two-dimensional WAXD patterns were circularly averaged to generate plots of diffracted intensity as a function of the azimuthal angle q. The plot of the diffracted intensity was analyzed by a deconvolution procedure, as described elsewhere [38]. The full spectrum was considered as a superposition of reflections due to the presence of different crystalline fractions. A total of 11 reflections were considered: 6 for the α phase (q = 0.99, 1.20, 1.31, 1.55, 2.03, 2.97), 1 for the mesomorphic form (q = 1.06), 2 for the β form (q = 1.14, 1.50), 1 for the γ phase (q = 1.44), and 1 for the amorphous halo. Each reflection was described by a Lorentzian function, and the area below each reflection, A_i, was evaluated. The amount of each crystalline fraction, ξi, was evaluated as

$$\xi i = \frac{A_i}{\sum_i^{11} A_i} \tag{1}$$

The overall crystallinity, Xc, was evaluated as the sum of the contribution of the crystalline fractions (α, β, γ, and mesomorphic phase).

The error of the measurement was estimated as ±3% on the percentage of each phase.

The Herman's orientation factors for (040) and (110) planes were calculated for each spectrum. The Wilchinsky [39] equation allowed evaluating the orientation for each spectrum. The method of Wilchinsky was applied to derive the c-axis orientation using the (110)α and (040)α reflections and the angle of 72.5° between b-axis and the (110)α plane with:

$$< cos^2\sigma > = 1 - 0.901 < cos^2\varphi_{040} > - 1.099 < cos^2\varphi_{110} > \tag{2}$$

The flow direction was taken as the reference direction. For a set of hkl planes, the average orientation, expressed as $\langle \cos^2\varphi \rangle_{hkl}$, was mathematically calculated using the equation

$$\left\langle cos^2\varphi \right\rangle_{hkl} = \frac{\int_0^{\pi/2} I(\varphi)\, cos^2\varphi \sin\varphi}{\int_0^{\pi/2} I(\varphi) \sin\varphi} \tag{3}$$

with φ being the azimuthal angle, and $I(\varphi)$ being the scattered intensity along the angle φ. Herman's orientation function, f, was defined as

$$f = \frac{3\left\langle cos^2\varphi \right\rangle_{hkl} - 1}{2} \tag{4}$$

with f having a value of −0.5 with the normal of the reflection plane being perpendicular to the reference direction (φ = 90°), a value of 1 with the normal of the refection plane being parallel to the reference direction (φ = 0°), and a value of 0 with the orientation being random.

The lamellar size was also evaluated from SAXS analyses. In particular, the scattered intensity was obtained as a function of the scattering vector q, which is given by:

$$q = \frac{4\pi}{\lambda} sen(\theta) \tag{5}$$

where θ is half of the scattering angle, and the wavelength λ = 1.54 Å. The measured scattering intensity could be transposed into the 1D scattering intensity using Lorentz correction. If the electron density differences in one direction are known, the average lamellar thickness l_c could be obtained from [40]:

$$lc = \frac{2\pi}{q_{I_{MAX}}} X_c \tag{6}$$

where $q_{I_{MAX}}$ corresponds to the maximum of the Lorentz corrected scattering intensity, and X_c is the crystalline volume fraction (see Section 2.3 *SAXS and WAXS*) [41].

2.4. HarmoniX AFM

Specimens cut along the flow thickness plane and chemically etched, according to the procedure described in literature [42], were analyzed by atomic force microscopy equipped with HarmoniX tool (Multimode Dimension V coupled with Nanoscope V, Veeco, Santa Barbara, CA, USA) for the simultaneous characterization of the morphology and mechanical properties at the nanometrical level. In the literature, this method was found in a good agreement with data obtained from conventional mechanical analyses (namely dynamic mechanical analysis and micro–indentation) for molded produced with the iPP adopted in this work [43,44]. Tests were performed with HMX probe silicon cantilevers (Bruker, Billerica, MA, USA) with nominal radii of c.a. 10 nm. Cantilevers were calibrated using a standard polystyrene/low-density polyethylene target. The calibration of the tip was performed according to the procedure reported elsewhere [45]. The adopted vertical frequency was 44 kHz, and the torsional frequency was 1130 kHz. The cantilever vibration-free amplitude was of 750 mV, in air. The force level was modulated, adopting the amplitude setpoint. The amplitude set point was used for feedback control, as a reference value; ca. 60% of the free amplitude was selected. Imaging was performed with 0.5 Hz scan rates, considering 20 harmonics. Vertical and torsional motions were applied to the cantilever. The vertical movement allowed the acquisition of the topography; in other words, the morphology developed along the sample thickness. The torsional movement due to the tip-sample approach/withdraw generated force curves that allowed the evaluation of the elastic modulus. The Bruker NanoScope software, version 7.30, allowed the elaboration of the data in order to obtain maps of the measured properties. In particular, adhesion (force required to pull-off the tip from the surface), dissipation (area under the negative part of the force withdrawal curve) and elastic modulus maps (evaluated from force curves through the Derjaguin-Muller-Toporov model) were acquired considering a scanning area of 10 μm × 10 μm. Elastic modulus profiles along the sample thickness were obtained, using the software NanoScope Analysis version 1.80, averaging the elastic modulus distribution over the considered scanning area.

2.5. Simulation of the Process

A simulation software, developed at the University of Salerno for the injection molding process [19,46], was adopted for the description of the thermomechanical history undergone by the polymer.

3. Results

Figure 2a shows the temperature evolutions recorded at 15 mm downstream from the gate (position P2 of Figure 1) with 150 °C cavity surface temperature and two values of the cavity surface heating time, t_h = 0.7 s and 6 s (A and B). The temperature evolution recorded during the passive test was also reported for comparison. Figure 2b shows the pressure evolutions of the same tests recorded in the same position. In both graphs, the time t = 0 corresponded to the time at which the melt contacted the cavity in position P2.

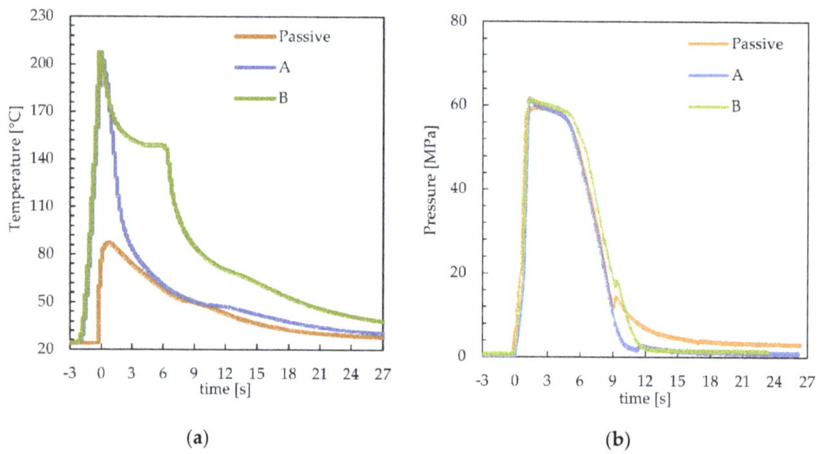

Figure 2. (**a**) Temperature and (**b**) pressure evolutions recorded during injection molding experiments performed with 150 °C cavity surface temperature and different heating times. The temperature and pressure evolution of the passive test were also reported.

For the cases A and B, a first temperature increase, from 25 °C to 150 °C (Figure 2a), was obtained activating the heating devices during the time, about 2 s, that the melt took to fill the sprue and the runner; afterward, the melt came in contact with the cavity surface, inducing an additional increase in the cavity surface temperature up to 210 °C. After the melt contacted with the cavity surface, the temperature started to decrease toward the value set for the heating device (150 °C) (kept active for 0.7 s for case A and 6 s for case B). At the heating device deactivation, the temperature started to decrease toward the temperature set for the whole mold, 25 °C. With reference to the passive case, the increase in temperature, due to the contact of the melt with the cavity surface, was significantly smaller than the cases in which the heating device was activated, also the temperature decrease rate was smaller with respect to those. In all the considered cases, the small inflection of the temperature evolutions, taking place within the temperature range 50–80 °C, was determined by the heat of crystallization that might partially overlap with the detachment of the polymer from the cavity surface (determined by the polymer shrinkage). The effect of the temperature evolution on the pressure evolution, in the same position where the temperature was recorded, is shown in Figure 2b. In the passive case, the pressure reached 23 MPa at the end of filling (0.7 s), and the increase in the cavity surface temperature induced a significant decrease of the filling pressure down to 16 MPa. For the passive and A cases, pressure evolutions in P2 after the filling stage were essentially the same, and both curves showed a change in the slope at 6 s due to the gate sealing [37]. When the cavity surface temperature was kept high for the whole holding stage (case B), the decrease in pressure delayed by some tenths of a second.

The evolution of the cavity surface temperature could be adopted to modulate the morphology and properties of the molded objects. Firstly, the morphology distribution along the sample thickness was strongly influenced by the evolution of the cavity surface temperature. Figure 3 shows optical micrographs of the slices cut in position P2, along the flow thickness plane, of the samples obtained with the operating conditions reported in Table 1.

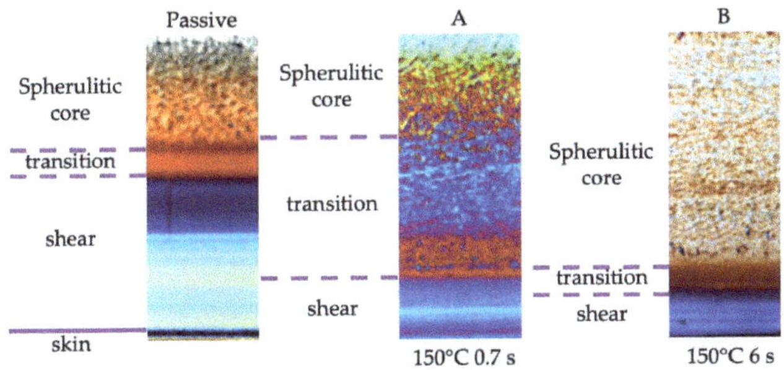

Figure 3. Optical micrographs of the molded samples obtained with different cavity surface temperature evolutions, from left to right: Passive, 150 °C kept for 0.7 s and 150 °C kept for 6 s (see Table 1 for the operating conditions). The slices were cut along the flow-thickness plane, and only half-thickness was shown.

Four regions could be detected along the thickness of the passive molding sample: the skin, close to the sample surface, an oriented layer named shear, a spherulitic *core* in the inner part of the sample, and a transitional region that could be identified between the spherulitic core and the shear layer. The transitional region was characterized by a continuous transition from the fibrillar to the spherulitic morphology. The samples A and B showed regions similar to those of the passive sample, with the exception of the skin. The cavity surface temperature evolution had a significant effect on the width of each layer. When the surface temperature was kept high, the shear layer's thickness decreased, and the transitional layer shifted toward the external part of the molding; when the heating time increased, the spherulitic core covered most of the sample thickness.

A deeper understanding of the morphology developed within the samples was achieved by the knowledge of distributions of both orientation and crystalline phase fractions within the molded samples. Both these characteristics were evaluated by WAXS analyses, performed by synchrotron acquisitions along the sample thickness of slices cut in the position P2, to which the morphologies reported in Figure 3 were referred.

Figure 4a shows the cross-section orientation distribution along the passive sample thickness. In particular, WAXS scattering images taken in different positions along the sample thickness (as shown on the topmost of the optical micrograph) and Herman's factors (as obtained from WAXS scattering images) were reported versus the normalized distance (normalized with respect to the whole sample thickness) from one sample surface.

The WAXS analyses showed that the orientation distribution, in terms of Herman's factor (see Figure 4a), was characterized by two maxima, located in the center of the shear layer regions ($d = 0.18$ and $d = 0.82$), by a minimum in the inner part of the sample and intermediate values at the sample surface. Figure 4b shows the crystalline phase fraction distributions (mesomorphic phase, α-phase, β-phase, and γ-phase) along the normalized distance from one sample surface. As expected for the adopted grade of iPP, the α-phase was the predominant phase, whereas only small fractions of other phases were detected. Interestingly, the distribution of the γ-phase was quite similar to the distribution of the orientation. This finding was consistent with the fact that the high orientation and high pressure induce the formation of the γ-phase [47,48]. Small fractions of the mesomorphic phase were concentrated at the sample surfaces. The β-phase fraction was found negligible over the whole sample thickness.

Figure 5 shows the distribution of mesomorphic, α, β, and γ-phases fractions evaluated from the WAXS analyses for the passive sample up to 0.1 normalized distance from one sample surface.

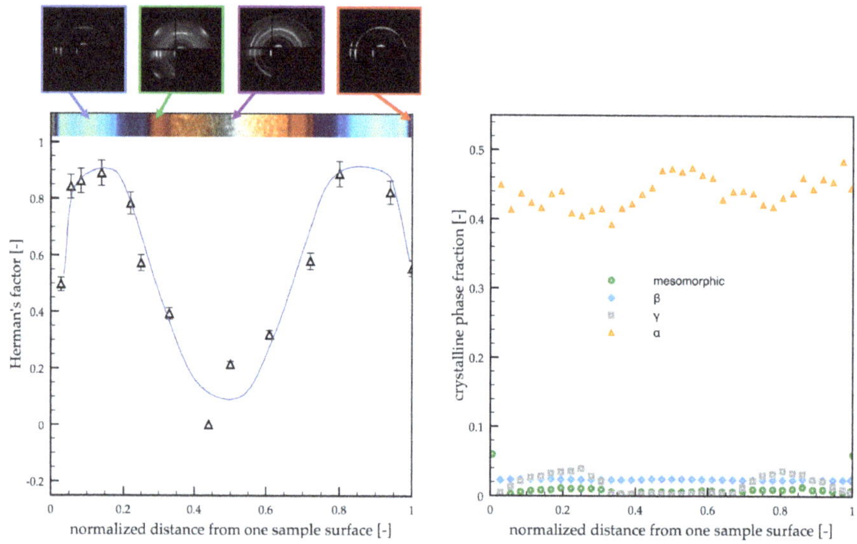

(**a**) (**b**)

Figure 4. (**a**) WAXS analyses of the passive molding in different positions along the sample thickness. Each analyzed position is indicated on the optical micrograph. The distribution of the orientation in terms of Herman's factor was also reported. (**b**) Distributions of mesomorphic, α, β, and γ-phase fractions along the sample thickness evaluated from the WAXS analyses.

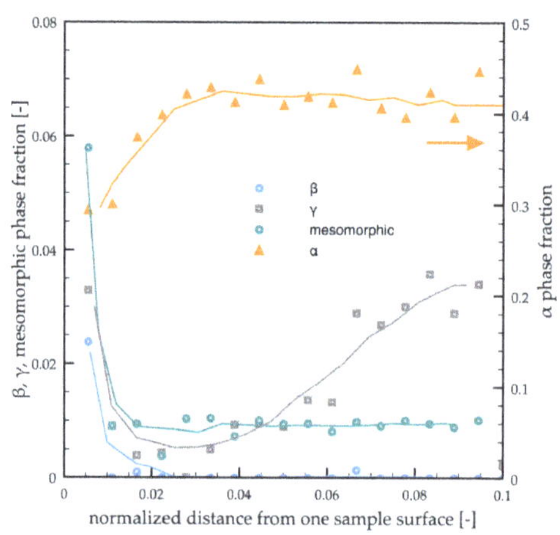

Figure 5. Distribution of mesomorphic, α, β, and γ-phases fractions evaluated from the WAXS analyses for the passive sample, up to 0.1 normalized distance from one sample surface.

The α-phase fraction increased from 0.30 to 0.45 in the interval 10–100 μm from sample surface (0.005–0.05 normalized distance). The mesomorphic phase fraction reached 0.06 in the first 10 μm from the sample surface, and it decreased down to 0.01 toward the inner parts of the sample (within the first 50 μm).

Similar analyses were performed on the samples obtained with a 150 °C cavity surface temperature, and 0.7 s heating time (case A in Table 1). Figure 6 shows the orientation distribution (in terms of

Herman's factor) along the sample thickness, the WAXS analyses in different positions along the sample thickness, and the optical micrograph of the sample along the flow thickness plane.

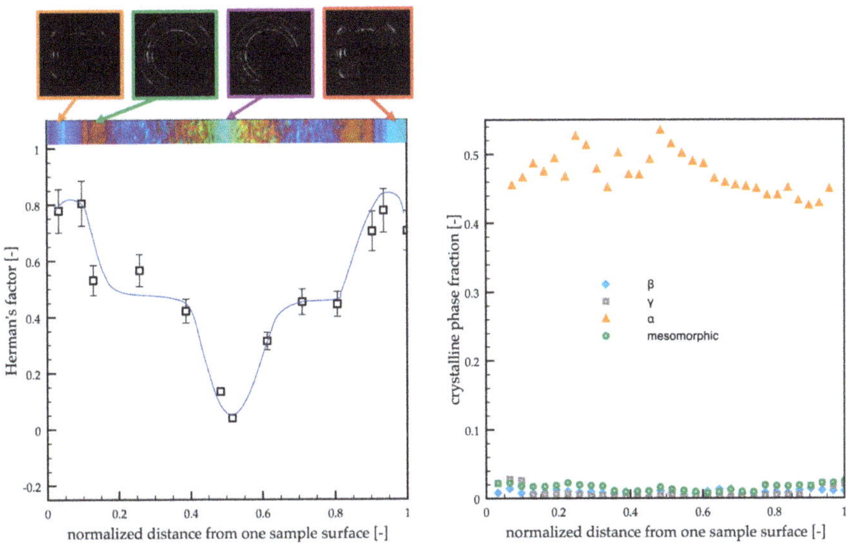

(a) (b)

Figure 6. (**a**) WAXS analyses of the A molding in different positions along the sample thickness. Each analyzed position is indicated on the optical micrograph. The distribution of the orientation in terms of Herman's factor was also reported. (**b**) Distributions of mesomorphic, α, β, and γ-phase fractions along the sample thickness evaluated from the WAXS analyses

The orientation of sample A was smaller with respect to the passive case. The maxima of orientation were located close to the sample surface, and the minimum was located in the sample core. In the regions between the core and the shear layer, the orientation was found almost constant and slightly lower with respect to the passive case. The distribution of the crystalline phase fractions (Figure 6b) showed again that the α-phase was the predominant one, only a small fraction of γ-phase was present, and it was mainly located close to the sample surface. The mesomorphic and the β-phase fractions were found negligible (smaller than 0.02).

Similar analyses were performed on the samples obtained with a 150 °C cavity surface temperature and 6 s heating time (case B, in Table 1). Figure 7 shows the orientation distribution (Herman's factor) and the crystalline phase fraction distributions along the sample thickness (Figure 7a,b, respectively). The maxima of orientation were located in very narrow regions close to the sample surface, and, differently from the passive sample, the reduction of the orientation at the sample surface was not observed. It is possible to observe a significant decrease in Herman's factor in the thin transition region. In the wide spherulitic core, the orientation was found essentially constant and comparable with the minimum values observed in the passive and A samples. The distribution of the crystalline phase fractions (Figure 7b) showed again that the predominant phase was the α-phase, only a small fraction of γ-phase was found and mainly located close to the sample surface, whereas neither the mesomorphic nor the β-phase were detected in a significant amount.

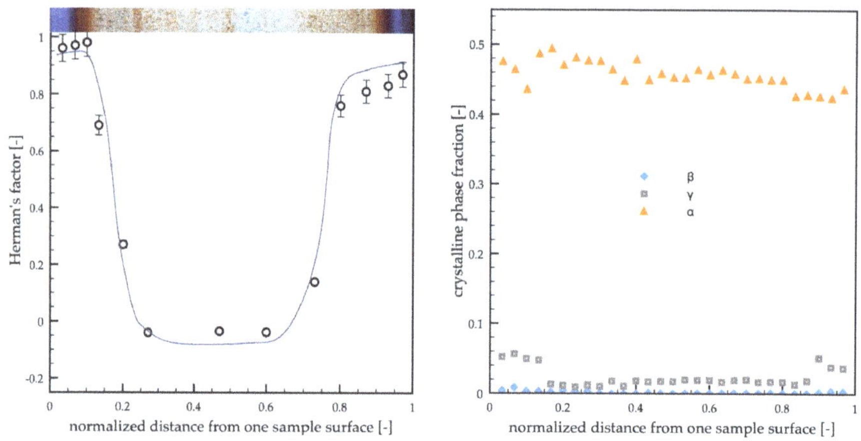

(a) (b)

Figure 7. (**a**) WAXS analyses of the B molding in different positions along the sample thickness. Each analyzed position is indicated on the optical micrograph. The distribution of the orientation in terms of Herman's factor was also reported. (**b**) Distributions of α, β, and γ-phase fractions along the sample thickness evaluated from the WAXS analyses.

4. Elastic Modulus Distributions

In order to assess the influence of the morphology and the crystalline structure on the properties of the molded samples, nanomechanical tests were performed. Figure 8 shows the AFM maps acquired on the passive sample along the thickness at two different distances from the sample surface: 150 μm (in the shear layer), 300 μm (in the transitional layer). Five maps were shown for each distance: the height that allowed analyzing the morphology of the sample, the phase, the adhesion, the dissipation, and the elastic modulus maps.

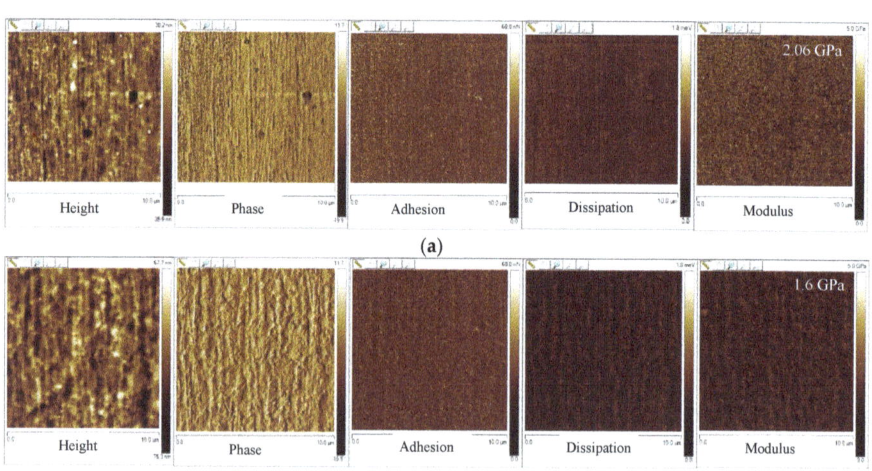

(a)

(b)

Figure 8. AFM maps of the height, phase, adhesion, dissipation, and elastic modulus, for the passive sample, acquired along the sample thickness at different distances from the surface: (**a**) 150 μm, namely the shear layer; (**b**) 300 μm, namely the transition layer.

The height maps showed the morphology of the passive sample in different areas along the thickness direction. At 150 µm distance from the surface, fibrillar morphology could be detected; at a higher distance from the sample surface (namely 300 µm), the distance between the fibrils increased, and the fibrils appeared to be thicker.

The phase-contrast microscopy, using AFM, is generally adopted to detect changes in composition or viscoelastic properties of polymers [49]. A viscoelastic material would strain periodically when subjected to periodic stress, such as the stress experienced during the AFM analysis in tapping mode. However, the strain response lagged the stress by a phase angle that is characteristic of the material. The presence of the regions with different values of phase lag could be attributed to the presence of regions characterized by elastic/viscoelastic behavior. The viscoelastic behavior could be attributed to the presence of parts with small crystalline fractions. This was confirmed by dissipation maps. The dissipation was the energy that was dissipated by the AFM tip during the withdraw from the sample. The dissipated energy and the adhesion were higher in the regions with a higher amount of viscous-viscoelastic material. Obviously, the elastic modulus was higher in the regions where the viscoelastic part was lower, i.e., in the shear layer.

Figure 9 shows the elastic modulus distributions obtained for the samples passive, A, and B.

The passive sample showed two maxima of the elastic modulus in the shear regions, the lowest values could be detected at the sample surface and in the transition regions, and the core was characterized by intermediate values of the elastic modulus. The samples A and B showed similar trends of the elastic modulus distribution. Furthermore, they were characterized by higher values of the elastic modulus with respect to the passive sample. The high values of the elastic modulus characteristic of the shear region moved toward the sample surface, accordingly with the reduction of the shear layer thickness due to the increase in the cavity surface temperature.

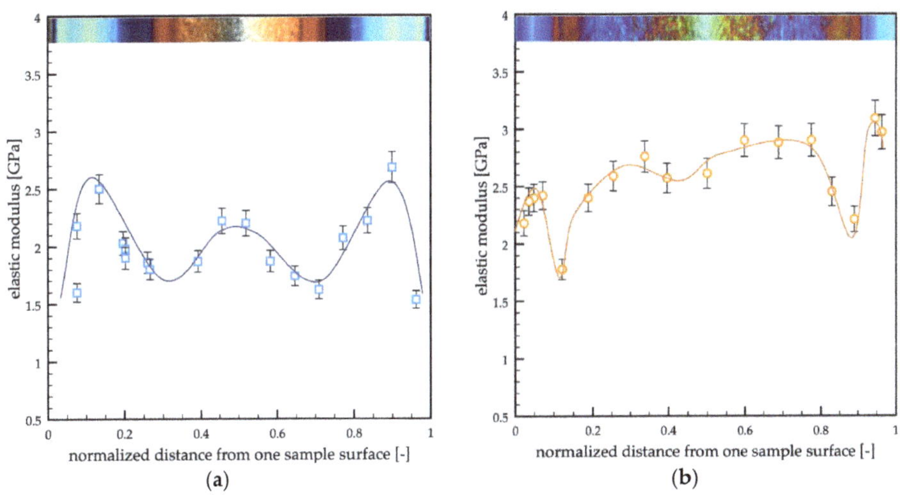

Figure 9. *Cont.*

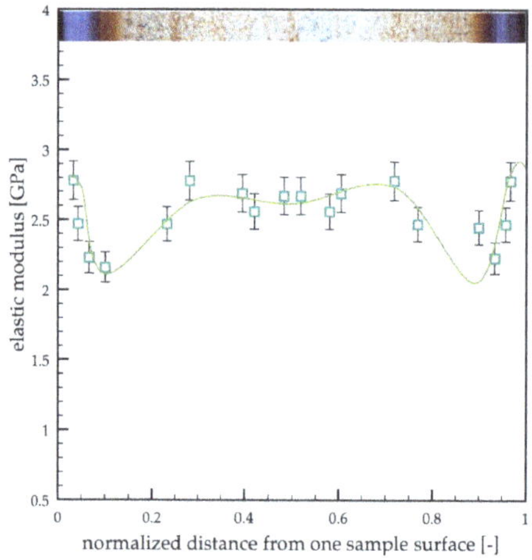

(c)

Figure 9. Elastic modulus distributions measured with the HarmoniX technique along the thickness of the (**a**) passive, (**b**) A, and (**c**) B sample.

5. Discussion

Figure 10 shows the WAXS analyses at different distances from the sample surface for the passive, A, and B cases. The orange borders delimit the regions with a high orientation level.

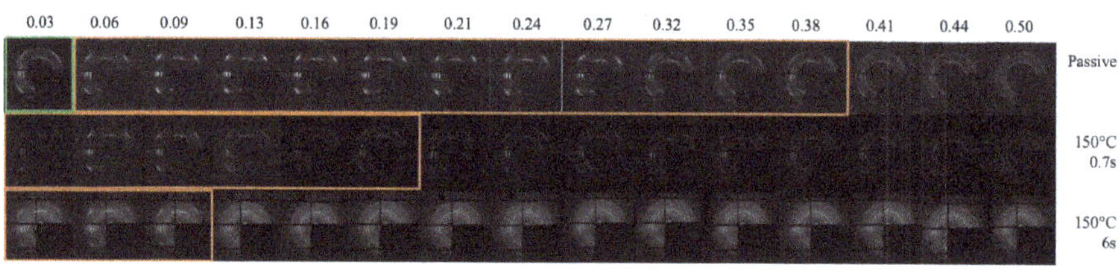

Figure 10. WAXS scattering distribution of the passive, A, and B moldings at different distances from one sample surface (the normalized distance is reported in the topmost of the figure). The higher orientation regions are delimited by orange borders, whereas the region with poor orientation (close to the surface of the passive sample) is delimited with the green borders.

The diffraction scattering intensity distribution consisted of five diffraction rings associated with different lattice planes of iPP, namely (110), (040), (130), (111), and (−131), from inner to outer circles, which were typical of the α-phase. An additional (300) lattice plane appeared in the transition layer, corresponding to the reflection of β-phase. γ-phase (117) diffraction ring was also observed. It was possible to detect traces of β-phase in sample A, where the β-phase ring was detectable up to 200 μm distance from the sample surface. The formation mechanism of β-phase was related to the shear flow, which could induce the formation of α-row structures. The surface of these row structures provided nucleation sites for the β-phase. The portion of β-phase depended on process conditions: the β-phase formed at low shear rates and low cooling rates, and its content in the moldings considered in this work was negligible. At a high shear rate, strong shear flow induced highly oriented structures, i.e., shish-kebabs, formed by α-phase [50–52]. The γ-phase ring was visible in the passive sample and in the sample A up to 450 μm from the sample surface. The γ-phase found in the regions close to the sample surface could be attributed to both the intense flow and to the fact that in these positions, the material experienced solidification under high pressure. The WAXS distribution showed on the left corner of the first line of Figure 10, highlighted with green color, was related to the skin region close to the surface of the passive sample; in this region, the mesomorphic phase fraction achieved 6%, whereas the mesomorphic phase fraction is negligible in the inner part of the sample. The observed distribution of the mesomorphic phase is consistent with the distribution reported in the literature [53]. The formation of the mesomorphic phase took place as a consequence of fast cooling experienced by the melt when contacted the cold surface of the mold [41,54,55]. As a further confirmation that the formation of the mesomorphic phase occurred under fast cooling conditions, the samples A and B, obtained with 150 °C cavity surface temperature, showed only a negligible fraction (smaller than 2%) of the mesomorphic phase since, in those conditions, the cooling rate was smaller than in the passive case.

The lamellar size distributions, *lc*, were evaluated from the SAXS analyses, and their results are shown in Figure 11. The information about the lamellar thickness allows understanding the structure developed at a smaller level with respect to the observed morphology (spherulites or fibrils).

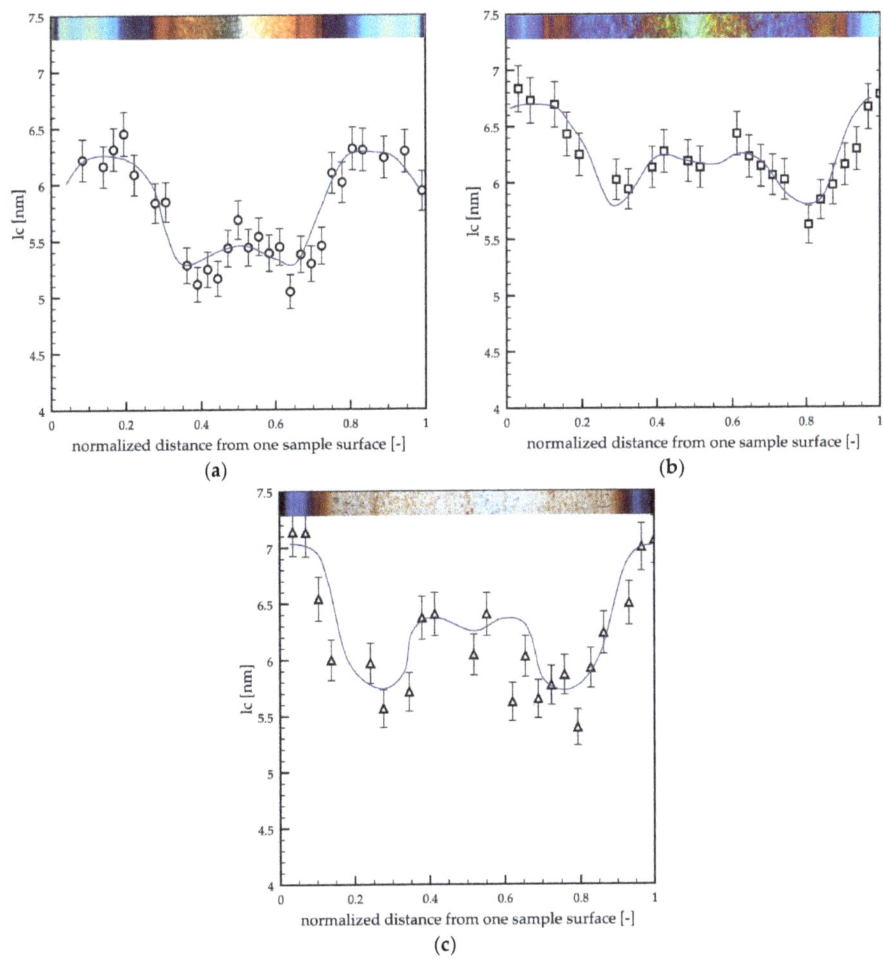

Figure 11. Lamellar size distributions along the thickness of the passive (**a**), A (**b**), and B (**c**) samples.

In the three cases analyzed in this work, the lamellar size, *lc*, was found to range between 5 and 7 nm. The passive sample showed slightly smaller values of the lamellar size with respect to the other two cases. In all the cases, *lc* was found larger at the sample surface, namely in the shear layers, and the values were almost constant in these regions. The width of the area in which the lamellar size was constant depended on the evolution of temperature on the cavity surface. In particular, the width of areas decreased with the increase in surface temperature and heating time. The transition regions were characterized by the smallest values of the lamellar size, whereas the core showed intermediate values.

The shear rate at the surface oriented (and stretched) the polymer chains, thus determining an entropy decrease within the system. Consequently, in the shear layer, the material crystallized at higher temperatures, with respect to the quiescent condition, and the formation of thicker lamellae was favored [56–59]. Sample B (obtained with high cavity surface temperature, 150 °C, kept for a long time, 6 s) showed a slight increase in the lamellar thickness, *lc*, especially in the core region, with respect to the passive and A samples. It seemed that the higher temperature, 150 °C, kept for a long time, 6 s, during the process, enhanced the structuring of the polymer chains, determining the increase of lamellar thicknesses.

A first attempt to find a relationship between structure and properties, especially mechanical property, was done by comparing elastic modulus distributions with the data obtained by SAXS and WAXS analyses for each molding condition. The WAXS analyses showed that the distributions of total crystallinity fractions were essentially homogeneous along the sample thickness, and the α-phase was the predominant crystalline phase, whereas only small fractions of the other phases were detectable. Furthermore, the increase in the cavity surface temperature induced a small increase in the crystallinity degree. Although small, the increase in the crystallinity degree could induce a small increase in the elastic modulus. Furthermore, the distributions of orientation, also obtained by WAXS analyses, appeared to be similar to the elastic modulus distribution. In particular, the maxima of the elastic modulus and orientation occurred at the same distance from the sample surface. The orientation was correlated with the stretch distribution that could be evaluated by the simulation software developed at the University of Salerno. Figure 12 shows the stretch distributions along the sample thickness calculated at the end of the process.

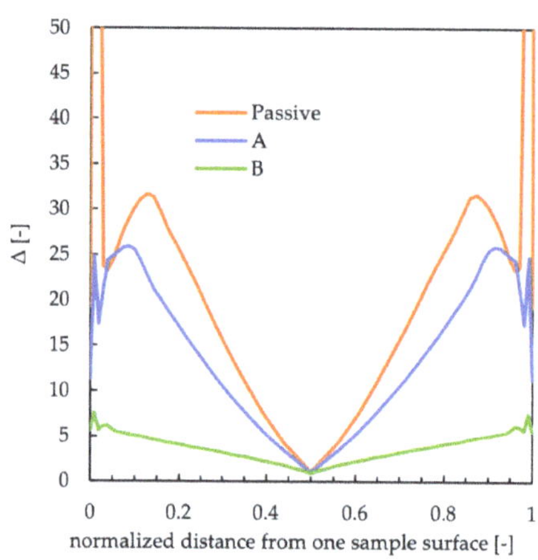

Figure 12. Distributions of the stretch, Δ, calculated by the simulation at the end of the process for the passive case and for cases A and B.

The stretch distribution of the passive sample was characterized by the presence of a maximum in the shear layer; beyond the shear layer, the stretch decreased down to the minimum in the core. The regions characterized by these values of stretch were composed of tightly packed fibrils. In the regions where the stretch decreased, a transitional layer characterized by a continuous transition from the fibrils to the spherulites was found. The formation of these areas was due to weak flows and slower cooling rates with respect to shear layers. The stretch distributions of the samples A and B are also shown in Figure 12. The stretch distribution of samples A and B showed a behavior similar to the one showed by the passive sample, with a maximum of the stretch in the shear layer. However, since the shear layer moved toward the sample surface, the maximum of the stretch was located closer to the sample surface with respect to the maximum of the passive sample. Sample B showed the lowest values of the stretch along the whole thickness according to the increase in the cavity surface temperature and heating time duration. These results confirmed the hypothesis previously introduced that the formation of the transitional areas was due to the weak flow, meaning small values of the stretch and slow cooling rates.

It seemed that in the shear layer, the stretch behavior was the main responsible for the high values of the elastic modulus. At the sample surface, even if the stretch assumed the highest value, the fast cooling rate, undergone by the molecules, prevented the molecular structuring, and thus lowest values of elastic modulus were found. The increase in the cavity surface temperature and heating time duration induced a decrease in the width of the regions with higher values of stretch and, thus, elastic modulus.

In the core region, elastic modulus distributions seemed to be not correlated to the orientation and stretch profiles. Indeed, in the core region, orientation and stretch assumed minimum values in all the considered cases, whereas the elastic modulus showed intermediate values. It seemed that the elastic modulus followed the behavior of the lamellar thickness that was correlated to the structuring during crystallization. Figure 13a,b shows the temperature evolution during the crystallization for two samples, passive and sample B, at several dimensionless distances from the sample surface, evaluated by the simulation software for injection molding developed at the University of Salerno.

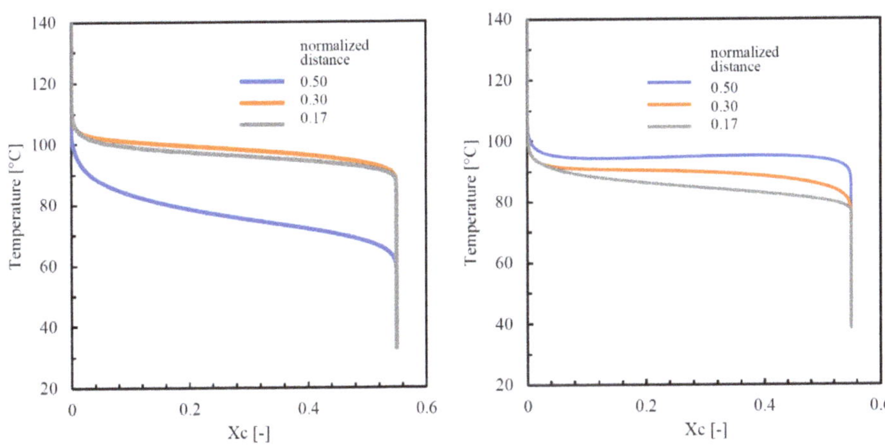

(**a**) (**b**)

Figure 13. The temperature during the crystallization for the samples passive (**a**) and B (**b**).

Figure 13a shows that, at the two distances (0.17 and 0.30 normalized distances) from the sample surface, the evolutions of the temperature during crystallization were very close, although the distances from the surfaces were different, this was because:

1. for given crystallization kinetics, crystallization temperature increased when cooling rate decreased (which happened as the distance from the surface increased);

2. for a given cooling rate, crystallization temperature decreased when the shear rate (and thus the molecular stretch) decreased (which happened as the distance from the surface increased).

As the distance from the surface increased, the effect of a decrease in cooling rate on the crystallization temperature was partially compensated by the decrease in shear rate, and thus crystallization took place at about the same temperature in the two positions 0.17 and 0.30.

In the core region, the crystallization started at lower temperatures, and the temperature kept on decreasing, from 100 °C to 60 °C, while the crystallization proceeded. Such a large difference in behavior with respect to the regions close to the sample surface was due to the fact that the crystallization in the core occurred after flow cessation, namely during the cooling, whereas in the regions close to the sample surface, the crystallization occurred during the filling and the first instants of the packing. Vice versa, when the temperature of the cavity surface was kept constant and high, at 150 °C, during filling and packing (test B, Figure 13b), most of the molecular stretch relaxed over the whole cross-section before crystallization and, consequently, the differences in crystallization temperature with the distances from the surface (0.17, 0.3, and 0.5) were essentially due to differences in the cooling rates. Thus, the sequence of the curves in Figure 13b followed the sequence of cooling rates, which decreased as the distance from the surface increased. Minor effects might be due to the residues of molecular stretch not yet completely relaxed. The evolution of temperature during crystallization was determined by a balance between the local cooling rate and the rate of heat generation due to the crystallization rate. In the core zone (the blue curve in Figure 13b), the crystallization temperature remained essentially constant and high; there (being far from the surface), the heat was lost under a slow rate, and the heat generation could slowly prevail, determining a small temperature increase. Thus, the time the polymer remained in the high range of the crystallization temperature, and available for structuring, was long in the whole cross-section, especially in the central zone. In previous work, the increase in the elastic modulus was attributed to the possibility of the polymer chains of structuring [44]. It appears that the longer is the time a molecule spends within the crystallization temperature range, the higher would be the structuring level and, therefore, the higher would be the elastic modulus. Obviously, this effect overlaps the effect of the molecular stretch. The higher values of the modulus in the core for the passive sample and the high values of the modulus along the whole thickness of sample B were consistent with the previous statements.

Summarizing, the increase in the elastic modulus in the core with respect to the transition areas was due to a decrease in the cooling rate that, on its turn, allowed better structuring of the molecules. The increase in the structuring level allowed an increase in the elastic modulus in the central part of the sample, which was responsible for the mechanical behavior of the whole sample.

6. Conclusions

Injection-molded samples of isotactic polypropylene were produced, modulating cavity surface temperature during the process, with the aim of tailoring the performance by the process. The samples were deeply characterized by adopting several techniques. Final properties distributions along sample thickness were obtained, adopting polarized optical microscopy, SAXS, WAXS, and HarmoniX atomic force microscopy. Crystallinity degree was found essentially constant along the sample thickness and close to 45% for all the analyzed samples; only a slightly increase of the crystallinity degree was found on increasing the cavity surface temperature. The α-phase was the predominant phase. Low fractions of mesomorphic phase were found only close to the sample surface, where it reached 6% for the sample obtained in conventional injection molding conditions, whereas, in the samples obtained with the cavity surface temperature modulation, the mesomorphic fractions were found to be negligible. The γ-phase fraction was found only in the shear layer, where it reached values smaller than 10%. The γ-phase fraction distribution was found quite similar to the distribution of the orientation. The β-phase fraction was found negligible for all the analyzed samples.

The morphology developed within the samples was accurately characterized by both optical microscopy and atomic force microscopy. Four areas were detected along the sample thickness: a skin, characterized by poorly oriented and structured features, a shear layer, a spherulitic core, and a transition layer between the shear and the spherulitic layers. The shear layer was composed of tightly packed fibrils. This area was characterized by the highest values of the orientation; Herman's factor was close to 1 for all the tests reported in this work. The elastic modulus found in the shear layer was the highest one, from 2.5 to 3.0 GPa, and it slightly increased with the cavity surface temperature and the heating time. The transition layer was characterized by fibrils that gradually evolve toward spherulitic structures. The orientation in this area gradually decreased from the values of the shear layer to the values of the spherulitic core. The elastic modulus showed the minimum values, from 1.7 to 2.2 GPa, in the transition. These values were close to the values found in the skin, 1.6 GPa. The spherulitic core was characterized by isotropic structures, the spherulites, which showed the smaller values of orientation, and Herman's factor was close to zero. The values of the elastic modulus in the spherulitic core were found to be intermediate between the values found in the shear and the values found in the transition layer. Interestingly, the values of the elastic modulus in the spherulitic core depended on the modulation of the cavity temperature; in particular, the higher were the cavity surface temperature and the heating time, the higher was the elastic modulus. The lamellar thickness, which was also evaluated from the SAXS analyses, showed distributions similar to the elastic modulus distribution. The maximum values (6.5–7.0 nm) were found in the shear, and the minimum values (5.0–5.5 nm) were found in the transition. The core showed intermediate values of lamellar thickness; furthermore, the lamellar thickness increased in the core with the cavity temperature and heating time.

The distributions of the mechanical properties and orientation were compared with the stretch the polymer chains experienced during the process, which was evaluated by the software developed at the University of Salerno for the injection molding process. In the shear and transitional layers, elastic modulus distribution showed a behavior similar to the orientation and stretch distributions. In the core layer, high values of elastic modulus seemed to be related to the increase in lamellar size. The interpretation of the aforementioned observation was given on the basis of the temperature evolution during the material crystallization and calculated using the aforementioned simulation software. In the core layer, even if the molecules did not experience high levels of orientation/stretch, they spent a long time within the crystallization temperature range, reaching a higher level of structuring and thus high values of elastic modulus. The skin layer, although characterized by the maximum level of stretch, showed small values of the elastic modulus. In this area, indeed, the fast cooling limited the structuring. Therefore, distributions of elastic modulus along the sample thickness seemed to be correctly interpreted by combining stretch and level of structuring.

Author Contributions: Formal analysis, V.S.; Investigation, S.L.; Methodology, G.T. and R.P.; Supervision, R.P.; Writing—original draft, S.L. and V.S.; Writing—review and editing, S.L., V.S., G.T., and R.P. All authors have read and agreed to the published version of the manuscript.

Funding: This research received no external funding.

Conflicts of Interest: The authors declare no conflict of interest.

References

1. Saipriya, I. Injection Molded Plastic Market Growth. Top Four Trends Driving Injection Molded Plastic Market over 2016–2023: U.S. to Remain a Prominent Revenue Contributor. Available online: https://gminsights.wordpress.com/2017/08/03/injection-molded-plastic-market/ (accessed on 28 January 2017).
2. Greener, J.; Wimberger-Friedl, R. *Precision Injection Molding*; Hanser Publications, Carl Hanser Verlag GmbH&Co.KG: Munich, Germany, 2006.
3. He, D.; Luo, Y.; Lu, S.; Liu, M.; Song, Y.; Lei, L. Microplastics in soils: Analytical methods, pollution characteristics and ecological risks. *TrAC Trends Anal. Chem.* **2018**, *109*, 163–172. [CrossRef]
4. Hahladakis, J.N.; Velis, C.A.; Weber, R.; Iacovidou, E.; Purnell, P. An overview of chemical additives present in plastics: Migration, release, fate and environmental impact during their use, disposal and recycling. *J. Hazard. Mater.* **2018**, *344*, 179–199. [CrossRef]
5. Cherif Lahimer, M.; Ayed, N.; Horriche, J.; Belgaied, S. Characterization of plastic packaging additives: Food contact, stability and toxicity. *Arab. J. Chem.* **2017**, *10*, S1938–S1954. [CrossRef]
6. Magalhães da Silva, S.P.; Lima, P.S.; Oliveira, J.M. Non-isothermal crystallization kinetics of cork-polymer composites for injection molding. *J. Appl. Polym. Sci.* **2016**, *133*. [CrossRef]
7. Schneider, D.; Hübner, C.; Bourbigot, S. New approach for the efficient attainment of flame retardancy using multi component injection molding. *AIP Conf. Proc.* **2019**, *2055*, 070007.

8. Furio, A.; Landi, G.; Altavilla, C.; Sofia, D.; Iannace, S.; Sorrentino, A.; Neitzert, H.C. Light irradiation tuning of surface wettability, optical, and electric properties of graphene oxide thin films. *Nanotechnology* **2017**, *28*, 054003. [CrossRef]
9. Eriksen, M.K.; Pivnenko, K.; Olsson, M.E.; Astrup, T.F. Contamination in plastic recycling: Influence of metals on the quality of reprocessed plastic. *Waste Manag.* **2018**, *79*, 595–606. [CrossRef] [PubMed]
10. Al-Salem, S.M.; Lettieri, P.; Baeyens, J. Recycling and recovery routes of plastic solid waste (PSW): A review. *Waste Manag.* **2009**, *29*, 2625–2643. [CrossRef] [PubMed]
11. Badia, J.D.; Ribes-Greus, A. Mechanical recycling of polylactide, upgrading trends and combination of valorization techniques. *Eur. Polym. J.* **2016**, *84*, 22–39. [CrossRef]
12. Zhou, Y.-G.; Turng, L.-S.; Shen, C.-Y. Morphological evolution and orientation development of stretched iPP films: Influence of draw ratio. *J. Polym. Sci. Part B Polym. Phys.* **2010**, *48*, 1223–1234. [CrossRef]
13. Lamberti, G. Flow-induced crystallization during isotactic polypropylene film casting. *Polym. Eng. Sci.* **2011**, *51*, 851–861. [CrossRef]
14. Wang, B.; Huang, H.-X.; Wang, Z.-Y. Process-induced phase and crystal morphologies in water-assisted injection molded polypropylene/polymeric β-nucleating agent blend parts. *Polym. Eng. Sci.* **2015**, *55*, 1698–1705. [CrossRef]
15. Ameli, A.; Kazemi, Y.; Wang, S.; Park, C.B.; Pötschke, P. Process-microstructure-electrical conductivity relationships in injection-molded polypropylene/carbon nanotube nanocomposite foams. *Compos. Part A Appl. Sci. Manuf.* **2017**, *96*, 28–36. [CrossRef]
16. Sun, X.; Kharbas, H.; Peng, J.; Turng, L.S. A novel method of producing lightweight microcellular injection molded parts with improved ductility and toughness. *Polymer (Guildf)* **2015**, *56*, 102–110. [CrossRef]
17. Guilong, W.; Guoqun, Z.; Huiping, L.; Yanjin, G. Analysis of thermal cycling efficiency and optimal design of heating/cooling systems for rapid heat cycle injection molding process. *Mater. Des.* **2010**, *31*, 3426–3441. [CrossRef]
18. Huang, C.T.; Hsien, I.S.; Tsai, C.H.; Chiou, Y.C.; Tang, C.C. The effects of various variotherm processes and their mechanisms on injection molding. *Int. Polym. Process.* **2011**, *26*, 265–274. [CrossRef]
19. Liparoti, S.; Speranza, V.; Pantani, R.; Titomanlio, G. Process Induced Morphology Development of Isotactic Polypropylene on the Basis of Molecular Stretch and Mechanical Work Evolutions. *Materials (Basel)* **2019**, *12*, 505. [CrossRef]
20. Liparoti, S.; Speranza, V.; Pantani, R. Replication of micro- and nanofeatures in injection molding of two PLA grades with rapid surface-temperature modulation. *Materials (Basel)* **2018**, *11*, 1442. [CrossRef]
21. Weng, C.; Wang, F.; Zhou, M.; Yang, D.; Jiang, B. Fabrication of hierarchical polymer surfaces with superhydrophobicity by injection molding from nature and function-oriented design. *Appl. Surf. Sci.* **2018**, *436*, 224–233. [CrossRef]
22. Baruffi, F.; Gülçür, M.; Calaon, M.; Romano, J.-M.; Penchev, P.; Dimov, S.; Whiteside, B.; Tosello, G. Correlating nano-scale surface replication accuracy and cavity temperature in micro-injection moulding using in-line process control and high-speed thermal imaging. *J. Manuf. Process.* **2019**, *47*, 367–381. [CrossRef]
23. Jiang, J.; Wang, S.; Sun, B.; Ma, S.; Zhang, J.; Li, Q.; Hu, G.-H. Effect of mold temperature on the structures and mechanical properties of micro-injection molded polypropylene. *Mater. Des.* **2015**, *88*, 245–251. [CrossRef]
24. Park, K.; Kim, Y.-S. Effect of Mold Temperature on Mechanical Properties of an Injection-Molded Part with Microfeatures. *J. Polym. Eng.* **2009**, *29*. [CrossRef]
25. Nagarajan, V.; Zhang, K.; Misra, M.; Mohanty, A.K. Overcoming the Fundamental Challenges in Improving the Impact Strength and Crystallinity of PLA Biocomposites: Influence of Nucleating Agent and Mold Temperature. *ACS Appl. Mater. Interfaces* **2015**, *7*, 11203–11214. [CrossRef] [PubMed]
26. Kuzmanović, M.; Delva, L.; Cardon, L.; Ragaert, K. The Effect of Injection Molding Temperature on the Morphology and Mechanical Properties of PP/PET Blends and Microfibrillar Composites. *Polymers (Basel)* **2016**, *8*, 355. [CrossRef] [PubMed]
27. Wang, W.; Zhao, G.; Guan, Y.; Wu, X.; Hui, Y. Effect of rapid heating cycle injection mold temperature on crystal structures, morphology of polypropylene and surface quality of plastic parts. *J. Polym. Res.* **2015**, *22*, 84. [CrossRef]
28. Shih, S.-Y.; Nian, S.-C.; Huang, M.-S. Comparison between single- and multiple-zone induction heating of largely curved mold surfaces. *Int. Commun. Heat Mass Transf.* **2016**, *75*, 24–35. [CrossRef]

29. Chen, S.-C.; Minh, P.S.; Chang, J.-A.; Huang, S.-W.; Huang, C.-H. Mold temperature control using high-frequency proximity effect induced heating. *Int. Commun. Heat Mass Transf.* **2012**, *39*, 216–223. [CrossRef]
30. Gao, S.; Qiu, Z.; Ma, Z.; Yang, Y. Development of high efficiency infrared-heating-assisted micro-injection molding for fabricating micro-needle array. *Int. J. Adv. Manuf. Technol.* **2017**, *92*, 831–838. [CrossRef]
31. De Meo, A.; De Santis, F.; Pantani, R. Dynamic local temperature control in micro-injection molding: Effects on poly(lactic acid) morphology. *Polym. Eng. Sci.* **2018**, *58*, 586–591. [CrossRef]
32. Xiao, C.-L.; Huang, H.-X.; Yang, X. Development and application of rapid thermal cycling molding with electric heating for improving surface quality of microcellular injection molded parts. *Appl. Therm. Eng.* **2016**, *100*, 478–489. [CrossRef]
33. The Nhan, P.; Do, T.T.; Anh Son, T.; Son Minh, P. Study on External Gas-Assisted Mold Temperature Control for Improving the Melt Flow Length of Thin Rib Products in the Injection Molding Process. *Adv. Polym. Technol.* **2019**, *2019*, 1–17. [CrossRef]
34. Pantani, R.; Nappo, V.; De Santis, F.; Titomanlio, G. Fibrillar morphology in shear-induced crystallization of polypropylene. *Macromol. Mater. Eng.* **2014**, *299*, 1465–1473. [CrossRef]
35. Pantani, R.; Speranza, V.; Titomanlio, G. Simultaneous morphological and rheological measurements on polypropylene: Effect of crystallinity on viscoelastic parameters. *J. Rheol. (N. Y. N. Y).* **2015**, *59*, 377–390. [CrossRef]
36. De Santis, F.; Pantani, R.; Titomanlio, G. Effect of shear flow on spherulitic growth and nucleation rates of polypropylene. *Polymer (Guildf)* **2016**, *90*, 102–110. [CrossRef]
37. Liparoti, S.; Sorrentino, A.; Titomanlio, G. Temperature and pressure evolution in fast heat cycle injection molding. *Mater. Manuf. Process.* **2018**, 1–9. [CrossRef]
38. Murthy, N.S.; Minor, H. General procedure for evaluating amorphous scattering and crystallinity from X-ray diffraction scans of semicrystalline polymers. *Polymer (Guildf)* **1990**, *31*, 996–1002. [CrossRef]
39. Wilchinsky, Z.W. Orientation in cold-rolled polypropylene. *J. Appl. Polym. Sci.* **1963**, *7*, 923–933. [CrossRef]
40. Caelers, H.J.M.; Govaert, L.E.; Peters, G.W.M. The prediction of mechanical performance of isotactic polypropylene on the basis of processing conditions. *Polymer (Guildf)* **2016**, *83*, 116–128. [CrossRef]
41. Liparoti, S.; Sorrentino, A.; Guzman, G.; Cakmak, M.; Titomanlio, G. Fast mold surface temperature evolution: Relevance of asymmetric surface heating for morphology of iPP molded samples. *RSC Adv.* **2015**, *5*, 36434–36448. [CrossRef]
42. White, H.M.; Bassett, D.C. On Row Structures, Secondary Nucleation and Continuity in Alpha- Polypropylene. *Polymer (Guildf)* **1998**, *39*, 3211–3219. [CrossRef]
43. Liparoti, S.; Sorrentino, A.; Speranza, V.; Titomanlio, G. Multiscale mechanical characterization of iPP injection molded samples. *Eur. Polym. J.* **2017**, *90*, 79–91. [CrossRef]
44. Liparoti, S.; Sorrentino, A.; Speranza, V. Micromechanical Characterization of Complex Polypropylene Morphologies by HarmoniX AFM. *Int. J. Polym. Sci.* **2017**, *2017*, 9037127. [CrossRef]
45. Liparoti, S.; Speranza, V.; Sorrentino, A.; Titomanlio, G. Mechanical Properties Distribution within Polypropylene Injection Molded Samples: Effect of Mold Temperature under Uneven Thermal Conditions. *Polymers (Basel)* **2017**, *9*, 585. [CrossRef]
46. Pantani, R.; Speranza, V.; Titomanlio, G. Effect of flow-induced crystallization on the distribution of spherulite dimensions along cross section of injection molded parts. *Eur. Polym. J.* **2017**, *97*, 220–229. [CrossRef]
47. Kalay, G.; Zhong, Z.; Allan, P.; Bevis, M.J. The occurrence of the γ-phase in injection moulded polypropylene in relation to the processing conditions. *Polymer (Guildf)* **1996**, *37*, 2077–2085. [CrossRef]
48. Anczykowski, B.; Gotsmann, B.; Fuchs, H.; Cleveland, J.P.; Elings, V.B. How to measure energy dissipation in dynamic mode atomic force microscopy. *Appl. Surf. Sci.* **1999**, *140*, 376–382. [CrossRef]
49. Scott, W. Use of phase imaging in atomic force microscopy for measurement of viscoelastic contrast in polymer nanocomposites and molecularly thick lubricant films. *Ultramicroscopy* **2003**, *97*, 151–169. [CrossRef]
50. Speranza, V.; Liparoti, S.; Pantani, R.; Titomanlio, G. Hierarchical structure of iPP during injection molding process with fast mold temperature evolution. *Mater. Rev.* **2018**, *12*, 424. [CrossRef]
51. Mi, D.; Xia, C.; Jin, M.; Wang, F.; Shen, K.; Zhang, J. Quantification of the Effect of Shish-Kebab Structure on the Mechanical Properties of Polypropylene Samples by Controlling Shear Layer Thickness. *Macromolecules* **2016**, *49*, 4571–4578. [CrossRef]

52. Wang, K.; Chen, F.; Zhang, Q.; Fu, Q. Shish-kebab of polyolefin by "melt manipulation" strategy in injection-molding: A convenience pathway from fundament to application. *Polymer (Guildf)* **2008**, *49*, 4745–4755. [CrossRef]
53. Pantani, R.; Coccorullo, I.; Speranza, V.; Titomanlio, G. Modeling of morphology evolution in the injection molding process of thermoplastic polymers. *Prog. Polym. Sci.* **2005**, *30*, 1185–1222. [CrossRef]
54. De Rosa, C.; Auriemma, F.; Galotto, N.G.; Di Girolamo, R. Mesomorphic form of isotactic polypropylene in stereodefective polypropylene: Solid mesophase or liquid-crystal like structure. *Polymer (Guildf)* **2012**, *53*, 2422–2428. [CrossRef]
55. Liparoti, S.; Titomanlio, G.; Sorrentino, A. Analysis of asymmetric morphology evolutions in iPP molded samples induced by uneven temperature field. *AIChE J.* **2016**, *62*, 2699–2712. [CrossRef]
56. Housmans, J.-W.; Gahleitner, M.; Peters, G.W.M.; Meijer, H.E.H. Structure–property relations in molded, nucleated isotactic polypropylene. *Polymer (Guildf)* **2009**, *50*, 2304–2319. [CrossRef]
57. Zhong, G.-J.; Li, Z.-M.; Li, L.-B.; Mendes, E. Crystalline morphology of isotactic polypropylene (iPP) in injection molded poly(ethylene terephthalate) (PET)/iPP microfibrillar blends. *Polymer (Guildf)* **2007**, *48*, 1729–1740. [CrossRef]
58. Marand, H.; Xu, J.; Srinivas, S. Determination of the Equilibrium Melting Temperature of Polymer Crystals: Linear and Nonlinear Hoffman−Weeks Extrapolations. *Macromolecules* **1998**, *31*, 8219–8229. [CrossRef]
59. Ryan, A.J.; Stanford, J.L.; Bras, W.; Nye, T.M.W. A synchrotron X-ray study of melting and recrystallization in isotactic polypropylene. *Polymer (Guildf)* **1997**, *38*, 759–768. [CrossRef]

polymers

MDPI

Article

Improvement of Peptide Affinity and Stability by Complexing to Cyclodextrin-Grafted Ammonium Chitosan

Andrea Cesari [1,†], Alessandra Recchimurzo [1,†], Angela Fabiano [2], Federica Balzano [1,*], Nicolò Rossi [2], Chiara Migone [2], Gloria Uccello-Barretta [1], Ylenia Zambito [2] and Anna Maria Piras [2,*]

1 Department of Chemistry and Industrial Chemistry, University of Pisa, via G. Moruzzi 13, 56126 Pisa, Italy; andrea.cesari@outlook.com (A.C.); alessandra.recchimurzo@phd.unipi.it (A.R.); gloria.uccello.barretta@unipi.it (G.U.-B.)

2 Department of Pharmacy, University of Pisa, via Bonanno Pisano 6, 56126 Pisa, Italy; angela.fabiano@unipi.it (A.F.); rossiniko94@hotmail.it (N.R.); chiaramigone@gmail.com (C.M.); ylenia.zambito@unipi.it (Y.Z.)

* Correspondence: federica.balzano@unipi.it (F.B.); anna.piras@unipi.it (A.M.P.)

† The authors contributed equally to this work.

Received: 23 January 2020; Accepted: 17 February 2020; Published: 19 February 2020

Abstract: Cyclodextrin-grafted polymers are attractive biomaterials that could bring together the host–guest complexing capability of pristine cyclodextrin and the pharmaceutical features of the polymeric backbone. The present paper is aimed at characterizing the potential application of ammonium–chitosan grafted with 2-methyl-β-cyclodextrin (N^+-rCh-MCD) as the functional macromolecular complexing agent for the oral administration of the neuropeptide dalargin (DAL). Specific NMR characterization procedures, along with UV and fluorescence techniques, as well as biological in vitro assessments have been performed. The results indicate that N^+-rCh-MCD forms water-soluble complexes with DAL, with a prevalent involvement of Tyr or Phe over Leu and Ala residues. The association constant of DAL with the polymeric derivative is one order of magnitude higher than that with the pristine cyclodextrin (K_a: 2600 M^{-1} and 120 M^{-1}, respectively). Additionally, N^+-rCh-MCD shields DAL from enzymatic degradation in gastrointestinal in vitro models with a three-fold time delay, suggesting a future pharmaceutical exploitation of the polymeric derivative. Therefore, the greater affinity of N^+-rCh-MCD for DAL and its protective effect against enzymatic hydrolysis can be attributed to the synergistic cooperation between cyclodextrin and the polymer, which is realized only when the former is covalently linked to the latter.

Keywords: chitosan; cyclodextrin; macromolecular complex; dalargin; enkephalin; chymotrypsin; NMR; oral peptide delivery; peptide stability

1. Introduction

Natural or synthetic polymers are widely used in pharmaceutical technology as drug delivery systems that are capable of protecting the active ingredients from possible degradation processes along the pathway to the target [1]. For this purpose, covalent conjugation or noncovalent supramolecular interactions between the drug and the controlled release platform can be exploited [2]. Moreover, polymeric materials can affect the absorption rate of active ingredients by acting on the permeability of epithelial barriers [3]. For these reasons, polysaccharides are suitable candidates, taking into account their biocompatibility, biodegradability, and relevant mucoadhesive properties. After exerting their role as drug transporters, polysaccharides are easily degraded into well-tolerated molecules by the metabolism [4]. In order to further improve their desirable properties such as drug affinity and mucoadhesivity, it is possible to modulate the physical and chemical properties of polysaccharides by modification of the constituent glycosidic units. Among amino-polysaccharides, which are made by amino-sugars monomers, chitosan (Ch) is facing fast-growing interest in the pharmaceutical industry [5,6]. It can be easily obtained by the partial deacetylation of the abundant polysaccharide chitin. Due to the high molecular weight of chitosan, and therefore a very high viscosity that hinders its use in biological applications, a first convenient modification is the degradation of chitosan, producing reduced molecular weight chitosan (rCh), which is carried out by chemical or enzymatic reactions [7]. Derivatizations of chitosan at –OH and/or $-NH_2$ groups, such as in *O*- and *N*-carboxymethyl chitosan, 6-*O*-sulfate chitosan, and *N*-methylenephosphonic chitosan have been reported [8]. The quaternarization of $-NH_2$ is one of the most common and relatively simple modifications, leading to a cationic polyelectrolyte that is soluble in water in a wide range of pH. Ammonium chitosan (N^+-rCh) has also shown enhanced mucoadhesivity provided by the interactions with negative charges located on the mucous membrane [9–11]. To further improve its mucoadhesion properties, ammonium chitosan modified with free or *S*-protected thiol groups have been proposed [12–16]. Such derivatives are capable of forming covalent bonds with the cysteine-rich subdomains of mucin, which is a key component of the mucosal layers. The grafting of cyclodextrins into the chitosan backbone constitutes another promising modification, which has not yet been fully explored [17].

Cyclodextrins are cyclic oligosaccharides that are endowed with cavities that are able to encapsulate specific guests, on the basis of a convenient size fitting. The covalent conjugation of cyclodextrins seems to play a specific role in the control of releasing and protecting active ingredients, compared with the simple physical mixture of cyclodextrin/chitosan [18]. Properties of cyclodextrin–chitosan conjugates can also be affected by modification of the macrocycle counterpart, as in the use of methoxy- or ethoxy-β-cyclodextrins derivatives, the water solubility of which is remarkably enhanced with respect to parent native cyclodextrin [19,20]. In particular, a degree of substitution (DS) of 0.5 (methyl per glucose unit, i.e., an average of one methyl per two units) guarantees the maximum improvement of water solubility [19]. Aliphatic groups make the internal surface of the toroidal cavity more water-repellent, favoring the inclusion of lipophilic molecules. Methyl-cyclodextrins grafted on chitosan are hence responsible for a considerable acceleration of dissolution processes of poorly water-soluble drugs [21].

The present work is aimed to determine the role of the cyclodextrin conjugation in ammonium-chitosan grafted with 2-methyl-β-cyclodextrin (N^+-rCh-MCD, Figure 1a) toward its binding ability to dalargin (DAL, Figure 1b), in comparison with the physical mixture of parent ammonium-chitosan and cyclodextrin (N^+-rCh/MCD). The hexapeptide dalargin has demonstrated efficacy in the treatment of ischemia, arrhythmias, gastric ulcer, and alcohol withdrawal syndrome [22–26]. However, its administration is limited to i.m./e.v. injections, since it undergoes fast enzymatic degradation and a loss of efficacy within the gastrointestinal tract, similar to most of the therapeutically active peptides. Several approaches to protect peptides from enzymatic degradation have been proposed. The most used one is the loading of peptides in nanosystems such as nanoparticles, liposomes, or self-emulsifying drug delivery systems [27–29]. However, the realization of these nanosystems is very complicated, and their industrial application may be limited by high production costs. For this reason, we have thought of a much simpler system consisting of N^+-rCh-MCD and DAL, which, in the aqueous physiological fluids, could give rise to the formation of a soluble macromolecular complex, in which N^+-rCh-MCD could protect DAL from enzymatic degradation. Indeed, it is known that cyclodextrins have the ability to stabilize peptides, thus protecting them from enzymatic degradation [30]. Therefore, the protective role of the cyclodextrin in the polymer against the degradation by α-chymotrypsin (CHT) has been evaluated with respect to the separate precursors. NMR proton selective relaxation rate measurements have been exploited as a non-invasive investigation tool in order to compare dalargin to polymer affinities. UV, fluorescence techniques, and biological in vitro assessments have been also performed.

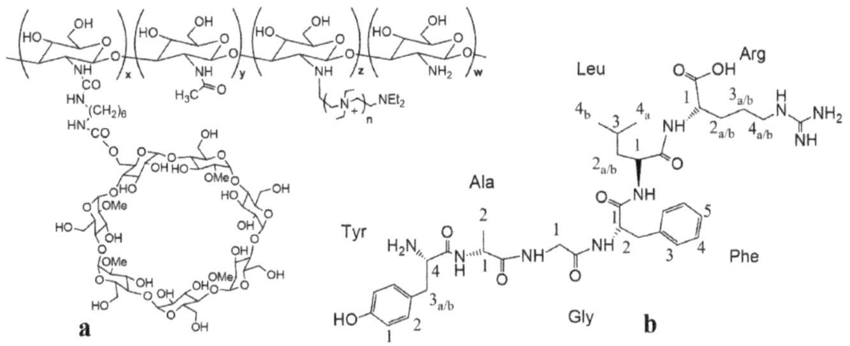

Figure 1. (**a**) 2-methyl-β-cyclodextrin conjugates of ammonium chitosan (N^+-rCh-MCD) and (**b**) dalargin (DAL).

2. Materials and Methods

2.1. Materials

Chitosan was purchased from Faravelli (molecular weight = 300 kDa) (Milan, Italy). Sodium nitrite, 2-diethylaminoethyl chloride hydrochloride (DEAE-Cl·HCl), α-chymotrypsin (CHT), dimethyl sulfoxide (DMSO), 1,6-hexamethylene diisocyanate (HMDI), and triethylamine (TEA) were purchased from Sigma-Aldrich (St. Louis, MO, USA); dalargin ((D-Ala2)-Leu-Enkephalin-Arg) was purchased from Bachem (Weil am Rhein, Germany); 2-methyl-β-cyclodextrin (DS = 0.5) was purchased from Roquette, Lestrem, France); standard pullulans (642 ÷ 6.10 kg/mol) were purchased from the American Polymers Standard Corporation (8680 Tyler Blvd., Mentor, OH, USA), and standard RC Dialysis membranes (MWCO 12500) were purchased from Spectra/Por® (Sigma, Los Angeles, CA, USA). NMR deuterated solvents, D_2O e NaOD (30%), were purchased from Deutero GmbH (Kastellaun, Germany).

HMDI and DMSO were distilled under reduced pressure (65 °C/0.2 mbar and 24 °C/0.2 mbar, respectively); TEA was refluxed over calcium hydride and distilled before use. MCD and DEAE-Cl·HCl were anhydrified under vacuum at 37 °C.

Cell line Caco-2 was purchased from the American Type Culture Collection (ATCC® HTB-37™) LGC standards, Milan, Italy) and propagated as indicated by the supplier; Minimum Essential Medium (MEM), non-essential amino acid, 0.01 M pH 7.4 Dulbecco's Phosphate Buffer (DPBS), bovine fetal serum (BCS), glutamine, antibiotics (penicillin/streptomycin), and Hanks balanced solution were purchased from Sigma, Milan, Italy; antimycotic was supplied from Invivogen, San Diego, CA, USA. Cell proliferation reagent WST-1 was provided by Roche diagnostic, Milan, Italy.

2.2. Analytical Methods

GPC (Gel permeation chromatography) and HPLC analyses were carried out with a Perkin-Elmer instrument (Perkin-Elmer, Waltham, MA, USA), consisting of a Series 200 LC-290 AT pump, Rheodyne injector with a loop of 20 μL, UV-VIS SPD–6 AV detector, and Turbochrom Navigator software (Turbochrom 6.1, Perkin-Elmer, Waltham, MA, USA). For GPC analyses, the standard pullulans were employed; the stationary phase was an X-stream H_2O Mixed Bed 50 × 4.6 mm column, the mobile phase was acetate buffer 0.5 M (pH = 2.9), and the wavelength of the detector was set to 237 nm. For HPLC, the suitable mobile phase result was H_2O:CH_3CN = 70:30, the column was a C18 Aquapore OD-300 (Perkin-Elmer, Waltham, MA, USA) 7 μm 4.6 × 250 mm, and the wavelength of the detector was set to 227 nm.

NMR analyses were conducted on Varian INOVA600 spectrometer (Varian inc, Palo Alto, CA, USA) operating at 600 MHz for ^1H and at 150 MHz for ^{13}C. The samples were analyzed in a solution of D_2O or phosphate buffer (50 mM, pH = 6.8). The temperature was controlled through Varian control unit (accuracy ±0.1 °C). 2D NMR spectra were obtained using standard sequences and the minimum spectral width in both dimensions. Two-dimensional (2D) gCOSY (gradient COrrelated SpectroscopY) maps were recorded with a relaxation interval of 1–2 s, 256–512 increments with 8–32 transients each, and 2K data points. TOCSY (TOtal Correlation SpectroscopY) maps were recorded by using a relaxation delay of 1 s, 256 increments of 4 transients, each with 2K points and a mixing time of 80 ms. 2D ROESY (Rotating-frame Overhauser Effect SpectroscopY) maps were recorded with a mixing time of 50–200 ms, relaxation delay of 1 s, 256–512 increments with 16–32 transients, and 2K data points. gHSQC (gradient Heteronuclear Single Quantum Coherence) and gHMBC (gradient Heteronuclear Multiple Bond Coherence) maps were obtained with 256 or 512 increments of 128–256 transients each and relaxation delay of 1.2 s. The gHMBC experiment was optimized for long-range J-coupling ^{1}H-^{13}C of 8 Hz. The mono-selective and/or bi-selective relaxation rates were measured in the approximation of the initial rate [31] through a 180° pulse set to one or two selective frequencies. After each incremental delay, a 90° non-selective pulse was applied to follow the exponential decay of the selected frequencies.

2.3. *Preparation of Quaternary Methyl-β-Cyclodextrin Ammonium–Chitosan Conjugate*

High molecular weight commercial chitosan (Ch) was depolymerized following a procedure previously reported [32]. Briefly, 5 g of Ch were dissolved in 2% acetic acid solution and reacted with $NaNO_2$ (0.1 M, 12.5 mL) at 30 °C for 4 h. Afterwards, the solution pH was adjusted to pH 7.8 by using NaOH 10 M, and the obtained reduced molecular weight chitosan (rCh) was recovered by filtration, washed with water, and vacuum dried. The rCh was alkylated on -NH_2 groups via a reaction of quaternarization optimized by Zambito et al. [2,11]: 2-methyl-β-cyclodextrin (MCD) was grafted on N^+-rCh, following a procedure [21] articulated into several steps: 350 mg of N^+-rCh were solubilized in 7 mL of DMSO under inert atmosphere (N_2) overnight. Sequentially, N^+-rCh (350 mg) and TEA (0.056 mL, 0.40 mol) were added to a solution of HMDI (0.7 mL, 4.4 mmol) in DMSO (8 mL), and the mixture was placed in an oil bath at 70 °C, under inert atmosphere. The reaction was maintained for 2 h and controlled steadily, until it became turbid. Thereafter, the polymer was precipitated in cold diethyl ether (0 °C) and washed thrice, eliminating the supernatant, and then redissolved in 3 mL of DMSO. MCD (2.38 g in 7 mL of DMSO) and 0.140 mL of TEA were directly used to dissolve the recovered ammonium–chitosan grafted with HMDI spacers; afterwards, the mixture was heated at 70 °C for 3 h. Finally, the solution was quenched by adding it to deionized water at 80 °C, drop by drop, under stirring, showing an initial effervescence. After 1 h, the mixture was dialyzed with deionized water for 3 days. The insoluble portion of product was removed through centrifugation (Sorvall, MTX, 20.000 rpm, 30 min, 4 °C, thermo Fisher Scientific In, Waltham, MA, USA) and the supernatant portion was dried, giving the cyclodextrin–polymer conjugate (N^+-rCh-MCD).

2.4. *Characterization of DAL Inclusion Complexes*

2.4.1. Complexes Stoichiometry (Job's Plot)

The complex stoichiometry of MCD and DAL, and polymers and DAL, were evaluated by the continuous variation method (Job's plot). Series of samples were prepared keeping constant the total molar concentration (275 μM) of the binding partners (DAL and MCD), but varying their mole fraction between 0 and 1. Similarly, complex stoichiometry was evaluated for MCD in the presence of N^+-rCh and for the MCD-grafted polymer (N^+-rCh-MCD); the dilution specifications are reported in Table S1 of the supplementary material). The absorbance of the solution sets was acquired by UV-VIS (Perkin Elmer Precisely lambda-25) at time 0 min, 1 h, and 24 h from solution mixing, with solutions kept at room temperature. The second derivative calculation was performed in order to lessen the polymers' nonspecific contribution. Job's plot was built on the difference between 275 and 316 nm absorbance values (ΔA). The plot displays the variation of $\Delta A \cdot \chi$ vs χ with $\chi = [DAL]/([DAL] + [MCD])$. The abscissa value corresponding to the curve maximum represents the complex stoichiometry as 1:2, 1:1, or 2:1 with $\chi = 0.33$, $\chi = 0.5$, or $\chi = 0.66$, respectively.

2.4.2. Evaluation of Complex Association Constant (Benesi–Hildebrand Method)

Complex association constant (K_a) was determined by spectrometric titration based on the Benesi–Hildebrand method [33] and performed both by UV-VIS and fluorescence spectrometry. Two sets of dilutions were prepared (Tables S2–S5, Supplementary material). In the case of UV spectrometry, 275 μM DAL was kept, whereas MCD concentrations varied, either as MCD alone or as conjugated (N^+-rCh-MCD). The solutions were prepared and analyzed after being equilibrated for 1 h at room temperature. UV spectra in the range 220–325 nm were collected, and the intensity of absorbance at 282 nm was normalized on the correspondent value at 247 nm. Diversely, for fluorescence spectroscopy evaluations, DAL concentration was kept at 70 μM, and samples were analyzed by using a PerkinElmer LS 45 fluorimeter, with excitation λ set at 270 nm and the emission spectrum collected between 340 and 270 nm. The variation of fluorescence intensity was evaluated at 304 nm, and it was normalized on 285 nm.

K_a was calculated from the ratio between the intercept and angular coefficient of the linear regression of a modified Benesi–Hildebrand equation (Equation (1)) [34]:

$$1/(I - I_0) = 1/\{(I' - I_0)\,K_a\}\;1/[MCD] + 1/(I' - I_0) \qquad (1)$$

where [MCD] and [DAL] referring to the total amount of MCD and DAL in solution, I and I_0 are the intensities in the presence and absence of MCD, respectively, and I′ is the value when the DAL molecules are all complexed with MDC.

2.5. Stability under Enzymatic Hydrolyses

Stock solutions of dalargin (1 mg/mL) either alone or as a 1:1 complex with MCD or polymers were mixed with α-chymotrypsin (0.36 mg/mL) solutions to obtain 0.50 mg/mL and 0.18 mg/mL (corresponding to 7.2 UI/mL) final concentrations, respectively [35]. All solutions were prepared in phosphate buffer (pH = 6.8, 0.4 M), and the enzymatic hydrolysis was performed under thermostated conditions (37 °C). All samples were analyzed after 5 min by interrupting the digestion with temperature shock (4 °C) and extensive dilution (1:20 with cold HPLC mobile phase). Residual DAL was expressed as the percentage of initial DAL as detected by the HPLC method.

DAL complexes were also prepared as solid powder by freeze-drying in a VirTis AdVantage wizard 2.0, SP Scientific, lyophilizator. Samples were frozen at a temperature of −40 °C and sublimated at a pressure of 30–40 mTorr, with an end point at 16 °C. The lyophilized samples containing DAL either alone or in a 1:1 complex were redispersed in PB and submitted to α-chymotrypsin hydrolysis, as already described.

2.6. Sample Preparation for NMR Studies

2.6.1. Affinity Studies

In the preparation of samples for the NMR analyses, the use of standardized solubilization procedures is crucial, as pointed out in our precedent work [12]. As a matter of fact, the conformational stabilization phenomena of polymeric material can occur over time, especially for high molecular weight polymers, thus affecting the NMR relaxation parameters. For these reasons, the relaxation rate measurements were repeated over time in order to avoid time-dependent effects on NMR parameters. DAL concentration was fixed at 0.50 mg/mL (the same concentration was used in HPLC degradation studies). On the basis of the knowledge of cyclodextrin content in the conjugated polymer, an N^+-rCh-MCD concentration of 3.60 mg/mL (corresponding to 2.50 mg of polymer into the NMR tube) was selected in order to set the 1:1 grafted MCD/DAL molar ratio. Accordingly, the ammonium–chitosan precursor concentration was 2.00 mg/mL (corresponding to 1.40 mg polymer into the NMR tube).

For NMR analyses, all polymers were dissolved in D_2O at 40 °C in vortex (600 rpm) for 2 h. After 1 h without stirring at room temperature, a solution of DAL was added to obtain the final ratio of 2.50 mg/0.33 mg for the mixture N^+-rCh-MCD/DAL or 1.40 mg/0.33 mg for the mixture N^+-rCh/DAL into the NMR tube. The corresponding mixture was kept at 25 °C for 2 h in vortex and then for another 2 h motionless before NMR measurements. The DAL/MCD mixture was kept under stirring at room temperature for 2 h and then kept for another further 2 h without stirring.

2.6.2. Enzymatic Hydrolyses

The same DAL/enzyme ratio and concentrations of components selected for HPLC were used in NMR analysis at 37 °C through the entire digestion process. DAL and α-chymotrypsin stock solutions were prepared in phosphate buffer (pH = 6.8) and mixed directly into the NMR tube containing predissolved polymers at 40 °C for 2 h.

2.7. Biological Evaluation

2.7.1. Cell Viability

Cell viability in the presence of either solutions of DAL or solutions of DAL and polymers/cyclodextrin was carried out using the human cell line Caco-2. Cells were grown in Minimum Essential Medium (MEM) supplemented with 1% (*v/v*) non-essential amino acid in 10% (*v/v*) fetal bovine serum, 2 mM glutamine, 100 U/mL penicillin, 100 g/mL of streptomicyn, and 0.2% (*v/v*) antimycotic. A subconfluent monolayer of Caco-2 cells was trypsinized using a 0.25% trypsin and 1 mM EDTA solution, centrifuged at 1000 rpm for 5 min, resuspended in growth medium, and counted. Caco-2 cells were seeded in each well of 96-well plates at a seeding density of 2×10^4 cells/well. Cells were incubated at 37 °C, 5% CO_2 for 24 h until 60%–70% confluence was reached. Then, the medium was then removed from each well, washed with Hanks balanced solution (HBSS), and replaced with HBSS containing the samples DAL, DAL/N^+-rCh-MCD, DAL/N^+-rCh, DAL/MCD, and DAL/N^+-rCh/MCD, all containing same amount of DAL (in the range 15–150 μg/mL) and prepared maintaining a 1:1 complex stoichiometry. After 3 h of incubation, DAL containing media were removed and substituted by WST-1 reagent solution (according to manufacturer instructions), for 4 h at 37 °C, 5% CO_2. Afterwards, formazan dye absorbance was quantified at 450 nm with the reference wavelength at 655 by using an Enspire 230 (Perkin-Elmer, Waltham, MA , USA) multilabel reader.

2.7.2. In Vitro Evaluation of Protection from Enzymatic Degradation

A Caco-2 monolayer was obtained by cell seeding (10^5 cells/well) on a polycarbonate membrane filter (pore size 0.4 μm, area 1.12 cm^2) in Corning Transwell, 12-well plates (Sigma-Aldrich, Saint Louis, MO, USA). Thereafter, 0.5 and 1.5 mL of culture medium was added to apical and basolateral compartments of the transwell inserts, respectively; the culture medium was replaced every two days for 24 days. As a rule, the experiments were performed with Caco-2 monolayers that have been grown on filter inserts for 21–29 days. By day 21, the monolayers have become differentiated with regard to the expression of many transport proteins, enzymes, and brush border hydrolases [36]. The TransEpithelial Electrical Resistance (TEER, Ωcm^2) across the Caco-2 monolayer was determined by measuring the potential difference between the apical and basolateral sides of the transwell using the Millicell-ERS (Millipore, Milano, Italy). TEER was measured prior to each experiment to ensure the confluence of the monolayers, during transport studies (every 30 min), and at 24 h from the samples' removal to assess monolayer stability recovery. Experiments were performed by using DAL-containing samples, which had either DAL alone or the presence of polymers and cyclodextrin. Caco-2 monolayers were gently rinsed twice with prewarmed HBSS and preincubated at 37 °C for 15 min with 0.5 mL and 1.5 mL of HBSS for the apical and basolateral side, respectively. Afterwards, the sample solutions (0.5 mL), all containing DAL 50 μg/mL, were applied to the apical side of each of the transwell inserts. At the end of the experiment (3 h), the apical chambers media were collected and analyzed by HPLC for DAL content evaluation.

3. Results and Discussion

3.1. NMR Characterization

Five regions were distinguished in the ^1H NMR spectrum of ammonium–chitosan N^+-rCh (Figure S1, Supplementary material). A quantitative NMR protocol developed for these specific systems [2] was applied in order to determine the degree of acetylation (8.8%), degree of derivatization (42.4%), and the ratio between the quaternary and neutral nitrogen of modified chains (n = 2.3).

Through GPC analysis, the molecular weights of the precursor (rCh, 134 kg/mol) and ammonium–chitosan (N^+-rCh, 184 kg/mol) were determined. Moreover, knowledge of the degree of acetylation and of the molecular weight of rCh made it possible to estimate the average number of units (800) in N^+-rCh.

The ^1H NMR spectrum of cyclodextrin-grafted ammonium–chitosan polymer (N^+-rCh-MCD) (Figure 2) roughly corresponds to the superimposition of N^+-rCh and MCD spectra. The degree of methylation of the cyclodextrin (DS = 0.5) was checked by NMR analysis of MCD (Figure S2, supplementary material). The use of a refined NMR quantitative analysis protocol (supplementary material) of N^+-rCh-MCD led to establishing that the N^+-rCh-MCD polymer contains 22% *w/w* of MCD and 18% *w/w* of spacer. Although the percentages of MCD and HMDI are almost equal, the conspicuous difference in their MWs indicates that there are numerous spacer chains bonded to the polymer only but unbound to the cyclodextrin. Excluding cross-linking phenomena, it was possible to estimate the minimum molecular weight of the conjugate, knowing the percentage by weight of the precursor and, therefore, the moles of conjugate, reporting a MW of 306 kg/mol, with 331 units of HMDI and 57 MCD, which was testimony to the fact that there is so much free spacer. DAL (Figure 1b) was characterized by ^1H e ^{13}C NMR spectroscopy (Figures S3–S5, Tables S6 and S7, supplementary material).

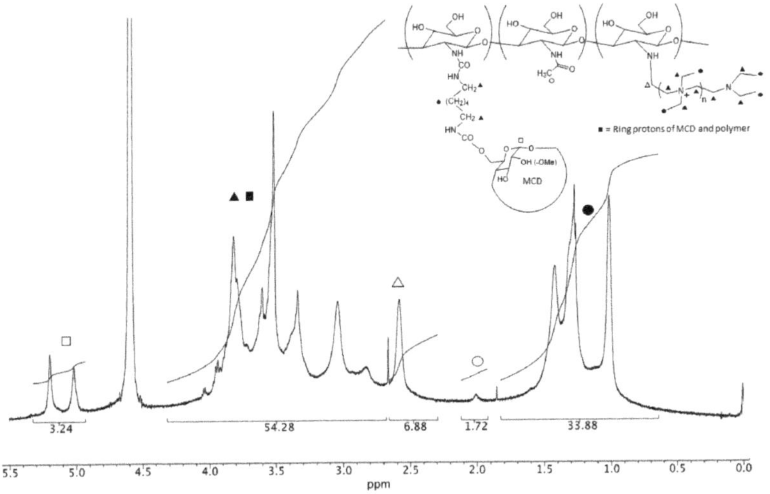

Figure 2. ^1H NMR spectrum (600 MHz, D_2O, 37 °C) of N^+-rCh-MCD (3.7 mg/mL). Black and white dots, squares and triangles refer to the corresponding protons marked on the structure.

3.2. Inclusion Complexes

3.2.1. Complexes Stoichiometry and Association Constant

The complex formation between DAL and MCD was investigated and compared to that of DAL with MCD conjugated to the quaternary ammonium chitosan (N^+-rCh-MCD) or in the presence of plain N^+-rCh (N^+-rCh / MCD). In all cases, the χ value was 0.5, corresponding to the 1:1 complex stoichiometry of DAL/MCD (Table S8, Supplementary material). The results evidenced that the quaternary ammonium polymer was not interfering with the formation of the complex neither when dissolved together with MCD nor when covalently linked to MCD. Additionally, χ values were not affected by the incubation time, indicating a prompt complexation between MCD and DAL (Table S8, supplementary material).

The determination of the association constant (K_a) for complexes of DAL/MCD and of DAL/N^+-rCh-MCD was performed both by UV and fluorescence spectrometry. Both techniques exploited the gradual variation of DAL absorbance or fluorescence in the presence of increasing amounts of host agent. These variations were better monitored by using DAL fluorescence, whereas UV spectrometry was poorly applicable to DAL/MCD complexes, which would have required an excessive amount of the binding partners to reveal the absorbance variation. Differently, this latter technique was successfully used only for the DAL/N^+-rCh-MCD complex. The calculated K_a values are reported in Table 1.

Table 1. Association constant (K_a) determined by UV absorbance or fluorescence spectrometry, for DAL complexing with 2-methyl-β-cyclodextrin (MCD) or N^+-rCh-MCD. Standard deviation (SD) and the coefficient of determination (R^2) of the linear regression are also displayed.

	Absorbance			Fluorescence		
	K_a (M^{-1})	SD	R^2	K_a (M^{-1})	SD	R^2
DAL/MCD	-	-	-	120	10	0.999
DAL/N^+-rCh-MCD	2326	406	0.999	2617	307	0.990

The good linearity of data regression confirmed the formation of a 1:1 complex in both cases, DAL/MCD and DAL/N^+-rCh-MCD. More importantly, the K_a evaluated for the peptide complexed to the N^+-rCh-MCD polymer was one order of magnitude higher than that calculated for MCD alone. Despite the two different spectroscopic techniques, the values are consistent with each other.

It is worth mentioning that the spectrometric titration performed for the evaluation of K_a evidenced a clear variation of DAL absorbance peak at 227 nm in terms of intensity and wavelength, as clearly displayed by the overlay of the II derivative absorbance spectra of DAL/N^+-rCh-MCD complexes (Figure 3A). Similarly, the greatest fluorescence variation of complexed DAL was revealed at the 304 nm emission band (Figure 3B). Both phenomena can be related to a main variation of the chemical environment surrounding Tyr residue, which appeared more pronounced for the DAL/N^+-rCh-MCD complex. As a matter of fact, the Tyr ABS peak at 227 nm is typically exploited for protein unfolding studies [37], and Tyr fluorescence is predominant with λ^{ex} 275 and λ^{em} peaks at 304 nm [38].

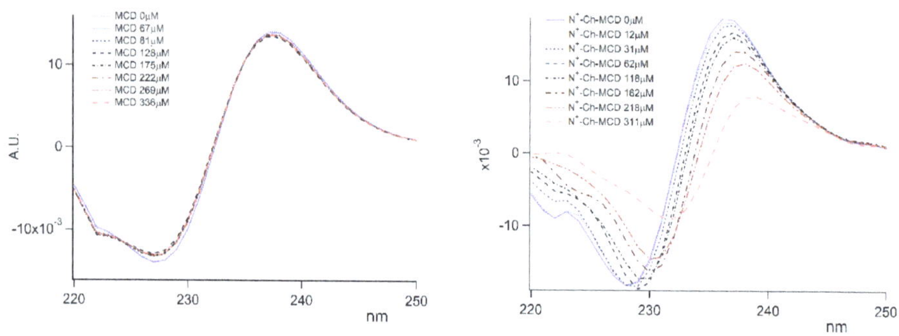

(A)

Figure 3. *Cont.*

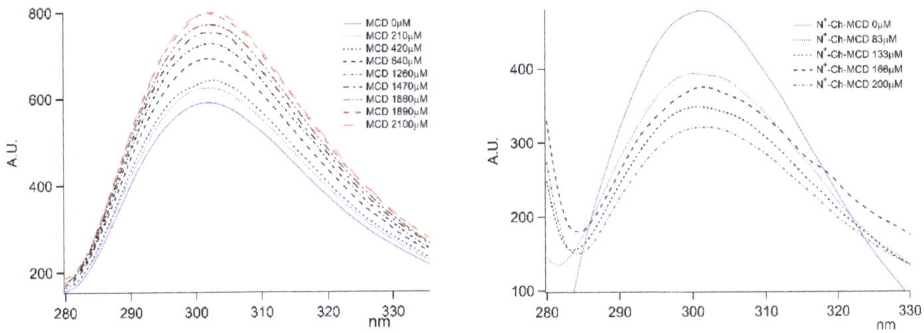

(B)

Figure 3. Complex titration by UV and fluorescence spectrometry. (**A**) Overlay of UV second derivative spectra; (**B**) Overlay of emission spectra collected by excitation at λ 275 nm.

3.2.2. NMR Interaction Studies and Complex Stereochemistry

The interaction between a small molecule (G) and a macromolecule (H) can be described by the simple equilibrium reported in Equation (2). In fast exchange conditions, the observable NMR parameters (P_{obs}) represent the weighted average between the bound (P_b) and free (P_f) states (Equation (3)).

$$H + G \rightleftharpoons HG \tag{2}$$

$$P_{obs} = \chi_f\, P_f + \chi_b\, P_b \tag{3}$$

where χ_f and χ_b are the molar fractions of the ligand in the free and bound states, respectively. By virtue of the marked difference between the molecular weights of H and G, it is necessary to operate with a large excess of G with respect to H in order to obtain an observable NMR signal. In these conditions, the molar fraction of the small molecule in its complexed form (χ_b) is negligible. Therefore, the choice of sensitive parameters is required to detect significant changes in the bound state with respect to the free state. Longitudinal relaxation rates represent an effective tool in the analysis of the interaction between small molecules and macromolecules. In particular, proton mono-selective relaxation rates ($R_i^{ms} = 1/T_i^{ms}$), measured by following the recovery of the magnetization of the selectively inverted spin *i*, are remarkably more sensitive than non-selective ones [39]. Considerable increases of R^{ms} are expected when the small molecule undergoes a change in molecular dynamic from fast ($\omega^2\tau_c^2 \ll 0.6$, where ω is the Larmor frequency and τ_c is the rotational correlation time) to slow motion regimes ($\omega^2\tau_c^2 \gg 0.6$), which is typical when bounded to a macromolecule. The cross-relaxation term (σ_{ij}), referring to the spin *i* and *j* at r_{ij} distance, is another susceptible parameter to complexation phenomena. σ_{ij} can be experimentally obtained from the difference between the bi-selective relaxation rate (detected by following the recovery of the magnetization of the nucleus *i* in conditions of simultaneous inversion of the spin *j*, $R_{ij}^{bs} = 1/\,T_{ij}^{bs}$) and mono-selective relaxation rate, as shown in Equation (4).

$$\sigma_{ij} = R_{ij}^{bs} - R_i^{ms} \tag{4}$$

In the two limit cases of fast and slow -motion regimes, the expression of σ_{ij} can be calculated by Equations (5) and (6), respectively.

$$\sigma_{ij} = 0.5\,\gamma^4\,\hbar^2\,r_{ij}^{-6}\,\tau_c \qquad (\omega^2\tau_c^2 \ll 0.6) \tag{5}$$

$$\sigma_{ij} = -0.1\,\gamma^4\,\hbar^2\,r_{ij}^{-6}\,\tau_c \qquad (\omega^2\tau_c^2 \gg 0.6) \tag{6}$$

Therefore, a negative σ indicates a slowdown of molecular motion, which is generated by the interaction of the active ingredient with the macromolecule.

Due to the extensive superimposition between DAL and the polymers' resonances, only the selective relaxation rates of phenylalanine and tyrosine aromatic protons and methyl moieties of leucine of DAL (0.5 mg/mL) were measured in solutions containing pure DAL or its mixtures with polymers/cyclodextrin.

The mono-selective and bi-selective relaxation of DAL protons is collected in Table 2, along with selected cross-relaxation terms. NMR measurements were performed both immediately after preparation of the solutions and after one week, with negligible variations in the relaxation parameters recorded.

Table 2. Mono-selective relaxation rates (R^{ms}, s^{-1}), bi-selective relaxation rates (R^{bs}, s^{-1}), and cross-relaxation terms (σ, s^{-1}) of the proton pair H_3/H_4^{Phe} and H_1/H_2^{Tyr} of DAL (600 MHz, 25 °C, D_2O, 0.68 mM).

	δ (ppm)	R_1^{ms} (s^{-1})				
		DAL	DAL/MCD	DAL/N$^+$-rCh	DAL/N$^+$-rCh-MCD	DAL/MCD/N$^+$-rCh
H_3^{Phe}	7.11	0.62	0.59	0.61	4.54	0.60
H_4^{Phe}	7.22	0.54	0.54	0.52	4.35	0.50
H_5^{Phe}	7.18	0.66	0.57	0.61	5.00	0.53
H_1^{Tyr}	6.77	0.36	0.32	0.34	4.54	0.34
H_2^{Tyr}	7.02	0.58	0.57	0.55	4.54	0.54
H_{4a}^{Leu}	0.78	1.69	1.65	1.67	4.17	1.67
H_{4b}^{Leu}	0.72	1.59	1.58	1.59	4.17	1.59
		R_1^{bs} (s^{-1})				
H_3^{Phe}	7.11	0.63	0.60	0.62	3.66	0.62
H_4^{Phe}	7.22	0.55	0.55	0.54	3.47	0.53
H_1^{Tyr}	6.77	0.39	0.35	0.36	3.92	0.37
H_2^{Tyr}	7.02	0.61	0.60	0.57	3.92	0.57
		σ (s^{-1})				
H_3/H_4^{Phe}	-	0.01	0.01	0.01	−0.88	0.02
H_1/H_2^{Tyr}	-	0.03	0.03	0.02	−0.62	0.03

The presence of an equimolar amount of MCD did not produce any significant variations in the relaxation parameters of DAL (Table 2), indicating minor drug/macrocycle interactions, according to the very low amount of bound molar fraction (7%) at the concentration of 0.68 mM. The same trend was found for the binary mixture DAL/N$^+$-rCh, with negligible variations with respect to free DAL. Importantly, the minimum impact of the presence of N$^+$-rCh also rules out possible interfering effects on relaxation parameters attributable to viscosity changes caused by the presence of the polymer, rather than to chemical interactions with the macromolecule. On the contrary, outstanding increments in relaxation rates were recorded when dalargin is mixed with ammonium–chitosan covalently conjugated with MCD (DAL/N+-rCh-MCD), especially for aromatic moieties. As an example, R_1^{ms} of H_3^{Phe} increases from 0.62 s^{-1} to a value of 4.54 s^{-1}. Cross-relaxation rates respectively decrease to the values of -0.88 s^{-1} and -0.62 s^{-1} for H_3^{Phe}/H_4^{Phe} and H_1^{Tyr}/H_2^{Tyr} proton pairs, due to the relevant slowing down of the molecular motion of DAL consequent to its interaction with N+-rCh-MCD. Interestingly, the simple physical mixture of ammonium–chitosan and cyclodextrin (DAL/N$^+$-rCh/MCD) produced negligible variations of the relaxation parameters compared to pure DAL, confirming the key role of the covalent grafting of cyclodextrin to the polymer (Table 1).

To obtain a parameter that reflects the different degree of involvement of DAL fragments in the interaction with polymers, normalized relaxation rates were calculated [$|R| = (R_1^{mix} - R_1^{free})/R_1^{free}$], showing a major involvement of tyrosine and phenylalanine group (Table 3). The particularly high value for the *ortho*-protons to hydroxyl group of tyrosine is due to the role of the polar group in the interaction with the polysaccharide material.

Table 3. Normalized relaxation rates ($|R| = (R_1^{mix} - R_1^{free})/R_1^{free}$) of selected protons of DAL in the binary mixture DAL/N$^+$-rCh-MCD.

	DAL/N$^+$-rCh-MCD						
	H_3^{Phe}	H_4^{Phe}	H_5^{Phe}	H_1^{Tyr}	H_2^{Tyr}	H_{4a}^{Leu}	H_{4b}^{Leu}
|R|	6.32	7.06	6.58	11.6	6.83	1.47	1.62

The mixture DAL/MCD was employed as a model for investigating the nature of the interaction of DAL with MCD. The study was carried out at 9.8 mM concentration, which guaranteed a significant bound molar fraction of DAL (about 40%) on the basis of the value of the association constant DAL/CD (120 M^{-1}). Such a kind of investigation was not possible for the mixture DAL/N$^+$-rCh-MCD due to the extensive overlapping of polymer and cyclodextrin signals. An extended conformation of DAL was found on the basis of the analysis of 1D ROESY spectra, where any significant dipolar interactions were not detected between protons bearing different amino acidic residues. As an example, the selective perturbation of Tyr protons produces negligible ROE effects on the adjacent Ala protons (Figure 4).

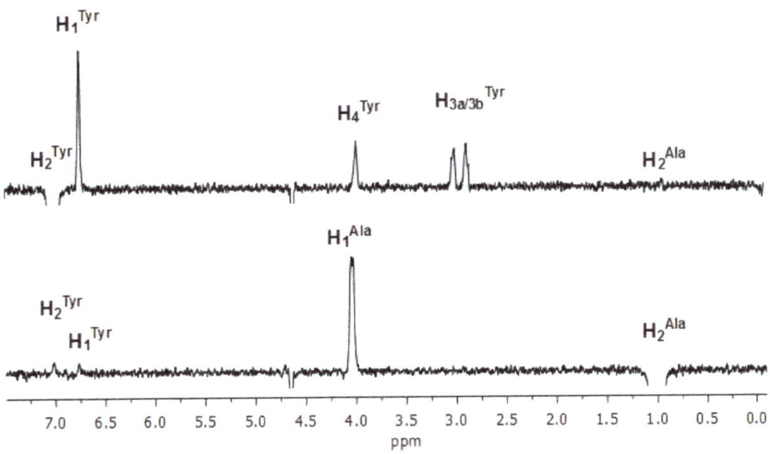

Figure 4. One-dimensional (1D) Rotating-frame Overhauser Effect SpectroscopY (ROESY) spectra (600 MHz, D_2O, 25 °C) of H_2^{Tyr} and H_2^{Ala} of DAL (9.8 mM).

In order to point out the nature of the DAL groups that were more involved in the interaction to MCD, complexation shifts were calculated ($|\Delta\delta| = |\delta^{DAL/MCD} - \delta^{MCD}|$, Hz). For this type of analysis, TOCSY traces were accurately analyzed: starting from the anomeric protons H_1 and $H_{1'}$ of MCD, resonances of the corresponding glucosidic rings protons were individuated. Complexation shifts were remarkably higher for $H_3/H_{3'}$ and H_5/H_5, which were located inside the hydrophobic cavity of MCD with a deep inclusion from the edge of larger diameter and supported by considerable values of $\Delta\delta$ for $H_5/H_{5'}$ (Table 4).

Table 4. MCD (600 MHz, D_2O, 25 °C, 9.8 mM) complexation shifts.

MCD	$\|\Delta\delta\|$ (Hz)	MCD	$\|\Delta\delta\|$ (Hz)
H_1	9.5	H_4	4.2
$H_{1'}$	4.1	$H_{4'}$	1.2
H_2	4.5	H_5	41.8
$H_{2'}$	6.1	$H_{5'}$	49.2
H_3	27.6	$H_6/H_{6'}$	3.2
$H_{3'}$	25.3		

Detailed information on the stereochemistry of the inclusion were obtained from the ROESY map, (Figure 5 and Figure S6, supplementary material), showing ROE effects mainly between the aromatic protons of DAL and the $H_3/H_{3'}$ and $H_5/H_{5'}$ internal protons of the cyclodextrin.

In particular, H_1^{Tyr} produces much more intense ROE effects on the $H_5/H_{5'}$ (narrower rim) compared with the $H_3/H_{3'}$ (wider rim), while H_2^{Tyr} showed an opposite trend, indicating that the phenolic group is embedded in the cavity of the cyclodextrin deeply and confirming what was observed with UV and fluorescence spectrometry.

Phenylalanine is also widely involved in the interaction with cyclodextrin, but the effects produced at $H_3/H_{3'}$ and $H_5/H_{5'}$ frequencies have comparable intensities, which is probably because of aromatic moieties mobility within the cavity of the cyclodextrin.

It is noteworthy that both methyl protons of leucine and alanine originated dipolar interactions with the internal $H_3/H_{3'}$ protons and external methoxy protons of the cyclodextrin, thus revealing superficial interactions at the wide rim of the cyclodextrin.

Therefore, taking into account that 1 to 1 complexation stoichiometry has been demonstrated, it can be concluded that dalargin may originate multiple interactions with the cyclodextrin, involving the deep inclusion of the aromatic groups of tyrosine or phenylalanine, as well as superficial interactions of the alkyl groups of leucine or alanine at the wide rim of the cyclodextrin.

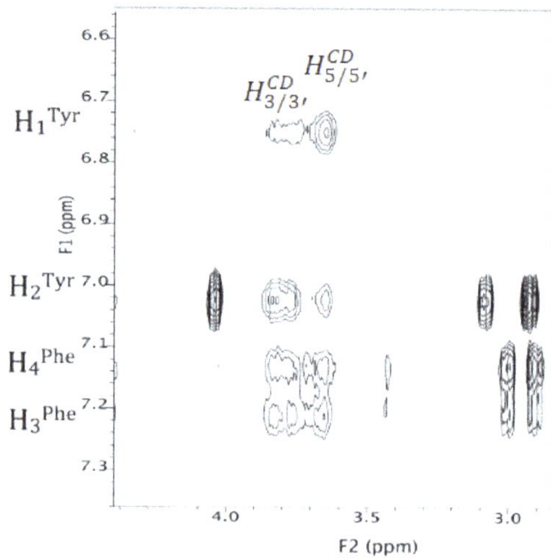

Figure 5. ROESY map (600 MHz, D_2O, 25 °C, 9.8 mM) of DAL/MCD mixture between 6.6–7.3 ppm and 2.8–4.4 ppm.

3.3. *Kinetic Studies of Dalargin Enzymatic Hydrolysis*

The protective role of N^+-rCh-MCD against DAL enzymatic hydrolysis was studied with two orthogonal methods (HPLC and NMR). α-Chymotrypsin (CHT) was chosen as the model enzyme for study of the degradation of DAL in the human body [35]. CHT belongs to the proteases and is capable of breaking peptide bonds with preference for tyrosine, phenylalanine, and leucine residues [40]. Pure DAL is stable in water at 37 °C for several days, as demonstrated by comparing ^1H NMR spectra recorded at different time after preparation: no differences were recorded neither in chemical shifts nor in the peaks' integrated areas.

The digestion of DAL by CHT was performed under controlled conditions to simulate the digestion occurring in the intestinal environment. A first characterization was performed on freshly prepared solutions of DAL, either alone or mixed with MCD, N^+-rCh /MCD, or N^+-rCh-MCD. The analysis that was carried out 5 min after the addition of the enzyme evidenced that the peptide underwent a rapid degradation, with a reduction equal to 80% of the initial content. The degradation process was merely slowed down in the presence of MCD, in an equal extent of MCD and MCD mixed with N^+-rCh. Diversely, when using the MCD-grafted polymer (N^+-rCh-MCD), DAL suffers only a 45% reduction. The results highlighted that only when MCD is covalently bound to the quaternary polymer was there a significant protection from the enzymatic attack (Figure 6).

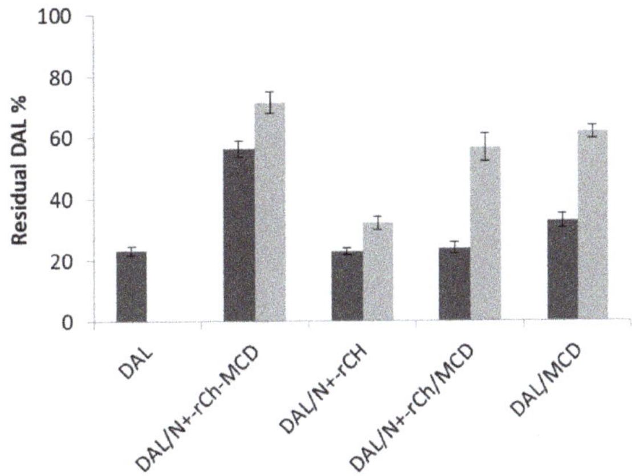

Figure 6. α-chymotrypsin (CHT) hydrolysis of DAL: Residual DAL % after 5 min of CHT activity on solutions (■); on lyophilized complexes (■). Error bars indicate the SD values of three independent experiments.

A similar digestion was conducted also on lyophilized samples that were freshly redispersed in buffer before adding the enzyme. Similarly, plain DAL was rapidly degraded, but an extended stability was observed for the DAL/MCD complex and even more for the complex DAL/N^+-rCh-MCD (Figure 4). Therefore, it was assumed that the conjugate was able to protect DAL from enzymatic digestion, and that the lyophilization process strengthened the interaction of DAL with the cyclodextrin, increasing its protection [41]. Despite the augmented stability due to the lyophilization process, the use of the conjugated polymer N^+-rCh-MCD led again to a significantly reduced digestion compared to the use of N^+-rCh/MCD mixture ($p < 0.01$).

By using the same DAL/CHT molar ratio selected for HPLC analysis, low and high frequency NMR spectral regions of dalargin, pure (Figure 7a) and in presence of CHT (Figure 7b, c), were monitored over time at 37 °C. In the binary mixture DAL/CHT, a rapid decrease of the original dalargin signals was observed already after 5 min with the simultaneous formation of new peaks, which was attributable to the digestion products. A comparison of integrated areas showed that the digestion was almost complete after 14 min. In order to obtain a quantitative parameter, we refer to the lifetime of the drug (t_f), the specific time in which non-degraded DAL reached 6% with respect to the hydrolyzed products. If the mixture is maintained at room temperature (25 °C), the t_f of DAL is extended up to 45 min, owing to a reduced activity of the enzyme.

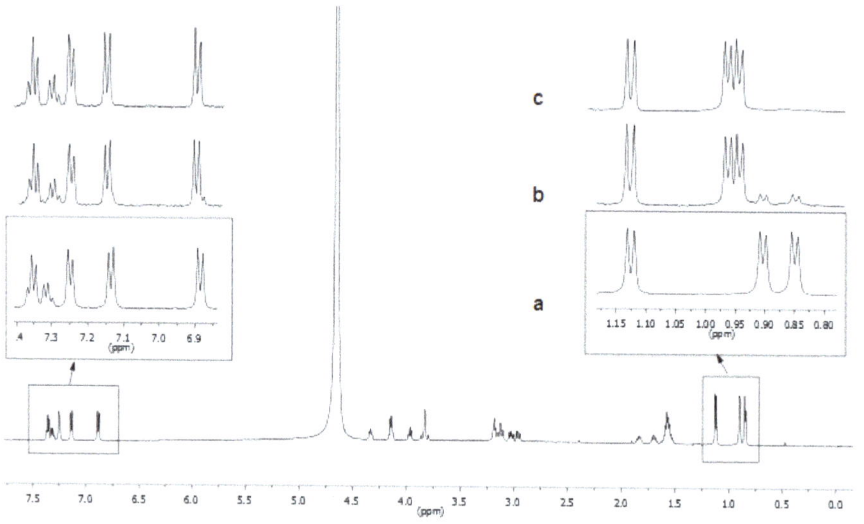

Figure 7. ^{1}H NMR spectra (600 MHz, D_2O, phosphate buffer, pH = 6.8, 37 °C) of pure DAL (0.68 mM, (**a**) and of the DAL/CHT after 5 min (**b**) and 14 min (**c**).

Effects due to the presence of MCD and/or the modified chitosans on the rates of hydrolysis were evaluated: as reported in Table 5, MCD, N^{+}-rCh, or their physical mixture MCD/N^{+}-rCh only slightly affected the hydrolysis rate. On the contrary, conjugated chitosan N^{+}-rCh-MCD produced remarkably prolonged t_f, confirming what was observed at a shorter digestion time with the HPLC technique. It clearly appeared that the covalent binding of MCD to the chitosan backbone was essential to generate a protecting effect against the DAL hydrolytic pathways.

Table 5. Complete hydrolyses times (t_f, min) of DAL (0.68 mM) in fresh mixtures and lyophilized mixture in the presence of CHT, as recorded by ^{1}H NMR (600 MHz, 37 °C, phosphate buffer, pH = 6.8).

Mixture	t_f (min)
DAL/CHT/MCD	16
DAL/CHT/N^{+}-rCh	18
DAL/CHT/N^{+}-rCh/MCD	17
DAL/CHT/N^{+}-rCh-MCD	61

3.4. Biological Evaluation

A quantitative evaluation of cell viability performed by WST-1 assay on a Caco-2 cell line (Figure 8) indicated a high tolerability for the complex of DAL with N^{+}-rCh-MCD, which was comparable to what was observed for plain DAL and DAL complexed with MCD. Differently, N^{+}-rCh polymer negatively affected cell viability, reducing it drastically at the tested concentration. Such behavior was shielded by the presence of MCD, especially when covalently conjugated as for the polymer N^{+}-rCh-MCD.

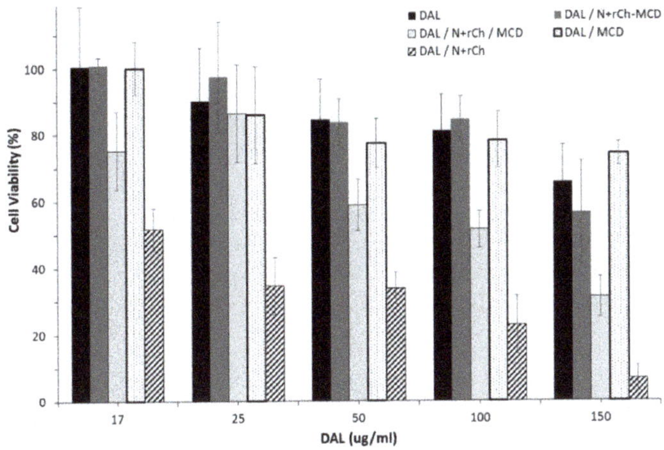

Figure 8. Caco-2 cells cell viability assay. Error bars indicating SD values of eight replicates.

DAL stability toward enzymatic degradation in the presence of N^+-rCh-MCD polymer significantly increased when tested on Caco-2 monolayers (Figure 9). After 3 h of incubation, DAL contents in the apical chamber of plain DAL, DAL/MCD, DAL/N^+-rCh, and DAL/N^+-rCh/MCD physical mixtures were not significantly different (in the range of 4.1–7.3 μg/mL, corresponding to 8%–15% of residual DAL). The MCD-bearing quaternary polymer (N^+-rCh-MCD) exhibited 16.7 μg/mL of DAL after 3 h of incubation, confirming its stronger stabilizing effect also in a more complex biological environment.

The effect of cyclodextrin and polymer derivatives on monolayers permeability was monitored by TEER measurements. TEER values exhibited the same trend of the untreated control layer until 1 h of incubation, followed by a plateau in the range of 86%–100% for the remaining contact time. On the contrary, N^+-rCh determined a significant reduction of TEER values after 2–3 h, reaching 64%. This finding is in agreement with our previous research studies [16], stressing that quaternary ammonium chitosan acts as a loosening/opening agent for intercellular tight junctions, resulting in the increased paracellular permeability of the monolayers, with the reversibility of the effect (Figure S7, supplementary material).

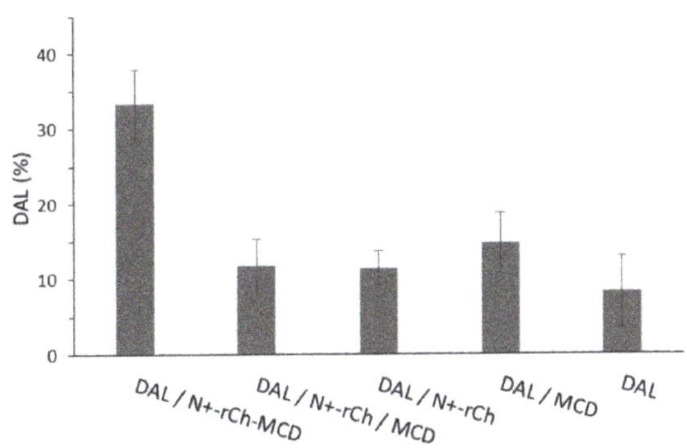

Figure 9. Percentage of DAL remaining in the apical chamber after 3 h of incubation on Caco-2 monolayers. Error bars indicate the SD values of three experiments.

4. Conclusions

MCD grafted to quaternary ammonium chitosan (N^+-rCh-MCD) has proved to be effective as a host macromolecule for the short, labile, and therapeutically active peptide DAL. This finding, along with the previously performed investigations [21,42], suggests the suitability of MCD-grafted quaternary chitosans as functional polymers for the oral administration of poorly bioavailable actives.

It has been demonstrated that the interaction of dalargin with MCD occurs through a deep inclusion in the oligosaccharide cavity, involving mainly Tyr or Phe aromatic groups or, with minor extent, the alkyl groups of the Leu and Ala units. Additionally, the covalent binding of MCD to the mucoadhesive chitosan derivative is not hampering the complexing ability of MCD, but rather further increasing the interaction with dalargin. The MCD grafting has a positive effect on polymer cytocompatibility, and it has been demonstrated that the polymer behaves as a shield versus the proteolytic attack of intestinal enzymes only when covalently linked to the MCD.

The mentioned features were described for water-soluble complexes of N^+-rCh-MCD/DAL, but the preliminary evaluation of lyophilized samples showed improved stabilization of the peptide, indicating the perspective of using a processed powder for the preparation of solid dosage forms.

Supplementary Materials: Available online at http://www.mdpi.com/2073-4360/12/2/474/s1.

Author Contributions: Conceptualization, F.B., G.U.-B., Y.Z., and A.M.P.; methodology, F.B., C.M., and A.M.P.; validation, F.B., G.U.-B., Y.Z., and A.M.P.; formal analysis, A.C., A.R., N.R., and C.M.; writing—original draft preparation, A.C., A.R., A.F., C.M., and A.M.P.; writing—review and editing, F.B., G.U.-B., Y.Z., and A.M.P.; supervision, F.B., G.U.-B., Y.Z., and A.M.P. All authors have read and agreed to the published version of the manuscript.

Funding: This research received no external funding.

Conflicts of Interest: The authors declare no conflict of interest.

Abbreviations

BCS	bovine fetal serum
Ch	chitosan
CHT	α-chymotrypsin
DAL	dalargin
DEAE-Cl HCl	2 diethylaminoethyl chloride hydrochloride
DMSO	dimethyl sulfoxide
DPBS	Dulbecco's phosphate buffer
HMDI	1,6-hexamethylene diisocyanate
MCD	2-methyl-β-cyclodextrin
MEM	minimum essential medium
N^+-rCh	reduced molecular weight ammonium chitosan
N^+-rCh-MCD	ammonium-chitosan grafted with 2-methyl-β-cyclodextrin
rCh	reduced molecular weight chitosan
TEA	triethylamine

References

1. Kosaraju, S.L. Colon targeted delivery systems: Review of polysaccharides for encapsulation and delivery. *Crit. Rev. Food Sci. Nutr.* **2005**, *45*, 251–258. [CrossRef] [PubMed]
2. Zambito, Y.; Uccello-Barretta, G.; Zaino, C.; Balzano, F.; Di Colo, G. Novel transmucosal absorption enhancers obtained by aminoalkylation of chitosan. *Eur. J. Pharm. Sci.* **2006**, *29*, 460–469. [CrossRef] [PubMed]
3. Liechty, W.B.; Kryscio, D.R.; Slaughter, B.V.; Peppas, N.A. Polymers for drug delivery systems. *Annu. Rev. Chem. Biomol. Eng.* **2010**, *1*, 149–173. [CrossRef] [PubMed]
4. Yadav, M.; Goswami, P.; Paritosh, K.; Kumar, M.; Pareek, N.; Vivekanand, V. Seafood waste: A source for preparation of commercially employable chitin/chitosan materials. *Bioresour. Bioprocess.* **2019**, *6*. [CrossRef]
5. Dash, M.; Chiellini, F.; Ottenbrite, R.M.; Chiellini, E. Chitosan—A versatile semi-synthetic polymer in biomedical applications. *Prog. Polym. Sci. (Oxford)* **2011**, *36*, 981–1014. [CrossRef]
6. Zhang, J.; Xia, W.; Liu, P.; Cheng, Q.; Tahirou, T.; Gu, W.; Li, B. Chitosan modification and pharmaceutical/biomedical applications. *Mar. Drugs* **2010**, *8*, 1962–1987. [CrossRef]

7. Harish Prashanth, K.V.; Tharanathan, R.N. Chitin/chitosan: Modifications and their unlimited application potential-an overview. *Trends Food Sci. Technol.* **2007**, *18*, 117–131. [CrossRef]
8. Rinaudo, M. Chitin and chitosan: Properties and applications. *Prog. Polym. Sci. (Oxford)* **2006**, *31*, 603–632. [CrossRef]
9. Aiello, F.; Balzano, F.; Carpita, L.; Fabiano, A.; Zambito, Y.; Uccello Barretta, G. Role of nanostructured aggregation of chitosan derivatives on [5-methionine]enkephalin affinity. *Carbohydr. Polym.* **2017**, *157*, 321–324. [CrossRef]
10. Uccello-Barretta, G.; Balzano, F.; Aiello, F.; Senatore, A.; Fabiano, A.; Zambito, Y. Mucoadhesivity and release properties of quaternary ammonium-chitosan conjugates and their nanoparticulate supramolecular aggregates: An NMR investigation. *Int. J. Pharm.* **2014**, *461*, 489–494. [CrossRef]
11. Zambito, Y.; Zaino, C.; Uccello-Barretta, G.; Balzano, F.; Di Colo, G. Improved synthesis of quaternary ammonium-chitosan conjugates (N+-Ch) for enhanced intestinal drug permeation. *Eur. J. Pharm. Sci.* **2008**, *33*, 343–350. [CrossRef]
12. Cesari, A.; Fabiano, A.; Piras, A.M.; Zambito, Y.; Uccello-Barretta, G.; Balzano, F. Binding and mucoadhesion of sulfurated derivatives of quaternary ammonium-chitosans and their nanoaggregates: An NMR investigation. *J. Pharm. Biomed. Anal.* **2020**, *177*. [CrossRef] [PubMed]
13. Fabiano, A.; Piras, A.M.; Uccello-Barretta, G.; Balzano, F.; Cesari, A.; Testai, L.; Citi, V.; Zambito, Y. Impact of mucoadhesive polymeric nanoparticulate systems on oral bioavailability of a macromolecular model drug. *Eur. J. Pharm. Biopharm.* **2018**, *130*, 281–289. [CrossRef] [PubMed]
14. Dünnhaupt, S.; Barthelmes, J.; Thurner, C.C.; Waldner, C.; Sakloetsakun, D.; Bernkop-Schnürch, A. S-protected thiolated chitosan: Synthesis and in vitro characterization. *Carbohydr. Polym.* **2012**, *90*, 765–772. [CrossRef] [PubMed]
15. Zambito, Y.; Felice, F.; Fabiano, A.; Di Stefano, R.; Di Colo, G. Mucoadhesive nanoparticles made of thiolated quaternary chitosan crosslinked with hyaluronan. *Carbohydr. Polym.* **2013**, *92*, 33–39. [CrossRef] [PubMed]
16. Zambito, Y.; Fogli, S.; Zaino, C.; Stefanelli, F.; Breschi, M.C.; Di Colo, G. Synthesis, characterization and evaluation of thiolated quaternary ammonium-chitosan conjugates for enhanced intestinal drug permeation. *Eur. J. Pharm. Sci.* **2009**, *38*, 112–120. [CrossRef]
17. Auzély-Velty, R.; Rinaudo, M. Chitosan derivatives bearing pendant cyclodextrin cavities: Synthesis and inclusion performance. *Macromolecules* **2001**, *34*, 3574–3580. [CrossRef]
18. Song, M.; Li, L.; Zhang, Y.; Chen, K.; Wang, H.; Gong, R. Carboxymethyl-β-cyclodextrin grafted chitosan nanoparticles as oral delivery carrier of protein drugs. *React. Funct. Polym.* **2017**, *117*, 10–15. [CrossRef]
19. Loftsson, T.; Brewster, M.E. Pharmaceutical applications of cyclodextrins. 1. Drug solubilization and stabilization. *J. Pharm. Sci.* **1996**, *85*, 1017–1025. [CrossRef]
20. Loftsson, T.; Duchêne, D. Cyclodextrins and their pharmaceutical applications. *Int. J. Pharm.* **2007**, *329*, 1–11. [CrossRef]
21. Piras, A.M.; Zambito, Y.; Burgalassi, S.; Monti, D.; Tampucci, S.; Terreni, E.; Fabiano, A.; Balzano, F.; Uccello-Barretta, G.; Chetoni, P. A water-soluble, mucoadhesive quaternary ammonium chitosan-methyl-β-cyclodextrin conjugate forming inclusion complexes with dexamethasone. *J. Mater. Sci. Mater. Med.* **2018**, *29*. [CrossRef] [PubMed]
22. Fleishman, M.U.; Zhivotova, E.U.; Lebedko, O.A.; Deigin, V.I.; Timoshin, S.S. Mechanisms for the effect of arginine-containing dermorphin analogue on proliferative processes in the gastric mucosa of albino rats. *Bull. Exp. Biol. Med.* **2007**, *144*, 309–311. [CrossRef] [PubMed]
23. Maslov, L.N.; Lishmanov, Y.B. The anti-arrhythmic effect of d-Ala 2, Leu 5, Arg 6-enkephalin and its possible mechanism. *Int. J. Cardiol.* **1993**, *40*, 89–94. [CrossRef]
24. Zhivotova, E.Y.; Fleishman, M.Y.; Sazonova, E.N.; Lebed'ko, O.A.; Timoshin, S.S. Gastroprotective effect of dalargin in gastropathy due to treatment with nonsteroid antiinflammatory drugs. *Bull. Exp. Biol. Med.* **2009**, *147*, 441–443. [CrossRef]
25. Zivotova, E.U.; Fleishman, M.U.; Lebedko, O.A.; Sazonova, E.N.; Timoshin, S.S. Effect of dalargin on DNA synthesis in the gastric mucosa of albino rats. *Bull. Exp. Biol. Med.* **2007**, *144*, 314–316. [CrossRef]
26. Aliautdin, R.N.; Petrov, V.E.; Ivanov, A.A.; Kreuter, J.; Kharkevich, D.A. Transport of the hexapeptide dalargin across the hemato-encephalic barrier into the brain using polymer nanoparticles. *Eksperimental'naia Klin. Farmakol.* **1996**, *59*, 57–60.

27. AboulFotouh, K.; Allam, A.A.; El-Badry, M.; El-Sayed, A.M. Role of self-emulsifying drug delivery systems in optimizing the oral delivery of hydrophilic macromolecules and reducing interindividual variability. *Colloids Surf. B Biointerfaces* **2018**, *167*, 82–92. [CrossRef]
28. Chen, Y.; Siddalingappa, B.; Chan, P.H.H.; Benson, H.A.E. Development of a chitosan-based nanoparticle formulation for delivery of a hydrophilic hexapeptide, dalargin. *Biopolym. Pept. Sci. Sect.* **2008**, *90*, 663–670. [CrossRef]
29. Martin-Serrano, Á.; Gómez, R.; Ortega, P.; Mata, F.J.D.L. Nanosystems as vehicles for the delivery of antimicrobial peptides (Amps). *Pharmaceutics* **2019**, *11*. [CrossRef] [PubMed]
30. Li, J.; Pan, D.; Yi, J.; Hao, L.; Kang, Q.; Liu, X.; Lu, L.; Lu, J. Protective effect of β-cyclodextrin on stability of nisin and corresponding interactions involved. *Carbohydr. Polym.* **2019**, *223*. [CrossRef] [PubMed]
31. Freeman, R.; Wittekoek, S. Selective determination of relaxation times in high resolution NMR. *J. Magn. Reson. (1969)* **1969**, *1*, 238–276. [CrossRef]
32. Mao, S.; Shuai, X.; Unger, F.; Simon, M.; Bi, D.; Kissel, T. The depolymerization of chitosan: Effects on physicochemical and biological properties. *Int. J. Pharm.* **2004**, *281*, 45–54. [CrossRef] [PubMed]
33. Benesi, H.A.; Hildebrand, J.H. A Spectrophotometric Investigation of the Interaction of Iodine with Aromatic Hydrocarbons. *J. Am. Chem. Soc.* **1949**, *71*, 2703–2707. [CrossRef]
34. Barman, S.; Barman, B.K.; Roy, M.N. Preparation, characterization and binding behaviors of host-guest inclusion complexes of metoclopramide hydrochloride with α- and β-cyclodextrin molecules. *J. Mol. Struct.* **2018**, *1155*, 503–512. [CrossRef]
35. Zupančič, O.; Rohrer, J.; Thanh Lam, H.; Grießinger, J.A.; Bernkop-Schnürch, A. Development and in vitro characterization of self-emulsifying drug delivery system (SEDDS) for oral opioid peptide delivery. *Drug Dev. Ind. Pharm.* **2017**, *43*, 1694–1702. [CrossRef] [PubMed]
36. Hubatsch, I.; Ragnarsson, E.G.E.; Artursson, P. Determination of drug permeability and prediction of drug absorption in Caco-2 monolayers. *Nat. Protoc.* **2007**, *2*, 2111–2119. [CrossRef] [PubMed]
37. Schmid, F.-X. Biological Macromolecules: UV-visible Spectrophotometry. In *eLS*; Wiley: Hoboken, NJ, USA, 2001.
38. Yang, H.; Xiao, X.; Zhao, X.; Wu, Y. Intrinsic fluorescence spectra of tryptophan, tyrosine and phenyloalanine. In Proceedings of the SPIE—The International Society for Optical Engineering, Jinhua, Suzhou, Chengdu, Xi'an, Wuxi, China, October and November 2016.
39. Valensin, G.; Kushnir, T.; Navon, G. Selective and nonselective proton spin-lattice relaxation studies of enzyme-substrate interactions. *J. Magn. Reson. (1969)* **1982**, *46*, 23–29. [CrossRef]
40. Gráf, L.A.; Szilágyi, L.A.; Venekei, I.A. Chymotrypsin. In *Handbook of Proteolytic Enzymes*; Academic Press, Elsevier: San Diego, CA, USA, 2013; Volume 3, pp. 2626–2633.
41. Loftsson, T.; Brewster, M.E. Cyclodextrins as functional excipients: Methods to enhance complexation efficiency. *J. Pharm. Sci.* **2012**, *101*, 3019–3032. [CrossRef]
42. Piras, A.M.; Fabiano, A.; Chiellini, F.; Zambito, Y. Methyl-β-cyclodextrin quaternary ammonium chitosan conjugate: Nanoparticles vs. macromolecular soluble complex. *Int. J. Nanomed.* **2018**, *13*, 2531–2541. [CrossRef]

polymers

MDPI

Article

Experimental Cold-Cured Nanostructured Epoxy-Based Hybrid Formulations: Properties and Durability Performance

Mariaenrica Frigione [1,*], Mariateresa Lettieri [2], Francesca Lionetto [1] and Leno Mascia [3]

1 Department of Innovation Engineering, University of Salento, 73100 Lecce, Italy; francesca.lionetto@unisalento.it
2 CNR-SPIN, 84084 Fisciano (Salerno), Italy; mariateresa.lettieri@cnr.it
3 Department of Materials, Loughborough University, Loughborough LE11 3TU, UK; l.mascia@lboro.ac.uk
* Correspondence: mariaenrica.frigione@unisalento.it; Tel.: +39-0832-297215

Received: 22 January 2020; Accepted: 18 February 2020; Published: 19 February 2020

Abstract: Different hybrid epoxy formulations were produced and cold-cured, monitoring the properties development during low temperature curing and aging. All systems were based on silane functionalized bis-phenol A (DGEBA) resins (Part A), cured at ambient temperature with two amine hardeners (Part B). The different components of the formulations were selected on their potential capability to bring about enhancements in the glass transition temperature. The durability of the produced hybrids was probed in comparison to the corresponding neat epoxies by monitoring changes in glass transition temperature (T_g) and flexural mechanical properties after exposure to different levels of humidity and immersion in water and at temperatures slightly higher than the local ambient temperature, in order to simulate the conditions encountered during summer seasons in very humid environments. The thermal degradation resistance of the hybrid systems was also evaluated by thermogravimetric analysis.

Keywords: cold-cure; epoxy resins; organic-inorganic hybrids; sol-gel

1. Introduction

Polymer-based nanocomposites have attracted a great interest over the last few decades due to their highly enhanced mechanical, electrical, thermal, barrier, and optical properties, derived from synergistic combinations of polymer matrix and inorganic nano-dimensional reinforcing and thermal resistant interpenetrating domains. From this advantageous combination new materials can be produced, exhibiting enhanced properties that can be tailored according to application requirements [1].

This class of polymeric nanocomposites is usually referred to as organic-inorganic (O-I) hybrid materials to denote a chemical and morphological structure consisting of nanoscale repeating units between the organic and inorganic constituents of nanoscale dimensions [2]. In such nanostructured systems, both the inorganic nano-dispersed component and the polymeric matrix are produced concurrently by the sol-gel method, consisting of hydrolysis and condensation reactions to produce three-dimensional interpenetrating networks. The very high surface area to volume ratio of the inorganic and organic components is considered a major factor responsible for the enhanced properties, which have successfully exploited in many applications, such as in the field of structural adhesives [3].

Many structural adhesive bonding applications make use of thermosetting resins (very often epoxy type), where curing reactions are carried out at cold, either "on field" or the products are too large to allow the use of high temperatures from typical sources of heat (ovens, lamps). Typical applications of cold-cured epoxies include: i) Structural adhesives employed either in construction (resins for injection) and in aeronautical/aerospace ("patch repair" technique) industries, and ii) matrices/adhesives for fiber-reinforced polymer (FRP) components to be used for the repairing and strengthening of infrastructures [4,5].

As well, testified by different papers that appeared in the last decade [6–11], cold-cured adhesives display intrinsic weaknesses, not only as a consequence of a hardening process carried out at cold and under uncontrolled temperature conditions but also due to adverse weathering effects on properties. Only moderate glass transition temperatures (T_g) can be achieved with such cold-cured resins: Never greater than 55–60 °C even after very long curing times (in the order of several weeks), which can be further reduced through the plasticization effect of liquid water or moisture in the atmosphere under outdoor service conditions. These aspects raise great concerns about the long-term performance of such cold-cured resins, even though these resin systems are already employed in many structural applications as cold-cured epoxy adhesives and matrices for fiber reinforced composites. For such applications, therefore, cold-cured polymer-based nanocomposite materials can offer significant advantages over commercial resins [3,12].

Nanomaterials are not "unknown" in the construction industry. For instance, it has been shown that concretes can be engineered by the incorporation of nanosized building blocks (e.g., nanoparticles and nanotubes) to enhance their performance [13–15]. It is also possible to graft molecules onto cement particles, cement phases, and aggregates (including nanosized additives) to provide surface functionality, which can be adjusted to promote specific interfacial interactions.

In this framework, nanostructured hybrid systems, based on common epoxy resins containing interpenetrating silica nanodomains, are expected to offer superior thermal (especially in terms of glass transition temperature), mechanical, and adhesive properties and greater durability against moisture, temperatures, hash environments, and fire over the commercial products. They are also expected to provide enhanced mechanical properties in comparison to the pure polymeric matrix due to the reinforcing effect of the inorganic nanodomains. Furthermore, the latter feature is likely to promote the adhesion to the typical substrates encountered in infrastructures (concrete, masonry, steel, wood, etc.). The use of such new cold-cured resins in the construction industry, therefore, would allow improvements in performance of existing constructions (buildings, bridges, galleries, etc.) by avoiding the demolishment and reconstruction of new ones. The use of nanomaterials as modifying agents, therefore, would be make it possible to extend the longevity of the structures for longer periods.

O-I hybrids, characterized by a morphology consisting of co-continuous silica-domains in the region of 5–20 nm of organic chains chemically linked to the inorganic phase, are obtained through the sol-gel method [16–20]. Usually, the first step in the production of the organic-inorganic hybrids is constituted by the silane-functionalization of the epoxy resin. Then, hydrolysis reactions of the functionalized epoxy with an appropriate alkoxysilane component takes place in presence of water. The final step of the synthesis of organic-inorganic hybrids is the simultaneous cross-linking of the (organic) epoxy-based component of the mixture, upon the addition of a suitable curing agent, and the condensation of the (inorganic) siloxane domains, leading to an in situ production of silica. In the case of cold-curing epoxy-based hybrids, the latter two processes both occur at ambient temperature.

In previous studies, different epoxy-based nano-silica hybrids were produced [21–23]. In this paper, the physical properties and durability characteristics of the hybrids developed in previous studies are reiterated, adding also unpublished results, highlighting at the same time differences between various new epoxy-based organic-inorganic hybrids. Although based on typical bisphenol epoxy resins, the O-I hybrid systems in this work differed for the nano-silica contents, as well as for the hardeners and the molar ratios. At the same time, changes are made regarding the alkoxysilane ingredients as precursor for the silica component and the procedure used for the pre-functionalization of the resin. Particularly novel is the use of a specific deep eutectic solvent (DES) as an auxiliary component for property enhancement: To the best of our knowledge, DES systems have not been used hitherto for epoxy hybrid compositions. The research investigated also the effects of different curing agents, both suitable for cure at ambient temperature, including the addition of an accelerator, and ammonium molybdate, as a siloxane network enhancer. The ingredients for the different hybrid systems were carefully selected with the view to reduce the curing times and improve the physical properties, especially the glass transition temperature for cure performed at ambient temperature. In particular, the present study compares the characteristics of different formulations, highlighting the effect of any single component on the final cured system.

2. Experimental

2.1. Materials and production of Hybrid and Non-Hybrid Formulations

Different experimental two-part formulations, whose compositions are detailed in Table 1, were produced using the following materials and procedures. In Scheme 1, the chemical structure of all the components employed are summarized.

Table 1. Compositions of the produced systems.

System	Resin (Part A)	Curing Agent (Part B)	Amine/Epoxy Ratio	SiO_2 Content (%)
B0	DGEBA	PACM + M851	0.75/1	-
BSi	DGEBA	PACM + M851	0.75/1	7.5%
BSiMo	DGEBA	PACM + M851 + $(NH_4)_2Mo_2O_7$	0.75/1	7.5%
Control DGEBA	DGEBA	PACM	0.75/1	-
Hybrid DGEBA	DGEBA	PACM	0.75/1	15.0%
Hyb-L-B	DGEBA	PACM	0.75/1	7.0%
Control Epoxy A	DGEBA	TETA	1:1	-
Hyb-L-A	DGEBA	TETA	1:1	5.1%

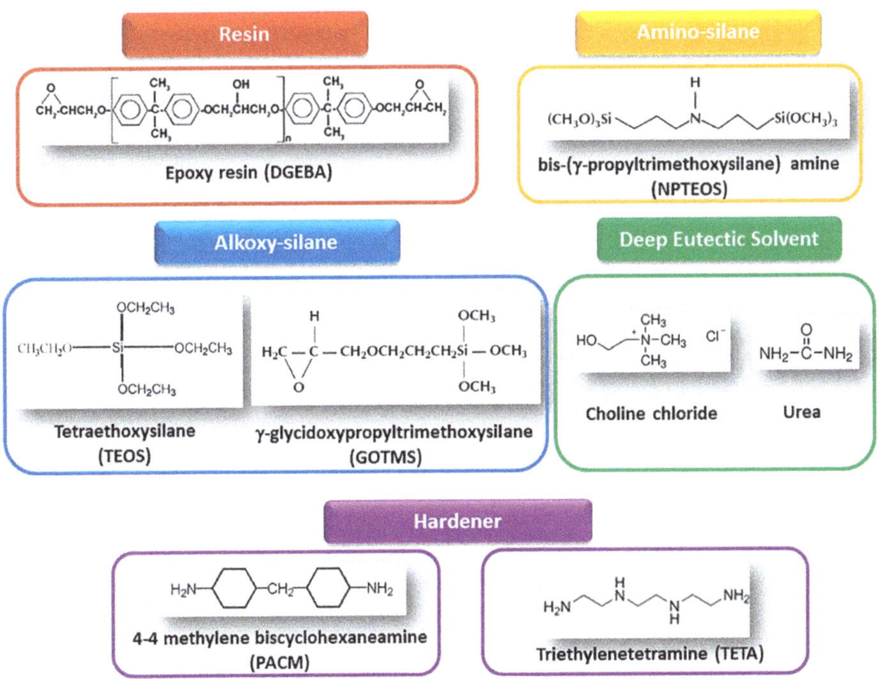

Scheme 1. Chemical structures of all the components employed.

In Scheme 2, the main steps that were followed for the production of the hybrids of this study are illustrated.

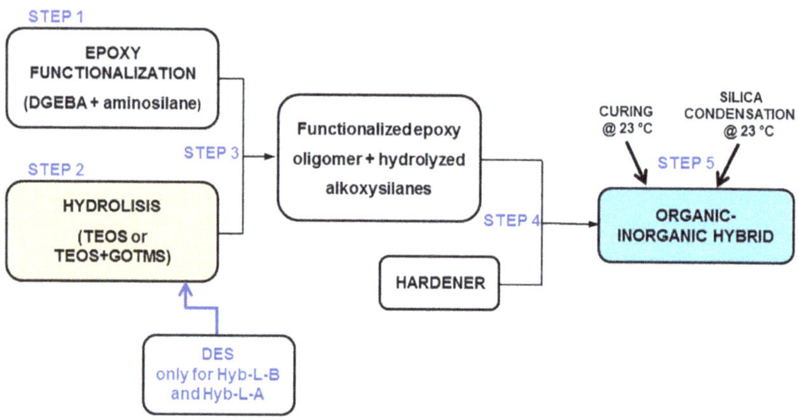

Scheme 2. Schematic presentation of steps involved in the production of the epoxy-silica hybrids.

All systems were based on typical bis-phenol A (DGEBA) resins. For the systems Control B0, BSi and BSiMo, a resin commercially known as NPEL 128E, obtained from Nan Ya Epoxy Resins was employed, characterized by an epoxy equivalent weight of 184–190 g. The remaining systems were produced with Epikote 828 (Resolution Performance Products), possessing the same epoxy equivalent value (i.e., 184–190 g).

Step 1: Silane functionalization of epoxy resin. The procedure used has been described elsewhere [24].

For the systems BiSi and BiSiMo, an amine-trialkoxysilane (supplied by Aldrich with a purity greater than 95%) was employed as a coupling agent, as well as an alkoxysilane possessing a high functionality, again supplied by Aldrich, as experimented in a previous study [24]. For the production of the Hybrid DGEBA, Hyb-L-A and Hyb-L-B systems, the bisphenolic resin was functionalized with bis-(γ-propyltrimethoxysilane) amine, commercially known as NPTEOS and purchased by Aldrich (purity>90%). The production of Hyb-L-A and Hyb-L-B systems is the subject of a patent [25]. The resulting compounds constituted Part A for each two-part hybrid formulation.

Step 2: Hydrolysis of alkoxysilane precursors.

The second step consisted of a prehydrolysis of alkoxysilane precursors. Only in the case of BiSi and BiSiMo hybrids, Tetra-ethoxysilane (TEOS, supplied by ACROS with a purity greater than 95%) was used in sufficient amounts to produce a 7.5 wt% nominal SiO_2 content, based on the total conversion of the alkoxysilane groups in the mixture [21]. For other hybrid systems (i.e., Hybrid DGEBA, Hyb-L-B, Hyb-L-A), a mixture of tetra-ethoxysilane (TEOS) and glycidoxypropyltrimethoxysilane (GOTMS) was used, both supplied by Aldrich (at a purity greater than 97%). A molar ratio 1:0.12 of TEOS:GOTMS was used in all the three cases. For the production of Hyb-L-A and Hyb-L-B systems, a deep eutectic solvent (DES) was added in the TEOS-GOTMS mixture. DES was obtained by mixing choline chloride (ChCl) and urea (U) in a 2:1 molar ratio; the two chemicals were supplied by Iolitec GmbH (Heilbronn, Germany) with a purity greater than 97%. The content of ChCl-U was set at 2.5 parts per hundred resins (phr), corresponding to 1.5% by weight of total mixture [23].

Step 3: Mixing of prehydrolyzed alkoxysilanes and functionalized epoxies.

This was carried out by mixing for 2 h the different solutions prepared in steps 1 and 2.

Step 4: Addition of the hardener.

Two hardeners were employed to cure the different hybrid systems, both suitable for the cure at ambient temperature. The first one was a cycloaliphatic amine, i.e., the 4,4′-methylenebis-cyclohexaneamine (PACM) supplied by Aldrich. PACM amine, although requiring somehow longer cold-curing times, is reported to produce superior elevated temperature performance. In all the formulations cold-cured by PACM, a molar ratio epoxy/amine = 1:0.75 was used. This choice was derived from previous studies carried out on cold-cured epoxy coatings, where the molar ratio 1:0.75 was identified as the optimum epoxy/amine ratio for property enhancement in relation to glass transition temperature range and the stiffness of systems cured at ambient temperature [26,27]. In the systems B0, BSi, and BSiMo, a small amount (<1%) of phenylen bis-methylamine M851 accelerator (supplied by Leuna Hartze) was also added, as a possible means to speed the curing reactions carried out at ambient temperature. In the BSiMo system, ammonium molybdate powder (with chemical formula $(NH_4)_2Mo_2O_7$ and obtained from ACROS) and dibutyl tin dilaurate catalyst (obtained from Aldrich with a purity greater than 95%) were mixed to PACM + M851 hardener at 6.0 and 1.0 wt%, respectively, with respect to the nominal SiO_2 content of the hybrid formulation. The addition of ammonium molybdate was expected to produce a denser siloxane network in the organic-inorganic system and to increase the T_g of the organic phase [28,29]. Molybdate salts, acting as inhibitors, have been previously added to the hybrid polymer coatings obtained via the sol-gel process for passivation of metals, steels, alloys [30–33]. It should be noted that BSi and BSiMo were proprietary systems produced by SAFE Marine Nanotechnologies (Ferentino, Frosinone, Italy), therefore some details in the compositions were not provided by the manufacturer.

The second curing agent selected for the curing at ambient temperature was an aliphatic amine, i.e., Triethylenetetramine, TETA, supplied by Elantas Italia S.r.l. (Italy) with the commercial name IG 824-K24. TETA curing agent was used without further modification. In all the systems cured with TETA, a stoichiometric (i.e., 1:1) molar ratio epoxy:amine was used.

The unmodified or modified curing agents constituted Part B of the hybrid systems, as well as for the control formulations.

Step 5: Cold-curing of hybrid (and non-hybrid) formulations.

Irrespective to the formulation examined (listed in Table 1), Part A and Part B were manually mixed at ambient temperature for a maximum of 30 min in order to avoid premature gelation. The liquid formulations were, then, poured in Teflon molds possessing different dimensions, depending on the standard tests to perform on them. The specimens were cured at ambient temperature in a controlled environment (23 ± 2 °C and 55% ± 5% relative humidity) for prolonged periods up to one year. The long-term cold-curing process was carried out on "free" specimens, i.e., they were all removed from the mold after few days of curing. During this last step of the process, the cross-linking reactions of the epoxy component and the nano-silica condensation of the inorganic phase are taken to an advanced stage, producing an interpenetrating network where organic and inorganic phases are interconnected at nanosize level.

The nominal SiO_2 contents of the cold-cured epoxy-siloxane hybrids are summarized in Table 1. For comparison purposes, an epoxy-based non-hybrid control formulation, representative of commercial cold-cured resins, was realized for each hybrid system.

2.2. *Characterization of Hybrid and Non-Hybrid Formulations in Standard Conditions*

The first parameter analyzed on all hybrid systems was the curing time. The effect of this parameter on different physical properties developed by organic-inorganic epoxy-silica formulations was, in fact, assessed, taking as reference the control non-hybrid systems.

As already explained, the main drawback of conventional cold-cured resins is the relatively "low" glass transition temperature, even after very long curing times. The evolution of the T_g was monitored by differential scanning calorimetry (DSC), using a Mettler Toledo DSC 822, on all the produced hybrid systems during the cold-cure process performed in a controlled environment. Curing times up to a year were analyzed. Dynamic DSC scans were carried out at a heating rate of 10 °C/min, from sub-zero temperature up to 300 °C. The specimens (with thicknesses not greater than 0.8 mm and an average mass of 10–14 mg) were placed in aluminum pans and analyzed under nitrogen atmosphere (flow rate: 60 mL/min), in order to avoid the occurrence of any undesirable oxidation reaction. The calorimetric runs were performed on three samples of each formulation and the results were averaged. The DSC analysis allowed to calculate also the residual heat of reaction after different cold-curing periods.

Flexural tests and dynamic mechanical thermal (DMTA) analysis were used to characterize the cold-cured hybrid materials. The flexural properties were measured in a three-point bending mode, using a LR5K Lloyd Instruments Machine according to the ASTM D790 standard [34] on rectangular specimens (100 × 10 × 4 mm^3), with a span/thickness ratio of 16:1, and at a rate of 1.7 mm/min. The systems analyzed were cured in controlled laboratory conditions for four months. The effect of shorter curing times was also assessed on some of the produced systems. At least five specimens were tested for each analyzed composition. According to Equations (1) and (2), the flexural strength (σ_{fl}) and flexural modulus (E_{fl}) were calculated as:

$$\sigma_{fl} = \frac{3 \times F_{max} \times L}{2 \times b \times h^2} \quad (1)$$

$$E_{fl} = \frac{F \times L^3}{4 \times b \times h^3 \times \delta} \quad (2)$$

indicating with F_{max} (N) the load at failure, L (mm) the span length, F (N) and δ (mm) the actual load and its displacement below the elasticity limit, respectively, b (mm) and h (mm) the width and the height of the tested specimen, respectively.

Dynamic mechanical thermal analysis was carried out on standard samples (40 × 10 × 1 mm^3) employing an ARES Rheometer (Rheometric Scientific) in the rectangular torsion configuration. The DMTA tests were performed in the scanning temperature mode, using a constant heating rate of 2 °C/min, in the range 30–150 °C, at a constant amplitude (1%) and frequency (1 Hz). Samples of Control and Hybrid DGEBA systems were analyzed in a DMTA module after a seven-month cold-cure carried out in controlled conditions. DES-based hybrids, and their respective controls, were cold-cured and aged for four months in the same controlled conditions. The tests were run in triplicate on each formulation and the results averaged for measurements of the storage and loss modulus values as functions of temperature. The glass transition temperature was measured as the maximum of the loss modulus (G″) curve.

Two scanning electron microscopies (a Zeiss EVO 40 SEM instrument and an ESEM, environmental scanning electron microscope, Mod. XL 30) were used to examine the internal morphology of fractured specimens. The specimens analyzed were fractured after a 1-min immersion in liquid nitrogen. The internal surface of some samples fractured during the flexural tests were also examined. The ESEM analyses were performed on samples without metallization, in "low vacuum" mode, with a pressure of 0.6 Torr, a beam accelerating voltage of 25 kV, and a working distance of 10 mm; secondary electron (GSE) detector was used. Energy-dispersive X-ray spectroscopy (EDS), coupled to the ESEM microscope, was applied to perform qualitative/quantitative elemental analyses on some of the produced hybrids. The spectra were collected in spots (live time 30 s) and the related results are reported as the average on five spectra. The EDS spectra were processed using the software Genesis Spectrum (version 6.2, EDAX Inc., Mahwah, NJ, USA).

2.3. Characterization of Hybrid and Non-Hybrid Formulations After Aging in Severe Environmental Conditions

One of the expected advantages of the hybrid cold-cured epoxy-based resins is their higher durability. In this study, this characteristic was determined by studying the effects of severe environmental conditions on both T_g values and mechanical properties.

The flexural properties were measured on the produced cold-cured (B0, BSi, BSiMo) specimens after their exposure to different levels of humidity (varying from 55% to 100%) for different times, up to three months. Before the exposure/immersion tests, the specimens were cold-cured in air for one month at ambient temperature and, then, dried to a constant mass for one additional month. The latter step was performed in a desiccator containing silica gel (corresponding to 10%–15% R.H.) for a total of cold-curing/aging period of two months. On the same aged specimens, the T_g values (from the DSC analysis) were also recorded as a function of exposure/immersion time.

Mechanical tests in flexural mode were also performed on cold-cured Hybrid DGEBA after exposure to moisture/immersion in water, as well as employing a testing temperature slightly higher (i.e., 50 °C) than the laboratory temperature, taking as a reference the control system. The choice of this test temperature was made based on a previous experiment performed during summer in Lecce (Italy) on a concrete component with the surface exposed to radiation of sun, in line with the typical applications of such cold-cured epoxy resins, such as adhesives for concrete in a Mediterranean climate. The temperature of this climatic condition for both inside and outside faces of the concrete component can surpass 50 °C with the temperature of air around 40 °C [35]. All the specimens of Hybrid DGEBA systems were cured at ambient temperature in a controlled environment (at 23 ± 2 °C and 55% ± 5% R.H.) for at least four months, in order to perform the experiments on a stable system.

For the mechanical tests carried out after different aging regimes, the specimens of Control and Hybrid DGEBA were exposed to a relative humidity of 75% ± 5% in a climatic chamber or immersed in distilled water for prolonged times (up to about six months), and subsequently tested in flexural mode. The results of flexural tests performed on the hybrid/control systems after the described aging procedures were, then, compared to those measured on unaged specimens cold-cured for the same time.

Finally, the thermal oxidation characteristics, as an indication of the fire behavior of the resins, was assessed by thermogravimetric analysis (TGA) on the hybrid/control Hyb-L-A and Hyb-L-B systems (four-month cure), employing a TGA/DSC1 (Stare System, Mettler Toledo). The samples, about 12–15 mg in weight, were placed in alumina pans and tested in air atmosphere from laboratory temperature (about 23 °C) up to 800 °C, at a heating rate of 10 °C/min. The TGA experiments were carried out in air to measure the decomposition temperatures of the produced systems to obtain an indication of the thermal oxidative resistance of the resins in a fire event.

3. Results and Discussion

3.1. Evolution of Thermal Properties during the Cold-Curing

The progress of the glass transition temperature displayed by the different hybrids was monitored during cold-curing times up to a year. In Figure 1, the average values of T_g (by DSC analysis) for all the systems examined are reported as a function of the curing time at ambient temperature and in the environmental conditions previously described. The general behavior of the cold-cured systems can be characterized in terms of T_g increments during the first stage of curing, irrespective of the type of formulation. In the later stages, when a stable state is achieved, on the other hand, the behavior is strongly dependent on the composition. After vitrification any further (small) increase in the degree of crosslinking can only be brought about by the diffusion of unreacted species present in the network, possessing sufficient mobility to reach neighboring reactive sites. However, the contribution to advancement of curing of these latter reactions is very limited and, practically, the T_g can be considered to have reached its ultimate value, even though the system may not be fully cured at this stage, which is manifested as residual heat of reaction in DSC thermograms [22].

The most notable features of these data are that the T_g values are significantly higher (as indicated in the different graphs reported in Figure 1) than those found for the corresponding control resins, alongside a reduction in the cure time required to reach a constant value for the T_g.

The lowest T_g values were registered for the systems BSi and BSiMo, reaching T_g values in the region of 65 °C after about a two-month curing time. The attainment of stable systems within a relatively short curing time (two months) confirms the advantageous role of the M851 accelerator in speeding up the crosslinking reactions. The residual heat of reaction (ΔH_{res}) of the B0 system was found to decrease during the first weeks of cure and remained unchanged after curing times longer than two months. This residual heat feature was substantially not visible for the BSi and BSiMo hybrids (reported for BSiMo in Figure 2a). No exothermic phenomenon ascribable to residual cross-linking was observed in the DSC curve (Figure 2a); in addition, in the temperature range between 75 and 180 °C, a large endothermic peak, most likely due to the evaporation of alcohol formed during the hybridization of the epoxy component, can hide the residual heat of reaction. The entrapment of alcohol could have been responsible for the modest increase in T_g of BSi and BSiMo hybrids (never greater than 65 °C); nevertheless, the T_g values result are still higher than that measured for B0 control system.

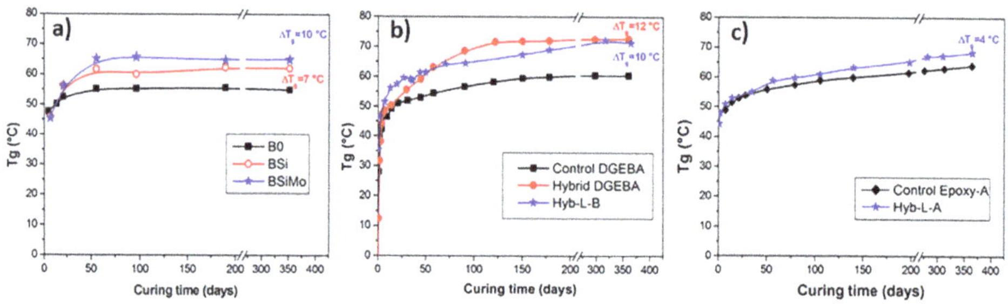

Figure 1. Glass transition temperature (T_g) values at different cold-curing times for hybrid and control epoxy systems: (**a**) B0, BSi, BSiMo; (**b**) Control bis-phenol A (DGEBA), Hybrid DGEBA, Hyb-L-B samples; (**c**) Control Epoxy A, Hyb-L-A samples.

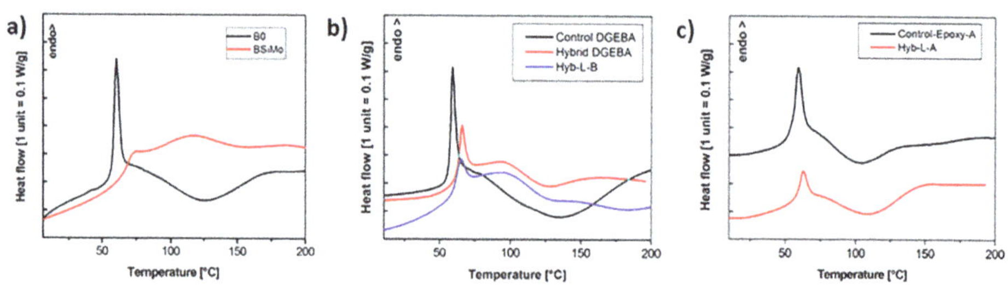

Figure 2. Typical differential scanning calorimetry (DSC) curves found for: (**a**) B0 and BSiMo; (**b**) Control DGEBA, Hybrid DGEBA, Hyb-L-B samples; (**c**) Control Epoxy-A, Hyb-L-A samples. All the systems were cold-cured for prolonged (greater than two months) curing times.

It is confirmed, furthermore, the positive effect of ammonium molybdate over the unmodified resin, reflected in the T_g values for the BSiMo hybrid being higher than those exhibited by the system not containing this component, i.e., BSi. The effect on T_g of the molybdate component was explained in terms of an acceleration of the development in T_g, as already observed in heat-cured hybrid systems [28]. From an applicative point of view, a crucial weakness noticed for these hybrid systems has been their short shelf life (a few months), even when they are stored at temperatures close to zero. This latter peculiarity, attributed to the effect of fast condensation reactions in presence of silanes [30,31,36], contributed to hinder a commercial exploiting of such systems.

The Hybrid DGEBA formulation, containing neither the accelerator for crosslinking reactions nor ammonium molybdate, developed a glass transition temperature even greater than 70 °C but after a slightly longer cold-curing time, i.e., about four months. The hardener employed to cold-cure this particular hybrid was the same employed in our previously studied systems (PACM). It is worth noting that T_g values at around 70 °C are higher over the commercially available epoxy-based resins reported in the literature [37–39]. As observable in Figure 2b, an exothermic peak, testifying that the cross-linking reactions were not completed during the cold-cure, is clearly visible also in the thermogram of Hybrid DGEBA, even if it is appreciably smaller if compared to the control resin (see Figure 3b). On the other hand, no evaporation of alcohol can be noticed from the same curves.

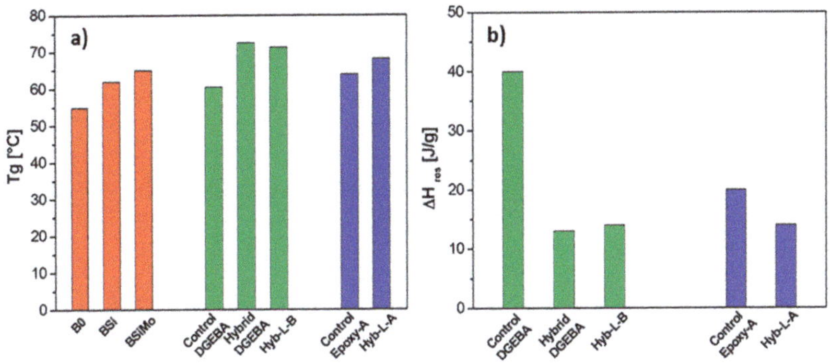

Figure 3. Final (**a**) glass transition temperatures (T_g) and (**b**) residual heat of reaction, (ΔH_{res}), measured on cold-cured systems: B0, BSi, and BSiMo; Control DGEBA, Hybrid DGEBA, Hyb-L-B; Control Epoxy-A, Hyb-L-A systems.

T_g values greater than 70 °C were achieved also for the Hyb-L-B system, again cured with the cyclohexaneamine PACM and containing a low amount nano-silica (7 wt%), after a prolonged cold-cure (about 9–10 months). By comparing this system to Hybrid DGEBA, as illustrated in Figure 3a, it can be concluded that the addition of choline chloride-urea brings about the same increase in T_g relative to the control resin at only half the level of hybridization of the resin, even though the development of T_g seems to be slower. It should be noted that the beneficial effect of DES is achieved with only small amounts (1.5 wt%). Moreover, when compared to BSi, BSiMo hybrids, the Hyb-L-B system is capable of developing a substantial greater T_g value, even though the SiO_2 is the same, as shown in Figure 3a. It is worth noting that the residual heat for cross-linking reactions, displayed in Figure 2b and in Figure 3b, is very similar to that measured on the system Hybrid DGEBA. Again, no endothermic peak, ascribable to evaporation of alcohol, can be observed in the DSC trace (Figure 2b).

The hybrid system containing DES and TETA aliphatic amine (i.e., Hyb-L-A) can be seen to display a T_g value approaching 70 °C after a very long curing time (about one year). It cannot be excluded, however, that the systems cured with TETA hardener, irrespective of whether it is used in the hybrid or its control neat epoxy, did not reach the stable T_g state (plateau) even after one year-curing, suggesting that it could continue to rise further at longer curing/aging times. The high T_g obtained in the Hyb-L-B system, however, can be partly attributed to rigidity of the aliphatic ring in the PACM molecule [23]. Once again, evaporation was not observed in DSC curves of Hyb-L-A system (Figure 2c); furthermore, it was possible to observe and measure a residual heat of reaction (Figure 3b), that was comparable with that measured for Hybrid DGEBA and Hyb-L-B systems.

In Figure 3, the final T_g (Figure 3a) and residual heat of reaction, ΔH_{res} (Figure 3b) measured on the cold-cured, control and hybrid, systems analyzed, are summarized.

From the previous discussion and the observation of the data reported in Figure 3a, it can be concluded that the produced cold-cured hybrid epoxy-silica resins are able to provide a noticeable increase in the glass transition temperature with respect to commercial cold-cured epoxies. This observed advantage seems to arise even in the presence of a low amount of silica (i.e., 7%) in the hybridized epoxy, but in the presence of a DES component. The chemical formulation of each hybrid, in terms of both the selected components and their amounts, has a relevant effect on the final T_g value, as well as on the time required to attain a stable system. The use of a catalyst makes it possible to achieve a stable system in shorter times at the expense of the final T_g of the hybrid.

3.2. Mechanical Characteristics and Morphology in Standard Conditions

In Table 2, the results of mechanical tests are summarized, performed in flexural mode. These show that in all cases the hybrids display general enhancements in both flexural strength and modulus in comparison to the respective control resin. The hybrid systems cured for two months in the presence of accelerator M851, i.e., BSi and BSiMo, however, displayed the lowest mechanical properties, even lower than those measured on the control formulations. This could be attributed to the presence of residual alcohol produced from the hydrolysis and condensation sol-gel reactions, bearing in mind the thickness of the specimens used in the tests (4 mm) and the DSC curves presented in Figure 2a. The T_g data in Figure 1, on the other hand, show that a stable high-T_g state is reached for the two-month cure period. In the latter case, the residual alcohol can escape through small openings in the nonhermitically sealed DSC cups used.

Table 2. Mechanical properties measured in flexural mode at ambient temperature on cold-cured hybrid/non-hybrid formulations.

System	Flexural Strength (MPa)	Flexural Modulus (GPa)	SiO_2 Content (%)
B0	46.8 ± 2.9	1.2 ± 0.1	-
BSi	16.8 ± 0.7	0.5 ± 0.1	7.5
BSiMo	20.9 ± 5.4	0.8 ± 0.1	7.5
Control DGEBA (two-month cure)	26.2 ± 3.8	2.9 ± 0.1	-
Control DGEBA (four-month cure)	30.9 ± 4.5	2.2 ± 0.1	-
Hybrid DGEBA (two-month cure)	47.0 ± 5.1	3.0 ± 0.2	15.0
Hybrid DGEBA (four-month cure)	75.2 ± 4.7	3.6 ± 0.1	15.0
Hyb-L-B	61.8 ± 2.4	2.6 ± 0.1	7.0
Control Epoxy A	42.5 ± 2.0	2.3 ± 0.1	-
Hyb-L-A	64.4 ± 4.1	2.1 ± 0.9	5.1

Nevertheless, the SEM examinations performed on the BSi system (Figure 4) revealed the presence of a homogenous structure through the entire thickness, with no occurrence of segregation of components or formation of voids. The systems produced in absence of the molybdate salt were completely transparent to the naked eye, while those containing the molybdate (BSiMo) were found somewhat opalescent, which is attributed to the presence of undissolved ammonium molybdate particles.

Appreciable improvements in mechanical properties, i.e., both in flexural strength (more than 60%) and stiffness (more than 140%) relative to the control resin, were found for Hybrid DGEBA cured for four months, due to the efficient reinforcement of the inorganic phase (siloxane domains). It should be noted that at the curing time used the T_g was in the plateau stage of the cure (see Figure 1b).

In order to assess the effect due to shorter cold-curing times, the mechanical properties of these hybrid/control systems were evaluated also after a two-month cold-curing period. The results of flexural tests, reported in Table 2, confirm that longer curing times, at least four months, are necessary for a full development of mechanical properties, particularly in the case of the Hybrid DGEBA system. Anyway, the superior performance of this hybrid epoxy-silica system, especially in terms of flexural strength, is evident even for curing times below the plateau stage is reached.

In Figure 4, SEM micrographs for Control DGEBA and Hybrid DGEBA specimens fractured in nitrogen are shown. These indicate that the control sample displays the typical topology of a brittle epoxy resin, characterized by distinct fracture lines, and the hybrid system exhibits features that are characteristic to organic-inorganic hybrids, consisting of a flat pattern with diffuse silica domains finely dispersed within the organic matrix [40,41].

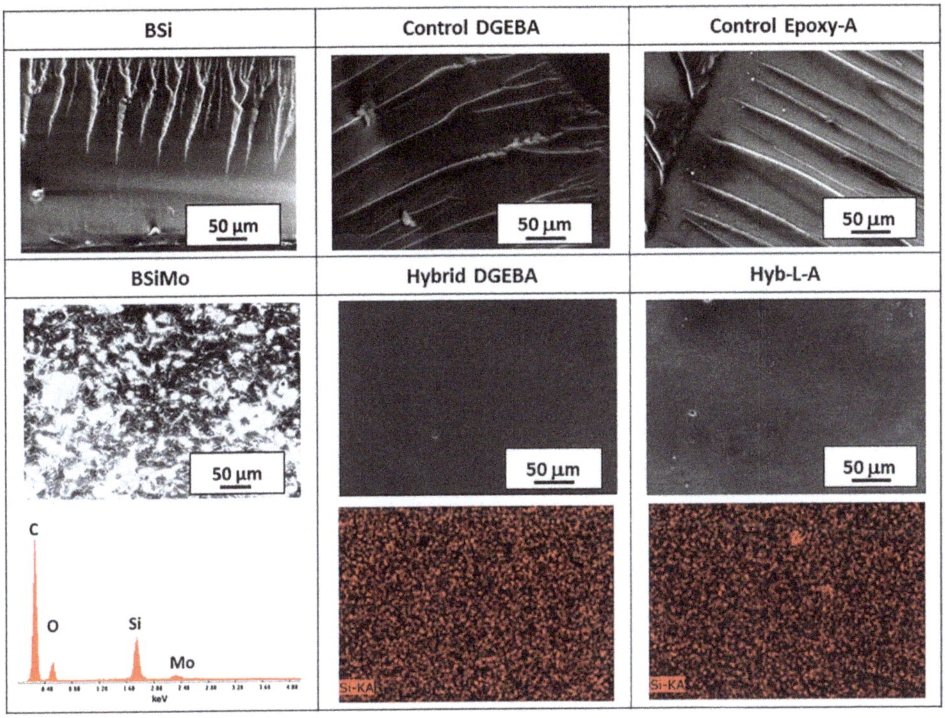

Figure 4. SEM micrographs of the different hybrid systems produced, compared with control systems. For some hybrid systems, EDS spectra or maps are also reported.

The hybrid systems cured for four months and containing the DES component exhibit a general increase in strength relative to the control resin, even twofold when a cycloaliphatic curing agent was employed (i.e., Hyb-L-B system). Only small increases in flexural modulus were observed in such hybrids relative to control resins and, again, the greatest increase was measured for the Hyb-L-B system. The flexural modulus value for the Hyb-L-B hybridized resin, on the other hand, was substantially higher than that obtained for the Hyb-L-A formulation, possibly due to a combination of factors, such as the rigid ring structure of the cycloaliphatic hardener and the lightly higher silica content in the Hyb-L-B system.

The SEM observations performed on control and hybrid systems fractured during the flexural tests are shown in Figure 4. Compared to the neat epoxies (Control DGEBA and Control Epoxy A), the hybrid systems show a significantly different pattern. This is to say that the control systems displayed the typical feature of brittle epoxy materials (i.e., long fracture lines) while no fracture lines were observed in either hybrid systems. Furthermore, in the latter case, no macroscopic phase separation between organic and inorganic phases are detectable, confirming that these are interconnected at nanoscale dimension level.

The EDS analyses performed on BSi and BSiMo revealed the presence of C, O, and Si as the main elements, while traces of Mo were found in BSiMo. In both samples, a homogeneous distribution of the constitutive elements was observed. The spectra, acquired in different points, resulted as repeatable and the content in each element (including Si), evaluated as weight percent, was uniform along the specimen.

As observed from the EDS map of the Silicon distribution in Hybrid DGEBA and Hyb-L-A samples in Figure 4, the Si signal is homogenously displayed in each part of the analyzed sample. The EDS analysis thus confirms that a homogenous distribution of silica is achieved also in these hybrids.

A comparison of flexural properties measured on the different hybrid systems produced suggests that the amount of silica (reported for each hybrid system in Table 2) has a significant effect on mechanical properties. A quantitative comparison, however, is not feasible since the procedures to obtain the different hybrids are somewhat different.

The results of DMTA tests performed on representative hybrid systems of the present work are presented in Figure 5.

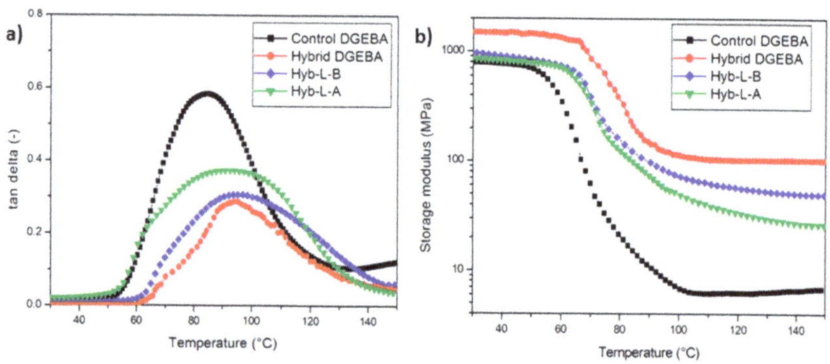

Figure 5. Results ((**a**) tanδ and (**b**) storage modulus) of the hybrids cold-cured in presence of PACM or TETA hardeners and of the Control DGEBA.

The storage modulus values at around typical ambient temperatures for potential outdoor applications of the Hybrid DGEBA system are expectedly higher than those recorded for the unmodified epoxy resin (Control DGEBA), as evident in Figure 5b. The increase in G′ modulus upon hybridization of epoxy is even more pronounced in the rubbery-plateau region (one order of magnitude), which is accompanied by a substantial depression of the viscous losses relative to Control DGEBA. Adding these characteristics to an approximately 10 °C increase in T_g (calculated as the peak of tanδ curve, Figure 5a, the performance enhancement exhibited by these Hybrid systems can be extended to applications with higher service temperatures, such as hot water tanks, coatings or matrices for fiber composites. The modulus at 100 °C for Hybrid DGEBA is equivalent to that of a typical medium density polyethylene with an expected incremented resistance to viscoelastic creep. It is well known, in fact, that MDPE has a Young modulus in the range 0.5–1.0 GPa [42]. The presence of co-continuous elastic siloxane domains produces a very efficient reinforcement, as well decreasing the viscoelastic nature of the system.

Very similar results can be observed also for the cold-cured hybrid systems containing the DES ChCl-U, i.e., Hyb-L-B and Hyb-L-A, shown in Figure 5a,b. The siloxane hybridization of the epoxy resin containing ChCl-U produced a synergistic effect for the enhancement of the storage modulus over the whole temperature range examined at half SiO_2 content with respect to the Hybrid DGEBA system. The increase in modulus G′ is again much more pronounced at high temperatures, i.e., the rubbery plateau region. The lower G′ values of Hyb-L-A can be ascribed to the less rigid structure of this system brought about by the aliphatic hardener compared to the cicloaliphatic hardener used for Hyb-L-B system. A rise in the T_g values, estimated as tanδ peak, is observed for Hyb-L-B and Hyb-L-A systems over the respective control resins (9 and 7 °C, respectively), alongside the depression of viscous loss at around the glass transition region. The behavior of tanδ curves indicates a hindering of the molecular motion of the organic chains brought about by the presence of silica nanodomains.

3.3. Properties and Performance of Hybrid Formulations in Severe Environmental Conditions

In order to assess the durability of the proposed systems when exposed to outdoor conditions, calorimetric and mechanical tests were performed on the produced hybrid systems after their exposure to environmental agents to simulate realistic weathering conditions.

The tests have revealed a peculiar behavior of the hybrid systems during aging in moisture/water environment. The expected deterioration in mechanical (flexural) properties, as well as in T_g, due to water plasticization, typically observed for commercial epoxy resins when aged in water or exposed to high levels of moisture [8,10,43], was not found in BSi and BSiMo cold-cured formulations, as illustrated in Figure 6. Instead, an increase of the glass transition temperatures, Young modulus, and the ultimate strength was recorded upon exposure to such environments. This unexpected behavior can be explained in terms of an increase in the network density of the siloxane nanodomains formed within the epoxy component of hybrids, resulting from a continuation of sol-gel reactions during the aging under high humidity conditions. An appreciable increase in the T_g of the hybrid material is also observed with increasing in the level of humidity in the atmosphere.

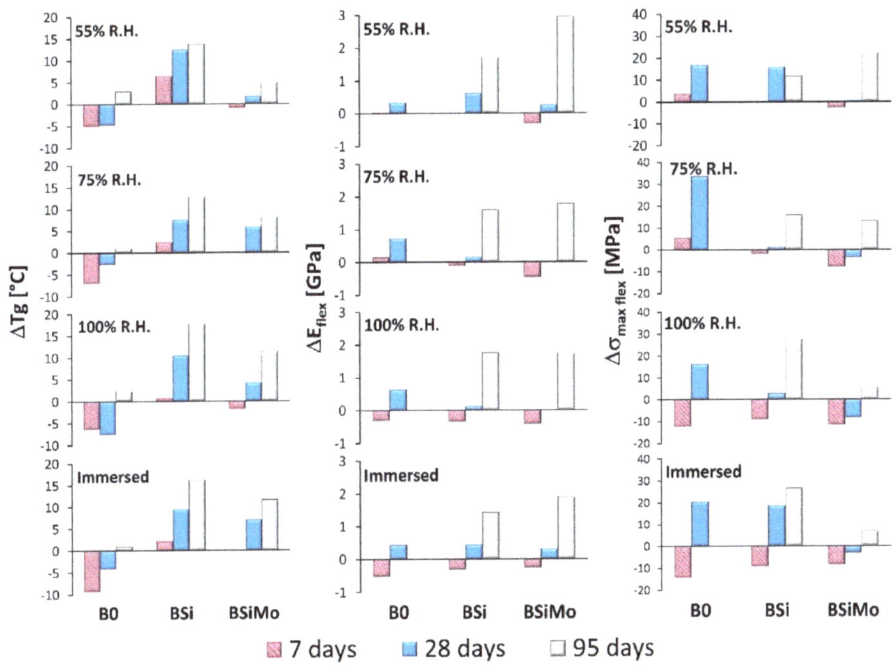

Figure 6. Variations in glass transition temperatures and in flexural mechanical properties (modulus and yield strength) for systems B0, BSi, and BSiMo subjected to different aging procedures for up to approximately three months.

After an initial decrease in T_g at shorter aging times upon immersion in water and/or exposure to 75% R.H., the Hybrid DGEBA system experienced a complete recovery of the attained T_g value after about six months aging, as shown in Table 3. Again, this can be ascribed to a continuation of sol-gel reactions activated by the absorbed water, causing further densification of the siloxane network. The superior mechanical performance of this hybrid system is confirmed from the results of the aging tests at 75% R.H. and immersion in water (see Table 3).

Table 3. Glass transition temperature (T_g) and mechanical properties (measured in flexural mode) on cold-cured Hybrid DGEBA system after different environmental exposures.

Environmental Exposure	Testing Temperature (°C)	Glass Transition (°C)	Flexural Strength (MPa)	Flexural Modulus (GPa)
75% R.H.	23	67.7 ± 1.5	66.3 ± 8.1	3.1 ± 0.1
Immersion in water	23	70.5 ± 1.8	58.5 ± 12.5	3.0 ± 0.3
None	50	-	45.9 ± 5.5	3.1 ± 0.1

The results show that the plasticization effects due to the absorbed water was appreciably reduced by the silica nanodomains formed through sol-gel reactions, which hinder the long-range molecular relaxations of the organic phase. This feature represents another important advantage of epoxy-silica hybrids over conventional epoxy systems.

In Table 3, it shown also that the cold-cured Hybrid DGEBA exhibit satisfactory mechanical properties even at 50 °C, which can be attributed to the acquired higher T_g (about 73 °C). It is worth noting that the strength and the stiffness measured on this hybrid at 50 °C remained greater than the same properties displayed by the control resin (see data reported in Table 2 for Control DGEBA). This feature represents another outstanding advantage of epoxy-silica hybrids over commercial cold-cured epoxies.

The behavior of the hybrid systems at very high temperatures is illustrated in Figure 7 with the results of thermo-gravimetric analysis performed on some of the produced hybrids.

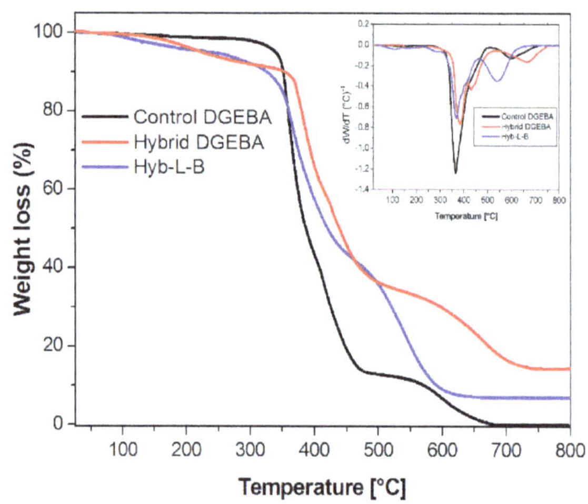

Figure 7. TGA curves of the hybrids cold-cured in presence of PACM amine and of the relative control. In the inset the derivative of the TGA curves obtained for the same systems is shown.

Compared to the neat epoxy resin, the hybrid systems in Figure 7 exhibit a significant weight loss between 100 and 300 °C, due to evaporation of volatiles (water and mostly alkanols) arising from the condensation reactions within the siloxane domains. In the case of Hyb-L-B system, the observed weight loss is also due to the thermal decomposition of ammonium cations associated with DES [23].

A high thermal resistance, also relative to the control resin, can be observed for both hybrid systems (i.e., Hybrid DGEBA and Hyb-L-B). The peak temperatures of the thermal degradation process measured for the three systems are respectively 366 °C for Control DGEBA, 383 °C for Hybrid DGEBA, and 368 °C for Hyb-L-B. The upward shift of peak degradation temperature is, therefore, very significant for the system with the highest silica content (15 wt% for Hybrid DGEBA).

The observed enhancement in the thermal stability can be again attributed to the presence of the inorganic siloxane component, which form siliceous interpenetrating barrier layers to both the infusion of oxygen and the outer diffusion of volatile pyrolysis products. In concordance with the aging results, the highest the level of hybridization, the better the thermal oxidation resistance.

4. Conclusions

Different organic-inorganic hybrid cold-cured epoxy-silica formulations were produced and cold-cured, aiming to provide the basis for the development of new systems as efficient cold-cured adhesives for structural applications. The study has evaluated the more relevant physical properties, as well as the durability in different exposure regimes. It was found that these novel O-I epoxy-based hybrids display significant advantages over the conventional epoxy resins used as structural adhesives or as matrices for fiber reinforced composites in terms of higher glass transition temperatures, better mechanical properties and enhanced durability in aqueous environments. The investigation has examined also the effects of different amine curing agents, as well as the use of a typical accelerator (M851), ammonium molybdate, and a deep eutectic solvent. These novel systems could provide a viable solution to the well-known durability issues of the conventional cold-cured epoxy systems employed for concrete repairing and structure strengthening applications. Ongoing studies are examining the possibility of using such cold-cured epoxy-silica hybrids in conjunction with precured fiber reinforced composites or other carbon fiber-based products suitable for retrofitting applications in structural engineering.

5. Patent

Some of the systems presented in the paper were developed in the framework of the patent: F. Lionetto, M. Frigione "Organic-inorganic hybrids polymerized in situ at room temperature", EP2977407 A1.

Author Contributions: Conceptualization, M.F.; methodology, M.L. and F.L.; investigation and formal analysis, M.L. and F.L.; data curation, M.L. and F.L.; writing—original draft preparation, M.F.; review and editing, L.M.; supervision, M.F.; projects administration, M.F.; funding acquisition, M.F. All authors have read and agreed to the published version of the manuscript.

Funding: This research was partly funded by PON 2007–2013 "IT@CHA—Italian Technologies for Advanced Applications in Cultural Heritage Assets" and by Apulia Regional projects "Applied Innovation Technologies for Diagnosis and Conservation of Built Heritage—A.I.Te.C.H." and "Protezione, consolidamento e pulitura di materiali lapidei caratteristici della Regione Puglia: Sperimentazione di prodotti a basso impatto ambientale e monitoraggio dei trattamenti".

Acknowledgments: The authors wish to thank: Maurizio Masieri for his technical support in performing the ESEM-EDS analyses on the BSi and BSiMo samples; and Luca Prezzi and SAFE Marine Nanotechnologies, Ferentino (FR), Italy for supplying some of the experimented materials (BSi and BSiMo set).

Conflicts of Interest: The authors declare no conflict of interest.

References

1. Zaferani, S.H. Introduction of polymer-based nanocomposites. In *Polymer-based Nanocomposites for Energy and Environmental Applications*; Jawaid, M., Khan, M.M., Eds.; Woodhead Publishing: Cambridge, UK, 2018; pp. 1–25.
2. Mascia, L.; Tang, T. Curing and morphology of epoxy resin-silica hybrids. *J. Mater. Chem.* **1998**, *8*, 2417–2421. [CrossRef]
3. Jojibabu, P.; Zhang, Y.X.; Prusty, B.G. A review of research advances in epoxy-based nanocomposites as adhesive materials. *Int. J. Adhes. Adhes.* **2020**, *96*, 102454. [CrossRef]
4. Karbhari, V.M. Building materials for the renewal of civil infrastructure. *Reinf. Plast.* **2005**, *49*, 14–25. [CrossRef]
5. Budhe, S.; Banea, M.D.; de Barros, S. Bonded repair of composite structures in aerospace application: A review on environmental issues. *Appl. Adhes. Sci.* **2018**, *6*, 3. [CrossRef]

6. Karbhari, V.M.; Ghosh, K. Comparative durability evaluation of ambient temperature cured externally bonded CFRP and GFRP composite systems for repair of bridges. *Compos. Part A Appl. Sci. Manuf.* **2009**, *40*, 1353–1363. [CrossRef]
7. Tatar, J.; Hamilton, H.R. Comparison of laboratory and field environmental conditioning on FRP-concrete bond durability. *Constr. Build. Mater.* **2016**, *122*, 525–536. [CrossRef]
8. Silva, P.; Fernandes, P.; Sena-Cruz, J.; Xavier, J.; Castro, F.; Soares, D.; Carneiro, V. Effects of different environmental conditions on the mechanical characteristics of a structural epoxy. *Compos. Part B-Eng.* **2016**, *88*, 55–63. [CrossRef]
9. Frigione, M. Durability problems of concrete structures rehabilitated with FRP. In *Eco-Efficient Repair and Rehabilitation of Concrete Infrastructures*; Pacheco-Torgal, F., Melchers, R.E., Shi, X., Belie, N.D., Tittelboom, K.V., Sáez, A., Eds.; Woodhead Publishing: Cambridge, UK, 2018; pp. 147–170.
10. Proia, A.; Matthys, S. Influence of environmental conditions on the glass transition temperature of epoxy used for strengthening applications. *Polym. Test.* **2019**, *79*, 106012. [CrossRef]
11. Pethrick, R.A. Design and ageing of adhesives for structural adhesive bonding–a review. *Proc. Inst. Mech. Eng. Part L: J. Mater. Des. Appl.* **2015**, *229*, 349–379. [CrossRef]
12. Morshed, S.A.; Young, T.J.; Chirdon, W.M.; Zhang, Q.; Tatar, J. Durability of wet lay-up FRP bonded to concrete with nanomodified epoxy adhesives. *J. Adhes.* **2018**, 1–26. [CrossRef]
13. Saloma; Nasution, A.; Imran, I.; Abdullah, M. Improvement of Concrete Durability by Nanomaterials. *Procedia Eng.* **2015**, *125*, 608–612. [CrossRef]
14. Rattan, A.; Sachdeva, P.; Chaudhary, A. Use of Nanomaterials in Concrete. *Int. J. Latest Res. Eng. Technol.* **2016**, *2*, 81–84.
15. Norhasri, M.S.M.; Hamidah, M.S.; Fadzil, A.M. Applications of using nanomaterial in concrete: A review. *Constr. Build. Mater.* **2017**, *133*, 91–97. [CrossRef]
16. Xenopoulos, C.; Mascia, L.; Shaw, S.J. Variables analysis in the gelation of alkoxysilane solutions for the production of polyimide-silica hybrids. *Mater. Sci. Eng. C* **1998**, *6*, 99–114. [CrossRef]
17. Mascia, L.; Prezzi, L.; Haworth, B. Substantiating the role of phase bicontinuity and interfacial bonding in epoxy-silica nanocomposites. *J. Mater. Sci.* **2006**, *41*, 1145–1155. [CrossRef]
18. Mitra, J.; Ghosh, M.; Bordia, R.K.; Sharma, A. Photoluminescent electrospun submicron fibers of hybrid organosiloxane and derived silica. *RSC Adv.* **2013**, *3*, 7591. [CrossRef]
19. Piscitelli, F.; Buonocore, G.G.; Lavorgna, M.; Verdolotti, L.; Pricl, S.; Gentile, G.; Mascia, L. Peculiarities in the structure-Properties relationship of epoxy-silica hybrids with highly organic siloxane domains. *Polymer* **2015**, *63*, 222–229. [CrossRef]
20. Liu, B.; Wang, H.; Guo, X.; Yang, R.; Li, X. Effects of an Organic-Inorganic Hybrid Containing Allyl Benzoxazine and POSS on Thermal Properties and Flame Retardancy of Epoxy Resin. *Polymers* **2019**, *11*, 770. [CrossRef]
21. Lettieri, M.; Lionetto, F.; Frigione, M.; Prezzi, L.; Mascia, L. Cold-cured epoxy-silica hybrids: Effects of large variation in specimen thickness on the evolution of the T_g and related properties. *Polym. Eng. Sci.* **2011**, *51*, 358–368. [CrossRef]
22. Lionetto, F.; Mascia, L.; Frigione, M. Evolution of transient states and properties of an epoxy silica hybrid cured at ambient temperature. *Eur. Polym. J.* **2013**, *49*, 1298–1313. [CrossRef]
23. Lionetto, F.; Timo, A.; Frigione, M. Cold-Cured Epoxy-Based Organic–Inorganic Hybrid Resins Containing Deep Eutectic Solvents. *Polymers* **2019**, *11*, 14. [CrossRef] [PubMed]
24. Prezzi, L.; Mascia, L. Network density control in epoxy–silica hybrids by selective silane functionalization of precursors. *Adv. Polym. Technol.* **2005**, *24*, 91–102. [CrossRef]
25. Lionetto, F.; Frigione, M. Organic-inorganic Hybrids Polymerized in Situ at Room Temperature. EP 2977407 A1, 20 July 2015.
26. Vanlandingham, M.R.; Eduljee, R.F.; Gillespie, J.W., Jr. Relationships between stoichiometry, microstructure, and properties for amine-cured epoxies. *Journal of Applied Polymer Science* **1999**, *71*, 699–712. [CrossRef]
27. Prezzi, L. Epoxy–silica hybrids for coating applications. Ph.D. Thesis, Loughborough University, Department of Materials, Loughborough, UK, 2003.
28. Mascia, L.; Prezzi, L.; Wilcox, G.D.; Lavorgna, M. Molybdate doping of networks in epoxy–silica hybrids: Domain structuring and corrosion inhibition. *Prog. Organ. Coat.* **2006**, *56*, 13–22. [CrossRef]

29. Chruściel, J.J.; Leśniak, E. Modification of epoxy resins with functional silanes, polysiloxanes, silsesquioxanes, silica and silicates. *Progr. Polym. Sci.* **2015**, *41*, 67–121. [CrossRef]
30. Balgude, D.; Sabnis, A. Sol–gel derived hybrid coatings as an environment friendly surface treatment for corrosion protection of metals and their alloys. *J. Sol-Gel Sci. Technol.* **2012**, *64*, 124–134. [CrossRef]
31. Figueira, R.B.; Silva, C.J.R.; Pereira, E.V. Organic–inorganic hybrid sol–gel coatings for metal corrosion protection: A review of recent progress. *J. Coat. Technol. Res.* **2015**, *12*, 1–35. [CrossRef]
32. Yasakau, K.A.; Ferreira, M.G.S.; Zheludkevich, M.L. Sol-Gel Coatings with Nanocontainers of Corrosion Inhibitors for Active Corrosion Protection of Metallic Materials. In *Handbook of Sol-Gel Science and Technology*; Klein, L., Aparicio, M., Jitianu, A., Eds.; Springer International Publishing: Cham, Switzerland, 2017; pp. 1–37. ISBN 978-3-319-19454-7.
33. Eduok, U.; Szpunar, J. Ultrasound-assisted synthesis of zinc molybdate nanocrystals and molybdate-doped epoxy/PDMS nanocomposite coatings for Mg alloy protection. *Ultrason. Sonochem.* **2018**, *44*, 288–298. [CrossRef]
34. *ASTM D790-03 Standard Test Methods for Flexural Properties of Unreinforced and Reinforced Plastics and Electrical Insulating Materials*; ASTM International: West Conshohocken, PA, USA, 2003.
35. Aiello, M.A.; Frigione, M.; Acierno, D. Effects of environmental conditions on performance of polymeric adhesives for restoration of concrete structures. *J. Mater. Civ. Eng.* **2002**, *14*, 185–189. [CrossRef]
36. Loureiro, M.V.; Ciriminna, R.; Lourenço, M.J.; Santos, L.F.; De Schrijver, A.; Bordado, J.C.; Pagliaro, M.; Marques, A.C. Organically-modified silica based microspheres for self-curing polyurethane one component foams. *Microporous Mesoporous Mater.* **2017**, *244*, 244–250. [CrossRef]
37. Moussa, O.; Vassilopoulos, A.P.; de Castro, J.; Keller, T. Early-age tensile properties of structural epoxy adhesives subjected to low-temperature curing. *Int. J. Adhes. Adhes.* **2012**, *35*, 9–16. [CrossRef]
38. Frigione, M.; Naddeo, C.; Acierno, D. Cold-Curing Epoxy Resins: Aging and Environmental Effects. I-Thermal Properties. *J. Polym. Eng.* **2001**, *21*, 23–51. [CrossRef]
39. Frigione, M.; Lettieri, M.; Mecchi, A.M. Environmental effects on epoxy adhesives employed for restoration of historical buildings. *J. Mater. Civ. Eng.* **2006**, *18*, 715–722. [CrossRef]
40. Matějka, L.; Dukh, O.; Kolařík, J. Reinforcement of crosslinked rubbery epoxies by in-situ formed silica. *Polymer* **2000**, *41*, 1449–1459. [CrossRef]
41. Kaushik, A.; Kaushik, J. Solvent Absorption Characteristics of Epoxy-Colloidal Silica Nanocomposites. *J. Reinf. Plast. Comp.* **2010**, *29*, 2821–2833. [CrossRef]
42. Mascia, L. *Polymers in industry from A to Z: A concise encyclopedia*; John Wiley & Sons: Hoboken, NJ, USA, 2012.
43. Savvilotidou, M.; Vassilopoulos, A.P.; Frigione, M.; Keller, T. Development of physical and mechanical properties of a cold-curing structural adhesive in a wet bridge environment. *Constr. Build. Mater.* **2017**, *144*, 115–124. [CrossRef]

polymers

MDPI

Article

Ionic Liquid as Dispersing Agent of LDH-Carbon Nanotubes into a Biodegradable Vinyl Alcohol Polymer

Valeria Bugatti [1,2], Gianluca Viscusi [1], Antonio Di Bartolomeo [3,*], Laura Iemmo [3], Daniela Clotilde Zampino [4], Vittoria Vittoria [1,2] and Giuliana Gorrasi [1,*]

1 Department of Industrial Engineering, University of Salerno, via Giovanni Paolo II, 132, 84084 Fisciano, Italy; vbugatti@unisa.it (V.B.); gviscusi@unisa.it (G.V.); vvittoria@unisa.it (V.V.)
2 Nice Filler s.r.l., via Loggia dei Pisani, 25, 80133 Napoli, Italy
3 Physics Department "E.R. Caianiello", University of Salerno, via Giovanni Paolo II, 132, 84084 Fisciano, Italy; liemmo@unisa.it
4 Institute for Polymers, Composites and Biomaterials (IPCB)–National Research Council, Via Paolo Gaifami, 18-95126 Catania, Italy; danielaclotilde.zampino@cnr.it
* Correspondence: adibartolomeo@unisa.it (A.D.B.); ggorrasi@unisa.it (G.G.); Tel.: +39-089-969189 (A.D.B.); +39-089-964146 (G.G.)

Received: 30 January 2020; Accepted: 23 February 2020; Published: 24 February 2020

Abstract: A Zn/Al layered double hydroxides (LDHs) hosting carbon nanotubes (80% of CNTs) was synthesized and dispersed into a commercial biodegradable highly amorphous vinyl alcohol polymer at different loading (i.e., 1; 3; 5; 10 wt %). In order to improve the degree of dispersion of the filler into the polymer matrix, an ionic liquid (IL) based on 1-hexadecyl-3-methylimidazolium dimethyl-5-sodiosulfoisophthalate was added to the composites' mixtures. Structural characterization of filler and polymeric composites was carried out. The analysis of thermal, mechanical and electrical properties of the composites, resulted improved compared to the unfilled material, allowed to hypothesize a good dispersion of the LDH-CNTs lamellar filler into the polymer matrix-assisted by the ionic liquid. This was demonstrated comparing electrical conductivity of composite at 5% of LDH-CNTs in the presence and in the absence of IL. The experimental results showed that the electrical conductivity of the sample with IL is four orders of magnitude higher than the one without IL. Furthermore, the percolation threshold of the whole system resulted very low—0.26% of LDH-CNTs loading, which is 0.21% of CNTs.

Keywords: layered double hydroxides; carbon nanotubes; ionic liquids; electrical properties

1. Introduction

Polymers reinforced with different forms of nanostructured carbon, such as fullerenes, graphene, nanoplatelets, nanotubes (CNTs), etc. are currently in increasing investigation. Due to their high aspect ratio and the surface of contact with the matrix, CNTs have definite advantages for the realization of composites with remarkable electrical conductivity and mechanical properties, such as hardness, impact resistance, vibration damping and photo-protectors [1–5]. However, the conductivity obtained so far is below the expectations based on the achievable mixing and high current carrying capability of CNTs. This is mainly due to the CNTs tendency to agglomerate, determining a poor three-dimensional structure of the conductive phase. The key parameters to control for the development of innovative CNTs-based composites with improved mechanical and electrical properties are: (i) The effective dispersion of CNTs (ii) the strong adhesion to the polymeric matrix leading to the formation of strong interfacial bonding. Dispersion is a determinant factor to reach high conductivity, and for this reason, chemical functionalization of CNTs was first tried. To achieve homogeneous dispersion and strong interfacial adhesion between CNTs and polymer matrix, a considerable number of studies have been carried out based on the chemical functionalization of CNTs [6], in which activated organic groups were grafted onto the surface of CNTs. These groups ought to promote the dispersion of CNTs in solvents, as well as in polymer matrix. Moreover, it is expected that the groups also improve the compatibility between CNTs and polymer matrix, resulting in the improvement of the interfacial property between CNTs and polymer matrix [7–11]. The simplest way to obtain an organic group on CNTs is their mild oxidation giving –COOH carbon nanotubes afterwards incorporated into resins. The presence of –COOH chemical groups on the nanotube surfaces was expected to improve the dispersion into the matrix, and therefore, to increase the electrical properties. Conversely, it was found that nanotube functionalization with –COOH results in a remarkable increase of the percolation threshold, as well as in a strong reduction of the conductance of the composite. This behavior was attributed to the fact that the presence of the –COOH groups catalyzes a fraction of homo-polymerization, creating a more adherent coating to the nanotubes that determines a higher tunneling resistance between interacting CNTs, and therefore, a higher percolation threshold and the lower limit value of the conductivity [12]. An alternative approach was the linking of the –COOH group to an anionic clay, zinc- and aluminum-based layered double hydroxide (LDH) [13]. CNTs and LDHs were simply mixed and incorporated into a silicone rubber matrix [14], produced as nano-hybrid from Mg/Al and incorporated into a poly (amide-imide) [15] or self-assembled to generate a three-dimensional CNTs-NiCoAl LDH hybrid by means of a urea-hydrolysis method [16]. Here we use the co-precipitation method to synthesize LDH modified with CNTs-COOH, based on Zn/Al, and its dispersion into a novel material, that is a highly amorphous ethylene-vinyl alcohol copolymer [17]. Different dispersants have been used to avoid CNTs aggregation, in particular solid substances and volatile organic solvents. Recently, ILs have been widely used as effective dispersants of nanofillers, such as CNTs [18–20], graphene [21], fullerenes [22], as a compatibilizer in polymer blends [23,24], and to enhance the dispersion of petroleum asfaltene [25]. In this paper, we report the fabrication of polymer composites based on a commercial biodegradable highly amorphous vinyl alcohol polymer and layered double hydroxides hosting carbon nanotubes. In order to improve the degree of dispersion of the filler into the polymer matrix, an ionic liquid, based on 1-hexadecyl-3-methylimidazolium dimethyl-5-sodiosulfoisophthalate was used.

2. Materials and Methods

2.1. Materials

$Zn(NO_3)_2{\cdot}6H_2O$, $Al(NO_3)_3{\cdot}9H_2O$, NaOH were purchased from Sigma-Aldrich (Milan, Italy). Carbon nanotubes, CNTs-COOH (3101 Grade), were purchased from Nanocyl S.A. (Sambreville, Belgium) 1-methylimidazole, 1-bromohexadecane, 1,3-dimethyl 5-sulfoisophthalate sodium salt, dichloromethane (DCM), ethyl acetate, tetrahydrofuran (THF), dimethyl sulfoxide-d6 (DMSO-d6), trans-2-[3-(4-tert-butylphenyl)-2-methyl-2-propenylidene]malononitrile (DCTB), were purchased by Sigma-Aldrich (Milan, Italy). All the chemicals were high purity products and were used as received. The polymer is a highly amorphous vinyl alcohol polymer (HAVOH), OKS 8049, kindly supplied from Nichigo G-Polymer (Nippon Gohsei Europe GmbH, Düsseldorf, Germany).

2.2. Preparation of ZnAl-CNTs by Co-Precipitation Method

CNTs-COOH were co-precipitated with $Zn(NO_3)_2$ and $Al_2(NO_3)_3$ giving a hydrotalcite-like solid (LDH), in which carbon nanotubes were linked to the hydrotalcite sheets by ionic bonds. 150 mL of an aqueous solution of $Zn(NO_3)_2{\cdot}6H_2O$ (2.23 g, 7.5 mmol) and $Al(NO_3)_3{\cdot}9H_2O$ (1.4 g, 3.75 mmol) were added to 150 mL of a CNTs-COONa solution (3 g, 7.5 mmol) under stirring and under nitrogen flow. The pH slowly reached the value of 9 by adding 1M NaOH. In the end, the precipitate was washed with distilled water and left in the oven at 125 °C for 5 h, under vacuum [26].

Elemental analyses for the detection of Zn and Al atoms were carried out by an atomic absorption spectrophotometer (Model Analyst 100, Perkin Elmer, Milan Italy) using solutions prepared by dissolving the samples in concentrated nitric acid. The C and H atoms were analyzed by an Elemental

Analyzer CHNS/O (Model Flash EA 1112, Thermo, Haan Germany), equipped by a thermoconductivity detector. The values of relative percentages and molar ratio of Zn, Al, C and H are reported in Table 1. The Zn/Al molar ratio of the starting solution of the Zn and Al nitrates was confirmed.

Table 1. Values for elemental analysis of ZnAl-CNTs.

	Zn	Al	C	H
%	7.2	1.4	60.4	0.82
Grams	7.2	1.4	60.4	0.82
Moles	0.11	0.05	5.03	0.82
Molar ratio	2.1	1.0	96.5	16

The chemical formula obtained from the elemental analysis was the following: $[Zn_{0.73}Al_{0.35}(OH)_2]C_{33.7}\ O_{8.9}{\cdot}1.75H_2O$ with the value of the molar fraction $x = M_{III} / M_{III} + M_{II}$ of 0.32 and molecular weight of 668.57 g/mol; the amount of nanotubes in LDH-CNTs is 80 wt % of the total weight. Therefore, almost all the aluminum is co-precipitated with the zinc ions to obtain a solid with the stoichiometry of two Zn(II) atoms for each Al(III) atom. This corresponds to an ideal arrangement of the brucite-like sheet with each aluminum atom surrounded by six zinc atoms [27].

2.3. *Preparation of Ionic Liquid (IL)*

The ionic liquid used is 1-hexadecyl-3-methylimidazolium dimethyl-5-sodiosulfoisophthalate ($HDmim^+$ $DIMSIP^-$) $C_{30}H_{48}N_2O_7S$. Its chemical structure is shown in Scheme 1.

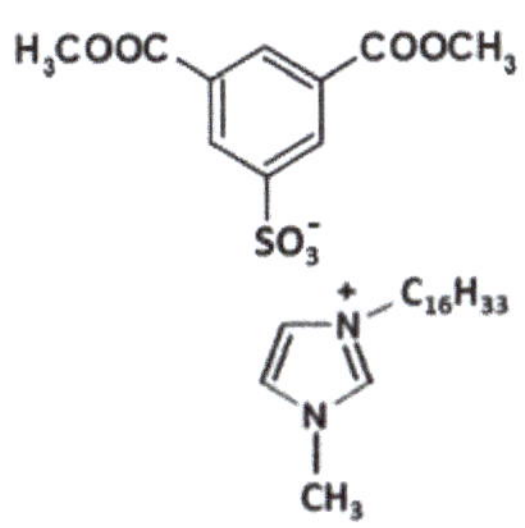

Scheme 1. Ionic liquid chemical structure.

It was synthesized in two steps.

The first one involved the alkylation of methyl imidazole with the formation of 1-hexadecyl 3-methyl imidazolium bromide. The second one was a metathesis reaction between the obtained IL bromide and 1,3-dimethyl 5-sulfoisophthalate sodium salt.

2.3.1. 1-Hexadecyl-3-methyl Imidazolium Bromide (Hdmim Br) Synthesis

The synthesis of 1-hexadecyl-3-methyl imidazolium bromide was carried out, and added to an ethyl acetate (4 mL) solution of 1-methylimidazole (1.56 mL, 20 mmol) 1-bromohexadecane (6.4 mL, 20 mmol). The mixture was stirred, under nitrogen, for 24 h at 65 °C. The compound was cooled down to room temperature, obtaining white gelatin, which was filtered and washed several times with ethyl acetate to remove unreacted initial compounds. The white solid obtained was dried under vacuum at 40 °C for 24 h (yield 95.0%).

1H NMR analysis: (200 MHz, DMSO-d6, δ ppm) 0.84 (t, 3H, CH_3–C_{15}), 1.23 (m, 26H, CH_2), 1.77 (m, 2H, CH_2–CH_2–N), 3.85 (s, 3H, CH_3–N), 4.15 (t, 2H, CH_2–N), 7.72 (s, 1H, CH in imidazolium ring), 7.78 (s, 1H, CH in imidazolium ring), 9.15 (s, 1H, N–CH–N in imidazolium ring).

2.3.2. 1-Hexadecyl-3-methyl Imidazolium Dimethyl-5-sulfoisophthalate (Hdmim DMSIP) Synthesis

1-hexadecyl-3-methyl imidazolium bromide (20 mmol, 7.77 g) dissolved in 40 mL of dichloromethane (DCM) and a solution of 1,3-dimethyl 5-sulfoisophthalate sodium salt (20.9 mmol, 6.03 g) dissolved in 130 mL of water were put together in a separating funnel and vigorously shaken (30 min) until no precipitate was present in the resulting two-phase mixture. The organic layer was separated, dried over magnesium sulfate, and the residual solvent was removed under reduced pressure. The obtained yellow viscous liquid was washed three times with ethyl acetate and dried under vacuum at 40 °C for 24 h (yield 92.0%). To verify the complete exchange of the bromide counter-ion, a silver nitrate test was performed on an aliquot of the organic layer. A new water solution containing the dimethyl-5sulfoisophthalate sodium salt was exchanged with the organic layer if the exchange was not complete.

1H NMR analysis (200 MHz, DMSO-d6, δ ppm): Signals of imidazolium ring and alkyl chain: 0.84 (t, 3H, CH_3–C_{15} chain), 1.22 (m, 26 H, CH_2), 1.75 (m, 2H, CH_2–CH_2–N), 3.83 (s, 3H, CH_3–N), 4.14 (t, 2H, CH_2–N), 7.69 (s, 1H, CH in imidazolium ring), 7.75 (s, 1H, CH in imidazolium ring), 9.10 (s, 1H, N–CH–N in imidazolium ring). Signals of benzene ring: 3.90 (s, 6H, CH_3–O), 8.37 (d, 2H, CH, orthoposition with respect to SO_{3-} substituents), 8.42 (d, 1H, CH, paraposition with respect to SO_{3-} substituents).

2.4. *Preparation of Composites*

Composites based on HAVOH with 1% of IL and 1; 3; 5 and 10 wt % of nano-hybrid LDH-CNTs were prepared by dissolving HAVOH in 30 mL of water at 100 °C for 30 min; after this time the powder of LDH-CNTs, in weight ratio (HAVOH: LDH-CNTs) 97:3, 95:5, 93:7 and 90:10 and the IL at 1 wt % were added. The solution containing the nano-hybrid LDH-CNTs, the HAVOH, and IL was then sonicated for 40 min. The mixtures obtained were slowly evaporated in Petri dishes. Films of pure HAVOH and HAVOH/LDH-CNTs/IL for each percentage of nano-hybrid were obtained in the same described experimental conditions. All films, having the same thickness ~300 μm, were dried in a vacuum oven at room temperature for 3 days. In order to better evaluate the effect of the IL on the degree of dispersion of the filler and the resulting physical properties, we prepared a composite based on HAVOH and 5 wt % of LDH-CNTs without IL, using the same experimental conditions described above.

2.5. *Methods of Investigation*

Scanning electron microscopy (SEM) was performed using an EVO MA 10 (Zeiss, Varese Italy) microscope, working in high-vacuum mode.

X-ray diffraction (XRD) patterns were taken, in reflection, with an automatic Bruker diffractometer equipped with a continuous scan attachment and a proportional counter, using nickel-filtered Cu Kα radiation (Kα = 1.54050 Å) and operating at 40 kV and 40 mA, step scan 0.05° of 2θ and 3 s of counting time (Karlsruhe, Germany).

Thermogravimetric analyses (TGA) were carried out in air atmosphere with a Mettler TC-10 thermobalance from 30 °C to 800 °C, at a heating rate of 10 °C/min (Mettler-Toledo GmbH, Greifensee, Switzerland).

FTIR was performed using a Bruker spectrometer (Vertex 70, Milan Italy). Spectra were obtained using 32 scans and a 4 cm^{-1}.

The elastic modulus, E (Mpa), of the samples was evaluated, in tensile mode, using a dynamometric apparatus INSTRON 4301 (Norwood, MA, USA). Experiments were conducted at room temperature on pure polymer and composites' films with a deformation rate of 2 mm/min. The specimens were 10 mm wide and ≅500 μm thick. The initial length of the samples was 10 mm. Elastic modulus was derived from the linear part of the stress-strain curves, giving to the samples a deformation of 0.2%. Data were averaged on five samples.

Electrical measurements were performed in a Janis ST-500 cryogenic probe station (Janis Research Company, LLC, 225 Wildwood Avenue, Woburn, MA, USA) connected to a Keithley 4200-SCS semiconductor parameter analyzer (Tektronix, Inc., 14150 SW Karl Braun Drive, P.O. Box 500 Beaverton, OR, USA). Direct current (DC) electrical measurements were performed at atmospheric pressure and room temperature. Films of ~2×2 cm^2 area and ~500–700 μm thickness were sputter coated with two Au pads, on the top and bottom surfaces. The Au pad had a thickness of 200 nm and area around 8×8 mm^2. The DC conductivity along the direction of the film thickness was obtained using a two-probe configuration by measuring the current resulting from the application of a voltage in the range 0–5 V.

3. Results

3.1. Filler Characterization

Figure 1 reports the SEM analysis of pristine LDH (a), pristine CNTs (b) and LDH-CNTs (c). The picture of pristine LDH reveals a layered plate-like morphology, typical of such kind of silicate. CNTs show a typical morphology of aggregate nanotubes densely packed, with numerous thinner bundles and single nanotubes coming out of the main bulk. For LDH-CNTs hybrid, it is evident the presence of CNTs among the LDH structure.

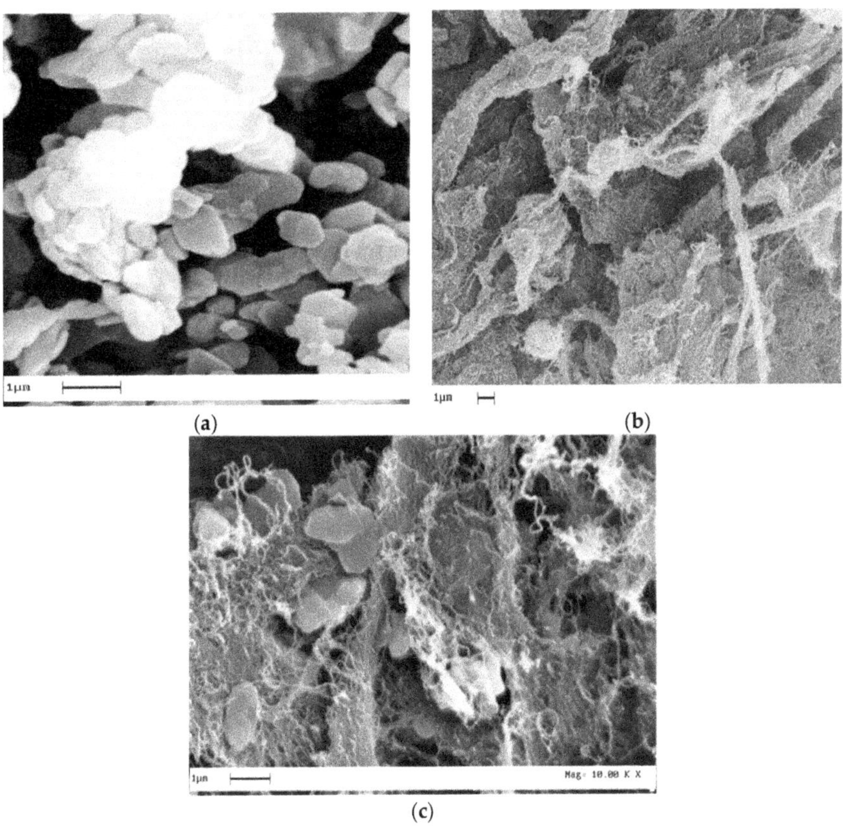

Figure 1. SEM analysis of (**a**) pristine layered double hydroxides (LDH), (**b**) pristine CNT and (**c**) LDH-CNT.

Figure 2 shows the TGA curves of the LDH in nitrate form, the CNTs and the hybrid LDH-CNTs. The decomposition temperature of the CNTs is around 550 °C, which indicates that the acid-treated CNTs are thermally stable up to this temperature. LDH-NO_3 displays three mass-loss steps [28]: The first one, centered at about 108 °C, is due to the loss of physically adsorbed water and water molecules around metal cations, the second one, at about 250 °C, is attributed to the loss of the nitrate anions, and the third one, above 400 °C, due to the de-hydroxylation process. For LDH-CNTs hybrids, three stages of mass loss were also observed in the TG curves. The first mass loss, at approximately 160 °C, is attributed to the removal of loosely bound water molecules from the LDH interlayer. The main mass loss, in the at about 500 °C, is primarily due to the oxidation of CNTs.

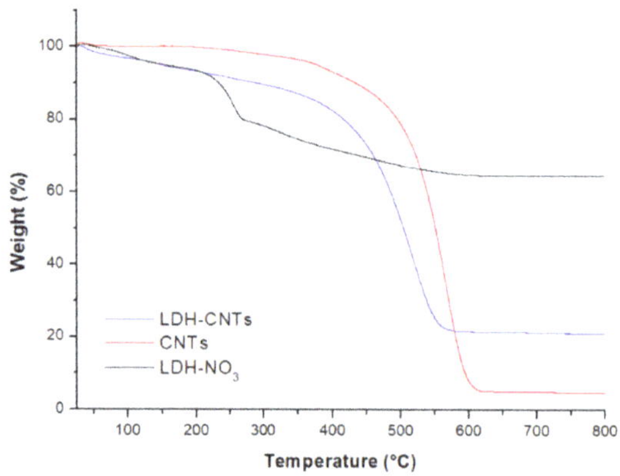

Figure 2. The TGA curves of LDH in nitrate form, the CNTs and the hybrid LDH-CNTs.

Figure 3 reports the XRD of LDH in nitrate form, the CNTs and the hybrid LDH-CNTs. The LDH-NO_3 shows the diffraction peaks at 9.8°, 19.8°, 29.8° of 2θ corresponding to the basal order reflections of (003), (006), (009), respectively. The peak at $2\theta = 25.4°$ for CNTs is typical of (002) basal reflection. The hybrid LDH-CNTs shows peaks of both pristine components, with an evident amorphization of the pristine LDH, after CNTs substitution with nitrate groups.

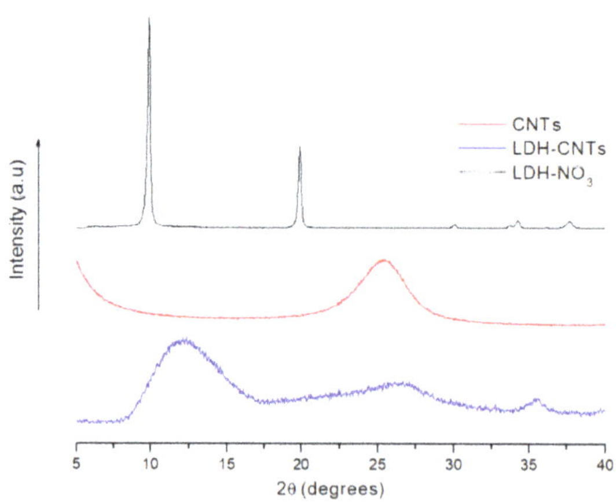

Figure 3. The XRD of LDH in nitrate form, the CNTs and the hybrid LDH-CNTs.

Figure 4 reports the FTIR of LDH in nitrate form, the CNTs, and the hybrid LDH-CNTs. For the LDH-NO_3 the band at around 3600 cm^{-1} is attributed to the stretching of the –OH bond of the hydroxyl groups and H_2O molecules. The weak band at 1630 cm^{-1} can be assigned to the H_2O bending vibration of the interlayer water. The strong peak around 1380 cm^{-1} is attributed to the anti-symmetric stretching mode of the nitrate anion present between the LDH layers. The bands observed around 839 and 670 cm^{-1} can be assigned to the weak out-of-plane symmetric deformation mode and the anti-symmetric deformation mode of nitrate, respectively. The FTIR spectrum of CNTs shows the evidence the absorption band of C=O at around 1600 cm^{-1}, and a strong absorption band centered at about 3462 cm^{-1}, attributed to the O–H stretching of carboxylic acid moieties from the surface of CNTs. The spectrum of the hybrid LDH-CNTs shows the same peak of C=O and O–H recorded in the pristine fillers, while the absence of the peak at 1380 cm^{-1} confirms the substitution of nitrate anion with the carboxyl groups on the LDHs' lamellae.

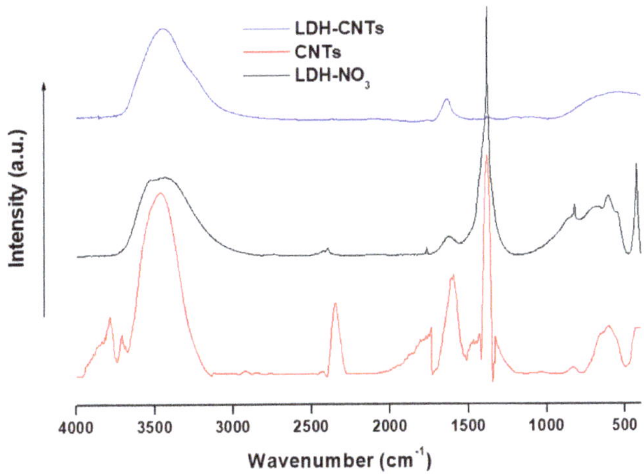

Figure 4. FTIR of LDH in nitrate form, the CNTs and the hybrid LDH-CNTs.

3.2. *Composites Characterization*

Figure 5 reports the XRD spectra of all the composites. Also, HAVOH with 1% of IL is reported. The amorphous structure of the polymer in composites is retained, also after the mixing with the filler in the presence of ionic liquid. The small peak at $2\theta = 11.3°$ is related to the LDH-CNTs.

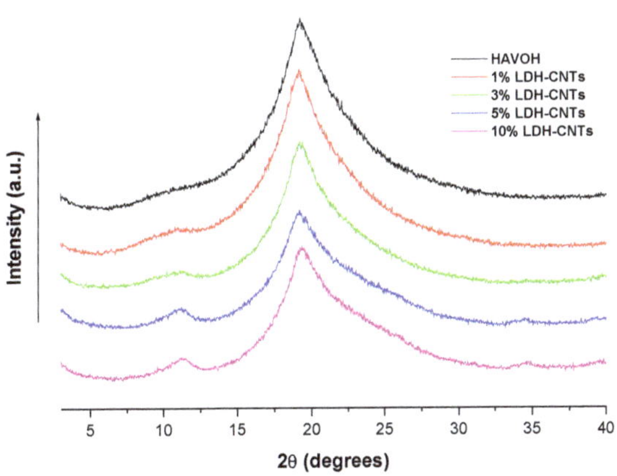

Figure 5. XRD spectra of HVOH and all the composites.

Figure 6 reports the TGA curves for all the composites and the unfilled polymer with 1% of IL. The first decomposition step is due to the loss of water, the second loss in weight is related either to the decomposition of hydroxyl groups or to a split of a part of the polymer chains, a third one is attributed to the decomposition of ethylene segments to carbon chains, and the last one to the volatilization of oxygenated low molecular weight compounds. The introduction of the filler anticipates the second decomposition step and delays the volatilization of low molecular weight compounds. The LDH lamellae represent a barrier to oxygen diffusion into the heated polymer, due to the accumulation of the oxides produced by thermal degradation of the LDH on the surface of the volatizing polymer. The presence of carbon nanotubes, bonded to the lamellae, could also act as physical barriers and/or entrap radicals formed during the decomposition process.

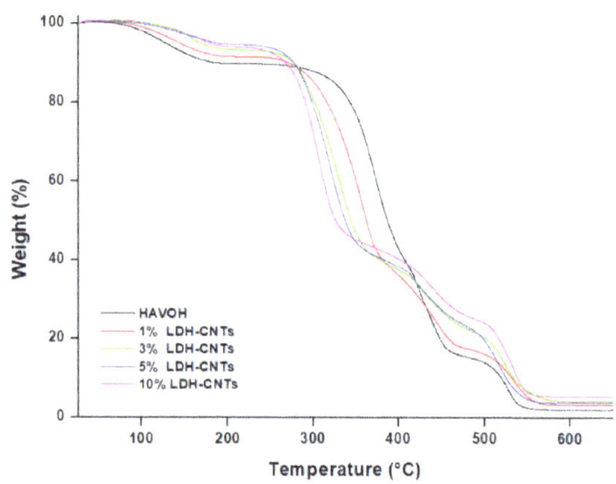

Figure 6. TGA curves for all the composites and the unfilled polymer.

Figure 7 reports the elastic modulus, E (MPa), as a function of filler loading. It increases up to 3% of filler, due to the reinforcing effect of the LDHs lamellae, and reaches a plateau value up to 10% of filler. Being the samples amorphous at any loading composition (see XRD), the reinforcement is only due to the filler dispersion. The degree of load transfer after 5% of filler loading is almost the same for all the composite, resulting in a constant response in mechanical behavior. The sample filled only with 5% of LDH-CNTs display an elastic modulus of 350 ± 20 MPa, higher than the composite filled with 5% of LDH-CNTs and 1% of IL. This is expected because of the plasticizing effect of IL in polymer matrices.

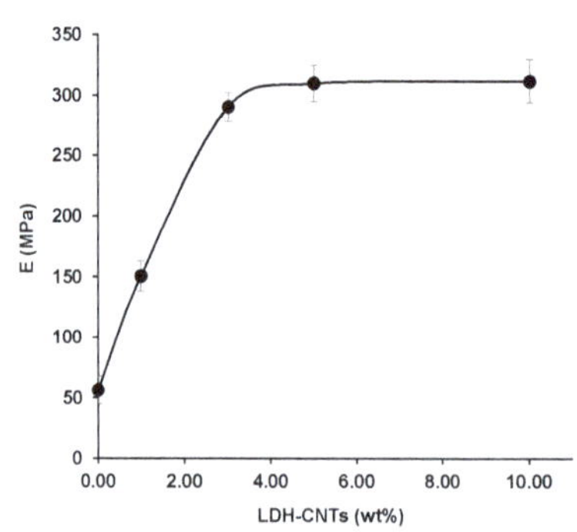

Figure 7. Elastic modulus, E (MPa), as a function of filler loading.

The DC electrical conductivity σ as a function of the filler loading is shown in Figure 8a. σ is obtained from the I-V curves, shown in Figure 8b, using the formula:

$$\sigma = G\left(\frac{t}{A}\right) \tag{1}$$

where $G = dI/dV$ is the sample conductance, that is the slope of the I-V curves of Figure 8b, while t and A are the sample thickness and Au pad area, respectively. We note that the ohmic behavior of the I-V characteristics, shown in Figure 8b, is typical of CNTs' filled films and is often found in defective insulators at low bias and high temperature owing to the hopping of carriers between close conducting regions [29,30]. In our sample, hopping can occur between the conductive network LDH-CNTs. The rapid growth of the conductivity with the increasing filler content, up to the maximum $\sigma_m = 5 \times 10^{-5} \frac{S}{m}$ at 10% loading, corresponds to the formation of a conductive pathway (percolation) throughout the sample and is typical of composite materials with well-dispersed and electrically high-conductive fillers. We also measured the electrical conductivity of the sample, based on HAVOH and 3% of LDH-CNTs without IL. The value of electrical conductivity resulted in 2×10^{-9} S/m, four orders of magnitude lower than the composite with IL. In order to support such hypothesis we performed SEM analysis on both these samples. Figure 9a reports the SEM image of the sample filled with LDH-CNT and ionic liquid, and Figure 9b the composite without IL. It is evident that the CNTs bonded, and not bonded, to LDH lamellae are better dispersed in the composite with IL (Figure 9a), creating a favored percolation pathway; whereas, in the case of the composite (with no ionic liquid), a separation of CNTs is evident, explaining the lower electrical conductivity.

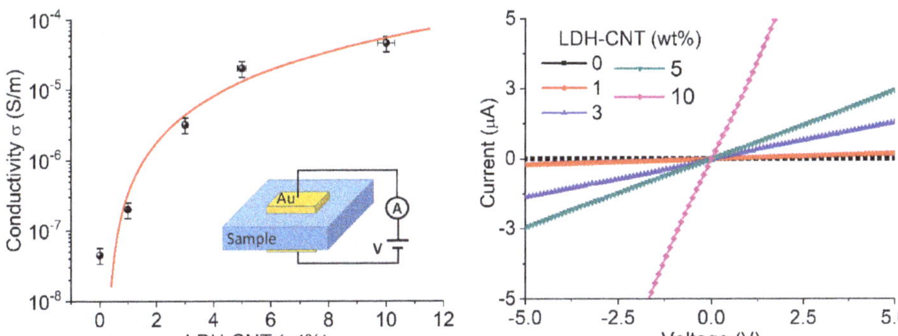

Figure 8. (**a**) Conductivity as a function of the filler loading. The inset shows the sample and measurement setup; (**b**) I-V characteristics for different filler loading.

The lowest concentration of the filler at which the insulating material is converted into a conductive composite is defined as the percolation threshold, which can be estimated from the power-law equation [31]:

$$\sigma = \sigma_0\left(\frac{\eta - \eta_c}{1 - \eta_c}\right)^s \tag{2}$$

where σ and σ_0 are the DC conductivities of the composite material and the filler, respectively, η is the volume fraction of filler loading, η_c is the weight fraction at percolation (also known as critical weight fraction), and s is the power law exponent.

The fit of Equation (2) to the data in Figure 9b (red line) yields $\eta_c \approx 0.26\%$, $\sigma_c \sim 0.01$ S/m and $s = 2.1$. The so-obtained percolation threshold can be considered a very good result when compared with typical values reported in the literature for CNTs based polymer composites [32], and confirms the very good dispersion of the filler into the considered material. It is worth to note that the CNTs amount into the nano-hybrid filler is 80%. This means that the percolation threshold, if referred to the only conductive nanoparticles, is even low.

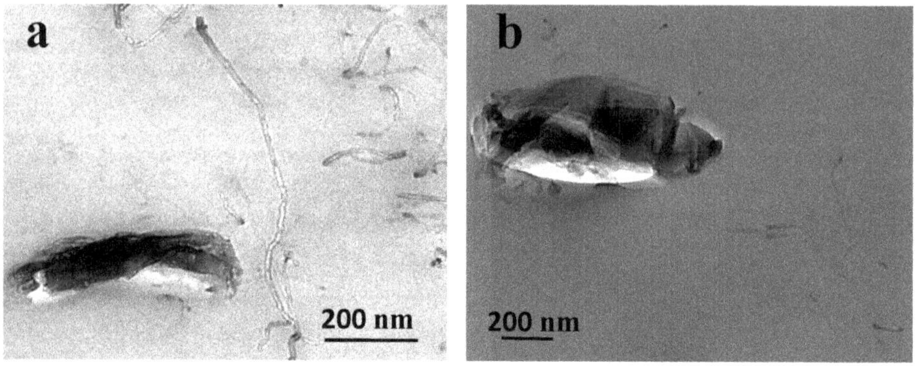

Figure 9. (**a**) SEM of the sample filled with 3 wt % of LDH-CNT and IL and (**b**) SEM of the sample filled with 3 wt % of LDH-CNT without IL.

4. Conclusions

The paper reports the preparation of layered double hydroxides (LDHs) hosting carbon nanotubes (80% of CNTs) synthesized by co-precipitation method. The obtained nano-hybrids, characterized by SEM, XRD, FTIR, TGA and elemental analysis, demonstrated the substitution of the nitrate anion with the carboxyl groups on the LDHs' lamellae. Such nano-hybrid was dispersed into a commercial biodegradable highly amorphous vinyl alcohol polymer at different loading (i.e., 1; 3; 5; 10 wt %). In order to improve the degree of dispersion of the filler into the polymer matrix, an ionic liquid (IL) based on 1-hexadecyl-3-methylimidazolium dimethyl-5-sodiosulfoisophthalate was added to the composites' mixtures. The structural characterization and the analysis of physical properties of the obtained composites were carried out and compared to the unfilled polymer. The thermal degradation analysis demonstrated that LDH lamellae produce a barrier effect to oxygen diffusion into the heated polymer, due to the accumulation of the oxides produced by thermal degradation of the LDH on the surface of the volatizing polymer. The presence of carbon nanotubes, bonded and adsorbed to the lamellae, could also act as physical barriers and/or entrap radicals formed during the decomposition process. Mechanical properties (i.e., elastic modulus) increased up to 3% of filler, due to the reinforcing effect of the LDH lamellae, and reached a plateau value up to 10% of filler. The electrical properties, evaluated in the range −5–5 V, showed a very low percolation threshold, at 0.26 wt % of nano-filler, that is 0.21 wt % of carbon nanotubes loading, as a probe of the very good dispersion of the filler into the considered biopolymer favored by the IL. The comparison between the electrical conductivity of the sample filled with 5% of LDHs-CNT with IL and without IL showed a difference of four orders of magnitude. The higher electrical conductivity of the composite with IL demonstrated the effect of such additive in favoring a better dispersion of LDH-CNTs into the polymer matrix.

Author Contributions: Conceptualization, G.G. and V.V.; methodology, G.G., D.C.Z. and A.D.B.; software, G.V., L.I.; validation, G.G., D.C.Z., V.B. and V.V.; formal analysis, V.B., L.I., A.D.B.; investigation, V.B., L.I., G.V.; resources, G.G., A.D.B.; data curation, V.B., L.I., D.C.Z., G.V.; writing—original draft preparation, V.B., G.G., A.D.B.; writing—review and editing, G.G.; visualization, V.V., L.I.; supervision, G.G.; project administration, G.G.; funding acquisition, G.G., A.D.B. All authors have read and agreed to the published version of the manuscript.

Funding: This research was funded by the Italian Ministry of Education University and Research, MIUR, Project Prin 2017, grant number 2017B7MMJ5_001.

Conflicts of Interest: The authors declare no conflict of interest.

References

1. Sahoo, N.G.; Rana, S.; Cho, J.W.; Li, L.; Chan, S.H. Polymer nanocomposites based on functionalized carbon nanotubes. *Prog. Polym. Sci.* **2010**, *35*, 837–867. [CrossRef]
2. Spitalsky, Z.; Tasis, D.; Papagelis, K.; Galiotis, C. Carbon nanotube–polymer composites: Chemistry, processing, mechanical and electrical properties. *Prog. Polym. Sci.* **2010**, *35*, 357–401. [CrossRef]
3. Dintcheva, N.Tz.; Arrigo, R.; Carroccio, S.; Curcuruto, G.; Guenzi, M.; Gambarotti, C.; Filippone, G. Multi-functional polyhedral oligomeric silsesquioxane-functionalized carbon nanotubes for photo-oxidative stable Ultra-High Molecular Weight Polyethylene-based nanocomposites. *Eur. Polym. J.* **2016**, *75*, 525–537. [CrossRef]
4. Gorrasi, G.; Milone, C.; Piperopoulos, E.; Lanza, M.; Sorrentino, A. Hybrid clay mineral-carbon nanotube-PLA nanocomposite films. Preparation and photodegradation effect on their mechanical, thermal and electrical properties. *Appl. Clay Sci.* **2013**, *71*, 49–54. [CrossRef]
5. Zakaria, M.R.; Md Akil, H.; Abdul Kudus, M.H.; Ullah, F.; Javed, F.; Nosbi, N. Hybrid carbon fiber-carbon nanotubes reinforced polymer composites: A review. *Compos. Part B Eng.* **2019**, *176*, 107313. [CrossRef]
6. Byrne, M.T.; Gun'ko, Y.K. Recent Advances in Research on Carbon Nanotube-Polymer Composites. *Adv. Mater.* **2010**, *22*, 1672–1688. [CrossRef]
7. Blake, R.; Coleman, J.N.; Byrne, M.T.; McCarthy, J.E.; Perova, T.S.; Blau, W.J.; Fonseca, A.; Nagy, J.B.; Gun'ko, Y.K. Reinforcement of poly(vinyl chloride) and polystyrene using chlorinated polypropylene grafted carbon nanotubes. *J. Mater. Chem.* **2006**, *16*, 4206. [CrossRef]
8. Blond, D.; Barron, V.; Ruether, M.; Ryan, K.P.; Nicolosi, V.; Blau, W.J.; Coleman, J.N. Enhancement of Modulus, Strength, and Toughness in Poly(methyl methacrylate)-Based Composites by the Incorporation of Poly(methyl methacrylate)-Functionalized Nanotubes. *Adv. Funct. Mater.* **2006**, *16*, 1608–1614. [CrossRef]
9. Yang, B.-X.; Pramoda, K.P.; Xu, G.Q.; Goh, S.H. Mechanical Reinforcement of Polyethylene Using Polyethylene-Grafted Multiwalled Carbon Nanotubes. *Adv. Funct. Mater.* **2007**, *17*, 2062–2069. [CrossRef]
10. Yang, B.-X.; Shi, J.-H.; Pramoda, K.P.; Goh, S.H. Enhancement of the mechanical properties of polypropylene using polypropylene-grafted multiwalled carbon nanotubes. *Compos. Sci. Technol.* **2008**, *68*, 2490–2497. [CrossRef]
11. Xing, W.; Yang, W.; Yang, W.; Hu, Q.; Si, J.; Lu, H.; Yang, B.; Song, L.; Hu, Y.; Yuen, R.K.K. Functionalized Carbon Nanotubes with Phosphorus- and Nitrogen-Containing Agents: Effective Reinforcer for Thermal, Mechanical, and Flame-Retardant Properties of Polystyrene Nanocomposites. *ACS Appl. Mater. Interfaces* **2016**, *8*, 26266–26274. [CrossRef] [PubMed]
12. Tsang, S.C.; Chen, Y.K.; Harris, P.J.F.; Green, M.L.H. A simple chemical method of opening and filling carbon nanotubes. *Nature* **1994**, *372*, 159–162. [CrossRef]
13. Mishra, G.; Dash, B.; Pandey, S. Layered double hydroxides: A brief review from fundamentals to application as evolving biomaterials. *Appl. Clay Sci.* **2018**, *153*, 172–186. [CrossRef]
14. Pradhan, B.; Srivastava, S.K. Layered double hydroxide/multiwalled carbon nanotube hybrids as reinforcing filler in silicone rubber. *Compos. Part A Appl. Sci. Manuf.* **2014**, *56*, 290–299. [CrossRef]
15. Mallakpour, S.; Dinari, M. Hybrids of Mg–Al-layered double hydroxide and multiwalled carbon nanotube as a reinforcing filler in the l-phenylalanine-based polymer nanocomposites. *J. Therm. Anal. Calorim.* **2015**, *119*, 1905–1912. [CrossRef]
16. Si, J.; Ping, P.; Xie, H.; Yang, W.; Lu, H. The influence of multiwalled carbon nanotubes-NiCoAl layered double hydroxide hybrids on fire safety performance of poly(ethylene-co-vinyl acetate) composites. *Polym. Compos.* **2018**, *39*, E835–E841. [CrossRef]
17. Yan, N.; Capezzuto, F.; Buonocore, G.G.; Lavorgna, M.; Xia, H.; Ambrosio, L. Gas-Barrier Hybrid Coatings by the Assembly of Novel Poly(vinyl alcohol) and Reduced Graphene Oxide Layers through Cross-Linking with Zirconium Adducts. *ACS Appl. Mater. Interfaces* **2015**, *7*, 22678–22685. [CrossRef]
18. Fukushima, T.; Aida, T. Ionic Liquids for Soft Functional Materials with Carbon Nanotubes. *Chem. Eur. J.* **2007**, *13*, 5048–5058. [CrossRef]
19. Hameed, N.; Church, J.S.; Salim, N.V.; Hanley, T.L.; Amini, A.; Fox, B.L. Dispersing single-walled carbon nanotubes in ionic liquids: a quantitative analysis. *RSC Adv.* **2013**, *3*, 20034. [CrossRef]
20. Polo-Luque, M.L.; Simonet, B.M.; Valcárcel, M. Functionalization and dispersion of carbon nanotubes in ionic liquids. *TrAC Trends Anal. Chem.* **2013**, *47*, 99–110. [CrossRef]

21. Bordes, E.; Morcos, B.; Bourgogne, D.; Andanson, J.-M.; Bussière, P.-O.; Santini, C.C.; Benayad, A.; Costa Gomes, M.; Pádua, A.A.H. Dispersion and Stabilization of Exfoliated Graphene in Ionic Liquids. *Front. Chem.* **2019**, *7*, 223. [CrossRef] [PubMed]
22. Fileti, E.E.; Chaban, V.V. Imidazolium Ionic Liquid Helps to Disperse Fullerenes in Water. *J. Phys. Chem. Lett.* **2014**, *5*, 1795–1800. [CrossRef] [PubMed]
23. Soares, B.G. Ionic liquid: A smart approach for developing conducting polymer composites. *J. Mol. Liq.* **2018**, *262*, 8–18. [CrossRef]
24. Soares da Silva, J.P.; Soares, B.G.; Livi, S.; Barra, G.M.O. Phosphonium–based ionic liquid as dispersing agent for MWCNT in melt-mixing polystyrene blends: Rheology, electrical properties and EMI shielding effectiveness. *Mater. Chem. Phys.* **2017**, *189*, 162–168. [CrossRef]
25. Rashid, Z.; Wilfredand, C.D.; Murugesan, T. Effect of hydrophobic Ionic Liquids on Petroleum Asphaltene Dispersion and Determination Using UV-Visible Spectroscopy. *AIP Conf. Proc.* **2017**, *1891*, 020118. [CrossRef]
26. Frunza, M.; Lisa, G.; Popa, M.I.; Miron, N.D.; Nistor, D.I. Thermogravimetric analysis of layered double hydroxides with chloramphenicol and salicylate in the interlayer space. *J. Therm. Anal. Calorim.* **2008**, *93*, 373–379. [CrossRef]
27. Oswald, H.R.; Asper, R. Bivalent Metal Hydroxides. In *Preparation and Crystal Growth of Materials with Layered Structures*; Lieth, R.M.A., Ed.; Springer Netherlands: Dordrecht, The Netherlands, 1977; pp. 71–140. ISBN 978-90-481-8344-9.
28. Rad, F.A.; Rezvani, Z. Preparation of cubane-1,4-dicarboxylate–Zn–Al layered double hydroxide nanohybrid: comparison of structural and optical properties between experimental and calculated results. *RSC Adv.* **2015**, *5*, 67384–67393. [CrossRef]
29. Giordano, C.; Filatrella, G.; Sarno, M.; Bartolomeo, A.D. Multi-walled carbon nanotube films for the measurement of the alcoholic concentration. *Micro. Nano. Lett.* **2019**, *14*, 304–308. [CrossRef]
30. Gorrasi, G.; Bugatti, V.; Milone, C.; Mastronardo, E.; Piperopoulos, E.; Iemmo, L.; Di Bartolomeo, A. Effect of temperature and morphology on the electrical properties of PET/conductive nanofillers composites. *Compos. Part B Eng.* **2018**, *135*, 149–154. [CrossRef]
31. Tunnicliffe, L.B.; Kadlcak, J.; Morris, M.D.; Shi, Y.; Thomas, A.G.; Busfield, J.J.C. Flocculation and Viscoelastic Behaviour in Carbon Black-Filled Natural Rubber: Flocculation and Viscoelastic Behaviour in Carbon Black-Filled. *Macromol. Mater. Eng.* **2014**, *299*, 1474–1483. [CrossRef]
32. Bauhofer, W.; Kovacs, J.Z. A review and analysis of electrical percolation in carbon nanotube polymer composites. *Compos. Sci. Technol.* **2009**, *69*, 1486–1498. [CrossRef]

polymers

MDPI

Article

Synthesis and Characterization of High-Performing Sulfur-Free Tannin Foams

Jonas Eckardt [1,2], Jonas Neubauer [1], Thomas Sepperer [1,2], Sandro Donato [3,4], Michela Zanetti [5], Nicola Cefarin [4], Lisa Vaccari [4], Marcel Lippert [1], Matthias Wind [1], Thomas Schnabel [1], Alexander Petutschnigg [1] and Gianluca Tondi [1,2,5,*]

1 Forest Products Technology & Timber Constructions Department, Salzburg University of Applied Sciences, Marktstrasse 136a, 5431 Kuchl, Austria; jonas.eckardt@fh-salzburg.ac.at (J.E.); jonas.neubauer@fh-salzburg.ac.at (J.N.); thomas.sepperer@fh-salzburg.ac.at (T.S.); mlippert.htw-m2019@fh-salzburg.ac.at (M.L.); mwind.htw-m2019@fh-salzburg.ac.at (M.W.); thomas.schnabel@fh-salzburg.ac.at (T.S.); alexander.petutschnigg@fh-salzburg.ac.at (A.P.)

2 Salzburg Center for Smart Materials, Jakob-Haringer-Strasse 2a, 5020 Salzburg, Austria

3 Physics Department, University of Calabria, Via Pietro Bucci, cubo 31C, 87036 Rende (CS), Italy; sandro.donato@fis.unical.it

4 Elettra-Sincrotrone Trieste S.C.p.A., Strada Statale 14-km 163,5 in AREA Science Park, 34149 Trieste, Basovizza, Italy; nicola.cefarin@elettra.eu (N.C.); lisa.vaccari@elettra.eu (L.V.)

5 Land, Environment, Agriculture and Forestry Department, University of Padua, Viale dell'Università 16, 35020 Legnaro (PD), Italy; michela.zanetti@unipd.it

* Correspondence: gianluca.tondi@unipd.it; Tel.: +39-049-827-2776

Received: 3 February 2020; Accepted: 1 March 2020; Published: 4 March 2020

Abstract: Tannin foams are green lightweight materials that have attracted industrial interest for the manufacturing of sandwich panels for insulation purposes. However, the dimensions of the cells and the presence of sulfur in the formulation developed until now have discouraged their upscaling. In this work, we present the synthesis and the characterization of the more promising small cell and sulfur-free materials. It was observed that, with respect to standard ones, foams catalyzed with nitric acid present similar physical properties and more phenolic character, which favors the absorption of ionic pollutants. Conversely, the foams blown with aliphatic solvents and surfactants present smaller pores, and higher mechanical and insulating properties, without affecting the chemical properties or the heating value. The combined foam produced with nitric acid as a catalyst and petroleum ether as a blowing agent result in sulfur-free and small cell material with overall improved features. These foams have been produced at $30 \times 30 \times 3$ cm^3, with high homogeneity and, to date, they represent the most suitable formulation for industrial upscaling.

Keywords: polyphenols; flavonoids; furanics; sulphur; bio-polymers; pentane; heating value

1. Introduction

The most commonly applied foams for the insulation of buildings in the European market are made of polystyrene or polyurethane [1–3]. Even if, nowadays, the interest in sustainable solutions is growing, the alternative foams under investigation are aerogels or bio-based polyurethanes [4,5]. Nevertheless, the tannin foams have been constantly monitored for a decade now, for their structure and morphology [6–8], and their chemical and physical properties [9–11]. Tannin foams have shown remarkable properties: they are based on natural resources [12,13]; they are very light, and can be produced with different heating systems [14,15]; they are fire and water resistant [16]; they can be modified and functionalized [17–19]; they can be easily combined with different top layers [20]; and they can even be reused at the end of their working life [21].

Despite the upscaling projects performed [22], there are still some technological drawbacks to be considered: (i) the consistent presence of sulfur in the formulation due to a sulfur-based catalyst (traces of sulfites due to the extraction process could still be present), and (ii) the morphological inhomogeneity occurring during the upscaling process, which affects the properties of the foams. For filling the remaining gap between the material properties and the industrial requirements, in the present work we aim the overcome these drawbacks by using alternative formulations.

2. Materials and Methods

2.1. Materials

The extract Weibull AQ, provided by the company Tanac (Montenegro, RS, Brazil), and furfuryl alcohol from Transfurans Chemicals (Geel, Belgium), were the basis for every formulation. The blowing agents used were diethylether (DEE), pentane (Pent), and petroleum ether (PE), while the catalysts were sulfuric acid (SA) and nitric acid (NA) from Roth (Karlsruhe, Germany). The surfactant Tween80 and the water pollutants methylene blue and sodium dodecyl sulfate (SDS) were purchased from VWR (Darmstadt, Germany).

2.2. Synthesis of the Tannin Foams

Tannin powder, furfuryl alcohol, water, surfactant, and blowing agent were mixed until a highly homogeneous emulsion was obtained with the proportions reported in Table 1. Then, the catalyst was added and vigorously stirred for 30 s. The formulation was poured into a preheated Medium Density Fiberboard (MDF) mold, and this was laid into a Höfer (Taiskirchen, Austria) press at 90 °C. The foams were blown and hardened in 5 min, and then they were removed from the mold and conditioned at 20 °C and 65% relative humidity for one week before testing.

Table 1. Formulation description in percentage by weight.

Formulations	Acronym	Tannin (%)	Furfuryl Alcohol (%)	Blowing Agent (%)	Tween80 (%)	Catalyst (%)
Standard	SA-DE	42.0	25.9	DEE 5.5	0.0	SA 5.8
Nitric Acid	NA-DE	42.0	25.9	DEE 5.5	0.0	NA 4.5
Pentane	SA-Pent	39.1	25.5	Pent 5.5	1.3	SA 5.7
Petroleum Ether	SA-PE	39.1	25.5	PE 5.5	1.3	SA 5.7
Nitric Acid-Petroleum ether	NA-PE	39.1	25.5	PE 5.5	1.3	NA 4.5

Note: The remaining part was water. Every formulation was prepared in tiles, five $10 \times 10 \times 3$ cm^3 and two $30 \times 30 \times 3$ cm^3.

2.3. Intrinsic Physical Characterization

2.3.1. Bulk Density, Skeletal Density, and Porosity

The bulk density (d) was calculated by measuring the length (l), width (w), height (h), and weight (m) for each cube, according to Equation (1).

$$d = m/(l \times w \times h) \quad (1)$$

The skeletal density of the different formulations was measured by He-pycnometry, using the gas pycnometer Quantachrome Instrument (Boynton Beach, FL, USA) Ultrapyc 1200e, and the porosity (P) of the material was calculated with Equation (2), where d is the bulk density and d_s is the skeletal density.

$$P = 1 - (d/d_s) \quad (2)$$

2.3.2. Cell Dimensions and Orthotropicity

The cell size dimensions were measured with a reflected light microscope (Nikon SMZ 1500, Tokio, Japan). For each formulation, at least 50 measurements in length and 100 in width were carried out in the direction parallel to the growth on three samples. The average foam cell diameter (Av.D) has been calculated with Equation (3), where D is the average of 150 measurements.

$$\text{Av.D} = (\pi/4) \times \text{D} \quad (3)$$

Orthotropicity (o) was calculated by using Equation (4), where Av.l is the average length and Av.w is the average width.

$$\text{o} = \text{Av.l}/\text{Av.w} \quad (4)$$

2.3.3. X-ray Tomography Acquisitions

From the center of each foam block, a cube of $3 \times 3 \times 3$ cm^3 was extracted. Those samples were imaged at the Tomolab laboratory, at the synchrotron facility of Elettra Sincrotrone Trieste.

The X-ray source was a sealed microfocus Hamamatsu X-ray tube (model L12161-07) with a Tungsten target, which can operate at voltages between 40 and 150 KV and at a maximum power of 75 W. The tube voltage was set to 40 keV and the current to 100 μA. The focal spot size was set to 5 μm. The detecting system consisted of a Hamamatsu CMOS camera (C12849-102, Hamamatsu, Japan) with a pixelsize of 6.5 μm and an active area of 20.48^2 pixels^2. The sample-to-detector distance was 120 mm, while the source-to-detector distance was 370 mm. Considering the magnification factor ~3, the equivalent pixelsize of the acquired projection was equal to 2.1 μm, and the corresponding imaged volume was approximatively equal to 80 mm^3. For each sample, 1800 equi-angular projections over 360°, with an exposure time of 2 s per projection, were acquired. Axial slices were reconstructed with an isotropic voxel size of 2.1 μm^3, using the standard FDK algorithm [23] for cone-beam geometry. Prior to volume rendering, digital slices were processed with a 3D median filter algorithm for noise reduction. For each foam, a volume of interest of nearly 5 mm^3 was selected for the 3D rendering done with Avizo®9.3.

2.3.4. Heating Value

The low heating value on a wet basis (PCI_W), and high heating value on a dry basis (PCS_0) of ground, tannin-based foam samples were calculated following the methodology described in the ISO standard 18,125 [24]. The high heating value of the samples as received has been measured using an IKA (Staufen, Germany) C-200 calorimeter. As indicated in the ISO standard, the average of two replications was calculated, and the repeatability limit did not differ by more than 140 J/g.

2.4. Intrinsic Chemical Characterization

Attenuated Total Reflectance Fourier-Transform Infrared (ATR FT-IR) Spectroscopy

The five foams were crushed into powder, dried, and finally scanned using a Perkin Elmer (Perkin-Elmer, Waltham, MA, USA) Frontier FT-IR spectrometer, equipped with an ATR Miracle accessory, with the same procedure applied in the previous work of the group [25]. Sixteen scans were performed for each sample, at a resolution of 4 cm^{-1}, in the range of 4000 to 600 cm^{-1}, and this was repeated in triplicate. Bio Rad KnowItAll (BioRad, California, CA, USA) software was used for normalizing and averaging the spectra.

2.5. Extrinsic Physical Characterization

2.5.1. Compression Strength

The compression strength was measured with the universal testing machine Zwick Roell Z250 (Zwick-Roell, Ulm, Germany) on samples 3 × 3 × 3 cm^3, according to the standard DIN 52,185 [26]. Nine specimens were tested parallel to the growth direction. The rate was set to 2 mm/min. If a clear break of the material appeared within 10% of specimen deformation, that value was taken as the result. If no clear break occurred within this range, the value at 10% of the deformation of the tested sample was taken as the maximum compression strength.

2.5.2. Thermal Conductivity

The thermal conductivity was measured with a λ-Meter Lambda-Messtechnik EP500e (Dresden, Germany) device on samples with dimension 25 × 25 × 3 cm^3. The measurements took place at three different temperatures (10, 25, and 40 °C) by applying 10K of temperature difference to the upper heating plate for each temperature.

2.5.3. Fire Resistance

Conditioned cubes of 2 cm per side of foam were held on a lab-made device laying on an analytical balance. The samples were exposed to 3 min of direct conical blue-flame at a distance of 3 cm from the Bunsen burner, and the weight was registered every 30 s during the burning phase (3 min), and during the glowing phase (2 min), before self-extinguishing. Five samples for each formulation were tested, and the percentage of mass loss was presented as average.

2.6. Extrinsic Chemical Characterization

2.6.1. Acid Recovery

The produced foams were pulverized and dried at 103 °C. Afterwards, 5 g of dry tannin-foam powder was added to 250 mL water and magnetically stirred for 3 min. The solution was then filtrated through a paper filter with a pore size of 12–25 µm and washed with a total amount of 750 mL of water. The pH-value of the solution was measured with a WTW inoLab pH 720 m (Xylem, New York, NY, USA). The process was done in triplicate.

2.6.2. Pollutant Absorption

The pollutant adsorption capacity test was set up by exposing 4 mg of the leached tannin foam powders to 20 ppm solutions of: (i) methylene blue, and (ii) sodium dodecyl sulfate for 48 h under stirring conditions, and 24 h under stationary conditions, at room temperature and in a dark environment, following the procedure reported by other sources [21,27,28].

These solutions were filtered with a vacuum glass filter of 16–40 µm (Por.3), and then the intensity of the signal of the solutions was measured with a Shimadzu UVmini 1240 UV-VIS spectrophotometer (Shimadzu, Kyoto, Japan) at 656 nm.

3. Results

The five foams, subject of the present study, have been characterized for their intrinsic and extrinsic properties from a physical and chemical point of view. In Figure 1, the different appearance of the standard (SA-DE) and the petroleum ether (NA-PE) foams are reported.

(a) (b)

Figure 1. Optical appearance of the standard (SA-DE) (**a**) and high performing petroleum ether (NA-PE) foams (**b**).

3.1. Intrinsic Physical Properties

The intrinsic physical properties that have been considered for the foams are summarized in Table 2.

Table 2. Bulk density, porosity average diameter, and orthotropicity of the tannin-based foams.

Foams	Bulk Density [kg/m^3]	Porosity [%]	Diameter [μm]	Orthotropicity
Standard (SA-DE)	62.3 (9.6)	95.5 (0.7)	172 (55)	1.41 (0.32)
Nitric acid (NA-DE)	68.4 (11.3)	95.3 (0.8)	174 (73)	1.53 (0.32)
Pentane (SA-Pent)	82.5 (6.7)	94.0 (0.5)	118 (36)	1.30 (0.25)
Petroleum ether (SA-PE)	83.8 (9.7)	93.9 (0.7)	113 (37)	1.31 (0.26)
Nitric Acid-Petroleum ether (NA-PE)	76.3 (5.4)	93.8 (0.4)	110 (33)	1.25 (0.18)

The measurements presented in this section are in line with the findings of previous studies of the group. Free expansion tannin foams reached density values down to 23 kg/m^3 [29]; porosities of 92.5% to 96.9% were found for the formaldehyde-reinforced foams [16]. A previous study showed generally higher cells dimensions (81 to 428 μm) and orthotropicity (1.46 to 1.99) [6]. However, by systematically comparing these foams, we can state that the foams produced with diethylether (standard and nitric acid) present a sensibly lower density and bigger diameter. As expected, the skeletal density (1380 kg/m^3) was similar for all the foams, with a consequently slightly higher porosity and orthotropicity for the surfactant-free foams. Generally, the foams with the surfactant presented more homogeneous cells dimensions, resulting in a more homogeneous material.

The computed tomography analysis highlighted a sensibly different morphology for the foams. In Figure 2, the structure of the cell wall is depicted for the standard foam (SA-DE), for the sulfur-free one (NA-DE), and for the one produced with aliphatic blowing agent (SA-Pent).

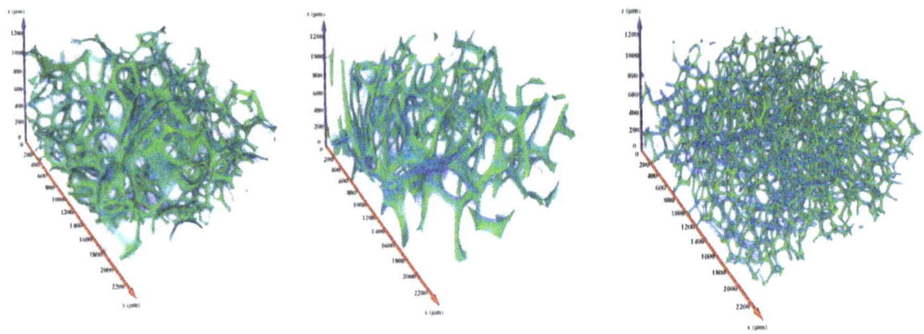

Figure 2. Three-dimensional Structures of the tannin foams (rendered volume ~5 mm^3): standard, nitric acid and pentane (left to right).

The features presented in Table 2 are confirmed by the CT analysis, so the SA-DE and the NA-DE are very similar: they exhibit very high porosity, lower pore homogeneity, and larger pore diameter than SA-Pent foams. Indeed, the foam formulated with aliphatic blowing agents is denser, characterized by more homogeneous and smaller pores. The porosity of the volumes represented in Figure 2 was calculated by performing a thresholding procedure, and then computing a ration between the walls and the total volume. The porosity values obtained were 95.9%, 95.4%, and 94.3%, for the standard, nitric and pentane formulations, respectively. These results are in good agreement with the experimental values of Table 2.

In view of a possible energy valorization of the tannin foams at the end of their life, their heating values were analyzed. The five foams presented similar calorific values, as reported in Table 3.

Table 3. Moisture content and heating values of the powdered foams.

Foams	Moisture Content [%]	PCI_w [MJ/kg]	PCS_0 [MJ/kg]
Standard (SA-DE)	12.50	18.89	22.77
Nitric acid (NA-DE)	9.69	20.30	23.57
Pentane (SA-Pent)	11.11	19.41	22.97
Petroleum ether (SA-PE)	10.55	19.22	22.61
Nitric Acid–Petroleum ether (NA-PE)	10.04	19.80	23.12

The foams showed a similar moisture content, and in particular the sulfur-free formulations (NA-DE and NA-PE) registered slightly lower values. These foams also showed slightly enhanced high heating values, which represents a further advantage for the end-life of these materials. As expected, these PCS_0 were intermediate between the mimosa tannin (=20.25 MJ/kg), and the polyfurfuryl alcohol (=27.89 MJ/kg).

3.2. Intrinsic Chemical Properties

In Figure 3, the FT-IR spectra of the foams are reported.

From the chemical viewing angle, the FT-IR spectra show similar absorption bands, but the relative intensity is affected by the used catalyst. The polymers produced with sulfuric acid have more intense spectral bands in the regions between: (i) 1220 and 1150 cm^{-1} and (ii) 920 to 820 cm^{-1}. The considered bands can be assigned to furanic signals [30]. In the foams catalyzed by nitric acid, a significant increment in the intensity of the bands between: (i) 1480 and 1220 cm^{-1} and (ii) 1150 and 1050 cm^{-1} are registered; both these regions are dominated by the absorption of phenolic components [31]. These observations suggest that the foams produced with nitric acid have lower furanic character than the standard foams catalyzed by sulfuric acid.

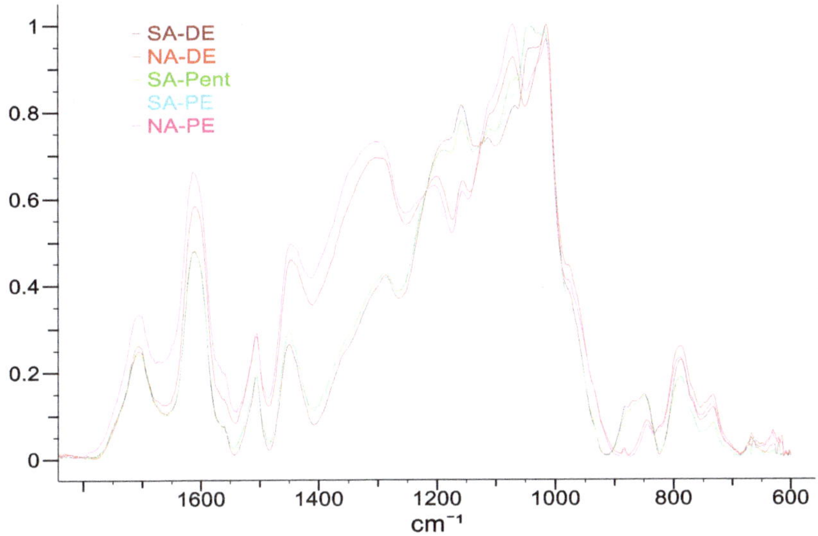

Figure 3. Attenuated Total Reflectance Fourier-Transform Infrared (ATR FT-IR) spectra of the five tannin-based foams between 1800 and 600 cm^{-1}.

3.3. Extrinsic Physical Properties

The five foams were tested for their mechanical, thermal and fire resistance properties.

3.3.1. Compression Strength

In Figure 4, the resistance against compression of the foams is summarized.

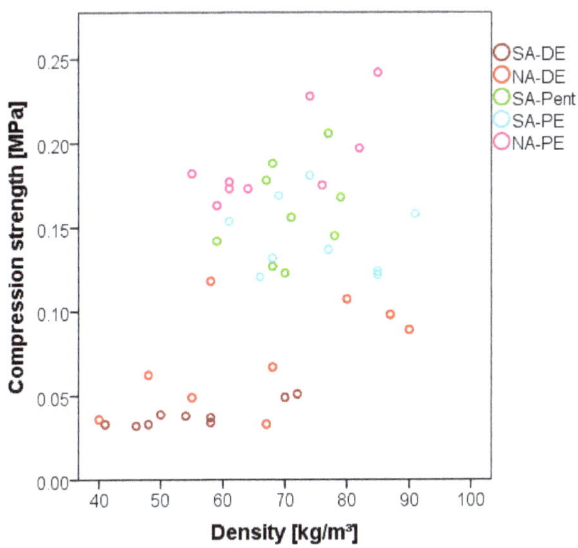

Figure 4. Compression strength of the foams against density.

The compression strength registered in this study can be compared with the previous studies on tannin foams: lightweight formaldehyde-free tannin foams of 50 kg/m^3 showed around 0.03 MPa, while heavier ones of 180 kg/m^3 reached up to 0.45 MPa [13,22].

The foams produced with aliphatic blowing agents show higher values of compressive strength independently of the density. Generally, the mechanical resistance of tannin foams has been reported to increase by increasing their density. Nevertheless, in the present study, a major role seems to be played by the dimension of the pores. In fact, foams with smaller and more homogeneous pores have a higher compression resistance. For instance, a pentane or petrol ether foam of around 70 kg/m^3 density (0.15 MPa) has around three times the strength of a standard or nitric acid foam with the same density (0.05 MPa).

3.3.2. Thermal Conductivity

The thermal conductivity of these foams is very similar for every formulation (Figure 5), and it ranges from 33 to 41 mW/m.K. These measurements enhance the results registered in previous studies, where the thermal conductivity was between 45 and 55 mW/m.K [22,32]. In particular, the foams with aliphatic blowing agents (SA-Pent, SA-PE, and NA-PE) show significantly improved insulation capacity.

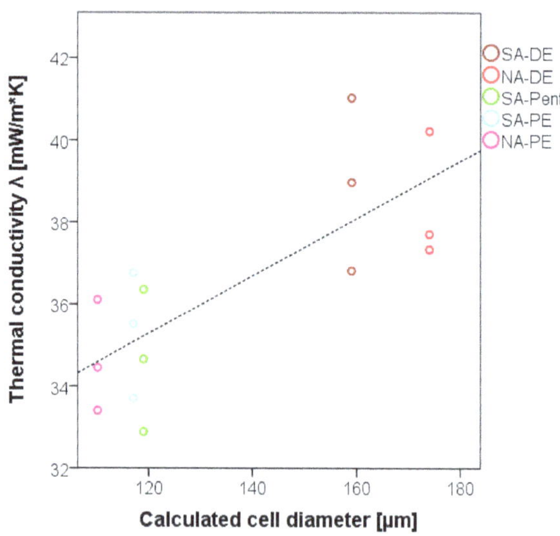

Figure 5. Thermal conductivity of the foams against cell diameter.

The effect of the foam density on the thermal conductivity is even more modest than the one it has on the compression strength, while the cell dimension is the dominant parameter: the smaller the cells are, the lower the thermal conductivity.

3.3.3. Fire Resistance

Tannin-furanic foams generally show a self-extinguishing behavior—which was already previously described for formaldehyde-reinforced foams of similar density—which extinguished in around 80 s [6], and for alkaline tannin foams, when the self-extinguishing occurred instantaneously [33]. The fire behavior of these formaldehyde-free formulations is presented in Figure 6.

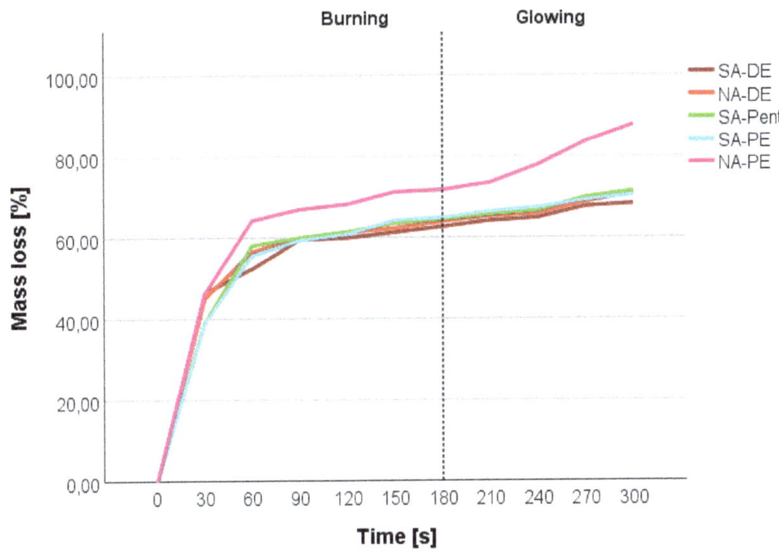

Figure 6. Mass loss of the tannin foams during direct flame exposure: burning and glowing phases.

The graphic shows the mass loss of the foams during the burning and the glowing phase. After an initial 1-minute steep mass loss, due to the release of solvents, of water, and the burning of the surfaces, the foams stabilize their mass loss rate, showing their reticence for burning and never keeping the flame during glowing. All foams show a similar trend, but the NA-PE appears to be more fire sensitive by showing around a 10% increase of mass loss in the initial burning phase.

Even if the difference might be statistically less important (an average of five samples), the different behavior of the NA-PE foam can be due to the presence of solvents in some closed pores. This presence may contribute to higher mass losses due to the further feed for the flame. Even in the final part of the glowing phase, there is a slightly higher slope in the mass loss, which can be attributed to the sudden increase of oxygen due to the opening of some closed regions.

3.4. Extrinsic Chemical Properties

The first extrinsic chemical property considered is the percentage of acid recovery. When the foams come into contact with water, the catalyst remaining on the foam surface of the foam is leached out, and it is possible to determine the pH value of the water, and hence the percentage of acid recovery. In Figure 7, the measured pHs are reported.

It can be seen that the pH of the solutions is always similar, and stays between 2.49 and 2.75. The use of nitric acid produces a less acidic environment, which could be due to different type and concentration of the acid, to its partial evaporation, but also to the different release after leaching. Numerically, 17.8% to 23.4% nitric acid could be recovered against the 29.6% to 39.7% for sulfuric acid. These higher values can be related to the lower phenolic character of the sulfuric acid catalyzed foams (SA-DE), suggesting the weaker interaction of the catalyst, which will be released more easily.

Further, the capacity of the material to act as a filter against organic and emerging pollutants such as methylene blue and sodium dodecyl sulfate [34,35] is evaluated and reported in Figure 8.

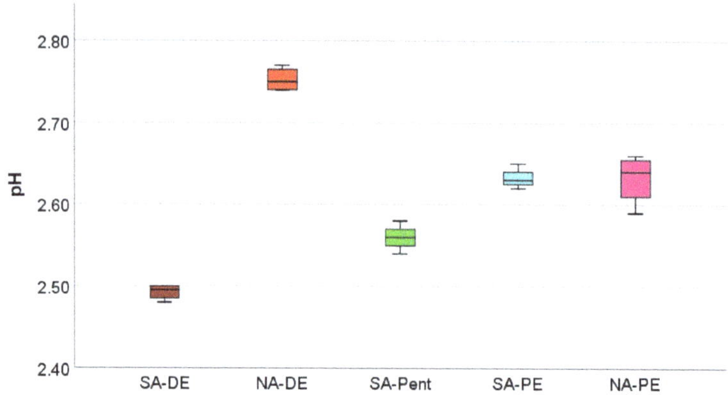

Figure 7. pH of the water after foam leaching.

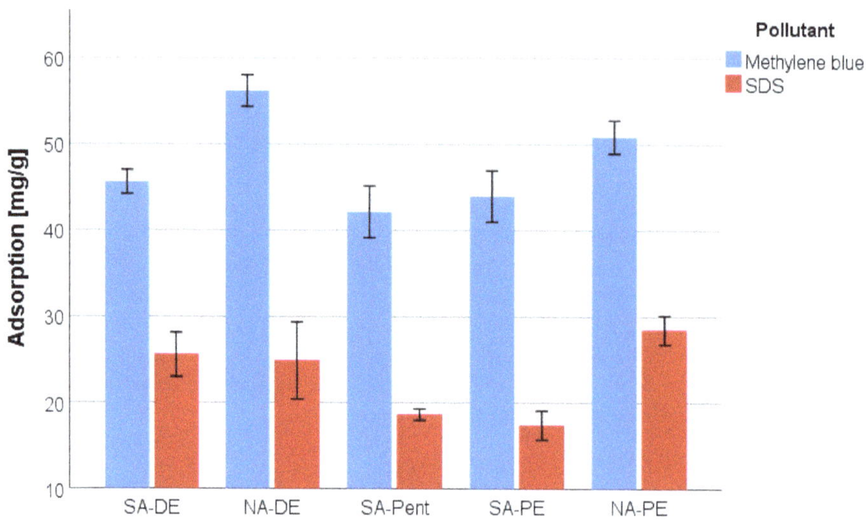

Figure 8. Pollutant adsorption of tannin foams.

It was observed that the foams are able to recover from 42% to 55% of methylene blue and from 15% to 27% of SDS. These results confirm the previous study of the group [21], but they highlight a marked increase in adsorption when nitric acid is used as a catalyst, carrying the absorption of methylene blue over 50% and SDS over 25%, increasing their absorption capacity.

According to the FT-IR analysis, the nitric acid-catalyzed foams seems to have a higher phenolic and a lower furanic character. Hence, the improved absorption of methylene blue and SDS with nitric acid confirms the higher importance of the tannin part in the absorption process.

4. Conclusions

An extended number of different formulations for the preparation of formaldehyde-free tannin foams are available, and some of them have gained industrial interest. In this work, we have presented sulfur-free foams produced using nitric acid as catalyst. They show less furanic character, less acid pH, and a higher absorption capacity for cationic and anionic pollutant by keeping the physical properties of the standard foams. Moreover, the foams obtained with the aliphatic blowing agent produce more homogeneous, smaller pores; more mechanically resistant; and are more insulating in respect to the standard ones, while maintaining the same intrinsic and extrinsic chemical properties. Finally, sulfur-free foams blown with petroleum ether have shown more homogeneous morphology, higher mechanical resistance, and lower thermal conductivity by keeping the higher phenolic character of the nitric acid catalyzed foams. Due to these properties, the NA-PE foam represents a very attractive sulfur-free product for the building insulation market, with higher chance to be industrially upscaled. Studies are currently running to further enhance the end-life opportunities for these materials.

Author Contributions: Conceptualization, G.T. and A.P.; Methodology, T.S. (Thomas Sepperer) and J.E.; Software, J.E.; Validation, J.E., J.N.; Formal Analysis, T.S. (Thomas Sepperer) and J.E.; Investigation, M.L., M.W., M.Z., J.E., J.N., N.C.; Resources, A.P. and L.V.; Data Curation, J.E., S.D.; Writing-Original Draft Preparation, G.T., J.E.; Writing-Review and Editing, G.T.; Visualization, J.E., S.D.; Supervision, G.T. and L.V.; Project Administration, T.S. (Thomas Schnabel), T.G., A.P. and L.V.; Funding Acquisition, L.V., G.T., M.Z. and A.P. All authors have read and agreed to the published version of the manuscript.

Funding: This research was funded principally by the European Regional Development Fund and Interreg V-A Italy Austria 2014–2020 through the Interreg Italy-Austria project ITAT 1023 InCIMa "Smart Characterization of Smart Materials".

Acknowledgments: The authors gratefully acknowledge also EFRE (European Funds for Regional Development), AWS (Austria Wirtschafts Service) and the region of Salzburg for the support in the development of the Salzburg Center for Smart Materials; and the "Competitive repositioning of the wood supply chain" (CORE-WOOD) project, implemented by Programme under the Regional Competitiveness and Employment Objective (POR Veneto), co-funded by the European Regional Development Fund (ERDF) 2014–2020 - axis 1, action 1.1.4. for supporting some specific activities.

Conflicts of Interest: The authors declare no conflict of interest.

References

1. Schiavoni, S.; Bianchi, F.; Asdrubali, F. Insulation materials for the building sector: A review and comparative analysis. *Renew. Sustain. Energy Rev.* **2016**, *62*, 988–1011. [CrossRef]
2. Gnip, I.; Vėjelis, S.; Vaitkus, S. Thermal conductivity of expanded polystyrene (EPS) at 10 C and its conversion to temperatures within interval from 0 to 50 C. *Energy Build.* **2012**, *52*, 107–111. [CrossRef]
3. Yan, D.; Xu, L.; Chen, C.; Tang, J.; Ji, X.; Li, Z. Enhanced mechanical and thermal properties of rigid polyurethane foam composites containing graphene nanosheets and carbon nanotubes. *Polym. Int.* **2012**, *61*, 1107–1114. [CrossRef]
4. Gama, N.V.; Soares, B.; Freire, C.S.; Silva, R.; Neto, C.P.; Barros-Timmons, A.; Ferreira, A. Bio-based polyurethane foams toward applications beyond thermal insulation. *Mater. Des.* **2015**, *76*, 77–85. [CrossRef]
5. Cuce, E.; Cuce, P.M.; Wood, C.J.; Riffat, S.B. Toward aerogel based thermal superinsulation in buildings: A comprehensive review. *Renew. Sustain. Energy Rev.* **2014**, *34*, 273–299. [CrossRef]
6. Tondi, G.; Pizzi, A. Tannin-based rigid foams: Characterization and modification. *Ind. Crop. Prod.* **2009**, *29*, 356–363. [CrossRef]
7. Pizzi, A.; Tondi, G.; Pasch, H.; Celzard, A. Matrix-Assisted Laser Desorption/Ionization Time-of-Flight Structure Determination of Complex Thermoset Networks: Polyflavonoid Tannin-Furanic Rigid Foams. *J. Appl. Polym. Sci.* **2008**, *110*, 1451–1456. [CrossRef]
8. Tondi, G.; Blacher, S.; Leonard, A.; Fierro, V.; Leban, J.M.; Celzard, A. X-ray microtomography studies of tannin-derived organic and carbon foams. *Microsc. Microanal.* **2009**, *15*, 384–394. [CrossRef]
9. Reyer, A.; Tondi, G.; Berger, R.J.F.; Petutschnigg, A.; Musso, M. Raman spectroscopic investigation of tannin-furanic rigid foams. *Vib. Spectrosc.* **2016**, *84*, 58–66. [CrossRef]
10. Tondi, G.; Pizzi, A.; Olives, R. Natural tannin based rigid foams as insulation for doors and wall panels. *Maderas* **2008**, *10*, 219–227. [CrossRef]
11. Tondi, G.; Oo, C.W.; Pizzi, A.; Trosa, A.; Thevenon, M.F. Metal Absorption of tannin based rigid foams. *Ind. Crop. Prod.* **2009**, *29*, 336–340. [CrossRef]
12. Meikleham, N.E.; Pizzi, A. Acid-and alkali-catalyzed tannin-based rigid foams. *J. Appl. Polym. Sci.* **1994**, *53*, 1547–1556. [CrossRef]

13. Link, M.; Kolbitsch, C.; Tondi, G.; Ebner, M.; Wieland, S.; Petutschnigg, A. Formaldehyde-free tannin foams and their use as lightweight panels. *Bioresources* **2011**, *6*, 4218–4228.
14. Kolbitsch, C.; Link, M.; Petutschnigg, A.; Wieland, S.; Tondi, G. Microwaves produced tannin-furanic foams. *J. Mater. Sci. Res.* **2012**, *1*, 84–91. [CrossRef]
15. Tondi, G.; Link, M.; Kolbitsch, C.; Petutschnigg, A. Infrared-catalysed synthesis of tannin-furanic foams. *Bioresources* **2014**, *9*, 984–993.
16. Tondi, G.; Zhao, W.; Pizzi, A.; Du, G.; Fierro, V.; Celzard, A. Tannin-based rigid foams: A survey of chemical and physical properties. *Bioresour. Technol.* **2009**, *100*, 5162–5169. [CrossRef]
17. Tondi, G.; Petutschnigg, A. Hydrophobic tannin foams. *Int. Wood Prod. J.* **2015**, *6*, 148–150. [CrossRef]
18. Tondi, G.; Johansson, M.; Leijonmark, S.; Trey, S. Tannin based foams modified to be semi-conductive: Synthesis and characterization. *Prog. Org. Coat.* **2015**, *78*, 488–493. [CrossRef]
19. Tondi, G.; Link, M.; Kolbitsch, C.; Gavino, J.; Luckeneder, P.; Petutschnigg, A.; Herchl, R.; Van Doorslaer, C. Lignin-based foams: Production process and characterization. *Bioresources* **2016**, *11*, 2972–2986. [CrossRef]
20. Tondi, G.; Link, M.; Kolbitsch, C. Sandwich panels with 100% natural tannin furanic foam core. *For. Prod. J.* **2015**, *65*, 33–38.
21. Sepperer, T.; Neubauer, J.; Eckardt, J.; Schnabel, T.; Petutschnigg, A.; Tondi, G. Pollutant Absorption as a Possible End-Of-Life Solution for Polyphenolic Polymers. *Polymers* **2019**, *11*, 911. [CrossRef] [PubMed]
22. Tondi, G.; Link, M.; Kolbitsch, C.; Lesacher, R.; Petutschnigg, A. Pilot plant up-scaling of tannin foams. *Ind. Crop. Prod.* **2016**, *79*, 211–218. [CrossRef]
23. Feldkamp, L.A.; Davis, L.C.; Kress, J.W. Practical cone-beam algorithm. *J. Opt. Soc. Am.* **1984**, *1*, 612–619. [CrossRef]
24. EN ISO 18125:2018. *Solid Biofuels–Determination of Calorific Value*; ISO: Geneva, Switzerland, 2018.
25. Tondi, G.; Petutschnigg, A. Middle infrared (ATR FT-MIR) characterization of industrial tannin extracts. *Ind. Crop. Prod.* **2015**, *65*, 422–428. [CrossRef]
26. DIN 52185. *Testing of Wood; Compression Test Parallel to Grain*; German Institute for Standardisation: Berlin, Germany, 1976.
27. Sánchez-Martín, J.; Beltrán-Heredia, J.; Delgado-Regaña, A.; Rodríguez-González, M.A.; Rubio-Alonso, F. Optimization of tannin rigid foam as adsorbents for wastewater treatment. *Ind. Crop. Prod.* **2013**, *49*, 507–514. [CrossRef]
28. Hayashi, K. A rapid determination of sodium dodecyl sulfate with methylene blue. *Anal. Biochem.* **1975**, *67*, 503–506. [CrossRef]
29. Eckardt, J. Tannin Basierende Schaumstoffe: Bewertung der Physikalischen Eigenschaften. Bachelor's Thesis, Salzburg University of Applied Sciences, Hallein, Austria, 2017.
30. Tondi, G.; Cefarin, N.; Sepperer, T.; D'Amico, F.; Berger, R.F.J.; Musso, M.; Birarda, G.; Reyer, A.; Schnabel, T.; Vaccari, L. Understanding the polymerization of polyfurfuryl alcohol: Ring opening & Diels-Alder reactions. *Polymers* **2019**, *11*, 2126.
31. Tondi, G.; Link, M.; Oo, C.W.; Petutschnigg, A. A Simple Approach to Distinguish Classic and Formaldehyde-Free Tannin Based Rigid Foams by ATR FT-IR. *J. Spectrosc.* **2015**. [CrossRef]
32. Tondi, G.; Petutschnigg, A. Tannin-Based Foams: The Innovative Material for Insulation Purposes. Handbook of Composites from Renewable Materials. *Struct. Chem.* **2016**, *1*, 93.
33. Basso, M.C.; Giovando, S.; Pizzi, A.; Lagel, M.C.; Celzard, A. Alkaline tannin rigid foams. *J. Renew. Mater.* **2014**, *2*, 182–185. [CrossRef]
34. Feddal, I.; Ramdani, A.; Taleb, S.; Gaigneaux, E.M.; Batis, N.; Ghaffour, N. Adsorption capacity of methylene blue, an organic pollutant, by montmorillonite clay. *Desalin. Water Treat.* **2014**, *52*, 2654–2661. [CrossRef]
35. Acero, J.L.; Benitez, F.J.; Real, F.J.; Teva, F. Removal of emerging contaminants from secondary effluents by micellar-enhanced ultrafiltration. *Sep. Purif. Technol.* **2017**, *181*, 123–131. [CrossRef]

polymers

MDPI

Article

Comparative Investigation on the Soil Burial Degradation Behaviour of Polymer Films for Agriculture before and after Photo-Oxidation

Francesco Paolo La Mantia [1,2,*], Laura Ascione [2], Maria Chiara Mistretta [1], Marco Rapisarda [3] and Paola Rizzarelli [3,*]

1 Dipartimento di Ingegneria, Università di Palermo, Viale delle Scienze, 90128 Palermo, Italy; mc.mistretta@gmail.com
2 Consorzio Interuniversitario Nazionale per la Scienza e Tecnologia dei Materiali (INSTM), Via Giusti 9, 55100 Firenze, Italy; ascione.laura@gmail.com
3 Istituto per i Polimeri, Compositi e Biomateriali, Consiglio Nazionale delle Ricerche, Via Paolo Gaifami 18, 95126 Catania, Italy; marcorapis7@gmail.com
* Correspondence: francescopaolo.lamantia@unipa.it (F.P.L.M.); paola.rizzarelli@cnr.it (P.R.)

Received: 26 February 2020; Accepted: 26 March 2020; Published: 30 March 2020

Abstract: Polymer films based on biodegradable polymers, polyethylene (PE) and modified PE with oxo-degradable additive were prepared by film blowing. Carbon black (1%) was added to all the films. Commercial biodegradable Ecovio® and Mater-Bi® samples were used. Mechanical properties, soil burial degradation and surface wettability were investigated, before and after UV irradiation. Chemical modifications induced by UV and soil degradation, or a synergic effect, were highlighted by Attenuated Total Reflection-Fourier Transform Infra-Red (ATR-FTIR). Photo-oxidized film samples with an elongation at break equal to 50% and 0.5 the initial value were selected for the soil burial degradation test at 30 °C. Weight loss measurements were used to follow biodegradation in soil. Predictably, the degradation in soil was higher for biodegradable polymer-based films than for the PE-based ones. UV irradiation increased surface wettability and encouraged the disintegration in soil of all the samples. In fact, photo-oxidation produced a molar mass reduction and hydrophilic end groups, thus increasing surface erosion and weight loss. This paper not only supplies new criteria to evaluate the performance of biodegradable films in agriculture, before and after lifetime, but also provides a comparative analysis on the soil burial degradation behaviour with traditional ones.

Keywords: biodegradable polymers; mulch films; soil burial test; Ecovio®; Mater-Bi®; polylactide; poly(butyleneadipate-*co*-butyleneterephtalate); polyethylene; polymer degradation; photo-oxidation

---

1. Introduction

Polymer films are widely applied in agriculture [1]. The main applications of plastic films are for greenhouses and mulching. The more extensively used polymers for these applications are polyethylenes (LDPE and LLDPE) and the copolymer poly(ethylene-vinyl acetate) (EVA). The importance of these films consists in the protection of the cultures and creation of a microclimate that strongly improves the yield of the plants. The principal problem of these films is their short, or relatively short, life and the consequent frequent necessity to eliminate them in huge amounts from the ground. Indeed, the photo-oxidation strongly worsens the mechanical and optical properties of the polymers and, nevertheless the strong stabilizing systems used, the lifetime is no more than three-four years. Anyway, at the end of the service life, all these films must be collected to avoid the dissemination in the ground with a negative environmental impact [2]. Recycling of these films is relatively simple because of the easy differentiated collection for the similar composition. However, the properties of the recycled materials are relatively poor for the photo-oxidation undergone during their use. The recycling is more complicated for the mulch films as their very low thickness (15–50 micron) implies many difficulties during their washing and the fragmentation. For these films, then, the use of biodegradable, compostable polymers could be a good opportunity [3,4]. The films are mixed with the soil and buried in the ground during the ploughing. The successive biodegradation process gives rise to compost that is a fertilizer for the ground.

Aliphatic polyesters are the most adopted biodegradable materials in food packaging as well as in agriculture and medical fields [5]. Diverse studies on commercial biodegradable polymers have been performed to verify their performance in mulching or irrigation pipes [4,6,7]. Nevertheless, both plastic mulch films and irrigation pipes, during their service life, are exposed to sunlight. Thus, the occurrence of photo-aging can modify their properties and performances as well as their biodegradation rates.

The degradation of plastic materials, traditional and biodegradable, involves different processes promoted by one or more environmental factors (i.e., heat, light, microorganisms) or chemicals. The degradation produces irreversible structural changes that are mostly unfavourable or, in some cases, necessary, as in biodegradation processes, or are further generated to determine polymer structure, such as in pyrolysis-gas chromatography-mass spectrometry (Py-GC/MS) studies [8,9].

The degradability in soil of poly(vinyl alcohol) [10], copolyesters [11] and poly(ester amide) [12] film samples has been investigated under controlled soil burial conditions. However, the influence of UV on degradation rates in soil has been limited investigated. In fact, despite extensive research on biodegradable materials, only some studies have been aimed on polymer degradation related to a synergistic effect of UV exposure and soil burial [6,7,13–16]. Briassoulis et al. evaluated the degradation behaviour of MaterBi®-based films and irrigation tubes under real field conditions [6,7]. Kijchavengkul et al. studied the degradation of an aliphatic-aromatic biodegradable polyester film under solar exposure and soil burial in a tropical area [13]. Recently, we investigated the photo-oxidative and soil burial degradation of irrigation tubes based on biodegradable polymer blends under controlled conditions [14]. Actually, laboratory tests supply more reproducible data for both degradation mechanisms and kinetics.

It should be always kept in mind that biodegradation is schematically a three step process: in the first one, macromolecular chains are depolymerized into monomers and oligomers; in the second stage, the monomers and oligomers are taken up as biomass; and finally in the third step, the respiration of biomass there consumes O_2 and produces CO_2 and H_2O (under aerobic conditions). The measurement of the plastic material consumption does not allow to ascertain whether the process has actually been completed or has prevented, for example, at depolymerisation. As a result, all the standardized methods for determining biodegradation are focused on the measurement of respiration, i.e., the conversion into CO_2 of the carbon initially present in the plastic through the use of the oxidant (O_2). On the other hand, most of the papers in the literature concerning polymer and composite biodegradation are based on weight loss [17–19] that is considered a measurement of plastic film biodegradability.

Aim of this work is to study the degradation behaviour in soil of a couple of biodegradable polymer systems after a photo-oxidative aging that simulates the real conditions of the service life of the film samples to be used in agriculture. For comparison, films of conventional polyethylene (PE) and PE-modified with an oxo-biodegradable compound have been used. A very interesting issue of the present work is that the films have not been aged under UV irradiation for the same time before the soil burial test. In particular, they have been aged until the elongation at break (EB) reached a given value, i.e., when the EB arrived at one half of the initial value, according to an international rule for the use of much films [20], as well as at the value of 50%. The elongation at break has been chosen to monitor the degradation of these films because it is a property of the polymers very sensitive to the changes of structure and morphology. In addition, it is very important for the use of these films as they must remain flexible during their use.

By blow film extrusion, we have prepared film samples based on biodegradable polymers (Ecovio® and Mater-Bi®), PE and PE modified with oxo-degradable additive. Soil burial tests were carried out and weight loss used to monitor biodegradation rate. Mechanical properties determined. Remarkably, the films have been photo-oxidized up to the EB reached the value of 50% and one half of the initial value. We compared mechanical properties, weight loss due to burial degradation in soil and surface wettability, before and after UV irradiation. Structural modifications induced by UV and soil degradation, or a synergic effect, were checked by Attenuated Total Reflection (ATR)-FTIR.

2. Materials and Methods

2.1. Materials

The samples investigated in this work were three films produced in our laboratory and a commercial one. Sample codes of the films, thickness and extrusion temperature are reported in Table 1. PE is a film made by a blend of 80% LDPE (FC 30, Versalis, San Donato Milanese, Italy) and 20% LLDPE (FG20, Versalis); this composition has been chosen because is used in the manufacturing of mulch films. PE-OXO is made by the same polymer blend by adding 1% wt/wt of an oxo-biodegradable masterbatch, d2w (Symphony Environmental, Hertfordshire, UK). B1 is a film prepared by ECOVIO F23B1 (Bayer, Leverkusen, Germany), a biodegradable blend poly(butyleneadipate-*co*-butyleneterephtalate) (PBAT) with small amount of polylactide (PLA), containing about 12% of insoluble, inert particles. This amount has been determined after a Soxhlet extraction in tetrahydrofuran. B2 is a commercial film based on Mater-Bi EF04P by Novamont. This biodegradable polymer system is a blend constituted by corn starch and a biodegradable thermoplastic copolyester, Origo BI (Novamont, Novara, Italy), containing about 8% of insoluble, inert particles. Carbon black (1%) was added to all the films. The first three films were produced in a Brabender single screw extruder (D = 19 mm, L/D = 25) equipped with a head for film blowing.

Table 1. Sample code, extrusion temperature of the three samples produced in laboratory scale and thickness for all the investigated films.

Sample Code	Film Thickness (μm)	Extrusion Temperature (°C)
PE	20	180
PE-OXO	20	180
B1	13	190
B2	15	—

2.2. Methods

2.2.1. Accelerated Weathering Tests

The films were exposed to accelerated weathering tests in a Q-UV Panel system (Q-Labs Corp., Westlake, OH, USA) containing eight UVB lamps with a peak at 313 nm. The accelerated weathering tests were carried out using continuously the following cycle: a period of UV irradiation for 8 h at 70 °C followed by a steam condensation (deionized water spray) for 4 h at 55 °C. Film samples were removed from the chamber at different aging intervals.

2.2.2. Mechanical Properties

Stress-strain curves were measured using a universal testing machine mod. 3365 (Instron, Norwood, MA, USA). The elastic modulus was measured at a speed of 1 mm/min until the deformation was 10%. Then, the crosshead speed was increased to 100 mm/min until the specimen breaks. The values of elastic modulus, E, tensile strength, TS, and elongation at break, EB, were calculated as average of 10 experimental values.

2.2.3. Attenuated Total Reflection-Fourier Transform Infra-Red (ATR-FTIR)

ATR-FTIR spectra were measured by using a Spectrum Two spectrometer (Perkin-Elmer, Waltham, MA, USA) with the Spectrum software. Spectra were recorded with 8 scans and 4 cm^{-1} resolution. In order to compare the different effects on the structures of photo-oxidation and degradation in soil, the carbonyl index (CI) was evaluated for the virgin and degraded samples. In particular, the CI was calculated from the ratio of the total area of the absorption bands between 1808–1550 cm^{-1} or 1801–1487 cm^{-1} (carbonyl region) and 3029–2749 cm^{-1} (stretching of the CH_2) respectively for the B1 and B2 samples, from the ratio of the total of the peaks in the range 1826–1550 cm^{-1} (carbonyl region) and that 2983–2662cm^{-1} (stretching of the CH_2 and CH_3 groups) for the PE based films.

2.2.4. Soil Burial Test

Tests were carried out at 30 ± 0.1 °C, under moisture-controlled conditions. Triplicate specimens of film samples were placed in darkened vessels containing a multi-layer substrate [12]. Filter paper and polyethylene samples were used, respectively, as a positive and negative control. Film portions of 2 cm × 2 cm were cut. Specimens of film (initial weight 7–8 mg, filter paper ~29 mg; Mettler Toledo MX5, d = 1 μg) were sandwiched between two layers of a mixture of milled perlite (50 g) and commercial soil (200 g), moistened with 100 mL of distilled water. The bottom and top layers were filled with 60 g of perlite moistened with 120 mL of distilled water. Perlite was used for increasing aeration to the soil and the amount of water retained. A flow of moistened air was supplied from the bottom of each vessel every 24 h for 15 min. Film samples were removed after regular intervals, brushed softly, washed with distilled water several times and dried under vacuum in the presence of P_2O_5 at room temperature, to constant weight [21]. The degree of degradation was evaluated by weight loss (WL) by using the following equation:

$$\text{WL}(\%) = (\text{Wi} - \text{Wt})/\text{Wi} \times 100, \quad (1)$$

where Wi is the initial weight of the sample and Wt is the weight after the established time.

2.2.5. Contact Angle

The surface wettability values of samples were measured at room temperature using a contact angle goniometer (OCA15EC, Dataphysics, Filderstadt, Germany). Static contact angle (CA) were determined dropping 2 μL of water from a micro syringe onto the film surfaces and analysing by software (SCA 20) the images taken by the connect video camera. To eliminate interference, the sample was previously equilibrated for 30 min and then CA was measured. The films were kept flat using a film sample holder that allowed their correct positioning and stretching. Five measurements were carried out for each sample in order to ensure repeatability of the experiments.

3. Results

3.1. Mechanical Properties

The values of the elongation at break (EB_0), tensile strength and elastic modulus of the virgin films are listed in Table 2. The dimensionless elongation at break vs. photo-oxidation time curves of all the samples are reported in Figure 1. The dimensionless values of the elongation at break were obtained by dividing the values of the elongation at break recorded for each photo-oxidation time by the corresponding initial value (EB_0).

Table 2. Mechanical properties of the virgin films.

Sample Code	Elastic Modulus, E (MPa)	Tensile Strength, TS (MPa)	Elongation at Break, EB_0 (%)
PE	206 ± 19	19.5 ± 1.1	586 ± 12
PE-OXO	183 ± 23	15.9 ± 1.4	582 ± 32
B1	186 ± 16	8.6 ± 2.2	460 ± 11
B2	129 ± 6	21.1 ± 2.0	392 ± 22

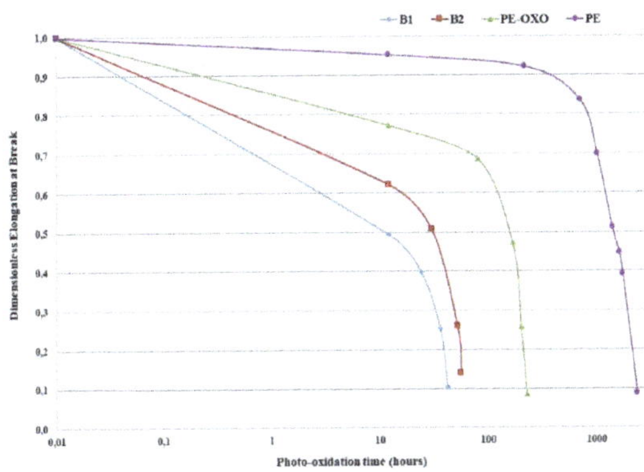

Figure 1. Dimensionless elongation at break vs. photo-oxidation time of all the samples.

Figure 1 highlights that the photo-oxidation affects the elongation at break of the two biodegradable polymers much more than that of the PE-based film samples. Of course, the presence of a pro-oxidant system strongly increases the kinetic of photo-oxidation of the PE matrix. As for the two biodegradable polymers, the photo-oxidation kinetics are slightly more rapid for the B1 samples.

In Figure 2 the photo-oxidation times at which all the films reach the elongation at break equal to 50% (EB_1) and to one half (EB_2) of the initial value (EB_0) are reported.

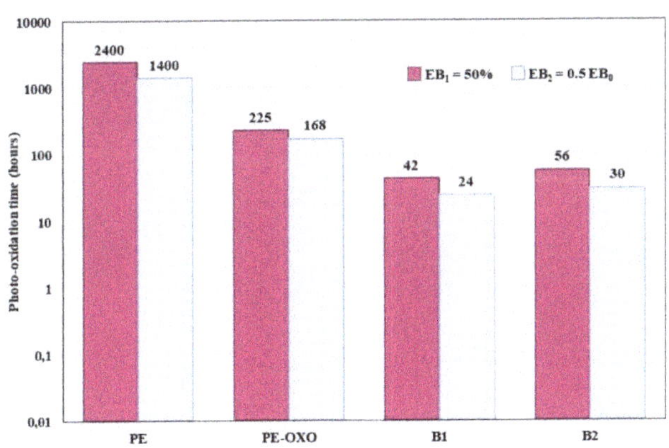

Figure 2. Photo-oxidation times (hours) of different samples when the elongation at break reaches the value of 50% (EB_1) and one half of the initial value (EB_2).

From this plot, it is well evident as the useful accelerated lifetime of the biodegradable films is significantly shorter than that of the PE samples. In particular, considering the values of the elongation at break 50% as the lowest value compatible with the use, the lifetime—as measured in these accelerated aging experimental conditions—of the PE sample is about 2400 h, while it is 42 and 56 h for the two biodegradable polymer systems. The presence of the pro-oxidant additive in the PE matrix considerably reduces the useful lifetime to about 225 h.

3.2. Soil Burial Degradation

Soil burial degradation tests were carried sandwiching the polymer films between two layers of a mixture of milled perlite and commercial soil to simulate soil degradation after their use lifetime. Perlite was used to increase the amount of water retained and accelerate degradation in soil.

Figure S1 summarizes the photo-oxidation and soil burial degradation intervals, together with some representative photographs of the film samples recovered after the burial test. Except for the B2 sample photo-oxidized (56 h), all the film portions appeared almost intact.

In Figure 3 the average weight losses of the virgin and photo-oxidized samples are reported vs. the soil burial time. Of course, the WL increases and kinetic of biodegradation decreases with burial degradation time. Additionally, the WL values increased because of UV exposure, as recently evidenced in soil burial degradation of photo-oxidized irrigation tubes based on biodegradable polymer blends [14]. Reasonably, UV exposure yielded a molecular weight (MW) reduction producing oligomeric chains, bearing hydrophilic chain ends, which are more easily removed from the film surface and susceptible to the attacks of microorganisms in the soil. In fact, it must be kept in mind that biodegradation proceeds in three stages. Whenever WL is used to monitor the degradation of polymer samples, just the first step is involved, i.e., macromolecular chain depolymerisation into monomers and oligomers that are eroded from the surface. UV irradiation enhances and accelerates the formation of monomers and oligomers, via random main chain scission, increasing consequently the WL rate.

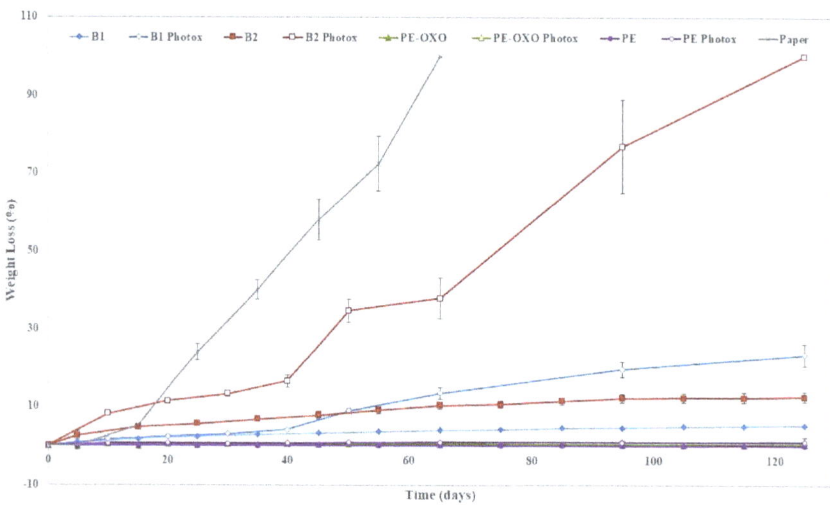

Figure 3. Average weight loss values (%) vs. degradation time for all the film samples—virgin and photo-oxidized.

The positive control (paper) and B2 sample photo-oxidized for 56 h were totally disintegrated (WL = 100%) respectively after 65 and 125 days of soil burial test (Figure 3). The complete disintegration in soil of the B2 sample photo-oxidized for 56 h suggests that the photo-oxidation is not limited to the surface but occurred throughout the matrix bulk, in agreement with the literature [13]. WL values of B1 (Figure 4a) are remarkably lower than that of B2 (Figure 4b) samples. The photo-oxidized samples of all the films show a more pronounced WL than the corresponding virgin samples (Figures 4 and 5). This holds also for the PE-based films, but the final values of the WL are almost negligible although the photo-oxidized PE-OXO film (Figure 5a) shows WL values slightly larger than that of the pure PE (Figure 5b). On the contrary, the weight losses of the two biodegradable samples are quite similar for the virgin films. WL curves of B1 and B2 photo-oxidized samples shows a comparable trend with an induction period (40 days) and then a rapid increase.

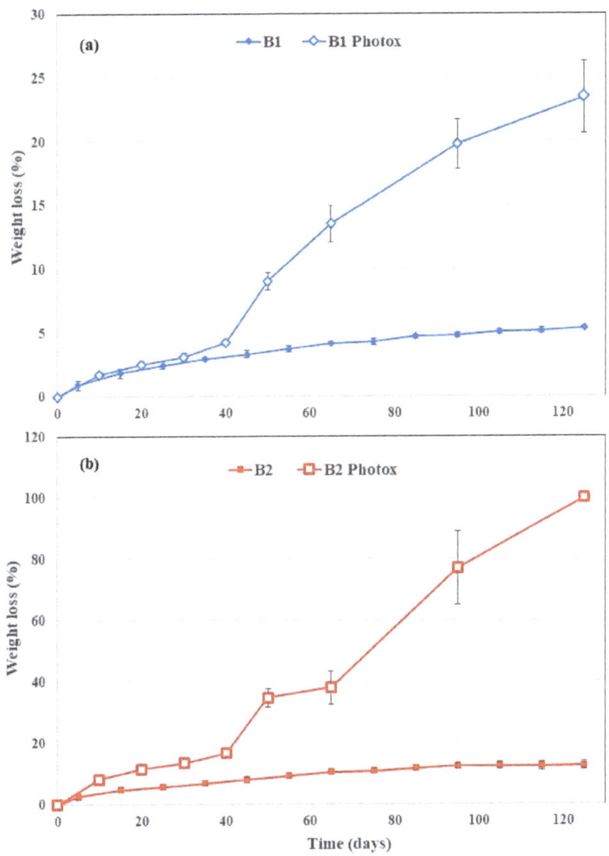

Figure 4. Average weight loss values (%) vs. degradation time for the film samples (**a**) B1 and (**b**) B2, virgin and photo-oxidized.

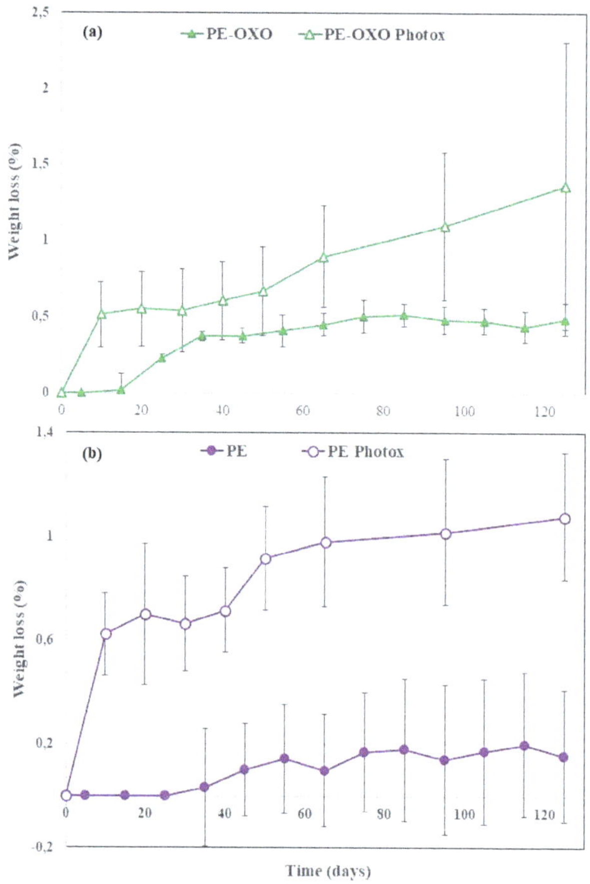

Figure 5. Average weight loss values (%) vs. degradation time for the film samples (**a**) polyethylene (PE)-OXO and (**b**) PE, virgin and photo-oxidized.

3.3. Wettability of the Film Surfaces

The wettability of the film surfaces submitted to UV irradiation was determined by contact angle measurements. In fact, chemical modifications on the surface of the thin films, induced by UV irradiation and/or soil degradation, increase plastic samples wettability, highlighted by CA decrease (Figures 6 and 7) and, consequently, the microbial susceptibility is encouraged. The UV induced increase in surface wettability is more pronounced for photo-oxidized and buried B1 (Figure 6) samples (CA ~68°) as well as for photo-oxidized B2 films (CA ~70°, Figure 6). CA decrease is reasonably related to the formation of hydrolytic chain ends. In photo-oxidised and buried B1 samples the CA is quite similar to that of B1 buried (Figure 6). This could be related to the surface erosion and the removal of the oligomeric species produced by hydrolysis.

In a similar way, the increase in the wettability of the irradiated PE (Figure 7) and PE-OXO (Figure 7) samples can be attributed to the formation of hydrophilic groups on the polymer surface by photo-oxidation, which occurs by the exposure of PE to UV irradiation in the presence of air [22,23]. The photo-degradation of PE samples, promoted by the oxo-degradable additive, induces a more marked decrease in the contact angle because of the introduction of carbonyl groups (Figure 7). However, wettability is higher in photo-oxidized and buried PE sample (CA ~79°, Figure 7) than in photo-oxidized and buried PE-OXO (CA ~91°, Figure 7). This can be reasonably due to the higher WL and thus faster surface erosion of oligomers, bearing hydrophilic chain ends, in the PE-OXO (Figure 5a) than in PE sample (Figure 5b).

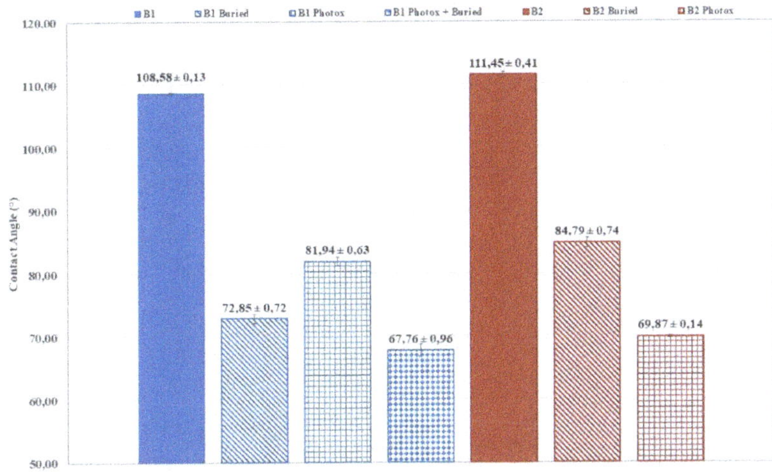

Figure 6. Average contact angle values for the film samples B1 and B2, virgin, buried (125 days) and photo-oxidized. After 125 days of soil burial test, photo-oxidized B2 samples were completely disintegrated and it was not possible to carry out the contact angle (CA) measurements.

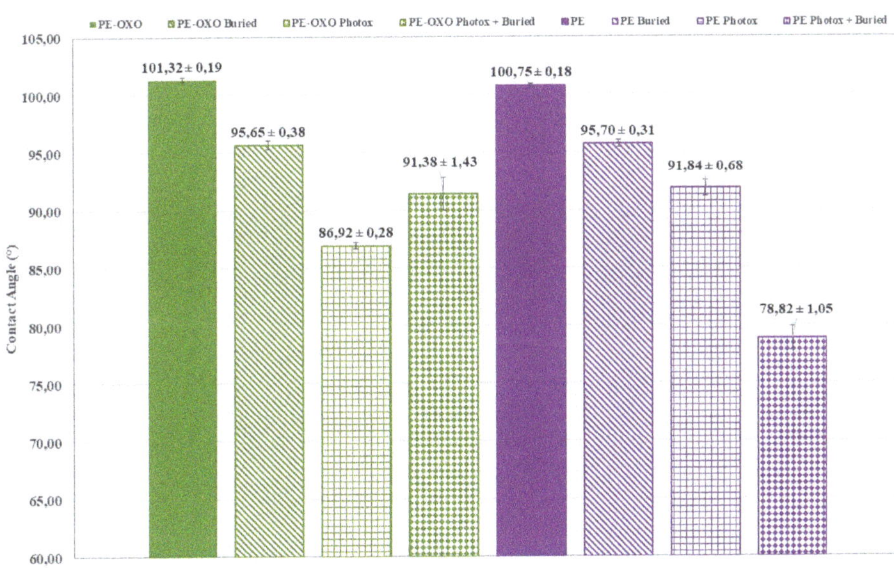

Figure 7. Average contact angle values for the film samples PE-OXO and PE, virgin, buried (125 days) and photo-oxidized.

3.4. UV and Soil Burial Induced Modifications

Both UV and soil burial degradation induced chemical modifications on the surface of all the polymeric films. Then, to check the structural changes caused by the different types of degradation, ATR-FTIR analysis was performed on virgin, photo-oxidized, soil buried and photo-oxidized + buried film samples. In Figures S2 and S3, ATR-FTIR spectra are reported as examples of the trend of variations in functional groups on the film surface. Overall, degradation produced a broadening and a change of the peak area more noteworthy in B2 (Figure S2b) and PE (Figure S3b) samples compared respectively with B1 (Figure S2a) and PE-OXO (Figure S3a). A significant decrease in peak intensity between photo-oxidized and photo-oxidized + buried B2 sample was more evident (Figure S2b). Remarkably, it is in accordance with the higher surface erosion of oligomers produced by UV irradiation and the resultant weight loss values increase.

The increase of the carbonyl groups was a marker of the photo-oxidation evolvement as well as of hydrolytic reactions taking place during soil degradation. Unsurprisingly, in agreement with the literature [13,14], the spectra of the two biodegradable polymer-based films show that both photo-oxidation and degradation in soil increase the CI (Figure 8a). Remarkably, the area ratio of photo-oxidized + buried B1 and B2 sample is lower than that photo-oxidized and similar to the not photo-oxidized and buried ones. This result suggests that the oligomers, bearing carbonyl groups, have been partially removed from the surface during the soil burial test. However, B1 films were clearly more vulnerable to photo-oxidation than to hydrolysis; photo-oxidation products were removed by degradation in soil and CI of "B1 photox + buried" was lower than "B1 photox" sample (Figure 8a). Compared to the virgin samples, the increase of CO functional groups, due to the combined photo-oxidation and soil degradation, was significantly more evident in the B2 than in the B1 samples; in fact, the UV irradiation did not accelerate the WL rate as it did in the B2 films (Figure 4). These results were in agreement with the degradation process and the choice of monitoring degradation in soil through weight loss. Actually, the B2 films were much more prone than the B1 ones, since monomers and low-molecular-weight oligomers were originated and removed in the erosion step faster than in the B1 sample.

The carbonyl index values show similar trends in PE-OXO and PE samples. Photo-oxidation provides an increase in the area ratio, significantly more marked for PE than for PE-OXO (Figure 8b), due to the formation of ketones, aldehydes and/or esters groups. The successive degradation in soil on photo-oxidized PE films produces a low CI decrease reasonably related to a limited surface erosion of oligomers highlighted by WL (Figure 5b). The small CI decrease in photo-oxidized PE-OXO and PE after soil burial test for 125 days is in agreement with the literature and was attributed to a minimal consumption of CO groups by microorganisms [23].

It has to be underlined that carbon back is a UV absorber and its photo-stabilizing action is due to the absorption of the ultraviolet energy that, then, is no more available for the formation of radicals and for the consequent propagation of the oxidation reactions [24]. This means that carbon black does not interfere with the photo-oxidative mechanisms of the different matrices and the effect is just to shift towards higher times the photo-oxidation increasing the induction time of the process. Carbon black slows down the molecular weight reduction and consequently the degradation process in soil. The shift towards higher time to starting photo-oxidation is relevant on a commercial point of view above all with biodegradable polymeric systems whose performance can be impaired by photo-oxidation processes more than the traditional ones [4].

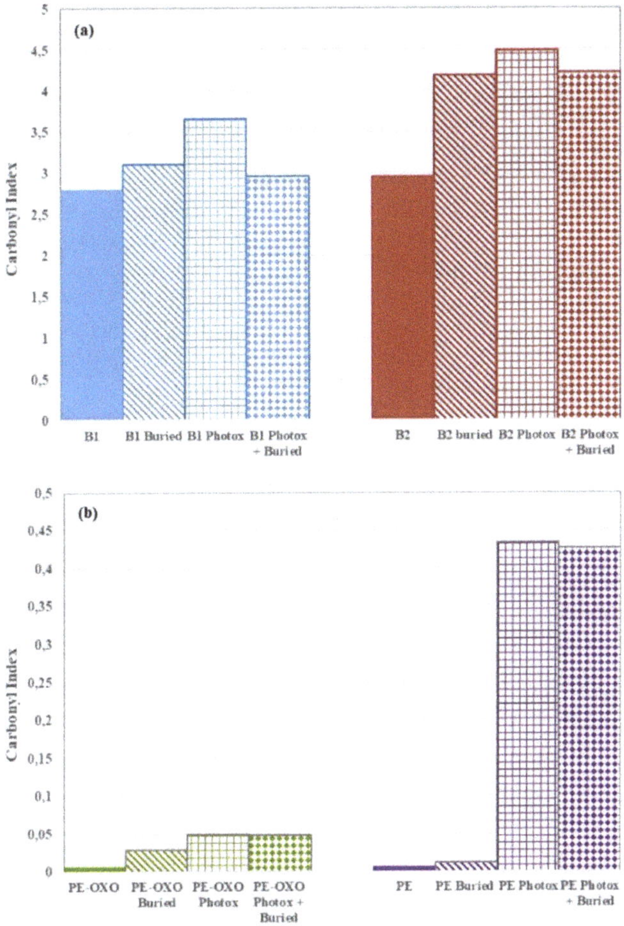

Figure 8. Carbonyl index values for (**a**) the two biodegradable and (**b**) PE based film samples. CI (B1) = A(1808–1550 cm^{-1})/A(3029–2749 cm^{-1}); CI (B2) = A(1801–1487 cm^{-1})/A(3029–2749 cm^{-1}); CI (PE and PE-OXO) = A(1826–1550 cm^{-1})/A(2983– 2662 cm^{-1}). B1 = Ecovio®-based and B2 = Mater Bi®-based films.

4. Conclusions

A comparative investigation on the degradation behaviour in soil of traditional and biodegradable polymer films for agriculture, before and after photo-oxidation, have been accomplished. The two polyethylene-based samples showed the better mechanical properties, while the Ecovio®-based films, named B1, displayed a lower value of the tensile strength and the Mater-Bi®-based films, called B2, had the lowest values of the elastic modulus and elongation at break. However, all the samples showed high values of the elongation at break that strongly decreases with the photo-oxidation time, in particular for the biodegradable films. Interestingly, the film samples have been aged under UV irradiation until the elongation at break reached the value of 50%, according to an international rule for the use of much films, as well as at one half of the initial value.

Soil burial degradation tests were carried sandwiching the samples, virgin and photo-oxidized, between two layers of a mixture of milled perlite and commercial soil to simulate degradation in soil after their use lifetime as mulch films. Weight loss measurements showed that the Mater-Bi®-based films were more susceptible to soil degradation than the Ecovio®-based ones were. Additionally, for all of the samples, soil degradation was accelerated by UV irradiation. The positive control (paper) was biodegraded in soil within 65 days, while photo-oxidized Mater-Bi®-based films were fully disintegrated in soil after 125 days at 30 °C. These results suggest that the photo-oxidation was not restricted to the surface but occurred throughout the bulk. Chemical structure changes induced by UV exposure affected wettability of films, as highlighted by contact angle decrease. A relevant increase intensity in carbonyl peaks and carbonyl index was highlighted in photo-oxidized biodegradable and PE film samples. Carbonyl index values of photo-oxidized biodegradable and PE film samples decreased after soil burial test because of surface erosion of monomers and oligomers.

This paper not only supplies new criteria to evaluate the performance of biodegradable films in agriculture, before and after lifetime, but also yields a comparative analysis on the soil burial degradation behaviour with traditional ones. However, accelerated weathering test does not provide real lifetime. Further studies will be addressed to establish a correspondence between accelerated weathering time and aging time in real conditions.

Supplementary Materials: The following are available online at http://www.mdpi.com/2073-4360/12/4/753/s1.

Author Contributions: Conceptualization, F.P.L.M. and P.R.; Data curation, L.A. and M.R.; Funding acquisition, F.P.L.M. and P.R.; Investigation, L.A., M.C.M. and M.R.; Project administration, F.P.L.M. and P.R.; Supervision, F.P.L.M. and P.R.; Visualization, F.P.L.M., L.A., M.R. and P.R.; Writing—original draft, F.P.L.M., L.A., M.R. and P.R. All authors have read and agreed to the published version of the manuscript.

Funding: This research was funded by Polieco (Italy) and Agriplast (Italy). POR FSE Sicily 2020 (project "Polymeric systems: innovative aspects and applications in the biomedical and agri-food fields – SPIN OFF of Polymers") and Call 11/2017 ("Strengthening employability in the R&D system and the emergence of research SPIN OFFS in Sicily") partially granted this work.

Acknowledgments: In this section you can acknowledge any support given which is not covered by the author contribution or funding sections. This may include administrative and technical support, or donations in kind (e.g., materials used for experiments).

Conflicts of Interest: The authors declare no conflict of interest.

References

1. Dubois, P. *Plastics in Agriculture*; Brighton, C.A., Ed.; Applied Science Publisher Ltd.: London, UK, 1978; pp. 1–176.
2. Briassoulis, D.; Babou, E.; Hiskakis, M.; Scarascia, P.P.; Guarde, D.; Dejean, C. Review, mapping and analysis of the agricultural plastic waste generation and consolidation in Europe. *Waste Manag. Res.* **2013**, *31*, 1262–1278. [CrossRef] [PubMed]
3. Siwek, P.; Domagala-Swiatkiewicz, I.; Bucki, P.; Puchalski, M. Biodegradable agroplastics in 21st century horticulture. *Polimery* **2019**, *64*, 480–486. [CrossRef]
4. Brodhagen, M.; Peyron, M.; Miles, C.; Inglis, D.A. Biodegradable plastic agricultural mulches and key features of microbial degradation. *Appl. Microbiol. Biotechnol.* **2015**, *99*, 1039–1056. [CrossRef] [PubMed]
5. Avérous, L.; Pollet, E. Biodegradable Polymers. In *Environmental Silicate Nano-Biocomposites. Green Energy and Technology*; Avérous, L., Pollet, E., Eds.; Springer: London, UK, 2012; pp. 13–39.
6. Briassoulis, D. Analysis of the mechanical and degradation performances of optimised agricultural biodegradable films. *Polym. Degrad. Stab.* **2007**, *92*, 1115–1132. [CrossRef]
7. Briassoulis, D.; Babou, M.; Hiskakis, M. Degradation behaviour and field performance of experimental biodegradable drip irrigation systems. *J. Polym. Environ.* **2011**, *19*, 341–361. [CrossRef]
8. Wypych, G. *Handbook of UV Degradation and Stabilization*, 2nd ed.; ChemTec Publishing: Scarborough, ON, Canada, 2015; pp. 1–420.
9. Rizzarelli, P.; Rapisarda, M.; Perna, S.; Mirabella, E.F.; La Carta, S.; Puglisi, C.; Valenti, G. Determination of polyethylene in biodegradable polymer blends and in compostable carrier bags by Py-GC/MS and TGA. *J. Anal. Appl. Pyrol.* **2016**, *117*, 72–81. [CrossRef]
10. Solaro, R.; Corti, A.; Chiellini, E. A new respirometric test simulating soil burial conditions for the evaluation of polymer biodegradation. *J. Envirom. Polym. Degrad.* **1998**, *6*, 203–208. [CrossRef]

11. Wu, H.; Wen, B.; Zhou, H.; Zhou, J.; Yu, Z.; Cui, L.; Huang, T.; Cao, F. Synthesis and degradability of copolyesters of 2,5-furandicarboxylic acid, lactic acid, and ethylene glycol. *Polym. Degr. Stab.* **2015**, *121*, 100–104. [CrossRef]
12. Rizzarelli, P.; Cirica, M.; Pastorelli, G.; Puglisi, C.; Valenti, G. Aliphatic poly(ester amide)s from sebacic acid and aminoalcohols of different chain length: Synthesis, characterization and soil burial degradation. *Polym. Degrad. Stab.* **2015**, *121*, 90–99. [CrossRef]
13. Kijchavengkul, T.; Auras, R.; Rubino, M.; Alvarado, E.; Montero, J.R.C.; Rosales, J.M. Atmospheric and soil degradation of aliphatic-aromatic polyester films. *Polym. Degrad. Stab.* **2010**, *95*, 99–107. [CrossRef]
14. Rapisarda, M.; La Mantia, F.P.; Ceraulo, M.; Mistretta, M.C.; Giuffrè, C.; Pellegrino, R.; Valenti, G.; Rizzarelli, P. Photo-oxidative and soil burial degradation of irrigation tubes based on biodegradable polymer blends. *Polymers* **2019**, *11*, 1489. [CrossRef] [PubMed]
15. Koutny, M.; Lemaire, J.; Delort, A.M. Biodegradation of polyethylene films with prooxidant additives. *Chemosphere* **2006**, *64*, 1243–1252. [CrossRef] [PubMed]
16. Stloukal, P.; Verney, V.; Commereuc, S.; Rychly, J.; Matisova-Rychlá, L.; Pis, V.; Koutny, M. Assessment of the interrelation between photooxidation and biodegradation of selected polyesters after artificial weathering. *Chemosphere* **2012**, *88*, 1214–1219. [CrossRef] [PubMed]
17. Emadian, S.M.; Onay, T.T.; Demirel, B. Biodegradation of bioplastics in natural environments. *Waste Manag.* **2017**, *59*, 526–536. [CrossRef] [PubMed]
18. Eubeler, J.P.; Zok, S.; Bernhard, M.; Knepper, T.P. Environmental biodegradation of synthetic polymers I. Test methodologies and procedures. *Trends Anal. Chem.* **2009**, *28*, 1057–1072. [CrossRef]
19. Rizzarelli, P.; Degli Innocenti, F.; Valenti, G.; Rapisarda, M. Biodegradation of green polymer composites: Laboratory procedures and standard test methods. In *Advanced Applications of Bio-Degradable Green Composites*; Inamuddin, M.P., Ed.; Materials Research Forum LLC: Millersville, PA, USA, 2020; pp. 1–44, ISBN 9781644900642. [CrossRef]
20. EN17033. *Plastics: Biodegradable Mulch Films for Use in Agriculture and Horticulture. Requirements and Test Methods*; European Standard, European Committee for Standardization: Brussels, Belgium, 1 January 2018.
21. Rizzarelli, P.; Puglisi, C.; Montaudo, G. Soil burial and enzymatic degradation in solution of aliphatic co-polyesters. *Polym. Degrad. Stab.* **2004**, *85*, 855–863. [CrossRef]
22. Suresh, B.; Maruthamuthu, S.; Kannan, M.; Chandramohan, A. Mechanical and surface properties of low-density polyethylene film modified by photo-oxidation. *Polym. J.* **2011**, *43*, 398–406. [CrossRef]
23. Gomes, L.B.; Klein, J.M.; Brandalise, R.N.; Zeni, M.; Zoppas, B.C.; Grisa, A.M.C. Study of Oxo-biodegradable Polyethylene Degradation in Simulated Soil. *Mater. Res.* **2014**, *17* (Suppl. 1), 121–126. [CrossRef]
24. Mistretta, M.C.; Botta, L.; Vinci, A.D.; Ceraulo, M.; La Mantia, F.P. Photooxidation of poypropylene/graphene nanoplatelets nanocomposites. *Polym. Deg. Stab.* **2019**, *160*, 35–43. [CrossRef]

polymers

MDPI

Article

Coincident Correlation between Vibrational Dynamics and Primary Relaxation of Polymers with Strong or Weak Johari-Goldstein Relaxation

Antonio Tripodo [1,†], Francesco Puosi [1,†], Marco Malvaldi [1], Simone Capaccioli [1,2] and Dino Leporini [1,2,*]

1 Dipartimento di Fisica "Enrico Fermi", Università di Pisa, Largo B.Pontecorvo 3, I-56127 Pisa, Italy; antonio.tripodo@df.unipi.it (A.T.); francesco.puosi@df.unipi.it (F.P.); marcoampelio@hotmail.com (M.M.); simone.capaccioli@unipi.it (S.C.)

2 Istituto per i Processi Chimico-Fisici-Consiglio Nazionale delle Ricerche (IPCF-CNR), via G. Moruzzi 1, I-56124 Pisa, Italy

* Correspondence: dino.leporini@unipi.it; Tel.: +39-050-2214937

† These authors contributed equally to this work.

Received: 26 February 2020; Accepted: 24 March 2020; Published: 31 March 2020

Abstract: The correlation between the vibrational dynamics, as sensed by the Debye-Waller factor, and the primary relaxation in the presence of secondary Johari-Goldstein (JG) relaxation, has been investigated through molecular dynamics simulations. Two melts of polymer chains with different bond length, resulting in rather different strength of the JG relaxation are studied. We focus on the bond-orientation correlation function, exhibiting higher JG sensitivity with respect to alternatives provided by torsional autocorrelation function and intermediate scattering function. We find that, even if changing the bond length alters both the strength and the relaxation time of the JG relaxation, it leaves unaffected the correlation between the vibrational dynamics and the primary relaxation. The finding is in harmony with previous studies reporting that numerical models *not showing secondary relaxations* exhibit striking agreement with experimental data of polymers also where the presence of JG relaxation is known.

Keywords: polymer melt; primary and secondary relaxations; Johari-Goldstein relaxation; bond reorientation; vibrational dynamics; molecular-dynamics simulations

1. Introduction

If polymers and liquids avoid crystallization during cooling or compression, they freeze into a microscopically disordered solid-like state, a glass [1]. On approaching the glass transition, molecular rearrangements occur via both the primary mode, referred to as structural or α relaxation, and the faster secondary (β) processes as evidenced by mechanical, electrical, and thermal properties of materials [2–4]. Although it has been at the focus of a large number of phenomenological and theoretical studies, as well as of experiments and simulations [4–10], there is still no definitive microscopic description available for the β relaxation. In particular, there is a special class of secondary relaxations involving translation or reorientation of the molecular unit as a whole, well different from the trivial ones, related to intramolecular degrees of freedom, such as the motion of pendant groups in polymers. This special class of β processes, called Johari Goldstein (JG) to honor the researchers that first noticed it [5], is universal in glass formers [4,6,7] and has been shown [6] to have strong relation to structural α-relaxation, being both governed by intermolecular interactions. A number of experimental results indicate that this β JG relaxation is sensitive to the thermodynamic variables underlying the glass transition [4,7], mimicking the α relaxation, being strongly pressure dependent and showing the invariance of the ratio τ_α/τ_β to variations of pressure and temperature, keeping τ_α constant. β JG, for these reasons, can be considered as the precursor to structural relaxation, having a slower dynamics due to cooperativity involving many body dynamics [6,8].

The purpose of the present paper is to investigate if the presence of well resolved secondary relaxation could disprove the universal correlation between the vibrational dynamics and the primary relaxation observed by experiments and numerical simulations in highly viscous liquids [11], for a recent mini review see Reference [12]. We are motivated by the fact that the timescale of the secondary relaxation ($\overline{\tau}_s$) is *intermediate* between the vibrational timescale (t^*), a few picoseconds, and the structural relaxation time of the primary relaxation ($\overline{\tau}_p$) which reaches about hundreds of seconds close to the glass transition:

$$t^* \ll \overline{\tau}_s < \overline{\tau}_p \tag{1}$$

The exact definitions of the symbols in Equation (1) will be given in Section 3. Here, we are interested in the secondary relaxation observed in the *main chain* of *linear* polymers without side groups. This class fits the definition of Johari-Goldstein (JG) relaxation [7]. There is consensus that in linear chains the JG process involves local motion of the polymer backbone.

Recently, the investigation of JG relaxation by using MD simulations has been facilitated by the development of coarse-grained models of linear polymers having fixed bond length and bond angles constrained to 120° [13–15]. We also mention studies of the JG relaxation in asymmetric diatomic molecules [15,16].

2. Model and Numerical Methods

We investigate two coarse-grained models of linear polymer chains. The melts are constituted by $N_c = 512$ linear chains made of $M = 25$ monomers each, resulting in a total number of monomers $N = 12800$. Non-bonded monomers at a distance r interact via a Lennard-Jones potential:

$$U^{LJ}(r) = \varepsilon\left[\left(\frac{\sigma^*}{r}\right)^{12} - 2\left(\frac{\sigma^*}{r}\right)^{6}\right] \tag{2}$$

where $\sigma^* = 2^{1/6}\sigma$ is the minimum of the potential, $U^{LJ}(r = \sigma) = -\epsilon$. The potential is truncated at $r = r_c = 2.5\sigma$ for computational convenience. Adjacent bonded monomers interact with each other via the harmonic potential $U^{bond}(r) = k_{bond}(r - l_0)^2$, where the constant k_{bond} is set to $2000\epsilon/\sigma^2$ to ensure high stiffness. We consider two distinct cases, corresponding to rest bond length l_0 set to either $l_0 = 0.48\sigma$ or $l_0 = 0.55\sigma$. The rationale behind our choice of the two bond lengths relies on the finding that previous Molecular-Dynamics (MD) simulations of the present model [14], investigating the rotational dynamics, revealed the steep *increase* of the separation between the primary and the JG relaxations by *decreasing* the bond length below $l_0^* = 0.5\sigma$ (corresponding to 2Å in Figure 3 of Reference [14]). Therefore, one anticipates that the JG relaxation is much more apparent if $l_0 = 0.48\sigma$ with respect to $l_0 = 0.55\sigma$. A bending potential $U^{bend}(\alpha) = k_{bend}\,(\cos\alpha - \cos\alpha_0)^2$, with $kbend = 2000\epsilon/\sigma^2$ and $\alpha_0 = 120°$, is introduced to maintain the angle α formed by two consecutive bonds nearly constant (see Figure 1 for typical chain conformations) [17].

All the data presented in this work are expressed in reduced MD units: length in units of σ, temperature in units of ϵ/k_B, where k_B is the Boltzmann constant, and time in units of $\tau_{MD} = (m\sigma^2/\epsilon)^{1/2}$. We set $\sigma = 1$, $\epsilon = 1$, $m = 1$ and $k_B = 1$.

Simulations were carried out with the open-source software LAMMPS [18,19]. Equilibration runs were performed at constant number of monomers N, constant vanishing pressure $P = 0$ and constant temperature T (NPT ensemble) [20]. For each state the equilibration lasted at least for $3\tau_{ee}$, being τ_{ee} the relaxation time of the end-to-end vector autocorrelation function [21–28]. Production runs have been performed within the NVT ensemble (constant number of monomers N, constant volume V and constant temperature T). Additional short equilibration runs were performed when switching from NPT to NVT ensemble. No signatures of crystallization were observed in all the investigated states.

As a final remark, we point out that our adoption of a monodisperse polymer model where all the chains have equal length is not expected to limit our conclusions. In fact, the vibrational scaling collapses in a single master curve both the MD results of monodisperse polymer models and the experimental data on polymers which, expectedly, refer to polydisperse systems [11].

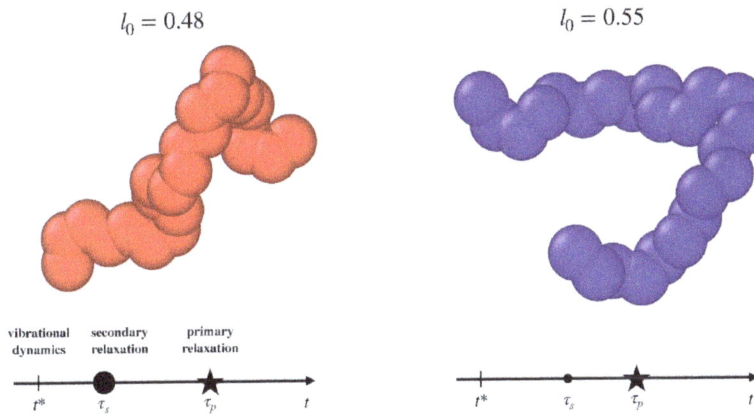

Figure 1. Pictorial view of the two kind of polymeric chains under consideration. L_0 denotes the bond length. The time axis signal, in a qualitative way, the location of the dynamical processes of interest. The size of the dot of the secondary relaxation is proportional to the strength of the relaxation. The exact definition of the symbols is given in Section 3.

3. Results and Discussion

3.1. Bond Reorientation

Since in linear chains the JG process involves local motion of the polymer chain backbone [4,7], we focus on the most elementary relaxation process, that is, the reorientation of the bond linking two adjacent monomers of the same chain. Let us define a suitable average bond correlation function $C(t)$ [29]. To this aim, we consider the unit vector along the m-th bond of the n-th chain at time t:

$$b_{m,n}(t) = \frac{1}{l_0}(r_{m,n}(t) - r_{m+1,n}(t)) \tag{3}$$

where $r_{m,n}$ (t) denotes the position of the m-th monomer in the n-th chain at time t. We define the correlation function $C(t)$:

$$C(t) = \frac{1}{N_c}\frac{1}{M-1}\sum_{n=1}^{N_c}\sum_{m=1}^{M-1}\langle b_{m,n}(t)\cdot b_{m,n}(0)\rangle \tag{4}$$

The brackets $\langle\cdots\rangle$ denote a suitable ensemble average. $C(t)$ decreases in time (by definition $C(0) = 1$), finally vanishing at long times when the bond orientation has explored all the unit sphere.

Figure 2 plots the correlation function $C(t)$ for the two polymer melts with chains having different bond length. In agreement with previous studies [13,14], it is seen that the chains having shorter bond length exhibits a characteristic two-step decay, signaling the presence of two distinct relaxation processes.

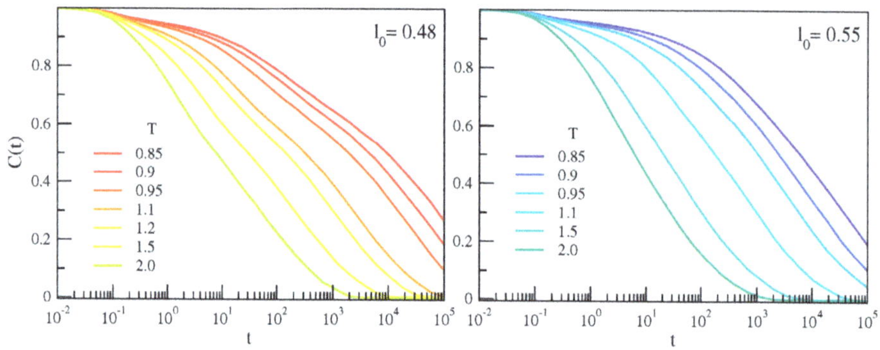

Figure 2. Temperature dependence of the bond correlation function of the chains with different bond length. If $l_0 = 0.48$, a two-step decay—evidencing two distinct relaxations—is observed.

To gain more insight, the correlation function $C(t)$ was analyzed under the assumption that it is represented by the weighted sum of two components, that is, a fast secondary process and a slower primary one:

$$C(t) = A_p f_p(t) + A_s f_s(t) \tag{5}$$

where $f_p(t)$ and $f_s(t)$ are two decaying functions with amplitudes A_p and A_s, respectively. The explicit form of $f_i(t)$ is taken as a stretched exponential:

$$f_i(t) = \exp\left[-\left(\frac{t}{\tau_i}\right)^{\beta_i}\right],\ i = p, s \tag{6}$$

where τ_i denotes the relaxation time and β_i the stretching exponent (as normally $\beta_i \leq 1$). We fit the MD data concerning the correlation function $C(t)$ with Equation (5) excluding the time window $t \leq \hat{t}$ where the decay of the function is controlled by the ballistic motion of monomers ($0.6 \leq \hat{t} \leq 1$ with $\hat{t}$ decreasing slightly by increasing T).

The analysis in terms of a sum of two stretched exponential decays, Equation (5) and Equation (6), conforms with most studies of primary and secondary relaxations in MD simulation [13–15,30,31] and in some experiments done in time domain [32,33]. Then, it facilitates the comparison between our results to those present in literature.

We have considered also the use of alternative functions to fit the secondary relaxation. For instance, we tested the Mittag Leffer function (with 2 fitting parameters like the stretched exponential), corresponding to the time-domain counterpart of the Cole-Cole function and usually used in studies of dielectric spectroscopy to describe the secondary process [34,35]. Yet, no significant differences have been found in terms of relaxation times and strengths with respect to the analysis with the sum of stretched exponentials.

Figure 3 shows the result of the best-fit of $C(t)$ with Equation (5) and Equation (6) for a representative state of the model. For comparison, the best-fit with a single stretched exponential decay is also shown. The residuals $C^{MD}(t) - C^{\text{fit}}(t)$ plotted in the inset of Figure 3 prove the poor performance of the single-relaxation curve to fit the MD data.

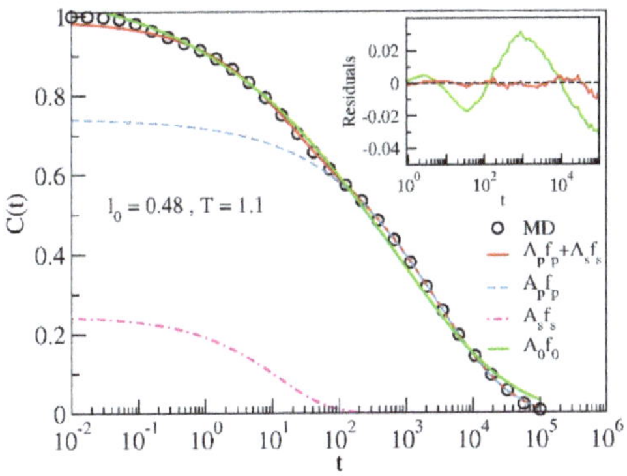

Figure 3. Illustrative example of the best-fit results via the double-relaxation function Equation (5). The best-fit with the single-relaxation function $A_0 f_0$ with f_0 in the form of Equation (6) is also shown. Inset: residuals of the best-fit with the double- and single- relaxation functions.

Figure 4 presents the temperature dependence of the parameters extracted from the best-fit of data with Equation (5) and Equation (6). It is seen that reducing the bond length has two major effects: (i) *increased* separation of the relaxation times τ_s and τ_p (ii) *enhanced* amplitude of the secondary relaxation with respect to the primary one. It is also worth noting that the primary relaxation is less stretched in the model where the secondary relaxation is more apparent (l_0 = 0.48) with respect to the case in which it is weak (l_0 = 0.55). The differences are small but significant, that is, larger than the errors on the stretching exponents. Stretched relaxation is usually associated with the presence of dynamical heterogeneities, namely the spatial distribution of mobilities, which may differ of orders of magnitude in regions only a few nanometers away. In this framework, our results suggest that the presence of a not negligible secondary relaxation process slightly decreases the degree of dynamical heterogeneity of the system. We plan to address this aspect in future works.

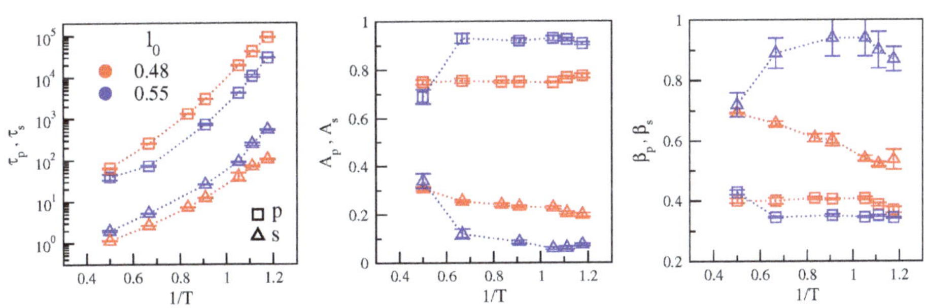

Figure 4. Plots of the temperature dependence of the best-fit parameters by using Equation (5) and Equation (6) (color codes as in Figure 1). From left to right: apparent relaxation times, relaxation strengths and stretching exponents.

To get rid of the temperature dependence of the stretching exponent β, we define the average relaxation time for the primary and the secondary relaxations, $\overline{\tau}_p$ and $\overline{\tau}_s$ respectively, as the time-integral of the corresponding relaxation function (f_p and f_s) leading to

$$\overline{\tau}_i = \frac{\tau_i}{\beta_i}\Gamma\left(\frac{1}{\beta_i}\right),\ i = p, s \tag{7}$$

The temperature dependence of the average relaxation times of the two polymer models are shown in Figure 5. For both systems the primary relaxation time (left panel in Figure 5) grows more than three orders of magnitude in the range of temperature from T = 2.0 to T = 0.85, exhibiting a non-Arrhenius upward concavity which is typical of fragile glassformers [36]. Conversely, the growth of the secondary relaxation time (right panel in Figure 5) is less pronounced (approximately two and a half decades in the same temperature range) and the deviations from an Arrhenius behavior are weaker, in particular for the system with stronger secondary relaxation (l_0 = 0.48, red symbols).

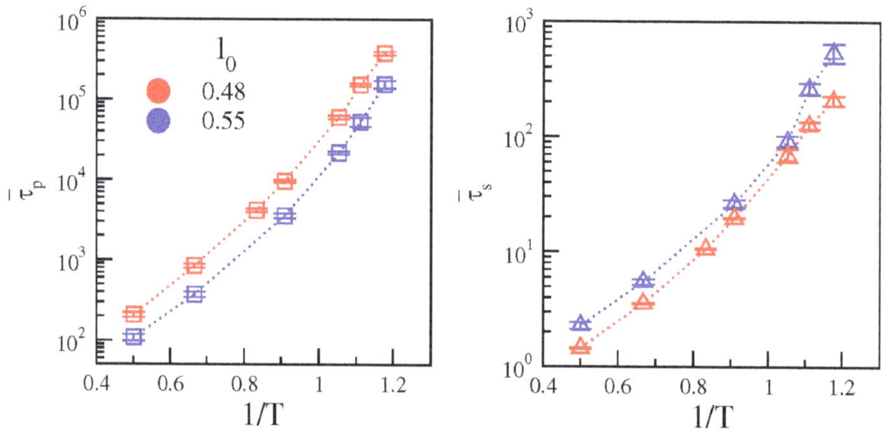

Figure 5. Arrhenius plots of the average relaxation times of the primary ($\overline{\tau}_p$, left) and secondary ($\overline{\tau}_s$, right) relaxations. Color code as in Figure 1.

3.2. Monomer Mobility

With the aim of further characterizing the dynamics of the two polymer melts with chains having distinct bond lengths, we now consider the monomer mobility, as quantified by the mean square displacement (MSD)

$$\langle r^2(t)\rangle = \frac{1}{N}\sum_{i=1}^{N}\langle \| x_t(t) - x_i(0)\|^2\rangle \tag{8}$$

where x_i (t) is the position of the i-th monomer at time t. Figure 6 shows MSD curves for the two systems at all the investigated temperatures. At very short times (ballistic regime) MSD increases according to $\langle r^2(t)\rangle \cong (3k_BT/m)t^2$. At later times a quasi-plateau region becomes apparent when the temperature is lowered. This signals the increased trapping of a particle in the cage of its neighbors. Once escaped from the cage, due to the presence of the chain connectivity, the monomers undergo a sub diffusive motion $\langle r^2(t)\rangle \propto t^\delta$ with $\delta < 1$ (Rouse regime) [21]. At very long times, monomers displace in a diffusive way ($\delta = 1$). Diffusion is hardly seen in our simulations since, due to the length of the chains, it occurs at the limit of the accessible timescales.

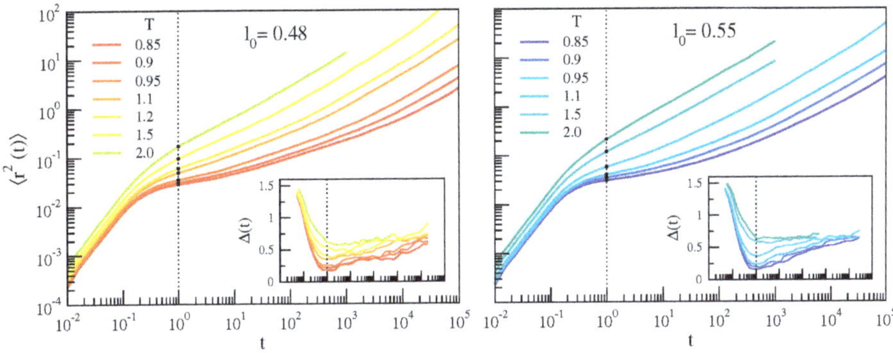

Figure 6. Monomer mean square displacement (MSD) of the two polymer melts with chains having bond lengths $l_0 = 0.48$ (left) and $l_0 = 0.55$ (right). Inset: corresponding MSD slope $\Delta(t)$, Equation (9). The vertical dashed lines mark the time $t^* \approx 1$ where $\Delta(t)$ reaches the minimum, locating the time where caging is more effective. t^* is found to independent of both the system and its physical state. The black circles indicate the values of $\langle u^2 \rangle$, Equation (10).

3.3. Cage Dynamics and Correlation with Primary and Secondary Relaxations

To identify a characteristic time-scale of the cage dynamics we consider the slope of MSD in the log-log representation

$$\Delta(t) \equiv \frac{\partial \log \langle r^2(t) \rangle}{\partial \log t} \tag{9}$$

The slope $\Delta(t)$ is plotted in the insets of Figure 6. $\Delta(t)$ tends to 2 at very short times, due to the ballistic motion. Then, it drops to a minimum value signaling effective trapping in the cage. Later, $\Delta(t)$ increases—staying less than 1 due to the Rouse sub-diffusion—with a progressive approach to the unit value characterizing the diffusive regime. The minimum occurs at $t^* \approx 1$ independently of both the system and its physical state. It offers a convenient definition of the monomer Debye-Waller (DW) factor $\langle u^2 \rangle$ as the MSD at t^* [11,12]:

$$\langle u^2 \rangle \equiv \langle r^2(t = t^*) \rangle \tag{10}$$

The position of $\langle u^2 \rangle$ is marked in Figure 6 with black circles. We remind that the DW factor provides a measure of the rattling motion of the monomer in the cage of the first neighbours. The temperature dependence of DW is reported in Figure 7. The dependence is *not* linear evidencing that, even at very short time scales, that is, $t^* \approx 1$ (corresponding to a few picoseconds [11], see also Reference [37]), the rattling of the monomer in the cage is *not* harmonic. Notably, the two systems exhibit different curves for $\langle u^2 \rangle$. In particular, at fixed temperature the system with chains having longer bond length exhibits larger DW factor.

We are now in a position to test the correlation between the vibrational dynamics and the relaxation. Recently, this correlation has been examined in great detail as far as the *primary* relaxation is concerned [11]. It was found that numerical models *not providing appreciable secondary relaxations* exhibit striking agreement with experimental data of polymers where the presence of JG relaxation is known, like 1,4 Polybutadiene, Polyvinylchloride and Polymethylmethacrylate [4,6,7,38]. This finding needs clarification since the secondary relaxation acts at an intermediate time scale between the fast vibrational dynamics and the primary relaxation and it could, in principle, interfere with the correlation.

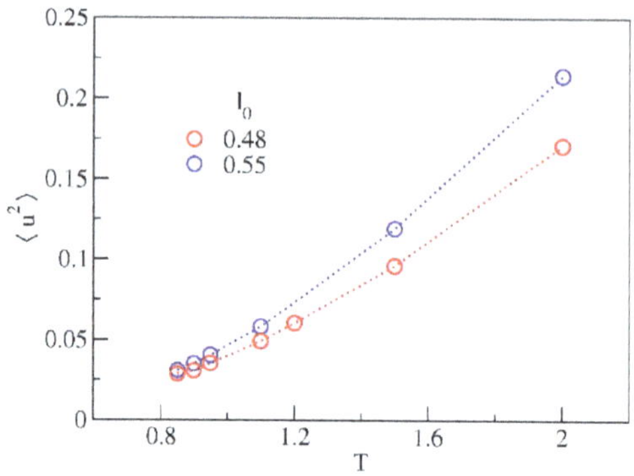

Figure 7. Temperature dependence of the Debye-Waller factor $\langle u^2 \rangle$. Color code as in Figure 1.

To address this aspect, Figure 8 plots the correlation between the average relaxation times and the DW factor. As major result of the present paper, it is clearly seen that the correlation between the fast vibrational dynamics and the primary relaxation is *not* affected by the presence of the secondary relaxation, in the sense that the data concerning the systems exhibiting either strong ($l_0 = 0.48$) or weak ($l_0 = 0.55$) secondary relaxation exhibit the same master correlation curve $\overline{\tau}_p$ vs. $\langle u^2 \rangle$. Notably, this coincidence takes place even if the correlation curves $\overline{\tau}_s$ vs. $\langle u^2 \rangle$ do depend on the bond length.

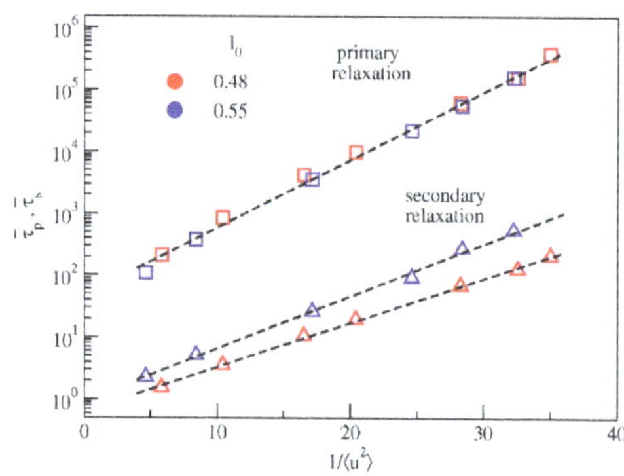

Figure 8. Average relaxation times $\overline{\tau}_p$ and $\overline{\tau}_s$ versus the inverse Debye-Waller factor. The correlation between the vibrational dynamics and the primary relaxation time $\overline{\tau}_p$ is unaffected by the changes of both the strength and the relaxation time of the secondary relaxation, see Figure 4. Color code as in Figure 1.

It seems proper to comment on a feature of Figure 8. The correlation between the primary relaxation and the DW factor of *distinct* systems is usually seen in terms of the reduced DW factor $\langle u^2 \rangle / \langle u_g^2 \rangle$ to get rid of specific system-dependent aspects, $\langle u_g^2 \rangle$ being the DW factor at the glass transition temperature [11]. Figure 8 shows that the correlation of the two systems under study is rather high by considering the unscaled DW factor. This suggests that, even if the glass transition of the two systems is *different* due to the different temperature dependence of the primary relaxation times, see Figure 5 (left), the DW factor at the glass transition of the two systems is *equal* within the errors.

3.4. Alternative Probe Functions of Secondary Relaxation

The present study reveals the JG relaxation via the bond correlation function $C(t)$, Equation (4). We pose the question of the JG sensitivity of alternative relaxation functions. We focus here on the familiar intermediate scattering function (ISF) [11,39] and the torsional autocorrelation function (TACF) [14]. ISF is defined as [11,39]:

$$F_s(q,t) = \frac{1}{N}\langle\sum_{i=1}^{N} e^{i\mathbf{q}\cdot(\mathbf{x}_i(t)-\mathbf{x}_i(0))}\rangle \quad (11)$$

In an isotropic liquid, ISF depends only on the modulus of the wavevector $q = \|\mathbf{q}\|$ and features the rearrangements of the spatial structure of the fluid over the length scale $\sim 2\pi/q$. As alternative relaxation function, the segmental relaxation has been characterized by TACF [14]:

$$TACF(t) = \frac{1}{N_c}\frac{1}{M-3}\sum_{n=1}^{N_c}\sum_{m=1}^{M-3}\frac{\langle|\theta_{m,n}(t)||\theta_{m,n}(0)|\rangle - \langle|\theta_{m,n}(0)|\rangle^2}{\langle|\theta_{m,n}(0)|^2\rangle - \langle|\theta_{m,n}(0)|\rangle^2} \quad (12)$$

where $|\theta_{m,n}(t)|$ is the modulus of the m-th dihedral angle of the n-th chain at a given time t [40].

Figure 9 compares the bond correlation function $C(t)$, Equation (4), with ISF and TACF at the lowest investigated temperature ($T = 0.85$, where the primary and the JG relaxation are expected to have the largest separation). A distinctive feature of $C(t)$ at that temperature is the presence of the step at $t \sim 4{\cdot}10^3$, signalling the secondary relaxation, see Figure 2, left. The step is clearly visible when $C(t)$ drops ~40% only. Figure 9, left, shows that the same step is also observed in TACF but when more than ~90% of the decay has been completed. Namely, it is harder to be observed. Figure 9, right, plots ISF for different wavevectors in a range including $q = q_{\max} \sim 2\pi/\sigma$, corresponding to the maximum of the static structure factor. It is seen that there is no evidence of the two-step decay observed in the bond correlation function $C(t)$ at the same temperature, see Figure 2, left. The results presented in Figure 9 are not unexpected. Previous MD studies [13] performed by using the same model of the present work (dubbed FRC model in Reference [13]) reported that ISF needs lower temperatures to reveal a two-step process in the relaxation. This suggests that ISF has lower JG resolution.

The ability to resolve in a limited time and temperature range both primary and secondary relaxation is related to their relative separation in time scale as well as relative strength. Concerning ISF, timescale separation is possible exploring high wavevector q, in order to be more sensitive to local and restricted dynamics, but the relative strength is also affected by this choice [41]. As recently shown [42], the microscopic density correlation function is dominated at short times by spatial fluctuations of some molecules within the cage formed by the nearest neighbours, that is, by rattling motions within the cage, while at longer times part of the relaxation strength is related to large spatial fluctuations extending at least up to the intermolecular distance, that is, outside the cage. The faster type of motion has been identified as the secondary, while the slower and more intense is the primary one. Timescale separation is usually larger for orientational dynamics, that entails for primary motions a much larger timescale than the intermediate scattering functions (as shown also in Figure 9). In contrast, for the secondary relaxation only local and fast rearrangements are probed by both observables (if the correct wavevector q is chosen). Again, $C(t)$ is much more sensitive than ISF to the presence of secondary relaxation because the relative strength is dominant. Indeed, secondary processes are characterized by rare (but not negligible) and fast rearrangements, resulting in large-amplitude angle motions, which bring a considerable decorrelation of the orientation function [15]. For these reasons orientational correlation function is usually much more sensitive to detect and resolve secondary motions with respect to what happen for intermediate structure function or density autocorrelation function. This has been demonstrated by experiments and numerical simulation studies [43–45].

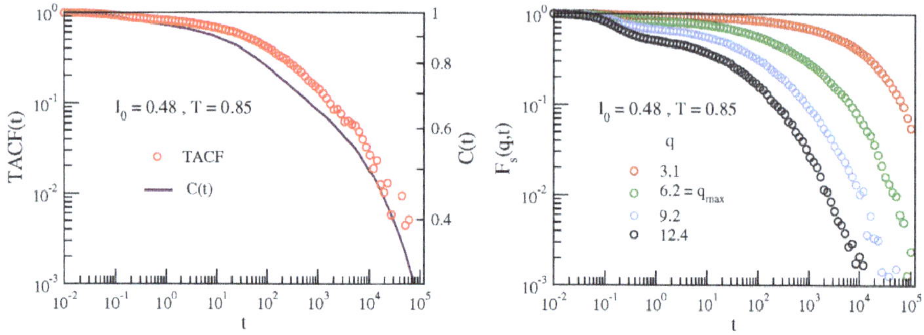

Figure 9. Left: comparison between the torsional autocorrelation function (TACF) and the bond-orientation correlation function, C(*t*) for the system with stronger secondary relaxation (l_0 = 0.48) at the lowest investigated temperature (T = 0.85). Notice the step observed in both functions at $t \sim 4{\cdot}10^3$, signalling the secondary relaxation, see Figure 2, left. The step is much more apparent in C(t), occurring when the latter is dropped of ~40% only. The same feature is observed in TACF when more than 90% of the decay has been completed. Right: corresponding intermediate scattering function (ISF) for different wavevectors in a range including $q = q_{\max}$, corresponding to the maximum of the static structure factor. No apparent two-step decay is seen.

4. Conclusions

We have studied by MD simulations two melts of polymer chains with different bond length, resulting in rather different strength of the secondary JG relaxation. Our interest is the correlation between the fast vibrational dynamics, as sensed by the Debye-Waller factor, and the slow primary relaxation in the presence of JG relaxation. Multiple relaxation processes were searched by using the bond-orientation correlation function, *C*(*t*) which proved higher JG sensitivity with respect to alternatives provided by TACF and ISF functions. We find that changing the bond length alters both the strength and the relaxation time of the secondary relaxation, as well as the correlation with the DW factor. On the other hand, it leaves unaffected the correlation between the vibrational dynamics and the primary relaxation. This finding is in harmony with previous studies reporting that numerical models *not showing secondary relaxations* exhibit striking agreement with experimental data of polymers where the presence of JG relaxation is known [11].

The present result fits into the context of the recent debate about the universality of the correlation between fast and slow degrees of freedom. Actually, a relation between viscous flow and vibrational properties in glass-forming materials has been found in recent studies examining the fragility and the nonergodicity factor, as obtained from scattering techniques in the glassy state [46] and extrapolated to the glass transition region [47]. These two quantities were found strongly correlated only once the effect of secondary relaxation processes, if present, was correctly accounted for. With regards to this issue, it is noteworthy to mention that DW factor as defined in Equation (10) is the best definition of cage vibration amplitude, devoid of any further decorrelation due to JG relaxation.

Author Contributions: Conceptualization, A.T., F.P., S.C. and D.L.; methodology, validation, formal analysis, investigation, A.T., F.P. and D.L.; software, A.T. and F.P.; writing—review and editing, A.T., F.P., M.M., S.C. and D.L.; supervision, F.P. and D.L.; funding acquisition, S.C. and M.M. All authors have read and agreed to the published version of the manuscript.

Funding: This research was funded by University of Pisa grant number PRA-2018-34 ("ANISE").

Acknowledgments: A generous grant of computing time from IT Center, University of Pisa and Dell EMC® Italia is gratefully acknowledged.

Conflicts of Interest: The authors declare no conflict of interest.

Abbreviations

ISF	Intermediate scattering function
JG	Johari-Goldstein
MD	molecular-dynamics
MSD	Mean square displacement
DW	Debye-Waller

References

1. Debenedetti, P.G. *Metastable Liquids*; Princeton University Press: Princeton, NJ, USA, 1997.
2. McCrum, N.G.; Read, B.E.; Williams, G. *Anelastic and Dielectric Effects in Polymeric Solids*; Dover Publications: New York, NY, USA, 1991.
3. Angell, C.A.; Ngai, K.L.; McKenna, G.B.; McMillan, P.; Martin, S.W. Relaxation in glassforming liquids and amorphous solids. *J. Appl. Phys.* **2000**, *88*, 3113–3157. [CrossRef]
4. Ngai, K.L. *Relaxation and Diffusion in Complex Systems*; Springer: Berlin, Germany, 2011.
5. Johari, G.P.; Goldstein, M. Viscous Liquids and the Glass Transition. II. Secondary Relaxations in Glasses of Rigid Molecules. *J. Chem. Phys.* **1970**, *53*, 2372. [CrossRef]
6. Ngai, K. Relation between some secondary relaxations and the relaxations in glass-forming materials according to the coupling model. *J. Chem. Phys.* **1998**, *109*, 6982–6994. [CrossRef]
7. Ngai, K.L.; Paluch, M. Classification of secondary relaxation in glass-formers based on dynamic properties. *J. Chem. Phys.* **2004**, *120*, 857–873. [CrossRef] [PubMed]
8. Capaccioli, S.; Paluch, M.; Prevosto, D.; Wang, L.M.; Ngai, K.L. Many-Body Nature of Relaxation Processes in Glass-Forming Systems. *J. Phys. Chem. Lett.* **2012**, *3*, 735–743. [CrossRef]
9. Cicerone, Q.Z.; Tyagi, M. Picosecond Dynamic Heterogeneity, Hopping, and Johari-Goldstein Relaxation in Glass-Forming Liquids. *Phys. Rev. Lett.* **2014**, *113*, 117801. [CrossRef]
10. Yu, R.R.; Samwer, K. Structural rearrangements governing Johari-Goldstein relaxations in metallic glasses. *Sci. Adv.* **2017**, *3*, 1701577. [CrossRef]
11. Larini, L.; Ottochian, A.; De Michele, C.; Leporini, D. Universal scaling between structural relaxation and vibrational dynamics in glass-forming liquids and polymers. *Nat. Phys.* **2008**, *4*, 42–45. [CrossRef]
12. Puosi, F.; Tripodo, A.; Leporini, D. Fast Vibrational Modes and Slow Heterogeneous Dynamics in Polymers and Viscous Liquids. *Int. J. Mol. Sci.* **2019**, *20*, 5708. [CrossRef]
13. Bedrov, D.; Smith, G.D. Molecular dynamics simulation study of the α and β-relaxation processes in a realistic model polymer. *Phys. Rev. E* **2005**, *71*, 050801. [CrossRef]
14. Bedrov, D.; Smith, G.D. Secondary Johari–Goldstein relaxation in linear polymer melts represented by a simple bead-necklace model. *J. Non-Crystal. Solids* **2011**, *357*, 258–263. [CrossRef]
15. Fragiadakis, D.; Roland, C.M. Participation in the Johari–Goldstein Process: Molecular Liquids versus Polymers. *Macromolecules* **2017**, *50*, 4039–4042. [CrossRef]
16. Fragiadakis, D.; Roland, C.M. Molecular dynamics simulation of the Johari-Goldstein relaxation in a molecular liquid. *Phys. Rev. E* **2012**, *86*, 020501. [CrossRef] [PubMed]
17. Larini, L.; Barbieri, A.; Prevosto, D.; Rolla, P.A.; Leporini, D. Equilibrated polyethylene single-molecule crystals: Molecular-dynamics simulations and analytic model of the global minimum of the free-energy landscape. *J. Phys. Condens. Matter* **2005**, *17*, L199–L208. [CrossRef]
18. Plimpton, S. Fast Parallel Algorithms for Short-Range Molecular Dynamics. *J. Comput. Phys.* **1995**, *117*, 1–19. [CrossRef]
19. Lammps. Available online: http://lammps.sandia.gov (accessed on 26 February 2020).
20. Prevosto, D.; Capaccioli, S.; Lucchesi, M.; Leporini, D.; Rolla, P. Pressure and temperature dependence of structural relaxation dynamics in polymers: A thermodynamic interpretation. *J. Phys. Condens. Matter* **2004**, *16*, 6597–6608. [CrossRef]
21. Doi, M.; Edwards, S.F. *The Theory of Polymer Dynamics*; Clarendon Press: Oxford, UK, 1988.

22. Giordano, M.; Leporini, D.; Martinelli, M.; Pardi, L.; Santucci, S.; Umeton, C. Nonlinear electron spin resonance techniques for the study of inhomogeneously broadened spectra. *J. Chem. Phys.* **1988**, *88*, 607–616. [CrossRef]
23. Andreozzi, L.; Giordano, M.; Leporini, D. Efficient characterization of the orientational ordering of ESR-active probes in supermolecular fluids. *Appl. Magn. Reson.* **1993**, *4*, 279–295. [CrossRef]
24. Leporini, D. Relationship between a nonlinear response and relaxation induced by colored noise. *Phys. Rev. A* **1994**, *49*, 992–1014. [CrossRef]
25. Andreozzi, L.; Giordano, M.; Leporini, D. Linear and non-linear electron spin resonance study of the rotational diffusion of a molecular tracer in supercooled o-terphenyl. *J. Non-Cryst. Solids* **1998**, *235*, 219–224. [CrossRef]
26. Andreozzi, L.; Faetti, M.; Giordano, M.; Leporini, D. Scaling between the rotational diffusion of tracers and the relaxation of polymers and glass formers. *J.Phys. Condens. Matter* **1999**, *11*, A131–A137. [CrossRef]
27. Andreozzi, L.; Faetti, M.; Giordano, M.; Leporini, D. Scaling of the Rotational Relaxation of Tracers in o-Terphenyl: A Linear and Nonlinear ESR Study. *J. Phys. Chem. B* **1999**, *103*, 4097–4103. [CrossRef]
28. Barbieri, A.; Gorini, G.; Leporini, D. Role of the density in the crossover region of *o*-terphenyl and poly(vinyl acetate). *Phys. Rev. E* **2004**, *69*, 061509. [CrossRef] [PubMed]
29. Barbieri, A.; Campani, E.; Capaccioli, S.; Leporini, D. Molecular dynamics study of the thermal and the density effects on the local and the large-scale motion of polymer melts: Scaling properties and dielectric relaxation. *J. Chem. Phys.* **2004**, *120*, 437–453. [CrossRef]
30. Smith, G.D.; Bedrov, D. A molecular dynamics simulation study of the α- and β-relaxation processes in 1,4-polybutadiene. *J. Non-Crystal. Solids* **2006**, *352*, 4690–4695. [CrossRef]
31. Fragiadakis, D.; Roland, C.M. Chain Flexibility and the Segmental Dynamics of Polymers. *J. Phys. Chem. B* **2019**, *123*, 5930–5934. [CrossRef]
32. Zorn, R.; Arbe, A.; Colmenero, J.; Frick, B.; Richter, D.; Buchenau, U. Neutron scattering study of the picosecond dynamics of polybutadiene and polyisoprene. *Phys. Rev. E* **1995**, *52*, 781–795. [CrossRef]
33. Gupta, S.; Mamontov, E.; Jalarvo, N.; Stingaciu, L.; Ohl, M. Characteristic length scales of the secondary relaxations in glass-forming glycerol. *Eur. Phys. J. E* **2016**, *39*, 40. [CrossRef]
34. Hilfer, R. Analytical representations for relaxation functions of glasses. *J. Non-Crystal. Solids* **2002**, *305*, 122–126. [CrossRef]
35. Pardo, L.C.; Lunkenheimer, P.; Loidl, A. Dielectric spectroscopy in benzophenone: The *beta* relaxation and its relation to the mode-coupling Cole-Cole peak. *Phys. Rev. E* **2007**, *76*, 030502. [CrossRef]
36. Angell, C. Relaxation in liquids, polymers and plastic crystals-strong/fragile patterns and problems. *J. Non-Crystal. Sol.* **1991**, *131*, 13–31. [CrossRef]
37. Ottochian, A.; Leporini, D. Scaling between structural relaxation and caged dynamics in $Ca_{0.4}K_{0.6}(NO_3)_{1.4}$ and glycerol: Free volume, time scales and implications for the pressure-energy correlations. *Philosophical Magazine* **2011**, *91*, 1786–1795. [CrossRef]
38. Arbe, A.; Buchenau, U.; Willner, L.; Richter, D.; Farago, B.; Colmenero, J. Study of the Dynamic Structure Factor in the β Relaxation Regime of Polybutadiene. *Phys. Rev. Lett.* **1996**, *76*, 1872–1875. [CrossRef] [PubMed]
39. Hansen, J.P.; McDonald, I.R. *Theory of Simple Liquids*, 3rd ed.; Academic Press: Cambridge, MA, USA, 2006.
40. Kremer, K.; Grest, G.S.; Carmesin, I. Crossover from Rouse to Reptation Dynamics: A Molecular-Dynamics Simulation. *Phys. Rev. Lett.* **1988**, *61*, 566–569. [CrossRef]
41. Arbe, A.; Richter, D.; Colmenero, J.; Farago, B. Merging of the α and β relaxations in polybutadiene: A neutron spin echo and dielectric study. *Phys. Rev. E* **1996**, *54*, 3853–3869. [CrossRef] [PubMed]
42. Caporaletti, F.; Capaccioli, S.; Valenti, S.; Mikolasek, M.; Chumakov, A.I.; Monaco, G. A microscopic look at the Johari-Goldstein relaxation in a hydrogen-bonded glass-former. *Sci. Rep.* **2019**, *9*, 14319. [CrossRef]
43. Richter, D.; Monkenbusch, M.; Arbe, A.; Colmenero, J.; Farago, B. Dynamic structure factors due to relaxation processes in glass-forming polymers. *Phys. B: Condensed Matter* **1997**, *241*, 1005–1012. [CrossRef]
44. Tölle, A. Neutron scattering studies of the model glass former *ortho*-terphenyl. *Rep. Prog. Phys.* **2001**, *64*, 1473–1532. [CrossRef]
45. Fragiadakis, D.; Roland, C.M. Dynamic correlations and heterogeneity in the primary and secondary relaxations of a model molecular liquid. *Phys. Rev. E* **2014**, *89*, 052304. [CrossRef]

46. Scopigno, T.; Ruocco, G.; Sette, F.; Monaco, G. Is the Fragility of a Liquid Embedded in the Properties of Its Glass? *Science* **2003**, *302*, 849–852. [CrossRef]
47. Scopigno, D.C.; Ruocco, G. Universal relation between viscous flow and fast dynamics in glass-forming materials. *Phys. Rev. B* **2010**, *81*, 100202. [CrossRef]

polymers

MDPI

Article

Preparation and Characterization of UV-LED Curable Acrylic Films Containing Biochar and/or Multiwalled Carbon Nanotubes: Effect of the Filler Loading on the Rheological, Thermal and Optical Properties

Valentina Strongone [1], Mattia Bartoli [2], Pravin Jagdale [3], Rossella Arrigo [1], Alberto Tagliaferro [2] and Giulio Malucelli [1,*]

1 Department of Applied Science and Technology, and local INSTM Unit., Viale Teresa Michel 5, 15121 Alessandria, Italy; valentina.strongone@polito.it (V.S.); rossella.arrigo@polito.it (R.A.)
2 Department of Applied Science and Technology, C.so Duca degli Abruzzi 24, 10129 Torino, Italy; mattia.bartoli@polito.it (M.B.); alberto.tagliaferro@polito.it (A.T.)
3 Italian Institute of Technology, Via Livorno 60, 10144 Torino, Italy; pravin.jagdale@iit.it
* Correspondence: giulio.malucelli@polito.it; Tel.: +39-0131-229369

Received: 11 March 2020; Accepted: 29 March 2020; Published: 2 April 2020

Abstract: UV-LED curable coatings represent an up-to-date attractive field due to the high curing efficiency even in the presence of high filler loadings, as well as to the absence of infrared wavelengths that may negatively impact on heat-sensitive substrates. The addition of carbonaceous materials, such as biochar (BC) and/or multiwalled carbon nanotubes (MWCNTs) could positively improve both the rheological and thermal properties. In this study we report on the synthesis and characterization of carbon-reinforced films containing nanometric (MWCNTs) and micrometric (BC) carbon-based materials. We analyze the rheological properties of the UV-LED curable dispersions, as well as the thermal and optical properties of the resulting films, establishing some correlations between filler dispersion/loading with the main observed properties.

Keywords: UV-LED curing; epoxy-acrylate resin; biochar; multiwalled carbon nanotubes; composites

1. Introduction

Photoinduced polymerization processes are very well-established curing methods that exploit high energy radiations for transforming a liquid system into a solid 3D network. This occurs in a very short time (i.e., a few seconds are enough for completing the curing process), with a low energy consumption (necessary just for triggering the curing reaction; unlike thermally-induced processes, there is no need to heat the bulk of the curable system and the process is carried out at room temperature) and without using solvents (as the recipes of photocurable systems usually already comprise reactive diluents that are able to adjust the viscosity while taking part to the curing reactions) [1–7]. As a result, photocuring processes have found several uses for different industrial applications, comprising varnishes, printing inks and protective coatings on a variety of substrates, including paper, wood, metals, plastics and fabrics. Besides, several high-tech and electronic applications have been thoroughly investigated and also developed at an industrial scale: they include the use of coatings on optical fibers, wave guides and optical recording media, and the fabrication of printed circuit boards [8,9].

UV radiation sources were the first to be designed and fabricated; they are still being employed for several advanced applications, mainly in the field of functional coatings and inks. However, they show some limitations concerning the low efficiency in photocurable pigmented systems or UV-curable composites containing high loadings of filler or reinforcing agent. In fact, all these additives, if the

UV-curable recipe is not well tailored and suitable mixtures of photoinitiators are employed, can absorb most of the incident radiation on the sample surface, hence lowering the quantum efficiency and leading to incomplete curing.

This drawback can be overcome by using UV-LED curing systems: unlike UV (i.e., mercury arc) lamps, UV-LED units deliver a monochromatic UV radiation (its wavelength is usually approximately about 400 nm or slightly shorter), displaying almost Gaussian distribution with a relatively narrow bandwidth. Besides, they are mercury-free and they do not generate ozone and therefore they determine a very low environmental impact. Furthermore, the "cold" radiation provided by UV-LED curing systems, mostly due to the absence of output in the infrared region, can be successfully exploited for applications on heat-sensitive substrates (such as wood and fabrics, among a few to mention) [10]. Last but not least, the electrical-to-optical conversion efficiency of UV-LED units is much higher with respect to standard UV lamps and the former can be switched off instantly, hence saving from around 50–75% of electrical power.

To the best of the authors' knowledge, the number of scientific articles dealing with the use of UV-LED curing processes for the preparation of thermosetting composite systems is at present very limited. Zhang and coworkers studied the effects of the UV-LED exposure dose on the cure homogeneity and interlaminar shear strength of composite laminates derived from prepregs made of E-glass fibers and a mixture of acrylic monomers and oligomers [11].

In another quite recent paper, Faes et al. [12] designed a novel 3D printing method that combines syringe extrusion and UV-LED curing: this way it was possible to produce high density ceramic components, after firing and sintering 3D printed UV-LED curable acrylic systems containing zirconia at different loadings (from 22.5 to 55 vol%).

In a recent work, it was possible to prepare UV-LED curable epoxy-acrylate nanocoatings containing intercalated phyllosilicates with enhanced thermal and thermomechanical features, as well as improved barrier properties towards oxygen diffusion [13].

As the incorporation of micro-to-nano-fillers in photocurable systems has clearly demonstrated its suitability for designing novel materials with improved mechanical, thermal, electrical and barrier properties with respect to the unfilled counterparts, in this work we investigated the effects of the presence of two selected carbon materials, namely multiwalled carbon nanotubes (MWCNTs) and biochar, on the rheological, thermal and optical properties of epoxy-acrylate UV-LED curable systems. Indeed, the scientific literature clearly highlights the exploitability of 1-3D carbonaceous fillers in various fields, as they can approve the properties of a plethora of nanocomposites [14–17].

The choice of these two carbon-based materials was driven by two main reasons: the first is that biochar is currently emerging as a cheap, functional material already useful for several applications (i.e., as storage for volatile nutrients, as an adsorber in functional clothing, as an insulating material in the building industry, as energy storage in batteries, among a few to mention) and represents a way for recovering wastes at the end of life, providing them with a new added-value [18]. The second reason refers to the shuttle effect that biochar, according to its structure and morphology, may exert on MWCNTs, favoring the dispersion of the latter into a polymer system.

In this work, each carbon-based filler was dispersed in bisphenol-A-ethoxylate-diacrylate (i.e., a commercially available resin, here used as the model system) at different loadings (ranging from 0.01 to 1.0 wt%), by means of a tip sonicator; further, the two additives were also premixed keeping 1:1 weight ratio and added to the resin, aiming at assessing the effect of their concurrent presence on the overall performances of the resulting UV-LED curable systems.

In particular, it was possible to establish some correlations between the rheological, thermal, optical properties of the composite systems and the structure and level of dispersions of the fillers used.

2. Materials and Methods

2.1. Materials

Exhausted coffee powder was selected as a real case study. It was collected from Bar Katia (Turin, Italy) supplied by Vergnano Spa (Torino, Italy) and corresponds to an Arabica mixture. The exhausted coffee was collected and dried at 105 °C for 72 h. Coffee samples (100 g) were pyrolyzed using a vertical furnace and a quartz reactor (heating rate: 15 °C/min) and kept at 800 °C for 30 min in argon atmosphere [19–21].

A commercially available epoxy-acrylate resin, Ebecryl 150 (bisphenol-A-ethoxylate-diacrylate hereinafter coded as EB150) was kindly supplied by Allnex (Brussels, Belgium). 2,4,6-Trimethylbenzoyl-diphenylphosphineoxide, herein after coded as TPO, was kindly supplied by IGM Resins (Mortara, Italy) and used as photoinitiator for the UV-LED curing.

MWCNTs (NC7000^{TM}, produced via the Catalytic Chemical Vapor Deposition (CCVD) process) were purchased from Nanocyl SA (Sambreville, Belgium). Carbon purity: 90%; surface area: 250–300 m^2/g, as indicated in the datasheet provided by the supplier.

2.2. Preparation of the UV-LED Curable Mixtures

Biochar was grinded using a mechanical mixer (Savatec BB90E) for 10 min at 300 rpm at room temperature in order to decrease the particle size according to Giorcelli et al. [20].

The carbonaceous materials (MWCNTs, Biochar) were dispersed in EB150 with final concentrations of 0.01, 0.10, 0.50 and 1.0 wt%, using a tip Sonics Vibra-cell ultrasonicator (Sonics and Materials, Inc., Newtown, CT, USA) for 15 min. In order to avoid the instantaneous temperature rise, ultrasounds were pulsed with cycles of 10 s alternating to 10 s pause to favor a better heat diffusion. Another set of samples was prepared mixing MWCNTs and biochar (1:1 weight ratio) using the same filler loadings employed for each single carbon material. Then, 6 wt% of TPO was added to the UV-LED curable mixture.

The stability of the resulting liquid dispersions was monitored by visual inspection up to 4 days after the ultrasonication process: no sedimentation of the fillers was observed during this time. Then, the dispersions were coated on glass plates using a wire-wound applicator: the thickness of the obtained coatings was about 150 µm. The coated glass plates were then subjected to the UV-LED curing process, using a Heraeus Noblelight UV-LED NC1 unit (Cambridge, UK), working in dynamic conditions (belt speed: 1 m/min), at 395 nm, with radiation intensity on the sample surface of about 4.8 W/cm^2.

2.3. Characterization Techniques

The completeness of the UV-LED curing reaction was assessed by means of a Perkin Elmer Spectrum 100 spectrometer (Shelton, Connecticut, USA) equipped with an attenuated total reflection (ATR) diamond probe. FTIR spectra were recorded at wavelengths from 700 to 4000 cm^{-1}, with 4 cm^{-1} resolution; 16 scans were collected.

The morphology of all the samples was investigated using a FESEM—Field Emission Scanning Electrical Microscope Zeiss Supra-40 (Carl Zeiss IMT, Oberkochen, Germany).

Rheological measurements were performed using an ARES (TA Instrument, New Castle, DE, USA) strain-controlled rheometer in parallel plate geometry (plate diameter: 25 mm). Strain sweep tests were carried out at 30 °C and $\omega = 1$ rad/s. The complex viscosity and storage and loss moduli were measured performing frequency scans from 10^{-1} to 10^2 rad/s at 30 °C. The strain amplitude was selected for each sample in order to fall in the linear viscoelastic region. At least two specimens for each system were tested and the results were averaged (standard deviation < 5%).

Differential scanning calorimetry (DSC) analyses were performed using a QA1000 TA Instrument apparatus (TA Instrument Inc., Waters LLC, New Castle, DE, USA). All the experiments were performed under dry N_2 gas (flow: 50 mL/min) using samples of about 10 mg in sealed aluminum pans. All the films underwent the following cycle: (1) heating from 0 to 160 °C at 10 °C/min; (2) cooling down to

0 °C at 10 °C/min and (3) heating from 0 to 160 °C at 10 °C/min. The glass transition temperature (T_g) was measured as the midpoint of heat capacity changes. Calibration was performed using indium as standard (T_m = 156.4 °C; ΔH_m = 28.15 J/g).

Thermogravimetric analyses (TGA) were performed using a Pyris1TGA apparatus (Perkin Elmer, Waltham, MA, USA; experimental error: ±0.5 wt%, ±1 °C). Samples (about 10 mg) were placed in alumina pans and runs were carried out in the range 50–700 °C, with a heating rate of 10 °C/min, under both N_2 and air flow (35 and 25 mL min^{-1}, respectively). $T_{5\%}$, $T_{10\%}$ (i.e., the temperatures, at which 5% or 10% weight loss, respectively, occurs) and T_{max} values were calculated; besides, the final residue at 700 °C was measured.

Thermal conductivity measurements were carried out with TPS 2500S apparatus (Hot Disk AB, Göteborg, Sweden), equipped with a Kapton sensor (radius 3.189 mm), using the transient plane source (TPS) method [22]. The test temperature (23.00 ± 0.01 °C) was controlled by a silicon oil bath (Haake A40, Thermo Scientific Inc., Austin, TX, USA) equipped with a temperature controller (Haake AC200, Thermo Scientific Inc., Austin, TX, USA).

UV-Vis spectroscopy measurements were performed on the UV-LED cured films by using a Shimadzu UV-Vis spectrophotometer UV2600 series (Shimadzu Italia Srl, Milano, Italy); wavelength range was set between 200 and 1200 nm.

3. Results and Discussion

As first point of discussion, a schematic overview on the radical polymerization mechanism of EB150 is shown in Figure 1.

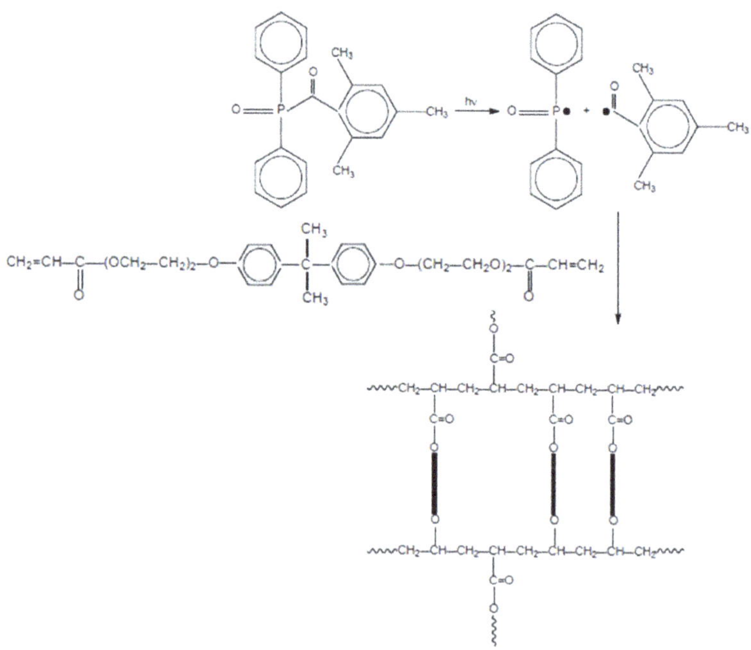

Figure 1. Scheme of UV-LED curing of EB150. The system undergoes a radical polymerization promoted by the generation of primary radicals derived from the photolysis of the photoinitiator as a consequence of the exposure to the UV-LED radiation.

The curing kinetics is very fast, due to the high efficiency of TPO photoinitiator and the high intensity of the UV-LED unit, whose emission spectrum well matches the absorption spectrum of

TPO; besides, the adopted experimental conditions for the photocuring process were suitable for achieving the completeness of the double bonds conversion, as revealed by FTIR-ATR spectra before and after the exposure to the UV-LED radiation. As an example, the typical spectra for the system containing 0.50 wt% of biochar, before and after UV-LED curing are shown in Figure 2: the complete disappearance of the band at 1630 cm^{-1} (attributed to the acrylic double bonds [23]) is evident.

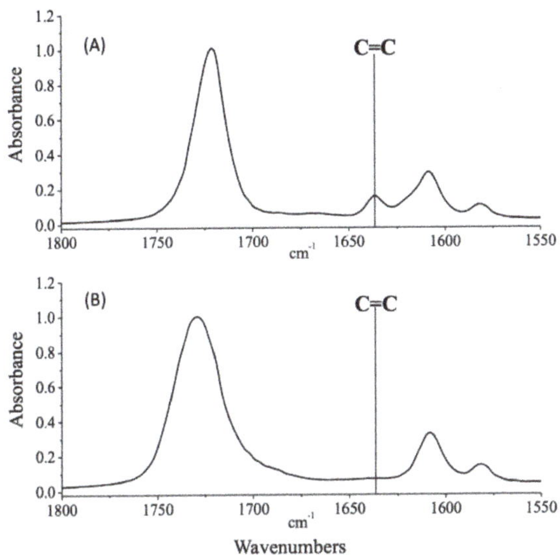

Figure 2. FTIR-ATR spectra of EB150 containing 0.50 wt% of biochar, before (**A**) and after (**B**) exposure to the UV-LED radiation. The disappearance of the band of the acrylic double bonds, located at about 1635 cm^{-1}, is a clear indication of the completeness of the UV-LED curing process.

The thickness of the obtained films estimated though FESEM technique, irrespective of their composition and filler loadings, was about 180 μm as shown in Figure 3.

Figure 3. FESEM capture of unfilled UV-LED cured EB150 film on the transversal plane.

3.1. Morphological Characterization of the Fillers

The carbonaceous fillers used were very different in both size and morphology, as clearly shown in FESEM images presented in Figure 4.

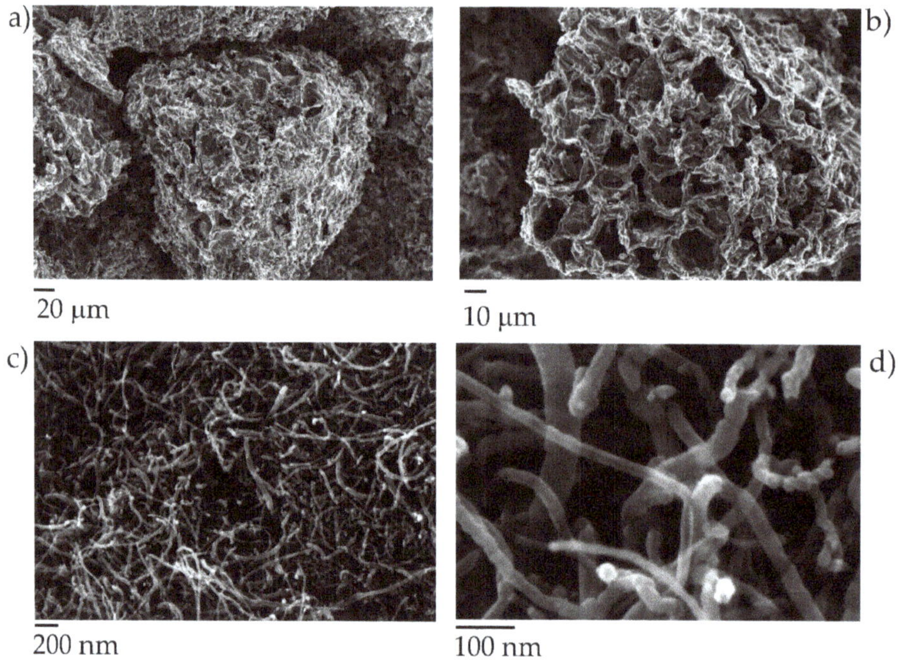

Figure 4. FESEM captures of BC (**a**,**b**) and multiwalled carbon nanotubes (MWCNTs; **c**,**d**) at different magnifications. It is worthy to note the highly porous structure of Biochar

BC was produced by carbonization of exhausted coffee following the procedure previously reported [14]. The resulting material showed a sponge-like porous structure with a grain size around 200–300 μm, average radius about 10–15 μm and a wall thick between 1 and 2 μm. According to Giorcelli et al. [20], the structure of BC is quite disordered and does not show residual functional groups (i.e., hydroxyl, quinoid and carboxyl functionalities). The employed MWCNTs are highly packed with an average diameter highly variable in the range from 10 to 50 nm with a length of several micrometers. This network structure deeply affects their dispersibility in organic matrix. Nonetheless, ultrasonic procedures can be effectively exploited to achieve good dispersion within the epoxy-acrylate resin [24]. Besides, using pulsed ultrasounds generated by the tip sonicator allowed achieving good dispersion without compromising the size of MWCNTs. On the other hand, ultrasonication always leads to a sized reduction of biochar, as proved by Bartoli et al. [25], who tested several biochars with different shape and size. However, an acceptable dispersion in EB150 was achieved for both the fillers and their mixtures as shown in Figure 5.

The unfilled UV-LED cured resin shows very smooth surfaces without the presence of any appreciable defects. The addition of MWCNTs at 1.0 wt% (Figure 5c–d) revealed some persistent aggregation but a good embedding of the nanofiller into the polymer matrix. The biochar-containing samples (Figure 5e–f) show a very good dispersion of the particles that, during sonication, underwent a disruptive process, showing surfaces very similar to the neat resin. Interestingly, when BC was mixed with MWCNTs, an improvement of MWCNTs dispersion was achieved as clearly presented in Figure 5g–h. This could be reasonably due to the shuttle effect of micrometric particles of biochar.

In particular, in the very first stage of the mixing process, MWCNTs strongly interact with BC surface through π–π stacking: a similar behavior was observed by Ismail et al. [26] using a mix of CNTs and carbon black. In a subsequent step, BC particle are reduced in size due to the sonication process and the surface interacting MWCNTs are homogeneously dispersed within the resin. This beneficial effect could explain the increased uniformity of filler dispersion, accompanied by the decrease of MWCNTs length.

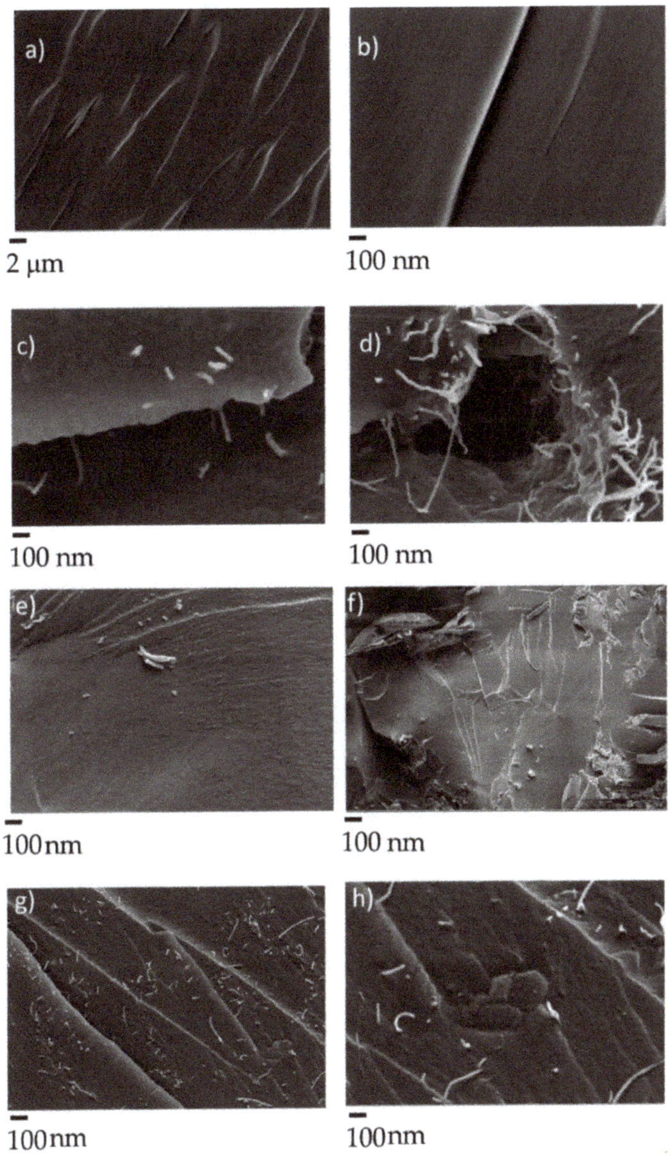

Figure 5. FESEM captures of UV-LED cured films (filler loading: 1 wt%): unfilled EB150 (**a**,**b**); EB150 + MWCNTs (**c**,**d**); EB150 + biochar (**e**,**f**) and EB150 + biochar/MWCNTs (**g**,**h**). All the filled UV-LED cured films show a good dispersion of the two carbonaceous fillers.

3.2. Rheological Behavior

Figure 6a–c shows the trends of complex viscosity as a function of frequency for EB150 and its mixtures with biochar, MWCNTs and biochar/MWCNTs at different filler loadings. First of all, it is worthy to note that the systems loaded with MWCNTs exhibit a high increase of complex viscosity compared with unfilled resin; the only exception is for the liquid dispersion containing 0.01 wt% of MWCNTs. This finding can be ascribed to the highly entangled structure taking place between π orbital system of MWCNTs and bisphenol A rings, hence to increasing interactions occurring between the oligomeric chains and the nanofiller. Conversely, 0.01 wt% of MWCNTs is too low for ensuring the occurrence of these interactions and the related complex viscosity curve is shifted downwards.

Conversely, for the systems containing biochar at different loadings, the complex viscosity curves shift towards lower values with increasing the filler loading. This finding clearly indicates that biochar exerts a lubricating action on the oligomeric chains, which is more pronounced as the filler loading increases.

Further, a balance in between these two opposite rheological behaviors is shown by the mixtures that combine together biochar and MWCNTs (Figure 6c). In this case, at low loadings (i.e., 0.10 wt%) the complex viscosity trend was mainly driven by the lubricating effect of biochar, as the complex viscosity values were lower or comparable with those of unfilled EB150. At variance, for higher fillers loadings, MWCNTs became prevalent on determining the rheological behavior of the dispersions: in fact, the complex viscosity curves approach those obtained for the systems containing MWCNTs only.

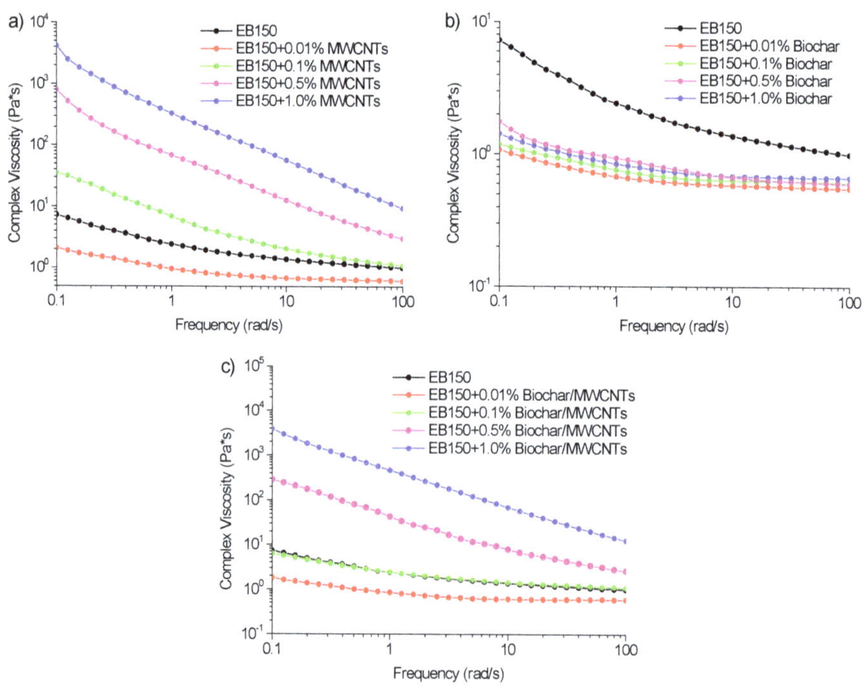

Figure 6. Complex viscosity vs. frequency curves for unfilled EB150 and EB150 + MWCNTs (**a**), EB150 + biochar (**b**) and EB150 + biochar/MWCNTs (**c**) at different loadings ranging from 0.01 to 1.00 wt%.

3.3. Properties of the UV-LED Cured Films

3.3.1. Thermal Properties

First of all, DSC analyses were performed on all the prepared systems, in order to (i) confirm the completeness of the UV-LED curing reaction (1st heating up—absence of exothermic peaks above the glass transition temperature) and (ii) measure the glass transition temperature (2nd heating up). Figures 7–9 plots the typical DSC traces for UV-LED cured EB150 and for all the prepared composite films, together with the T_g values measured as the midpoint of heat capacity steps.

First of all, it is worthy to note that no exothermic effects are detectable in any 1st heating up for UV-LED cured EB150 and its composites: this finding further supports the FTIR-ATR results as far as the completeness of the curing reaction is considered. Therefore, it can be concluded that the experimental conditions adopted for the UV-LED curing (i.e., radiation intensity on the films surface and belt speed) are adequate for promoting the formation of a fully cured network, even in the presence of the fillers. Conversely, the first heating up traces show the enthalpy relaxation (i.e., the endothermic peaks shown in Figures 7–9—1st run), which superimposes to the specific heat step associated with the glass transition of the polymer network. The presence of this phenomenon is likely to be attributed to the speed of the UV-LED curing process that freezes the macromolecules in a non-equilibrium thermodynamic state. Then, the 2nd heating up allows the free macromolecule segments between cross-links to rearrange in a more stable conformational structure and enthalpy relaxation disappears. This phenomenon has been already observed for similar systems containing different ZnO fillers [27].

From an overall point of view, the presence of the different carbon fillers at different loadings did not significantly affect the T_g values of the UV-LED cured polymer network, which were usually within 70 and 77 °C.

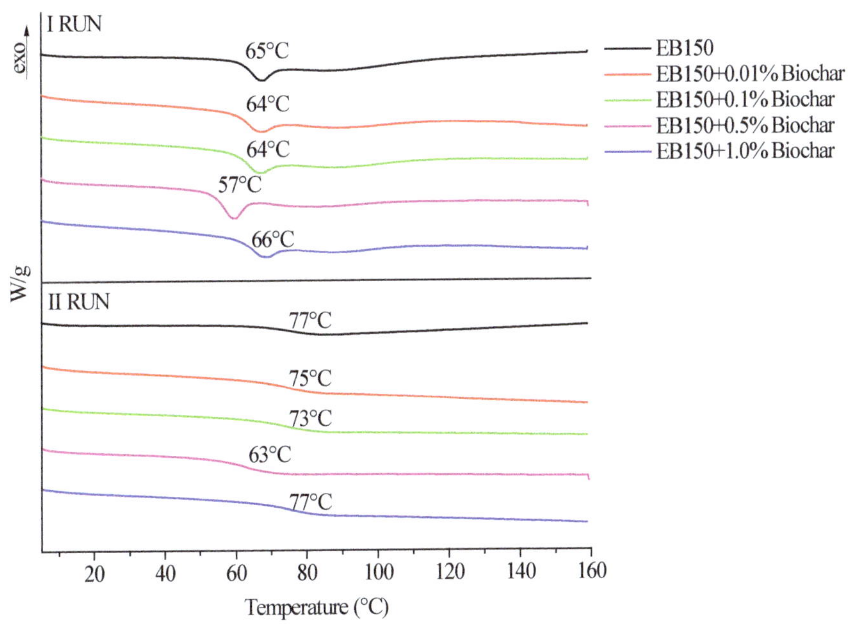

Figure 7. Differential scanning calorimetry (DSC) traces (1st and 2nd heating up) for UV-LED cured EB150 and its composites with different biochar loadings.

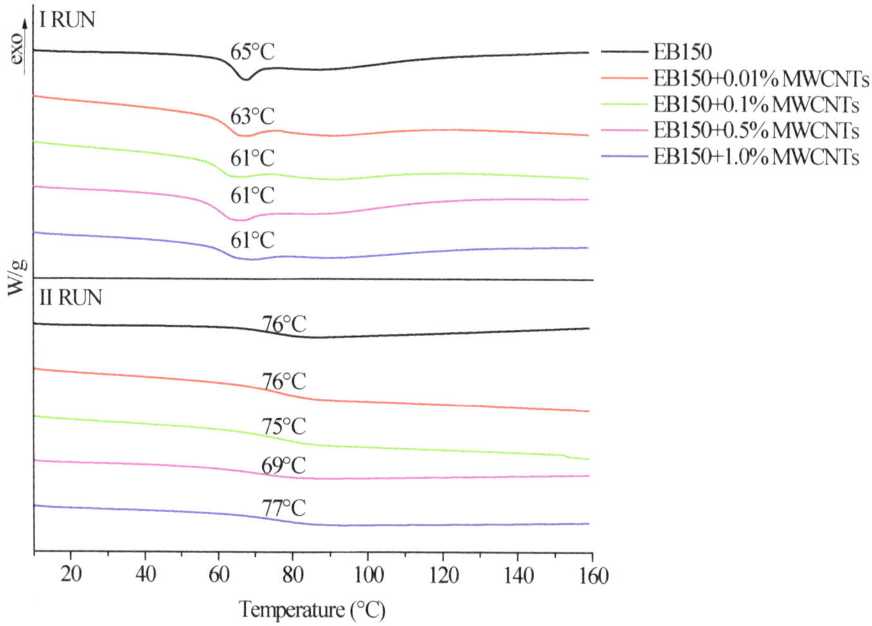

Figure 8. DSC traces (1st and 2nd heating up) for UV-LED cured EB150 and its composites with different MWCNTs loadings.

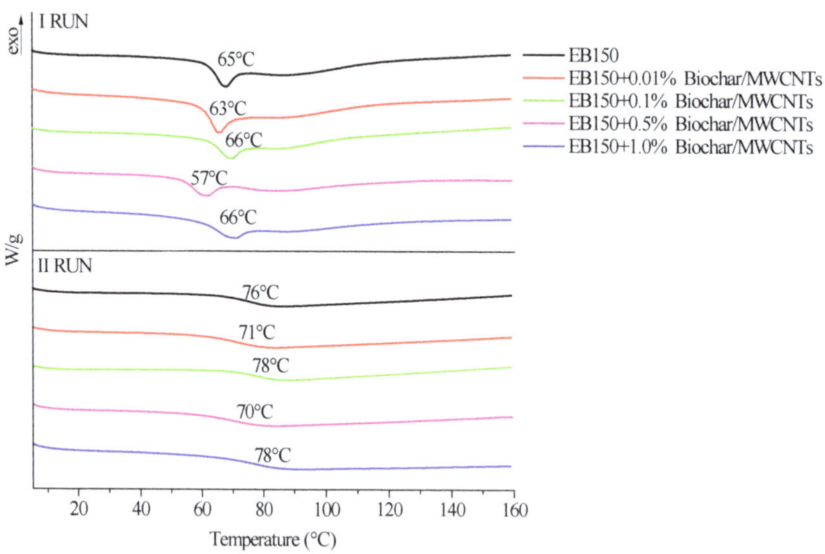

Figure 9. DSC traces (1st and 2nd heating up) for UV-LED cured EB150 and its composites with different MWCNTs loadings.

Then, all the UV-LED cured films were analyzed through thermogravimetric (TG) analysis carried out in nitrogen and air, in order to investigate the effect of the carbon-based fillers on the thermal and thermo-oxidative stability of the UV-LED cured resin, respectively. The collected data are summarized in Table 1.

Table 1. Thermal and thermo-oxidative stability of the UV-LED EB150 film and its composites *.

Filler	Filler Loading (wt%)	Atmosphere: N_2 $T_{10\%}$ (°C)	$T_{50\%}$ (°C)	T_{max} (°C)	Residue @T_{max} (wt%)	Residue 700 °C (wt%)	Atmosphere: Air $T_{10\%}$ (°C)	$T_{50\%}$ (°C)	T_{1max} (°C)	T_{2max} (°C)	T_{3max} (°C)	Residue 700 °C (wt%)
-	0.00	395	439	448	34.6	3.6	349	439	382	438	557	0.3
MWCNTs	0.01	393	439	447	35.5	3.6	341	437	382	436	556	0.2
	0.10	394	440	449	33.7	3.7	342	440	383	439	559	0.3
	0.50	392	434	441	37.5	4.1	331	441	378	440	565	0.8
	1.00	395	438	444	39.0	5.0	335	434	379	438	558	0.8
Biochar	0.01	393	438	445	36.5	3.4	346	440	382	437	559	0.3
	0.10	393	438	447	34.1	3.2	342	438	380	438	557	0.4
	0.50	391	437	445	37.4	5.1	350	437	387	437	557	0.4
	1.00	383	432	441	36.2	4.8	340	439	373	439	557	0.7
Biochar/MWCNTs	0.01	394	439	448	34.4	3.6	346	440	382	437	559	0.3
	0.10	391	437	445	36.7	3.5	342	438	380	438	557	0.4
	0.50	390	438	443	40.3	4.8	339	437	377	438	559	0.6
	1.00	388	434	440	38.9	4.7	340	439	373	439	557	0.7

* experimental error: ±0.5 wt%, ±1 °C.

In nitrogen atmosphere, degradation occurs according to a single step. Comparing the behavior of the unfilled cured resin with that of its composites, the presence of a filler, irrespective of the type and loading, did not affect the thermal stability of the cured polymer network. In fact, the changes of $T_{10\%}$, $T_{50\%}$ and T_{max} (i.e., the temperatures, at which 10% and 50% weight loss occurs and corresponding to the maximum weight loss rate in derivative—dTG—curves, respectively) in the presence of the different fillers are practically negligible, despite an increase of the residues of the filled systems at the end of the test as compared to UV-LED cured EB150.

In air atmosphere, the degradation occurs according to three successive steps. The first one, taking place at about 375 °C, can be related to the decomposition of low molecular weight structures, such as dimer and trimers present in the polymer network. Then, at higher temperatures (i.e., about 440 °C), the main degradation of the polymer network occurred. Finally, during the last stage (at about 560 °C), the degradation products of the previous step were further oxidized, giving rise to the formation of CO and CO_2. Comparably to what was already observed in nitrogen, the presence of the different carbonaceous filler did not affect the thermo-oxidative stability of the UV-LED cured networks, apart from an increase of the residues at the end of the tests.

Figure 10a–c shows the thermal conductivity values as a function of the filler type and loading.

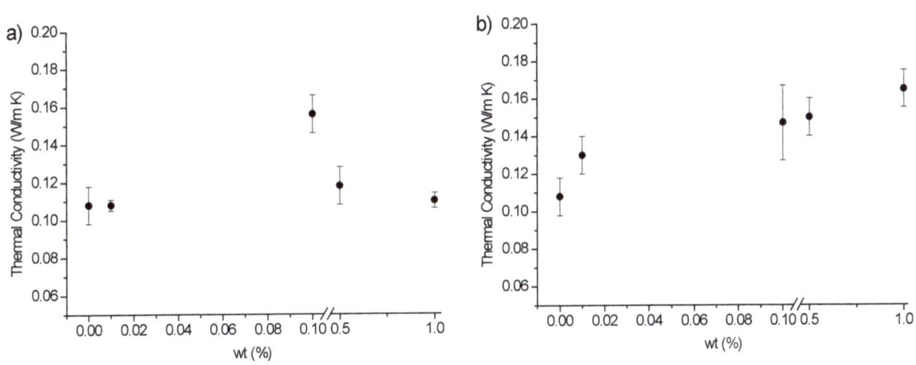

Figure 10. *Cont.*

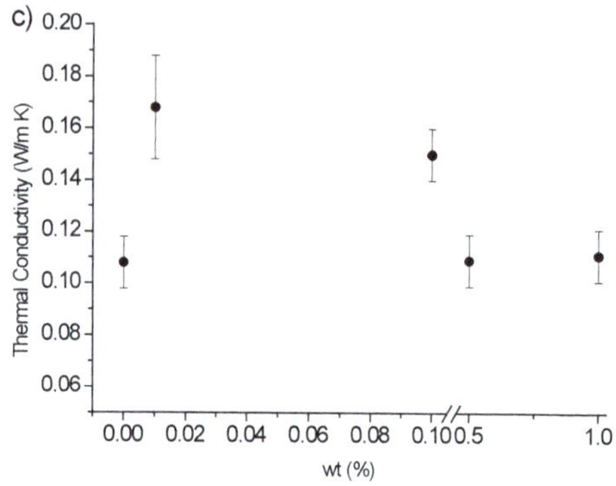

Figure 10. Thermal conductivity of UV-LED cured EB150 and its composite films as a function of filler loading: (**a**) with MWCNTs, (**b**) with biochar and (**c**) with biochar/MWCNTs. Different trends can be observed, according to the filler type and loading.

The thermal conductivity data clearly shows a different effect provided by MWCNTs and by biochar. More specifically, the addition of increasing amounts of MWCNTs increased the thermal conductivity that reached a maximum corresponding to 0.10 wt% loading; then, for higher filler loadings, it monotonically decreased. Conversely, the addition of biochar to the UV-curable system determined a monotonically increase of the thermal conductivity, which approached 0.165 $Wm^{-1}K^{-1}$ for the highest filler loading. Clustering phenomena occurring with increasing the nanofiller concentration might explain the trend of the thermal conductivity of the films containing MWCNTs: these phenomena have already been observed in FESEM analyses. The latter have also pointed out that biochar shows a very homogeneous dispersion within the polymer network, irrespective of its loading: as a consequence, the thermal conductivity shows a monotonic increasing trend. Finally, the concurrent presence of the two carbon fillers was affected by the different morphology of biochar and MWCNTs in UV-LED cured EB150. In particular, at low fillers loadings, the prevailing effect is exerted by biochar, so that thermal conductivity increases; then, the worse dispersion and clustering tendency of MWCNTs becomes predominant and the thermal conductivity significantly decreases approaching that of unfilled EB150 [28].

3.3.2. Optical Properties

The transmittance over a wide wavelength range of self-standing composite films was evaluated, assessing the effect of the different fillers and of their loadings on the optical properties of UV-LED cured blank counterparts. To this aim, UV-Vis spectroscopy measurements were performed; the obtained data are shown in Figure 11a–c.

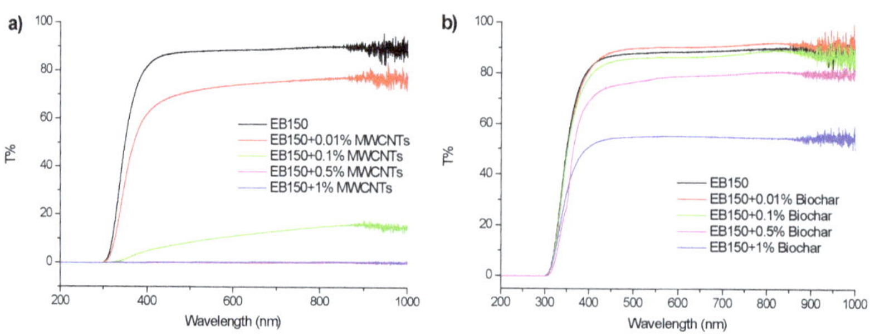

Figure 11. *Cont.*

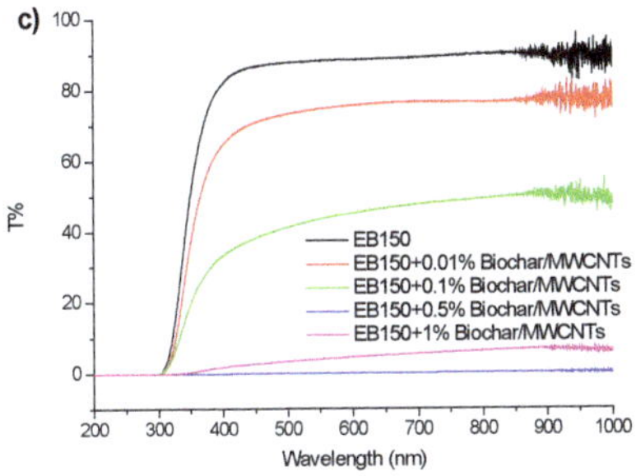

Figure 11. UV-Vis spectra of films containing (**a**) MWCNTs, (**b**) BC and (**c**) BC/MWCNTs in concentrations ranging from 0.01 to 1.00 wt%.

From an overall point of view, it is worthy to note that the ease of dispersibility of the fillers within the polymer network significantly affected the transmittance of the resulting composite films, especially when the filler loading was quite high. In particular, unlike MWCNTs, for which the transparency was maintained at very low filler loadings (i.e., 0.01 wt%) only because of clustering phenomena, biochar-containing films were able to keep 50% transmittance at the highest loading (i.e., 1.0 wt%). Finally, the combination of the two fillers did not provide significant enhancements in transmittance, probably due to the light scattering of MWCNTs that limit the beneficial effects provided by biochar.

4. Conclusions

In this work, UV-LED curing was exploited for designing new epoxy-acrylate films containing different amounts (ranging from 0.01 to 1.0 wt%) of carbon-based micro and nanofillers, namely biochar and MWCNTs. The UV-LED curing performed in dynamic configuration allowed achieving the completeness of the double bond conversion, even in the presence of the highest fillers' loadings. The rheological behavior of the UV-LED curable dispersions, as well as the morphology, the thermal and optical properties of the resulting UV-LED cured films were thoroughly investigated and correlated with the type of filler and its loading. In particular, unlike MWCNTs, which beyond 0.01 wt% were prone to clustering, biochar, irrespective of the used concentration, turned out to easily disperse into the UV-LED curable resin, as assessed by FESEM analyses. Besides, the presence of the fillers, regardless of the type and loading, did not affect the thermal and thermo-oxidative stability of the cured films; furthermore, T_g values were almost unchanged. According to their ease of dispersibility within the polymer network, MWCNTs and biochar showed a different effect on the thermal conductivity of the prepared films: in particular, increasing loadings of biochar providing a monotonic increasing trend of this parameter.

Finally, all the prepared UV-LED films exhibited a high transparency only at low filler loadings; however, the very homogeneous distribution of biochar in the cured resin allowed maintaining high transmittance (about 50%) even at 1.0 wt%, over a wide wavelength range.

Table 2 summarizes all these findings for an easier comparison of the obtained results for the different systems investigated.

Table 2. Main outcomes from the investigated systems.

	BC	MWCNTs	BC + MWCNTs
Completeness of the UV-LED curing process	Achieved	Achieved	Achieved
Dispersibility in EB150	Good	Average; tendency to form aggregates beyond 0.01 wt%	Good; shuttle effect on MWCNTs exerted by BC
Rheological behavior of the liquid dispersions	Lubricating effect at any loading	High increase of complex viscosity beyond 0.01 wt%	Driven by BC at low loadings
Effect of the filler on T_g	Negligible	Negligible	Negligible
Effect on thermal stability	Negligible	Negligible	Negligible
Effect on thermal conductivity	Monotonic increasing trend	Maximum achieved at 0.10 wt% filler loading	Prevailing effect of BC at low filler loadings
Effect on transmittance	50% transmittance kept at 1 wt% loading	High transmittance only below 0.01 wt% loading	Scattering phenomena provided by MWCNTs

Author Contributions: Conceptualization G.M.; methodology, V.S., A.T., M.B.; formal analysis, V.L., P.J.; investigation, V.S., P.J. R.A.; resources, G.M., A.T.; data curation, V.L., P.J., R.A.; writing—original draft preparation, M.B.; writing—review and editing V.S., P.J., G.M., M.B., A.T.; supervision, G.M. All authors have read and agreed to the published version of the manuscript.

Funding: This research received no external funding.

Conflicts of Interest: The authors declare no conflict of interest.

References

1. Dietliker, K.K.; Oldring, P.K.T. *Chemistry and Technology of UV and EB Formulation for Coatings, Inks and Paints*; SITA Technology: London, UK, 1991; Volumes 1–5.
2. Pappas, S.P. *Radiation Curing Science and Technology*; Plenum Press: New York, NY, USA, 1992.
3. Fouassier, J.P.; Rabek, J.F. *Radiation Curing in Polymer Science and Technology*; Chapman and Hall: London, UK, 1993; Volumes 1–5.
4. Roffey, C. *Photogeneration of Reactive Species for UV Curing*; Wiley: New York, NY, USA, 1997.
5. Hoyle, C.E.; Kinstle, J.F. *Radiation Curing of Polymer Materials*; ACS Symposium Series 417; American Chemical Society: Washington, DC, USA, 1990.
6. Fouassier, J.P. *Photoinitiator, Photopolymerization and Photocuring*; Hanser: Munich, Germany, 1995.
7. Scranton, A.C.; Bowman, C.N.; Peiffer, R.W. *Photopolymerization Fundamentals and Applications*; ACS Symposium Series 673; American Chemical Society: Washington, DC, USA, 1997.
8. Davidson, S. *Exploring the Science, Technology and Applications of UV and EB Curing*; SITA Technology: London, UK, 1999.
9. Decker, C. Photoinitiated crosslinking polymerisation. *Prog. Polym. Sci.* **1996**, *21*, 593–650. [CrossRef]
10. Landry, V.; Blanchet, P.; Boivin, G.; Bouffard, J.F.; Vlad, M. UV-LED curing efficiency of wood coatings. *Coatings* **2015**, *5*, 1019–1033. [CrossRef]
11. Zhang, X.; Duan, Y.; Zhao, X.; Li, D. UV stepwise cured fabrication of glass fiber/acrylate composites: Effects of exposure dose on curing uniformity and interlaminar shear strength. *J. Comp. Mater.* **2016**, *50*, 1395–1401. [CrossRef]
12. Faes, M.; Vleugels, J.; Vogeler, F.; Ferraris, E. Extrusion-based additive manufacturing of ZrO_2 using photoinitiated polymerization. *CIRP J. Manufact. Sci. Tech.* **2016**, *14*, 28–34. [CrossRef]
13. Malucelli, G. Synthesis and Characterization of UV-LED Curable Nanocomposite Coatings. *Curr. Org. Chem.* **2017**, *21*, 1–8. [CrossRef]
14. Lee, Y.S.; Lee, S.Y.; Kim, K.S.; Noda, S.; Shim, S.E.; Yang, C.M. Effective Heat Transfer Pathways of Thermally Conductive Networks Formed by One-Dimensional Carbon Materials with Different Sizes. *Polymers* **2019**, *11*, 1661. [CrossRef] [PubMed]

15. Cseri, L.; Baugh, J.; Alabi, A.; AlHajaj, A.; Zou, L.; Dryfe, R.; Budd, P.; Szekely, G. Graphene oxide—Polybenzimidazolium nanocomposite anion exchange membranes for electrodialysis. *J. Mater. Chem. A* **2018**, *6*, 24728–24739. [CrossRef]
16. Liang, X.; Dai, F. Epoxy Nanocomposites with Reduced Graphene Oxide-Constructed Three-Dimensional Networks of Single Wall Carbon Nanotube for Enhanced Thermal Management Capability with Low Filler Loading. *ACS Appl. Mater. Interfaces* **2020**, *12*, 3051–3058. [CrossRef] [PubMed]
17. Roy, N.; Sengupta, R.; Bhowmick, A.K. Modifications of carbon for polymer composites and nanocomposites. *Prog. Polym. Sci.* **2012**, *37*, 781–819. [CrossRef]
18. Schmidt, A.P. 55 Uses of Biochar. *Ithaka J.* **2012**, *1*, 286–289.
19. Arrigo, R.; Jagdale, P.; Bartoli, M.; Tagliaferro, A.; Malucelli, G. Structure–Property Relationships in Polyethylene-Based Composites Filled with Biochar Derived from Waste Coffee Grounds. *Polymers* **2019**, *11*, 1336. [CrossRef] [PubMed]
20. Giorcelli, M.; Bartoli, M. Development of coffee biochar filler for the production of electrical conductive reinforced plastic. *Polymers* **2019**, *11*, 1916. [CrossRef] [PubMed]
21. Jagdale, P.; Ziegler, D.; Rovere, M.; Tulliani, J.M.; Tagliaferro, A. Waste Coffee Ground Biochar: A Material for Humidity Sensors. *Sensors* **2019**, *19*, 801. [CrossRef]
22. Log, T.; Gustafsson, S.E. Transient plane source (TPS) technique for measuring thermal transport properties of building materials. *Fire Mater.* **1995**, *19*, 43–49. [CrossRef]
23. Socrates, G. *Infrared and Raman Characteristic Group Frequencies: Tables and Charts*, 3rd ed.; Wiley: New York, NY, USA, 2004.
24. Cheng, Q.; Debnath, S.; Gregan, E.; Byrne, H.J. Ultrasound-assisted SWNTs dispersion: Effects of sonication parameters and solvent properties. *J. Phys. Chem. C* **2010**, *114*, 8821–8882. [CrossRef]
25. Bartoli, M.; Giorcelli, M.; Rosso, C.; Rovere, M.; Jagdale, P.; Tagliaferro, A. Influence of Commercial Biochar Fillers on Brittleness/Ductility of Epoxy Resin Composites. *Appl. Sci.* **2019**, *9*, 3109. [CrossRef]
26. Ismail, H.; Ramly, A.; Othman, N. The effect of carbon black/multiwall carbon nanotube hybrid fillers on the properties of natural rubber nanocomposites. *Polym. Plast. Technol.* **2011**, *50*, 660–666. [CrossRef]
27. Malucelli, G.; Fioravanti, A.; Francioso, L.; De Pascali, C.; Signore, M.A.; Carotta, M.C.; Bonanno, A.; Duraccio, D. Preparation and characterization of UV-cured composite films containing ZnO nanostructures: Effect of filler geometric features on piezoelectric response. *Prog. Org. Coat.* **2017**, *109*, 45–54. [CrossRef]
28. Antar, Z.; Feller, J.-F.; Noel, H.; Glouannec, P.; Elleuch, K. Thermoelectric behaviour of melt processed carbon nanotube/graphite/poly (lactic acid) conductive biopolymer nanocomposites (CPC). *Mater. Lett.* **2012**, *67*, 210–214. [CrossRef]

polymers

MDPI

Article

Electrically-Conductive Polyketone Nanocomposites Based on Reduced Graphene Oxide

Esteban Alejandro Araya-Hermosilla [1,*], Marco Carlotti [1], Francesco Picchioni [2], Virgilio Mattoli [1,*] and Andrea Pucci [3,4,*]

1 Center for Micro-BioRobotics, Istituto Italiano di Tecnologia Viale Rinaldo Piaggio 34, 56025 Pontedera (PI), Italy; marco.carlotti@iit.it
2 Department of Chemical Product Engineering, ENTEG, University of Groningen, Nijenborgh 4, 9747AG Groningen, The Netherlands; f.picchioni@rug.nl
3 Dipartimento di Chimica e Chimica Industriale, Università di Pisa, Via Moruzzi 13, 56124 Pisa, Italy
4 CISUP, Centro per l'Integrazione della Strumentazione dell'Università di Pisa, Lungarno Pacinotti 43, 56126 Pisa, Italy
* Correspondence: esteban.araya@iit.it (E.A.A.-H.); virgilio.mattoli@iit.it (V.M.); andrea.pucci@unipi.it (A.P.); Tel.: +39-050-883417 (E.A.A.-H.); +39-050-883417 (V.M.); +39-0502219270 (A.P.)

Received: 25 March 2020; Accepted: 13 April 2020; Published: 16 April 2020

Abstract: In this work, we investigated the functionalization of polyketone 30 (PK30) with glycyl-glycine (Gly-Gly) via the Paal–Knorr reaction with the aim of homogenously dispersing two types of reduced graphene oxide (rGO, i.e., lrGO and hrGO, the former characterized by a lower degree of reduction in comparison to the latter) by non-covalent interactions. The functional PK30-Gly-Gly polymer was effective in preparing composites with homogeneously distributed rGO characterized by an effective percolation threshold at 5 wt. %. All the composites showed a typical semiconductive behavior and stable electrical response after several heating/cooling cycles from 30 to 115 °C. Composites made by hrGO displayed the same resistive behaviour even if flanked by a considerable improvement on conductivity, in agreement with the more reduced rGO content. Interestingly, no permanent percolative network was shown by the composite with 4 wt. % of lrGO at temperatures higher than 45 °C. This material can be used as an ON–OFF temperature sensor and could find interesting applications as sensing material in soft robotics applications.

Keywords: reduced graphene oxide; polyketone functionalization; electrically-conductive nanocomposites

1. Introduction

Graphene has called high attention thanks to its excellent mechanical properties, thermal conductivity, and electronic transport properties [1–4]. However, the lack of high quality samples, elevated cost of production, poor dispersibility in solvents, and irreversible aggregation in several media have limited its use [5]. Instead, graphene oxide (GO) has replaced graphene in many applications due to low cost of production and dispersibility in water and polar organic solvents [6,7]. As graphene, it possesses a 2D structure but some of the carbons atoms lost the sp^2 character being involved in the covalent linkage with hydroxyl, epoxide, and carbonyl groups generated during the oxidation of the graphene layer [8–10]. These differences are enough to affect the conjugation properties of the former and thus GO behaves as an insulator [11,12]. The electric properties of GO can be recovered by reduction treatments (producing reduced graphene oxide (rGO)), which partially restores the graphitic network of sp^2 carbons [13,14]. Notably, the residual functional groups remained in the rGO structure make its dispersion easier in organic solvents [15] and increase the number of effective interactions also with several polymer matrices thus helping their homogeneous distribution in the solid matrix [16,17]. One can therefore take advantage of these features for the preparation of functional polymer nanocomposites [18–20] where rGO can provide enhanced electrical and mechanical properties [21–25]. In this sense, these materials have found a variety of applications such as supercapacitor electrodes [26], chemical sensor [27], and antibacterial scaffolds [28]. However, despite the better results when compared to graphene, obtaining good dispersions of rGO into the polymer matrices is still challenging: the absence of highly interacting functional groups in several polymers causes rGO aggregation, which limits the ultimate application of the derived composites [29].

Conversely, functional polymers able to provide strong and reliable non-covalent interactions with rGO, can result in composites characterized by enhanced mechanical, electrical, and thermal properties even at low filler content [30]. Our group has recently reported the high exfoliation ability of polyketone functionalized with aromatic [31] and hydroxyl pendant groups [32] towards multiwalled carbon nanotubes (MWCNTs): the resulting nanocomposites showed an effective percolation network at low MWCNTs concentration and high stable electrical response after several heating/cooling cycles, making them promising materials for temperature sensors. Based on these findings, we decided to investigate the dispersion of rGO in a polyketone functionalized by the Paal–Knorr reaction (Figure 1). Polyketones are a promising starting polymer for the synthesis of functional polymers through the chemical modification approach due to the presence of highly reactive 1,4-di-carbonyl moieties that react with primary amines yielding water-resistant *N*-substituted pyrrole units [33]. The Paal–Knorr is a versatile reaction as it can be carried out in bulk, without catalysts, with high yields and with water being the only by-product [34]. In addition, it may be also performed without solvent or in various organic solvents depending on the physical and chemical properties of the amine compound. It is tolerant to a number of primary amines making this easy synthesis a fast, cheap, and appealing approach to create polymers with almost any desired pendant functional group [31,32,35,36]. The resulting functional polymers find use in different applications such as self-healing materials [37,38], emulsions with adhesive properties [39,40], and coating materials [41].

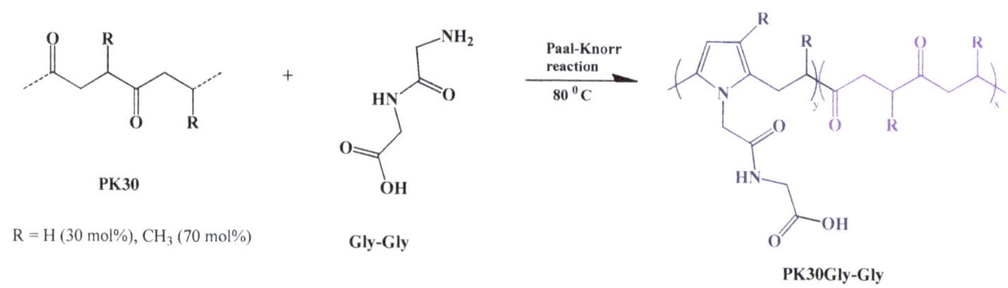

Figure 1. Scheme of Paal–Knorr reaction held by polyketone 30 (PK30) and glycyl-glycine (Gly-Gly).

The present work focuses on the synthesis of a functional polymer by chemical modification of alternating aliphatic polyketone with glycylglycine (Gly-Gly) via the Paal–Knorr reaction. The amide and carboxylic groups of Gly-Gly, along with the formation of the pyrrole groups, enable the polymer matrix to interact with rGO via non-covalent interactions (e.g., H-bonding and π–π interactions), thus promoting the formation of the nanocomposite without any chemical conversion of the functional moieties [42,43]. We, therefore, prepared a series of nanocomposites comprising different amount and types of rGO and characterized their final morphology and electric properties also in terms of the resistive behavior as a function of temperature.

2. Experimental

2.1. Materials and Methods

Aliphatic polyketones composed by ethylene, propylene, and CO were synthesized according to a reported procedure [44,45] yielding a polyketone with the aliphatic part comprised of 30 mol% ethylene and 70 mol% propylene (PK30, Mw 4670 Da). Glycyl-glycine (Gly-Gly) (Sigma Aldrich, Milan, Italy, 97%) and 2,5-hexanedione (Sigma Aldrich, Milan, Italy, 98%) were used as received. Different reduced graphene oxide (rGO), i.e., lrGO and hrGO, respectively, were kindly provided by Abalonyx (Oslo, Norway). Briefly, rGO was prepared by introducing the dry graphene oxide powder into a quartz tube tubular oven (about 1 g/min). The graphene oxide flashes in the hot zone and is then transported out of the tube by means of a continuous flow of air and collected on a filter. Residence time in the hot zone is estimated to be about 2 s. The oven temperature was set at 430 and 625 °C, for lrGO and hrGO, respectively. Elemental analysis: lrGO = C 68.84%, H 1.41%, O 29.14%, and N 0.48%; hrGO = C 74.64%, H 1.09%, O 23.12%, and N 0.64%.

2,5-hexanedione was prepared as a model compound and aimed at the correct assignment of ^{1}H-NMR signals after the Paal–Knorr synthetic process with Gly-Gly. The reaction between stoichiometric amounts of Gly-Gly and 2,5-hexadione was carried out in a 100 mL round-bottomed flasks equipped with a magnetic stirrer, a reflux condenser and a Heat-On™ Block System (Radleys Shire Hill, Saffron Walden, Essex, United Kingdom). First, 1.45 g of 2,5-hexadione (0.013 mol) and 1.72 g of Gly-Gly (0.013 mol) were placed in the flask. The reaction was carried out in ethanol at 80 °C under stirring (700 rpm) for 48 h in order to ensure reaction completeness. The solvent was removed at reduced pressure and the recovered material introduced in an oven at 40 °C for 48 h for complete dryness. ^{1}H-NMR (400 MHz δ, $CDCl_3$) = 2.2 ppm (s, 6H), 4.0 ppm (d, 2H, J = 5.8), 4.5 ppm (s, 2H), 5.8 ppm (t, 1H, J = 5.7 Hz), 5.87 ppm (s, 1H) [46].

2.2. Polyketone Modification

The functionalization of PK30 with Gly-Gly was carried out to reach 30% of conversion of polyketone di-carbonyl groups (Table 1). First, 47.192 g of PK30 were dissolved in 75 mL of ethanol in a 250 mL round-bottomed flasks equipped with a magnetic stirrer, a reflux condenser and a Heat-On™ Block System. Afterward, 14.21 g of Gly-Gly were added to the polymer solution. The reaction was carried out at reflux for 36 h. The polymer was filtered to collect the unreactive amine, re-dissolved in ethanol and filtered again and the procedure repeated three times. The solvent was removed at reduced pressure, and then placed in an oven at 40 °C for 48 h for complete dryness. The carbonyl conversion (C_{co}), i.e., the molar fraction of 1,4-dicarbonyl units reacted via the Paal–Knorr reaction, was calculated on the basis of elemental analysis as the following:

$$C_{co} = \frac{y}{y+x} \times 100\% \quad (1)$$

where x and y are the di-ketone and pyrrolic moles after conversion, respectively. y was determined as follows:

$$y = \frac{wt(N)}{A_m(N)} \quad (2)$$

where $wt(N)$ are the grams of N of the product as determined by elemental analysis and $A_m(N)$ is the atomic mass of N. x was then determined as follows:

$$x = \frac{g_{prod} - y \times M_w^y}{M_w^{pk}} \quad (3)$$

where g_{prod} is the gram of the product, M_w^y the molecular weight of the pyrrolic unit and M_w^{pk} the molecular weight of a 1,4 di-ketone unit. From the ratio between C_{co} and the corresponding amount in alimentation (C_{co}^{feed}), the conversion efficiency η con be calculated as the following:

$$\eta = \frac{C_{co}}{C_{co}^{feed}} \times 100\% \quad (4)$$

where C_{co}^{feed} corresponds to:

$$C_{co}^{feed} = \frac{Mol_{amine}}{Mol_{d-CO}} \times 100\% \quad (5)$$

with Mol_{amine} are the moles of amine and Mol_{d-CO} the moles of di-carbonyl units in alimentation.

Table 1. Elemental analysis of pristine PK30 and PK30 after functionalization with Gly-Gly.

Sample	x(%)	y(%)	C_{co}(%) [1]
PK30	100	-	-
$PK30_x$Gly-Gly_y	70.8	29.2	29.2

[1] Di-carbonyl conversion (CO %) obtained from EA.

2.3. Preparation of rGO/PK30-Gly-Gly Composites

The rGO/PK30-Gly-Gly composites were prepared by mixing PK30-Gly-Gly and lrGO or hrGO at different weight concentration (wt. %). First, PK30-Gly-Gly was dissolved in a round bottom flask containing 20 mL of chloroform under vigorous stirring at 50 °C for 20 min. The corresponding 4–8 wt. % for lrGO or 5–6 wt. % for hrGO amounts of rGO with respect to PK30-Gly-Gly was added to 20 mL of chloroform and tip sonicated for 5 min in an ice bath to avoid the loss of solvent. Afterward, the dispersion was added to the polymer solution and the system vigorously stirred at 50 °C for 24 h. After cooling to room temperature, the solvent was removed and the recovered rGO/PK30-Gly-Gly composite dried completely at 40 °C for 24 h.

2.4. Characterization

The elemental composition of rGO and polymers was determined by using an Elementar Vario Micro Cube for nitrogen, carbon and hydrogen.

^{1}H-NMR spectra were recorded at room temperature in $CDCl_3$ solution with a Bruker Avance DRX 400 spectrometer (Bruker, Billerica, MA, USA), using the residual solvent peak as internal reference.

ATR–FT–IR spectra were recorded by means of a Perkin-Elmer Spectrum One (San Francisco, CA, USA), within the 4000–650 cm^{-1} and averaged over 32 scans.

Differential scanning calorimetry (DSC, TA Instruments, New Castle, DE, USA) was carried out by means of a TA DSC250 under N_2 in agreement with previous studies [32,35].

Gel Permeation Chromatography (GPC) measurements were performed with an HP1100 Hewlett-Packard (Hewlett-Packard, Wilmington, Philadelphia, PA, USA) and in agreement with procedures previously reported [32,47].

Scanning transmission electron microscopy (STEM, Thermo Fisher Scientific, Hillsboro, Oregon, USA) was performed on rGO samples using a FEI Quanta 450 ESEM equipped with a field emission gun. Particle analysis was performed using the public domain ImageJ 1.52k software (National Institutes of Health, Bethesda, MD, USA).

Thermal degradation of rGO was analyzed via thermogravimetric analysis (TGA) with a Mettler Toledo TGA/SDTA851 instrument (Mettler Toledo, Columbus, OH, USA) under nitrogen flux (80 mL/min). All samples were tested in agreement with procedures previously reported [48,49].

Raman spectroscopy has been performed using a Horiba Jobin Yvon Xplora ONE confocal Raman microscope (Horiba Scientific, Horiba Italy, Roma, Italy). The wavelength of the excitation laser was 542 nm and the power of the laser was kept below 1 mW without noticeable sample heating.

Microscopic morphology of the composite samples was observed by Scanning Electron Microscopy, by using a Dual Beam FIB/SEM Helios Nano-Lab 600i (Thermo Fisher Scientific, Hillsboro, OR, USA), 10 kV accelerating voltage and variable magnification. For SEM imaging, the samples were prepared by Au deposition (layer about 40nm) using AC sputtering.

The electrical measurements of the lrGo and hrGO solid dispersions were carried out according to literature reports [50,51].

The temperature dependent resistivity measurements were performed on square 6 mm × 6 mm samples with 1.05 mm of thickness, connected with two copper electrodes to the opposite edge of the square: the temperature control was obtained by placing the sample on a thin ceramic plate substrate (0.3 mm) equipped with a gold resistance (on the bottom side, not in direct contact with sample) connected to a controllable power supply module. The temperature was monitored with a k-thermocouple brought in contact with the sample; the resistance was continuously measured using a High precision multimeter (Model 187, Fluke, Fluke Corporation, Everett, WA, USA) connected with the copper electrodes. Measurements were performed modulating the temperature by changing the current applied to the ceramic heater and waiting for the temperature stabilization.

3. Result and Discussion

3.1. Polyketone Functionalized with Gly-Gly Groups via the Paal–Knorr Reaction

We prepared PK30-Gly-Gly by chemical modification of polyketone 30 with Gly-Gly via the Paal–Knorr reaction. The functionalization of polyketone yielded a C_{co} of 29% as established from elemental analysis data (Table 1). The total di-carbonyl efficiency can be calculated by elemental analysis ($x + y$) using the relative content of nitrogen in the products. Moreover, PK30 shows a weight average molecular weight of 4670 with a dispersity index of 1.7, that are supposed to be mostly unchanged the functionalization process in agreement with the recent literature [47]. Nevertheless, attempts at measuring the molecular weight of the functionalized PK30 were not successful due to significant aggregation phenomena among the macromolecules during chromatography.

The PK30 functionalization with Gly-Gly is confirmed by ATR–FT-IR and ^{1}H-NMR spectroscopies. Figure 2 shows the ^{1}H-NMR spectra of PK30 before (B) and after (A) functionalization. The signals of the pyrrole rings closely correspond to those of the model compound that we described in the experimental section (Section 2.1). The success of the grafting process is indicated by the proton signals at 5.7 (H1), 2.0 (H2), and 3.97 (H3) ppm that we attributed to the pyrrole ring formed during the Paal–Knorr reaction, the methyl group, and methylene groups adjacent to it, respectively [34]. The methylene proton of the functional group was assigned at 4.15 (H6) and 4.51 ppm (H4), and the proton of the amide group at 4.75 ppm (H5), whereas the remaining signals between 2.3 and 2.9 ppm were attributed to the unreacted PK aliphatic part comprised of 30 mol% ethylene and 70 mol% propylene.

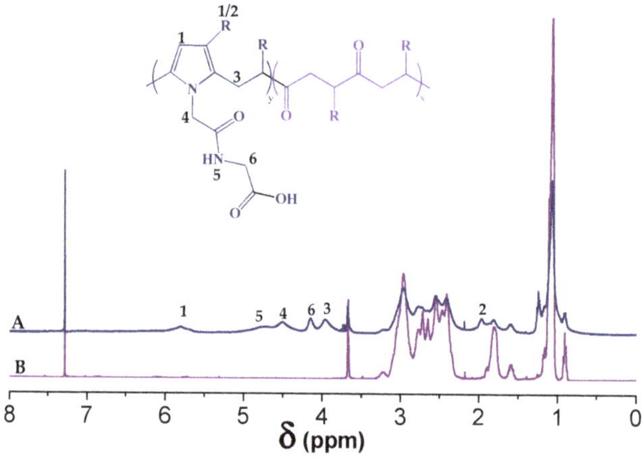

Figure 2. ^{1}H-NMR spectra of (**A**) PK30 functionalized with Gly-Gly and (**B**) pristine PK30 in $CDCl_3$.

Figure 3 shows the ATR–FTIR spectrum of PK30 and PK30-Gly-Gly. After the Paal–Knorr reaction has occurred, the intensity of the carbonyl group signal (1700 cm^{-1}) decreases, due to the disappearance of the 1,4-dicarbonyl moieties. In addition, two peaks appeared for PK30-Gly-Gly very close to 1700 cm^{-1} that correspond to the carbonyl groups of the carboxylic acid at 1745 cm^{-1} and the amide moiety at 1670 cm^{-1}. The appearance of these two peaks, together with the decrease in intensity of the signal associated with the PK carbonyl indicates the successful modification of the starting polymer. We also assigned the wide weak peak from 3650 to 2000 cm^{-1} in PK30-Gly-Gly to the hydrogen bonding of the carboxylic groups. Moreover, the peak associated to the pyrrole units was found at 3100 cm^{-1} (C=C–H) whereas the N–H and C–N stretching of the secondary amide can be found at 3397 and 1209 cm^{-1}, respectively. The weak signal at 1540 cm^{-1} may correspond either to stretching of C=N and C=C bonds or to both. Finally, stretching bands of aliphatic C–H of PK backbone and functional groups appeared between 2969 and 2873 cm^{-1} [46].

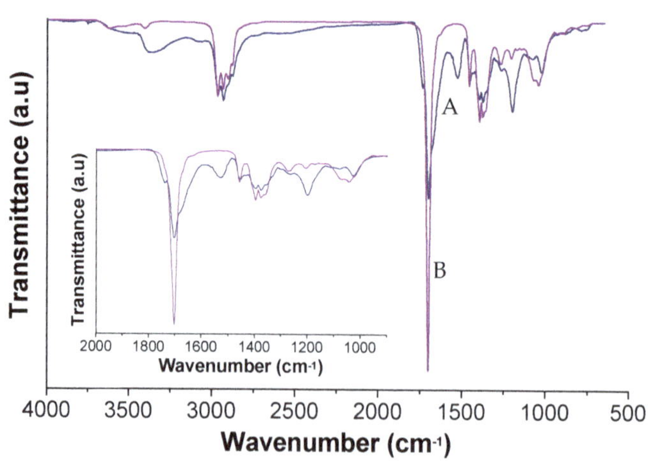

Figure 3. FT-IR spectra of (**A**) PK30 functionalized with Gly-Gly and (**B**) pristine PK30.

Another evidence that the chemical functionalization was successful can be found in the higher glass transition temperature (T_g) we measured for PK30-Gly-Gly (31.6 °C, Figure S1) compared to its precursor PK30 (−17.7 °C). This is possibly due to a combination of effect: a) the formation of pyrrole units after the Paal–Knorr reaction and the presence of hydrogen bonds between the Gly-Gly groups of different polymer chains [35,37], both contributing in the rigidity enhancement of the polymer matrix.

3.2. Preparation and Characterization of rGO/PK30-Gly-Gly Composites

We used PK30-Gly-Gly as supporting polymer matrix for two types of rGO (i.e., lrGO and hrGO) that were obtained from GO by means of thermally reduced processes at different temperatures, i.e., 430 and 625 °C, for lrGO and hrGO, respectively. PK30-Gly-Gly (Figure 1) was specifically designed as to show the presence of pyrrole rings along the backbone and the amide + carboxylic acid as pendant groups thus possibly allowing a strong and effective (H-bonding) interactions with graphitic fillers [32,52]. In particular, lrGO was characterized by a lower degree of reduction in comparison to hrGO, according to the different reduction temperature. This was confirmed by the different carbon content as determined by elemental analysis. Notably, on passing from lrGo to hrGO, the amount of carbon increased from 68.84% to 74.64%. Moreover, thermogravimetric analysis (TGA, Figure S2) evidenced a different degradation behaviour of the two types of rGO as well as a distinct residual mass. The weight loss is directly correlated to the amount of residual oxygen-containing groups, thus rGO with a larger reduction extent should display a lower weight loss. Indeed, lrGO had a residual mass of 68.9% in comparison with the 74.9% of hrGO, in strict agreement with the elemental analysis. Moreover, lrGO showed an inflection point at 218 °C that corresponds to the thermal decomposition of the oxygen-containing residual groups [53], which resulted much less pronounced in the hrGO sample. The surface size of the flakes for the two rGOs was evaluated by STEM (Figure S3) and found to be 2.95 ± 0.35 and 2.52 ± 0.74 μm^2 for lrGO and hrGO, respectively. Raman spectroscopy is a versatile technique used for the structural characterization of graphitic materials including graphene, graphene oxide, and reduced graphene oxide. It is well reported that the most prominent features in a Raman spectrum of rGO are the G and D bands (Figures S4 and S5). The G-band at about 1580 cm^{-1} is an intrinsic feature of graphene and related to the planar vibration of carbon atoms in most sp^2 graphitic materials. Conversely, the disorder-induced D-band at about 1340 cm^{-1} is attributed to the scattering from defects breaking the basic symmetry of the graphene sheet [54,55]. The large contribution at about 2900 cm^{-1}, is possibly attributed to the combination of the first overtone of the D band (2D band) and D + G band. Notably, the ratio of D and G bands peak intensities (I_D/I_G) is a common index for the extent of defects on different qualities of rGO. On passing from lrGO (Figure S4) to hrGO (Figure S5) it was found that the I_D/I_G ratios of rGO samples changed from 1.09 to 1.20 even if the contribution of the 2D + G + D bands did not substantially change. This result was attributed to the combination of two opposite phenomena: the first is the increased reduction extent on passing from lrGO to hrGO that favors the restoration of a larger amount of ordered graphitic structure, thus enhancing the I_D/I_G ratio. The second is the loss of carbon content that possibly occurs for thermally-activated reduction processes at higher temperature that, in turn, adversely affects the graphitic layer extent [56]. This last hypothesis is in agreement with the average surface size determined by STEM experiments.

The different reduction extent between lrGO and hrGO was also reflected on their electrical conductivity. Aliquots (100 μL) of rGO chloroform dispersions were drop-cast onto gold plated electrodes supported on a Kapton® film and the electrical resistance was measured after the complete evaporation of the solvent. Electrical resistances, determined as the average of three distinct depositions, of 81.7 ± 15.41 MΩ and 38.11 ± 2.11 MΩ were measured for lrGO and hrGO, respectively, thus confirming that the larger reduction degree is associated to a higher graphitic conductive extent.

rGO/PK30-Gly-Gly composites were then prepared by solvent-aided mixing of the polymer with different amounts of rGO and characterized by means of various techniques. DSC experiments of PK30-Gly-Gly composites comprising lrGO or hrGO showed that they have a higher T_g in comparison to PK30-Gly-Gly alone (Figure S1 shows the experimental curves used to derive the data). Figure 4 shows the T_g increase for the lrGO/PK30-Gly-Gly composites as a function of the filler concentration. The rGO/PK30-Gly-Gly composites with hrGO displayed a similar trend with T_g values of 52.7 °C and 53.5 °C for the 5 and 6 wt. % content, respectively. According to the literature, the increase of the T_g in nanocomposite is tightly related to the enhanced system viscosity due to the interfacial interaction between the polymeric matrix and the filler [57,58]: in our case the pendant functional groups and the pyrrole rings of the polymer both effectively interact with the rGO, thus limiting the polymer mobility [59–61]. Overall, these results prove the good affinity between the filler and the polymer.

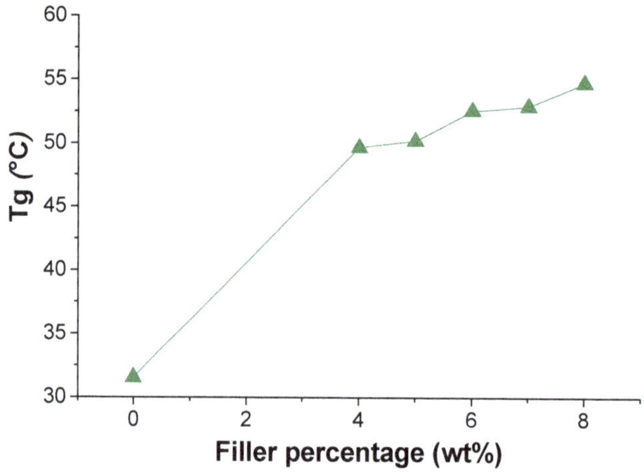

Figure 4. Glass transition temperature (differential scanning calorimetry (DSC)) of pristine PK30-Gly-Gly and of the respective composites containing lower degree of reduction graphene oxide (lrGO) as the filler.

We also evaluated the morphology of the composite material and the dispersion degree of rGO by SEM microscopy (Figure 5). Pictures in Figure 5 show a constant morphological change of the material as the amount of lrGO increase. The composite with less content of lrGO (4 wt. %) displays a smoother surface as also revealed at higher magnifications (Figure 5A,A1). As soon as the lrGO content increases a more porous structure is evidenced, especially at the highest 7 and 8 wt. % (Figure 5C,D). Nevertheless, the SEM micrographs do not evidence the presence of a significant phase separation between the composite components and lrGO appears as homogeneously distributed graphitic filler within the PK30-Gly-Gly matrix. More than that, at the highest magnification lrGO appears homogeneously distributed within the interacting PK30-Gly-Gly thus possibly suggesting the formation of effective percolation pathways (Figure S6). The composites comprising 5 and 6 wt. % of hrGO show similar morphology (Figure 6) and appear similar to those prepared with the same concentration of lrGO as well (Figures 5B,C and S7).

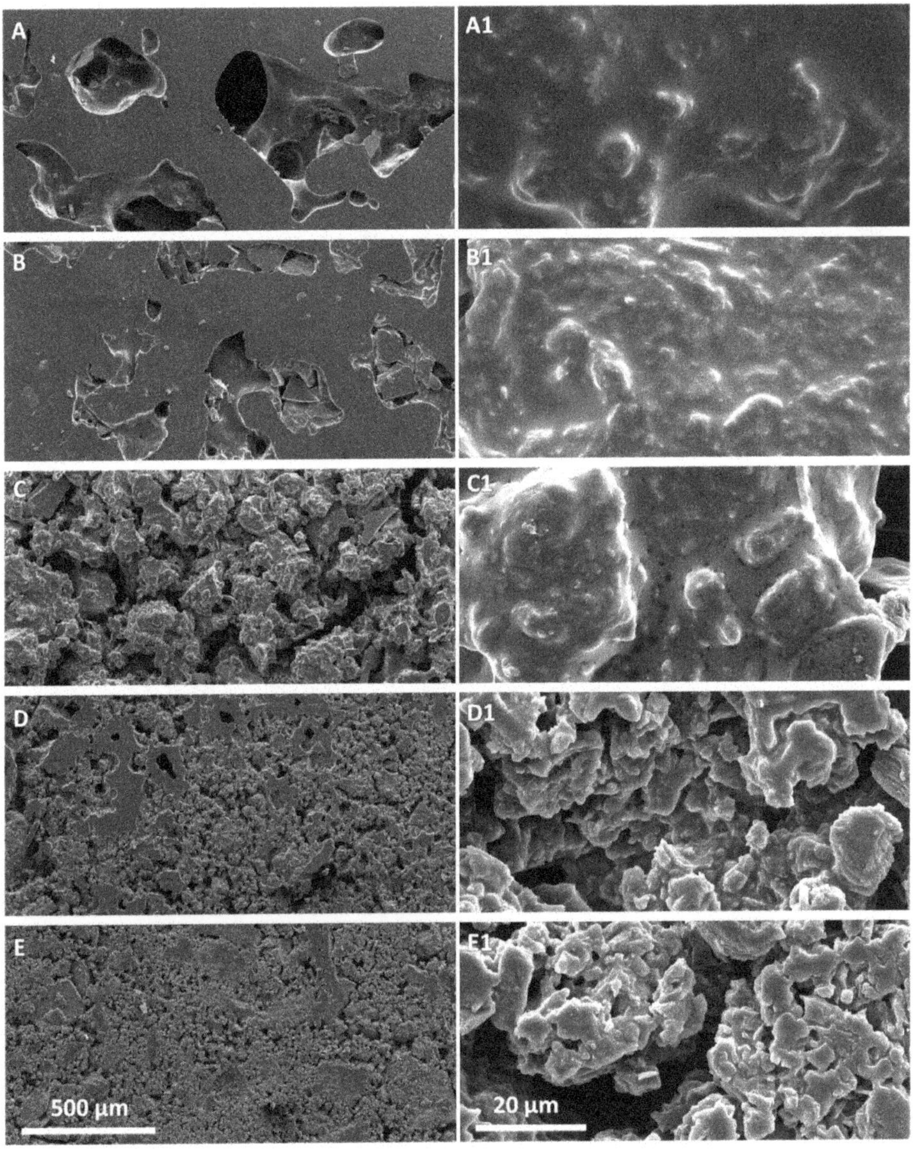

Figure 5. Scanning Electron Microscopy (SEM) micrographs at different magnification of the nanocomposite composed by PK30-Gly-Gly and lrGO at different weight percentage. (**A**,**A1**) 4 wt. %, (**B**,**B1**) 5 wt. %, (**C**,**C1**) 6 wt. %, (**D**,**D1**) 7 wt. %, and (**E**,**E1**) 8 wt. %. Left pictures scale bar 500 μm, right pictures scale bar 20 μm.

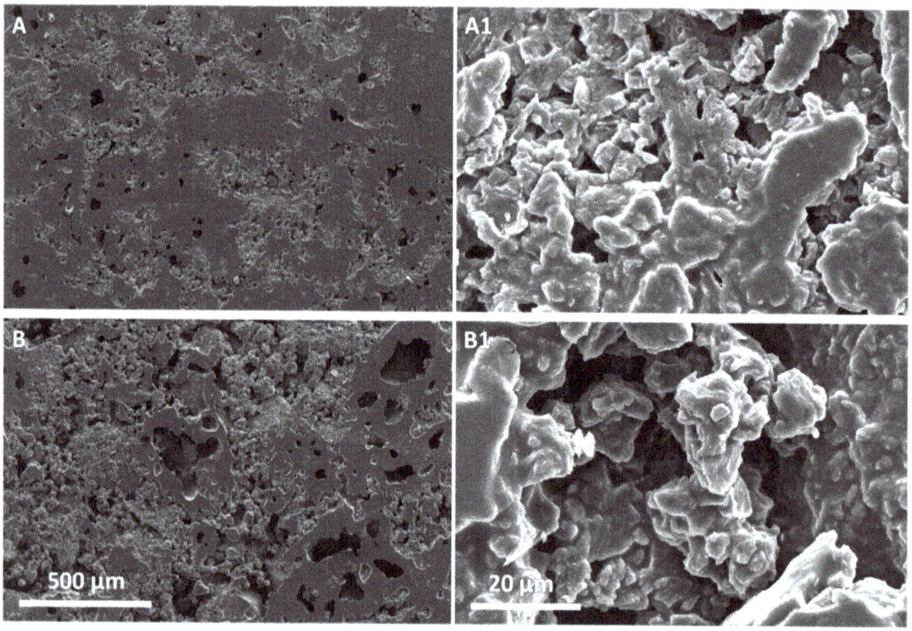

Figure 6. SEM micrographs at different magnification of the nanocomposite composed by PK30-Gly-Gly and hrGO at different weight percentage. (**A**,**A1**) 5 wt. % and (**B**,**B1**) 6 wt. %. Left pictures scale bar 500 µm, right pictures scale bar 20 µm.

Figure 7 summarizes the resistivity of the samples composed by PK30-Gly-Gly and lrGO. As expected, the resistance decreased when increasing the rGO content dispersed into the polymer. While the composite with 4 wt. % of lrGO displayed a surface resistivity higher than 500 MΩ/sq at 30 °C, an effective percolation pathway was reached with 5 wt. % of lrGO. Moreover, significant decrease in terms of surface resistivity occurred on going from 5 to 8 wt. % of the graphitic filler up to values of 250 kΩ/sq.

To evaluate the effect of different degrees of reduction in the rGO filler, we tested samples containing 5 and 6 wt. % of hrGO with a similar procedure. The results are summarized in the Figure 7. Unlike the previous case, the increment in rGO from 5 to 6 wt. % did not produce any evident variation in surface resistivity that was of 500 kΩ/sq for both composites. Nevertheless, these values appeared remarkably lower that those measured from the corresponding lrGO/PK30-Gly-Gly composites, in agreement with the more reduced rGO content in the hrGO sample.

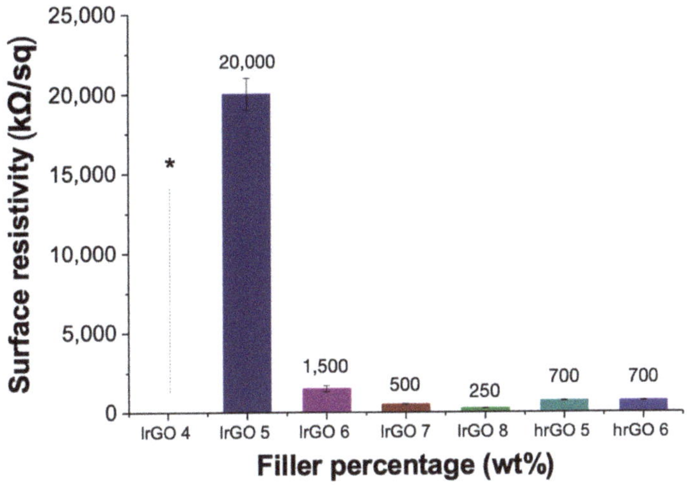

Figure 7. Surface resistivity of the nanocomposites composed by PK30-Gly-Gly and lrGO or higher degree of reduction graphene oxide (hrGO) at different weight percentage at 30 °C. * ≥ 500 MΩ/sq. Sample thickness of 1.05 mm.

3.3. Composite Resistance Sensitivity to Temperature

As a semiconductor, the conductivity of rGO increases upon increasing the temperature. Figure 8 shows the resistivity vs. temperature plot for the different lrGO/PK30-Gly-Gly composites (6, 7, and 8 wt. %). For each sample, the resistance lowered upon heating. In addition, the composites displayed a marked reduction in resistivity after 50 °C, which one can relate to the T_g of the same composites (Figure 4). Also, the composite with 4 wt. % started to show resistivity values from 45 °C (Figure S8). This indicates that above the T_g, the polymeric matrix allows for an increased mobility of the rGO flakes, thus reaching a temporal network where the electrons can flow through the material [62].

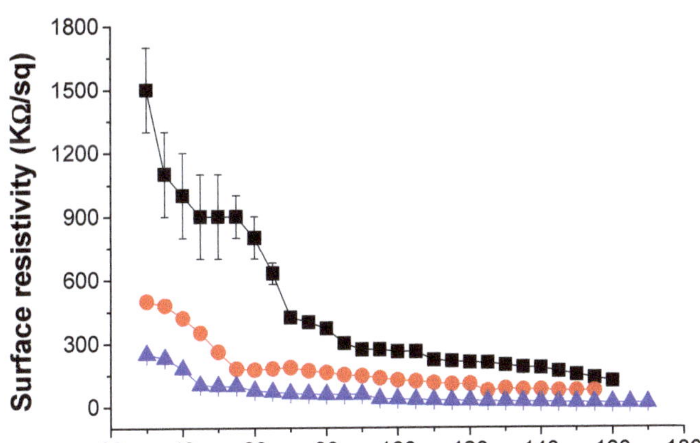

Figure 8. Surface resistivity of the nanocomposite composed by PK30-Gly-Gly and lrGO at different temperatures and filler concentration. (■) 6 wt. %. (●) 7 wt. %. (▲) 8 wt. %. Sample thickness of 1.05 mm.

An important characteristic to evaluate is the material fatigue upon heating and cooling cycles. In Figure 9, we show the resistivity variation of the lrGO/PK30-Gly-Gly composite during five heating and cooling cycles from 30 to 115 °C. In general, we observed that the resistivity of the materials was reliably restored across every cycle. Indeed, the composites with lrGO at concentrations of 4 wt. % returns to be an electrical insulator as the heating system is switch off. This evidence suggests that these composites could be used as temperature sensors over the investigated operating range, more preferably between room temperature and 80 °C, i.e., where the surface resistance variation was maximum.

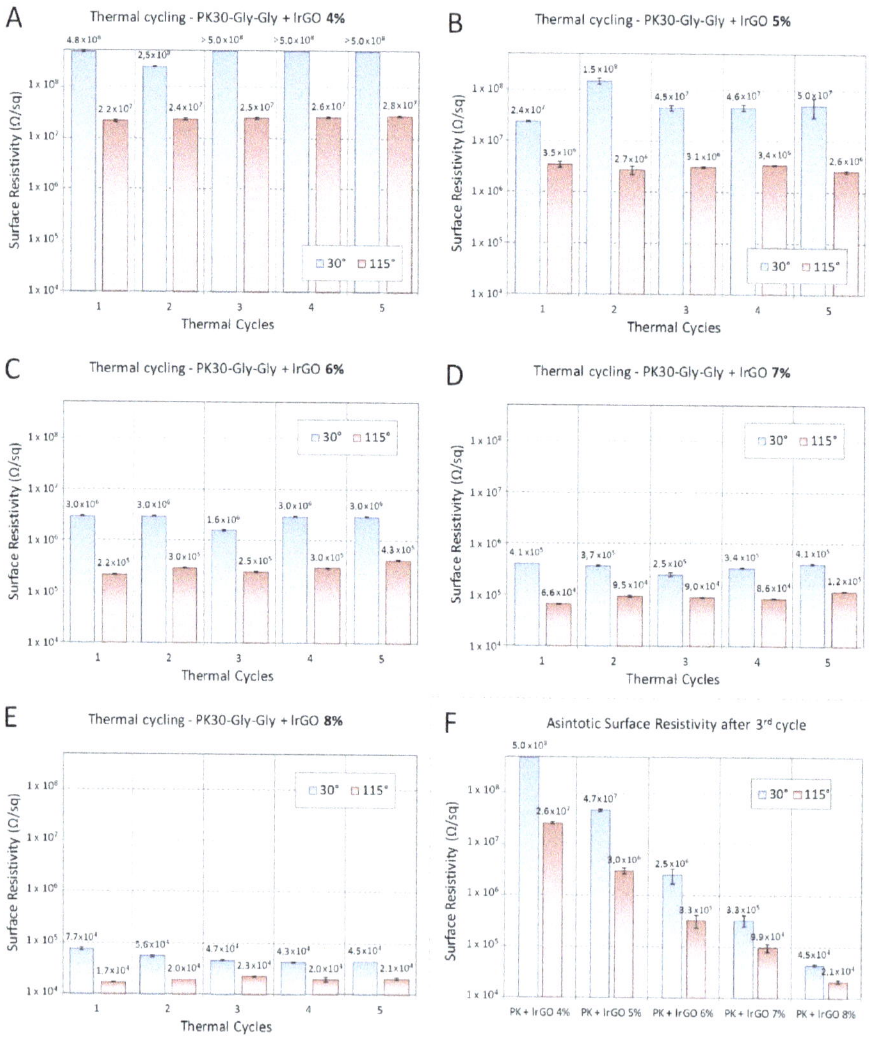

Figure 9. Surface resistivity measured at 30 °C (light blue) and 115 °C (light red) upon repeated cycles of the nanocomposite composed by PK30-Gly-Gly and lrGO. (**A**) 4 wt. %, (**B**) 5 wt. %, (**C**) 6 wt. %, (**D**) 7 wt. %, and (**E**) 8 wt. %. Sample thickness of 1.05 mm. (**F**) Asymptotic surface resistivity as average over last three thermal cycles.

The same experiment was repeated for the hrGO/PK30-Gly-Gly composites (Figures 10 and 11). Despite the variation of the filler, these composites possess the same behaviour of the ones comprising lrGO.

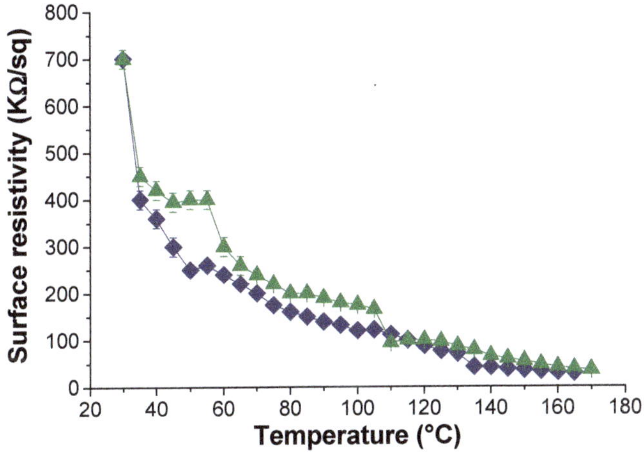

Figure 10. Resistivity of the nanocomposite composed by PK30-Gly-Gly and hrGO at different temperatures. 5 (♦) and (▲) 6 wt. % of filler concentration.

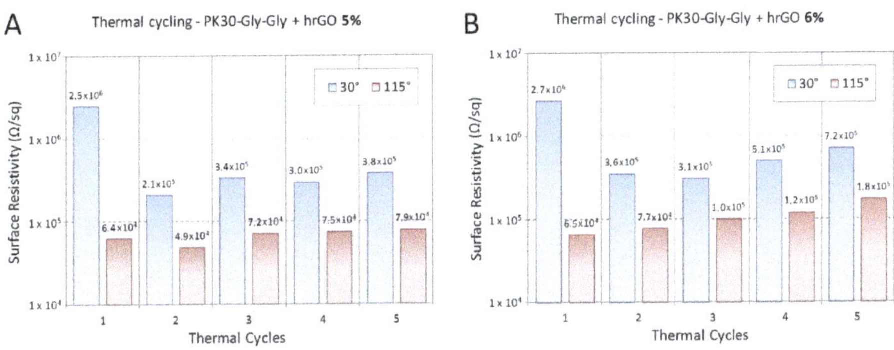

Figure 11. Surface resistivity measured at 30 °C (light blue) and 115 °C (light red) upon repeated cycles of the nanocomposite composed by PK30-Gly-Gly and hrGO. 5 (**A**) and (**B**) 6 wt. % of filler. Sample thickness of 1.05 mm.

4. Conclusions

We have shown the straightforward preparation of polymeric nanocomposites comprising rGO using the novel functionalized PK30-Gly-Gly polyketone bearing pyrrole rings in the backbone and amide and carboxylic acid as pendant groups. We synthesized the polymer by the chemical modification of an alternating aliphatic polyketone via the Paal–Knorr reaction. The former was able to interact with rGO by effective non-covalent interactions, thus facilitating the exfoliation process without damaging the one-dimensional arrangement of rGO. We also easily modulated the electrical resistance of the composite by increasing the concentration of rGO in the mixing process. Indeed, resistivity measurements supported by SEM investigations demonstrated that the effective percolation pathway is achieved from an rGO concentration of 5 wt. %. Heating and cooling cycles showed that the conductive network is preserved, suggesting a high stability of the rGO dispersion within the polymeric matrix. Interestingly, we found that the composite with 4 wt. % lrGO (below the percolation threshold) can act as an ON–OFF system with temperature, allowing a low resistance only when the temperature is higher than the T_g of the composite and otherwise behaving as an insulator. Overall, these results support the use of functionalized polyketones as the matrix for rGO nanocomposites. Such materials may have a predominant role in the development of the next generation of soft robotics devices, in which one can modulate mechanical and electrical properties by using diverse inputs such as temperature and deformation.

Supplementary Materials: The following are available online at http://www.mdpi.com/2073-4360/12/4/923/s1, Figure S1: DSC first cycle after thermal history erases of PK30Gly-Gly29 and its respective composite with lrGO and hrGO. The first curves are not shown since they are carried out to remove the thermal history of the polymer and composites; Figure S2: TGA analysis of a) lrGO and b) hrGO; Figure S3: STEM micrograph of a) lrGO and b) hrGO; Figure S4: Raman spectrum of lrGO; Figure S5: Raman spectrum of hrGO; Figure S6: SEM picture of the PK30-Gly-Gly nanocomposite containing 7 wt% lrGO; Figure S7: SEM picture of the PK30-Gly-Gly nanocomposite containing 6 wt% of hrGO; Figure S8: Surface resistivity of the nanocomposite composed by PK30-Gly-Gly and lrGO at different temperatures and filler concentration. (A) 4 wt%. (B) 5 wt%. Sample thickness of 1.05 mm.

Author Contributions: Conceptualization, A.P., F.P. and V.M.; methodology, E.A.A.-H.; data curation, V.M. and M.C.; writing—original draft preparation, E.A.A.-H.; writing—review and editing, A.P., V.M. and M.C.; funding acquisition, V.M. All authors have read and agreed to the published version of the manuscript.

Funding: This research received no external funding.

Acknowledgments: CISUP—Centre for Instrumentation Sharing—University of Pisa is kindly acknowledged for STEM measurements. Michela Meucci is kindly acknowledged for rGO characterizations. The assistance provided by Rune Wendelbo and Blerina Gjoka form Abalonyx and Marco Scatto from Nadir Srl is greatly appreciated.

Conflicts of Interest: The authors declare no conflict of interest.

References

1. Novoselov, K.S.; Geim, A.K.; Morozov, S.V.; Jiang, D.; Zhang, Y.; Dubonos, S.V.; Grigorieva, I.V.; Firsov, A.A. Electric Field Effect in Atomically Thin Carbon Films. *Science* **2004**, *306*, 666. [CrossRef]
2. Kuilla, T.; Bhadra, S.; Yao, D.; Kim, N.H.; Bose, S.; Lee, J.H. Recent advances in graphene based polymer composites. *Prog. Polym. Sci.* **2010**, *35*, 1350–1375. [CrossRef]
3. Geim, A.K.; Novoselov, K.S. The rise of graphene. *Nat. Mater.* **2007**, *6*, 183–191. [CrossRef] [PubMed]
4. Bonaccorso, F.; Lombardo, A.; Hasan, T.; Sun, Z.; Colombo, L.; Ferrari, A.C. Production and processing of graphene and 2d crystals. *Mater. Today* **2012**, *15*, 564–589. [CrossRef]
5. Lin, L.; Peng, H.; Liu, Z. Synthesis challenges for graphene industry. *Nat. Mater.* **2019**, *18*, 520–524. [CrossRef]
6. Ranjan, P.; Agrawal, S.; Sinha, A.; Rao, T.R.; Balakrishnan, J.; Thakur, A.D. A Low-Cost Non-explosive Synthesis of Graphene Oxide for Scalable Applications. *Sci. Rep.* **2018**, *8*, 12007. [CrossRef]
7. Paredes, J.I.; Villar-Rodil, S.; Martínez-Alonso, A.; Tascón, J.M.D. Graphene Oxide Dispersions in Organic Solvents. *Langmuir* **2008**, *24*, 10560–10564. [CrossRef]
8. Tarcan, R.; Todor-Boer, O.; Petrovai, I.; Leordean, C.; Astilean, S.; Botiz, I. Reduced graphene oxide today. *J. Mater. Chem. C* **2020**, *8*, 1198–1224. [CrossRef]
9. Ray, S.C. Chapter 2—Application and Uses of Graphene Oxide and Reduced Graphene Oxide. In *Applications of Graphene and Graphene-Oxide Based Nanomaterials*; Ray, S.C., Ed.; William Andrew Publishing: Oxford, UK, 2015; pp. 39–55.
10. Liu, W.; Speranza, G. Functionalization of Carbon Nanomaterials for Biomedical Applications. *C J. Carbon Res.* **2019**, *5*, 72. [CrossRef]
11. Mann, J.A.; Dichtel, W.R. Noncovalent Functionalization of Graphene by Molecular and Polymeric Adsorbates. *J. Phys. Chem. Lett.* **2013**, *4*, 2649–2657. [CrossRef]
12. Dasari Shareena, T.P.; McShan, D.; Dasmahapatra, A.K.; Tchounwou, P.B. A Review on Graphene-Based Nanomaterials in Biomedical Applications and Risks in Environment and Health. *Nano Micro Lett.* **2018**, *10*, 53. [CrossRef] [PubMed]
13. Park, S.; An, J.; Piner, R.D.; Jung, I.; Yang, D.; Velamakanni, A.; Nguyen, S.T.; Ruoff, R.S. Aqueous Suspension and Characterization of Chemically Modified Graphene Sheets. *Chem. Mater.* **2008**, *20*, 6592–6594. [CrossRef]
14. Alharbi, T.M.D.; Alghamdi, A.R.M.; Vimalanathan, K.; Raston, C.L. Continuous flow photolytic reduction of graphene oxide. *Chem. Commun.* **2019**, *55*, 11438–11441. [CrossRef] [PubMed]

15. Fan, H.; Wang, L.; Zhao, K.; Li, N.; Shi, Z.; Ge, Z.; Jin, Z. Fabrication, Mechanical Properties, and Biocompatibility of Graphene-Reinforced Chitosan Composites. *Biomacromolecules* **2010**, *11*, 2345–2351. [CrossRef]
16. Gupta, B.; Kumar, N.; Panda, K.; Kanan, V.; Joshi, S.; Visoly-Fisher, I. Role of oxygen functional groups in reduced graphene oxide for lubrication. *Sci. Rep.* **2017**, *7*, 45030. [CrossRef]
17. Menes, O.; Cano, M.; Benedito, A.; Giménez, E.; Castell, P.; Maser, W.K.; Benito, A.M. The effect of ultra-thin graphite on the morphology and physical properties of thermoplastic polyurethane elastomer composites. *Compos. Sci. Technol.* **2012**, *72*, 1595–1601. [CrossRef]
18. Feng, H.; Li, Y.; Li, J. Strong reduced graphene oxide–polymer composites: Hydrogels and wires. *Rsc Adv.* **2012**, *2*, 6988–6993. [CrossRef]
19. Layek, R.K.; Samanta, S.; Nandi, A.K. The physical properties of sulfonated graphene/poly(vinyl alcohol) composites. *Carbon* **2012**, *50*, 815–827. [CrossRef]
20. Huang, X.; Qi, X.; Boey, F.; Zhang, H. Graphene-based composites. *Chem. Soc. Rev.* **2012**, *41*, 666–686. [CrossRef]
21. Luceño-Sánchez, A.J.; Díez-Pascual, M.A. Grafting of Polypyrrole-3-carboxylic Acid to the Surface of Hexamethylene Diisocyanate-Functionalized Graphene Oxide. *Nanomaterials* **2019**, *9*, 1095. [CrossRef]
22. Cano, M.; Khan, U.; Sainsbury, T.; O'Neill, A.; Wang, Z.; McGovern, I.T.; Maser, W.K.; Benito, A.M.; Coleman, J.N. Improving the mechanical properties of graphene oxide based materials by covalent attachment of polymer chains. *Carbon* **2013**, *52*, 363–371. [CrossRef]
23. Castell, P.; Cano, M.; Maser, W.K.; Benito, A.M. Combination of two dispersants as a valuable strategy to prepare improved poly(vinyl alcohol)/carbon nanotube composites. *Compos. Sci. Technol.* **2013**, *80*, 101–107. [CrossRef]
24. Liang, J.; Huang, Y.; Zhang, L.; Wang, Y.; Ma, Y.; Guo, T.; Chen, Y. Molecular-Level Dispersion of Graphene into Poly(vinyl alcohol) and Effective Reinforcement of their Nanocomposites. *Adv. Funct. Mater.* **2009**, *19*, 2297–2302. [CrossRef]
25. Wang, Z.; Guo, Y.; Yan, L.; Bian, J.; Liu, H.; Huang, H.; Lin, H.; Sude, M.; Lijun, W.; Gu, Z. Mechanical properties and morphologies of polypropylene composites synergistically reinforced-toughened by styrene–butadiene rubber and graphene oxide nanosheets. *J. Thermoplast. Compos. Mater.* **2019**, *33*, 413–431. [CrossRef]
26. Zhang, K.; Zhang, L.L.; Zhao, X.S.; Wu, J. Graphene/Polyaniline Nanofiber Composites as Supercapacitor Electrodes. *Chem. Mater.* **2010**, *22*, 1392–1401. [CrossRef]
27. Yuan, W.; Huang, L.; Zhou, Q.; Shi, G. Ultrasensitive and Selective Nitrogen Dioxide Sensor Based on Self-Assembled Graphene/Polymer Composite Nanofibers. *Acs Appl. Mater. Interfaces* **2014**, *6*, 17003–17008. [CrossRef]
28. Pal, N.; Dubey, P.; Gopinath, P.; Pal, K. Combined effect of cellulose nanocrystal and reduced graphene oxide into poly-lactic acid matrix nanocomposite as a scaffold and its anti-bacterial activity. *Int. J. Biol. Macromol.* **2017**, *95*, 94–105. [CrossRef]
29. Tang, L.-C.; Zhao, L.; Guan, L.-Z. 7 Graphene/Polymer Composite Materials: Processing, Properties and Applications. In *Advanced Composite Materials: Properties and Applications*; De Gruyter: Berlin, Germany, 2017.
30. Ramanathan, T.; Abdala, A.A.; Stankovich, S.; Dikin, D.A.; Herrera-Alonso, M.; Piner, R.D.; Adamson, D.H.; Schniepp, H.C.; Chen, X.; Ruoff, R.S.; et al. Functionalized graphene sheets for polymer nanocomposites. *Nat. Nanotechnol.* **2008**, *3*, 327–331. [CrossRef]
31. Migliore, N.; Polgar, M.L.; Araya-Hermosilla, R.; Picchioni, F.; Raffa, P.; Pucci, A. Effect of the Polyketone Aromatic Pendent Groups on the Electrical Conductivity of the Derived MWCNTs-Based Nanocomposites. *Polymers* **2018**, *10*, 618. [CrossRef]
32. Araya-Hermosilla, R.; Pucci, A.; Araya-Hermosilla, E.; Pescarmona, P.P.; Raffa, P.; Polgar, L.M.; Moreno-Villoslada, I.; Flores, M.; Fortunato, G.; Broekhuis, A.A.; et al. An easy synthetic way to exfoliate and stabilize MWCNTs in a thermoplastic pyrrole-containing matrix assisted by hydrogen bonds. *Rsc Adv.* **2016**, *6*, 85829–85837. [CrossRef]
33. Zhang, Y.; Broekhuis, A.A.; Stuart, M.C.A.; Picchioni, F. Polymeric amines by chemical modifications of alternating aliphatic polyketones. *J. Appl. Polym. Sci.* **2008**, *107*, 262–271. [CrossRef]

34. Araya-Hermosilla, E.; Roscam Abbing, M.; Catalán-Toledo, J.; Oyarzun-Ampuero, F.; Pucci, A.; Raffa, P.; Picchioni, F.; Moreno-Villoslada, I. Synthesis of tuneable amphiphilic-modified polyketone polymers, their complexes with 5,10,15,20-tetrakis-(4-sulfonatophenyl)porphyrin, and their role in the photooxidation of 1,3,5-triphenylformazan confined in polymeric nanoparticles. *Polymer* **2019**, *167*, 215–223. [CrossRef]
35. Araya-Hermosilla, R.; Lima, G.M.R.; Raffa, P.; Fortunato, G.; Pucci, A.; Flores, M.E.; Moreno-Villoslada, I.; Broekhuis, A.A.; Picchioni, F. Intrinsic self-healing thermoset through covalent and hydrogen bonding interactions. *Eur. Polym. J.* **2016**, *81*, 186–197. [CrossRef]
36. Araya-Hermosilla, E.; Catalán-Toledo, J.; Muñoz-Suescun, F.; Oyarzun-Ampuero, F.; Raffa, P.; Polgar, L.M.; Picchioni, F.; Moreno-Villoslada, I. Totally Organic Redox-Active pH-Sensitive Nanoparticles Stabilized by Amphiphilic Aromatic Polyketones. *J. Phys. Chem. B* **2018**, *122*, 1747–1755. [CrossRef]
37. Araya-Hermosilla, R.; Broekhuis, A.A.; Picchioni, F. Reversible polymer networks containing covalent and hydrogen bonding interactions. *Eur. Polym. J.* **2014**, *50*, 127–134. [CrossRef]
38. Araya-Hermosilla, R.; Fortunato, G.; Pucci, A.; Raffa, P.; Polgar, L.; Broekhuis, A.A.; Pourhossein, P.; Lima, G.M.R.; Beljaars, M.; Picchioni, F. Thermally reversible rubber-toughened thermoset networks via Diels–Alder chemistry. *Eur. Polym. J.* **2016**, *74*, 229–240. [CrossRef]
39. Hamarneh, A.I.; Heeres, H.J.; Broekhuis, A.A.; Sjollema, K.A.; Zhang, Y.; Picchioni, F. Use of soy proteins in polyketone-based wood adhesives. *Int. J. Adhes. Adhes.* **2010**, *30*, 626–635. [CrossRef]
40. Zhang, Y.; Broekhuis, A.A.; Picchioni, F. Aqueous polymer emulsions by chemical modifications of thermosetting alternating polyketones. *J. Appl. Polym. Sci.* **2007**, *106*, 3237–3247. [CrossRef]
41. Toncelli, C.; Schoonhoven, M.-J.; Broekhuis, A.A.; Picchioni, F. Paal–Knorr kinetics in waterborne polyketone-based formulations as modulating cross-linking tool in electrodeposition coatings. *Mater. Des.* **2016**, *108*, 718–724. [CrossRef]
42. Layek, R.K.; Nandi, A.K. A review on synthesis and properties of polymer functionalized graphene. *Polymer* **2013**, *54*, 5087–5103. [CrossRef]
43. Wang, Q.H.; Hersam, M.C. Room-temperature molecular-resolution characterization of self-assembled organic monolayers on epitaxial graphene. *Nat. Chem.* **2009**, *1*, 206–211. [CrossRef] [PubMed]
44. Drent, E.; Keijsper, J.J. Polyketone polymer Preparation with tetra Alkyl Bis Phosphine Ligand and Hydrogen. U.S. Patent 5225523 A, 6 July 1993.
45. Mul, W.P.; Dirkzwager, H.; Broekhuis, A.A.; Heeres, H.J.; van der Linden, A.J.; Guy Orpen, A. Highly active, recyclable catalyst for the manufacture of viscous, low molecular weight, CO–ethene–propene-based polyketone, base component for a new class of resins. *Inorg. Chim. Acta* **2002**, *327*, 147–159. [CrossRef]
46. Silverstein, R.M.; Webster, F.X.; Kiemle, D.J.; Bryce, D.L. *Spectrometric identification of organic compounds*; Wiley India Private Limited: New Delhi, India, 2015.
47. Macedo, R.; Lima, G.; Orozco, F.; Picchioni, F.; Moreno-Villoslada, I.; Pucci, A.; Bose, K.R.; Araya-Hermosilla, R. Electrically Self-Healing Thermoset MWCNTs Composites Based on Diels-Alder and Hydrogen Bonds. *Polymers* **2019**, *11*, 1885. [CrossRef] [PubMed]
48. Manafi, P.; Ghasemi, I.; Karrabi, M.; Azizi, H.; Manafi, M.R.; Ehsaninamin, P. Thermal stability and thermal degradation kinetics (model-free kinetics) of nanocomposites based on poly (lactic acid)/graphene: The influence of functionalization. *Polym. Bull.* **2015**, *72*, 1095–1112. [CrossRef]
49. Yang, J.; Huang, Y.; Lv, Y.; Zhao, P.; Yang, Q.; Li, G. The intrinsic thermal-oxidative stabilization effect of chemically reduced graphene oxide on polypropylene. *J. Mater. Chem. A* **2013**, *1*, 11184–11191. [CrossRef]
50. Biver, T.; Criscitiello, F.; Di Francesco, F.; Minichino, M.; Swager, T.; Pucci, A. MWCNT/perylene bisimide water dispersions for miniaturized temperature sensors. *Rsc Adv.* **2015**, *5*, 65023–65029. [CrossRef]
51. Criscitiello, F.; Scigliano, A.; Bianco, R.; Beccia, M.R.; Biver, T.; Pucci, A. Perylene bisimide metal complexes as new MWCNTs dispersants: Role of the metal ion in stability and temperature sensing. *Colloids Surf. A Physicochem. Eng. Asp.* **2017**, *516*, 32–38. [CrossRef]
52. Araya-Hermosilla, R.; Pucci, A.; Raffa, P.; Santosa, D.; Pescarmona, P.P.; Gengler, Y.N.R.; Rudolf, P.; Moreno-Villoslada, I.; Picchioni, F. Electrically-Responsive Reversible Polyketone/MWCNT Network through Diels-Alder Chemistry. *Polymers* **2018**, *10*, 1076. [CrossRef]
53. Ossonon, B.D.; Bélanger, D. Synthesis and characterization of sulfophenyl-functionalized reduced graphene oxide sheets. *Rsc Adv.* **2017**, *7*, 27224–27234. [CrossRef]

54. Cançado, L.G.; Jorio, A.; Ferreira, E.H.M.; Stavale, F.; Achete, C.A.; Capaz, R.B.; Moutinho, M.V.O.; Lombardo, A.; Kulmala, T.S.; Ferrari, A.C. Quantifying Defects in Graphene via Raman Spectroscopy at Different Excitation Energies. *Nano Lett.* **2011**, *11*, 3190–3196. [CrossRef]
55. Muhammad Hafiz, S.; Ritikos, R.; Whitcher, T.J.; Md. Razib, N.; Bien, D.C.S.; Chanlek, N.; Nakajima, H.; Saisopa, T.; Songsiririthigul, P.; Huang, N.M.; et al. A practical carbon dioxide gas sensor using room-temperature hydrogen plasma reduced graphene oxide. *Sens. Actuators B Chem.* **2014**, *193*, 692–700. [CrossRef]
56. Díez-Betriu, X.; Álvarez-García, S.; Botas, C.; Álvarez, P.; Sánchez-Marcos, J.; Prieto, C.; Menéndez, R.; de Andrés, A. Raman spectroscopy for the study of reduction mechanisms and optimization of conductivity in graphene oxide thin films. *J. Mater. Chem. C* **2013**, *1*, 6905–6912. [CrossRef]
57. Du, F.; Scogna, R.C.; Zhou, W.; Brand, S.; Fischer, J.E.; Winey, K.I. Nanotube Networks in Polymer Nanocomposites: Rheology and Electrical Conductivity. *Macromolecules* **2004**, *37*, 9048–9055. [CrossRef]
58. Wu, D.; Wu, L.; Zhang, M. Rheology of multi-walled carbon nanotube/poly(butylene terephthalate) composites. *J. Polym. Sci. Part B Polym. Phys.* **2007**, *45*, 2239–2251. [CrossRef]
59. Yu, K.; Wang, M.; Wu, J.; Qian, K.; Sun, J.; Lu, X. Modification of the Interfacial Interaction between Carbon Fiber and Epoxy with Carbon Hybrid Materials. *Nanomaterials* **2016**, *6*, 89. [CrossRef] [PubMed]
60. Bhattacharya, M. Polymer Nanocomposites—A Comparison between Carbon Nanotubes, Graphene, and Clay as Nanofillers. *Materials* **2016**, *9*, 262. [CrossRef]
61. Liao, K.-H.; Aoyama, S.; Abdala, A.A.; Macosko, C. Does Graphene Change Tg of Nanocomposites? *Macromolecules* **2014**, *47*, 8311–8319. [CrossRef]
62. Potts, J.R.; Dreyer, D.R.; Bielawski, C.W.; Ruoff, R.S. Graphene-based polymer nanocomposites. *Polymer* **2011**, *52*, 5–25. [CrossRef]

polymers

Article

Wear Resistant Nanocomposites Based on Biomedical Grade UHMWPE Paraffin Oil and Carbon Nano-Filler: Preliminary Biocompatibility and Antibacterial Activity Investigation

Michelina Catauro [1,*], Cristina Scolaro [2], Giovanni Dal Poggetto [3], Severina Pacifico [4] and Annamaria Visco [2,5,*]

1 Department of Engineering, University of Campania "Luigi Vanvitelli", Via Roma 29, I-81031 Aversa, Italy
2 Department of Engineering, University of Messina, C.da Di Dio, 98166 Messina, Italy; cscolaro@unime.it
3 Ecoricerche, Srl, Via Principi Normanni, 81043 Capua (CE), Italy; giogiodp@hotmail.it
4 Department of Environmental, Biological and Pharmaceutical Sciences and Technologies; University of Campania "Luigi Vanvitelli", Via Vivaldi 43, 81100 Caserta, Italy; severina.pacifico@unicampania.it
5 Istituto per i Polimeri, Compositi e Biomateriali - CNR IPCB, Via Paolo Gaifami 18, 9-95126 Catania, Italy
* Correspondence: michelina.catauro@unicampania.it (M.C.); avisco@unime.it (A.V.); Tel.: +39-082/5010360 (M.C.); Tel.: +39-090-676-5249/3808 (A.V.)

Received: 25 March 2020; Accepted: 18 April 2020; Published: 22 April 2020

Abstract: In the present paper, we investigate the effectiveness of nanocomposites (composed of ultra-high molecular weight polyethylene (UHMWPE) mixed with carbon nano-filler (CNF) and medical grade paraffin oil (PO), from the biological point of view. Wear measurements were carried out without (air) and with lubricant (distilled water, natural, and artificial lubricant), and antibacterial activity and cytotoxicity were evaluated. The results highlighted that the presence of CNF is important in the nanocomposite formulation because it reduces the wear rate and prevents oxidative degradation during its processing. An amount of 1.0 wt % of CNF is best because it reaches the optimal distribution within the polymeric matrix, resulting in the best wear resistant, bio-active, and anti-bacterial nanocomposite among all investigated samples.

Keywords: UHMWPE; nanocomposites; wear tests; FTIR; bio-activity

1. Introduction

Biomedical grade ultra high molecular weight polyethylene (UHMWPE) has high wear/abrasion resistance toughness and biocompatibility features and has been the standard material for joint replacement (JR) in artificial knees and hips prosthesis for more than half a century to date [1].

However, UHMWPE is considered the weakest part of the artificial joint: during its use, plastic debris is produced from the wear mechanism. Debris gives rise to adverse reactions, which lead to osteolysis and aseptic loosening of the prosthesis causing the failure of the entire joint, and hence the revision surgery [2]. After implantation in the human body, indeed, UHMWPE undergoes degrading mechanisms in the physiological environment and for the sliding friction against the other metallic or ceramic harder components of the artificial mobile prostheses [3–5]. For this reason, the research of the last years is highly focused on the improvement of mechanical resistance of this material during its use in vitro [6–9]. Highly cross-linked ultra-high-molecular-weight polyethylene was observed to be the most clinically promising in total joint arthroplasty owing to its high wear resistance [10].

Cross-linking of UHMWPE can be initiated by free radicals, which can be introduced by various methods such as exposure to ionizing radiation (i.e., gamma rays, electron beam) or the incorporation of chemical cross-linking agents such as peroxides or silanes [11–14]. Then, an anti-oxidation treatment is required to avoid material oxidation: thermal quenching, thermal annealing, or blending with antioxidant molecules in an opportune amount [15,16]. Vitamin E has been considered as an important antioxidant for UHMWPE, preventing its oxidative degradation, and increasing its wear resistance and fatigue [17]. Anyway, if on the one hand, vitamin E has been effective as a free radical scavenger, on the other hand, it could reduce the cross-linking efficiency [18]. Recently, Oral et al. 2019 combined the consolidation and cross-linking of UHMWPE in one step, such an opportunity to manufacture highly wear and oxidation-resistant joint implant-bearing surfaces with much improved toughness. The presence of Vitamin E in the UHMWPE formulation is important for its oxidative stabilization [19].

With the aim to enhance the mechanical properties of UHMWPE, many other alternative methods have been used, such as the mixing with reinforcements (particles or fibres) [20]. It is known that graphite acts as a lubricant to increase the wear resistance in friction components. For this reason, UHMWPE reinforced with carbon nanofibers (codified as CFR-UHMWPE, or "Poly II") has been used in orthopedic implants for total hip or total knee arthroplasty (THA/TKA), in 1970 [21]. However, this composite has been withdrawn because of its decreasing in crack resistance, of the fibers pull off on the surface, and other applicative problems. However, these nanocomposites are reconsidered in these last years, owing to the great innovations in the incorporation methods developed in the new experimentation, and also because of the cytocompatibility of the carbon nanofiller [22]. As an example, Puertolas et al. [23] evaluated the influence of carbon nanotubes (CNTs) and graphene as reinforcement fillers in UHMWPE matrix, while Yousef et al. [4] highlighted that carbon nanofiller can lead to an improvement in the wear behavior of biomedical grade polyethylene.

In our previous papers [4,24], we have shown the wear resistance features of nanocomposites made by UHMWPE mixed with 1.0 wt % of carbon nano-filler (CNF) and 2.0 wt % of medical grade paraffin oil (PO). In particular, the carbon nanofiller acts as lubricant while the paraffin oil reduces the typically high viscosity of UHMWPE, favoring its process ability and the filler dispersion [4]. These nanocomposites are innovative because they exhibited an appreciable wear resistance compared with that of graphene-filled nanocomposite, with the advantage of low cost [24].

In the present paper, we investigate the effectiveness of these nanocomposites (which had a wear behavior enhancement) from the biological point of view (antibacterial activity and cytotoxicity). Biocompatibility or cytotoxicity aspects need to be evaluated to determine if adverse reactions could have occurred as the body's response.

In the nanocomposites, UHMWPE is mixed with PO and CNF. UHMWPE is commonly classified as bio-inert material. PO is a mineral oil and is compatible with the human body; for this reason, it is used in medical applications as well as in cosmetics [25]. Carbon filler instead needs special attention because its particles could migrate, being dangerous for the human body. In particular, if carbon filler contains dangling carbon bonds on their surface, they are considered highly reactive. Similarly, if carbon filler contains traces of residual catalysts, these could be dangerous and reactive as well. Consequently, the high wear resistance of an UHMWPE based nano-composites containing carbon fillers is the key-factor to limit the migration of the filler. In fact, the more resistant to wear the material, the lower the carbon filler load that could migrate outside in the surrounding areas of the nanocomposite. An improvement of wear resistance is very important to minimize the danger of adverse effects on implants. For these reasons, and because to date there are no studies on this aspect of these nanocomposites related to their biological features, we present the wear and biological (bioactivity and anti-bacterial) features of these materials.

2. Materials and Methods

The control sample (code UP) was made of pure medical grade UHMWPE (code GUR1020 Ticona®, Sulzbach, Germany, molecular weight = $2–4 \times 10^6$ g/mol and density = 0.93 g/cm^3) mixed with 2.0 weight% of pharmaceutical grade Paraffin oil (Sella Pharmaceutical and Chemical Laboratory, Schio (Vi), Italy), added to UHMWPE to favor its mixing and processing.

Nanocomposites were obtained by mixing pure UHMWPE (code: U) with paraffin oil (2.0 weight%—code: P) and with carbon nanofiller (0.1–0.5–1.0–1.5–2.0 weight%—code: C) in a ball milling for 30 min at 20 Hz. The nanocomposites were identified with the code UPC with a number indicating the weight% of carbon nano-filler (code CNF). The mixing with ball mill was useful in order to induce a high degree of the filler dispersion inside the polymeric matrix, which improves both the mechanical and thermal features of the nanocomposites [5].

CNF powder was obtained by milling short carbon fibers supplied by Zoltek (Bridgeton, MO, USA) in a ball milling (mod. Retsch-MM301, 30 cycles of 10 min at 50 rpm, Retsch, Haan, Germany).

Polymeric sheets were prepared by compression molding (mod. PM 20/200, Campana S.R.L, Veduggio (MB), Italy) in a laboratory press at 200 °C for 20 min, at pressure of 20 MPa. The geometry was 60 mm × 60 mm, 2 mm thick.

More details about the material's preparations have already been explained in detail in our previously published paper [5].

2.1. Materials Characterization

A pin-on-disc wear tester was used to perform wear resistance measurements in different lubricants: no lubricant (air), distilled water, artificial (simulated synovial fluid), and natural lubricant (bovine serum) [24]. Both artificial and natural lubricants were chosen in order to approximate the biological conditions of a human joint. The polymeric samples employed in this test had square shapes: 20 mm × 20 mm, and 2 mm thick. The pin was a ruby corundum grinding stones (code: M.2145 and 3 mm in diameter), at room temperature. The use of such hard material allowed us to obtain the "accelerate wear test", in which we observed a noteworthy weight loss and improvement in the accuracy of the final result. The pin-on-disc system gives a circular shape wear trajectory with a testing load of 30 N, spin rate of 60 rpm, and test time of 120 min. Anyway, more details are given in a previous paper [4]. For each sample, the specific wear rate Ws (mm^3/Nm) was calculated [26]:

$$W_s = \frac{\Delta m}{\mathcal{P} x F n x L} \tag{1}$$

where Δm (mg) is the mass loss of the specimen, $\mathcal{P}$ (g/mL) is the density (see values listed in Table 1), Fn (N) is the normal load, and L (m) is the total sliding distance. The mass loss was evaluated by a high sensitivity electronic weighing balance (mod. Explorer pro EP 214C, ASTM D1505 International standard, OHAUS Corporation, Parsippany, NJ, USA, accuracy: 10^{-4} g).

Three tracks of different diameter were evaluated after the two hours in each sample and the Ws value was the average of these three tracks. Besides, the wear test was performed in four different media: no lubricant (air), distilled water, artificial lubricant (or simulated synovial fluid), and natural lubricant (or bovine serum). The final value of the specific wear rate Ws was determined with the average of the Ws values of n.9 polymeric sheets (for each nanocomposite obtained) in the four different media.

Data presentation and statistical analysis: the mean differences and standard deviations of specific wear rate of all the samples, with the CNF filler amount, in different media (air, distilled water, and artificial and natural lubricant), were calculated. The media (air, distilled water, and artificial and natural lubricant) and the CNF (%) nanocomposites were the independent variables. The data were first verified with the D'Agostino & Pearson test for the normality of the distribution and the Levene test for the homogeneity of variances. The data were normally distributed and homogenous; therefore, they were statistically analyzed using two-way analysis of variance (ANOVA) and Bonferroni post hoc test for multiple comparisons at a level of significance set at $p < 0.05$ (Prism 8.4.1; GraphPad Software, Inc., La Jolla, CA, USA).

Artificial lubricant was a simulated synovial fluid (SSF), formed by dissolving 0.3 wt % of hyaluronic acid in a phosphate buffered saline solution at pH = 7.4. The electrolyte concentration was as follows: Na^+ 153.1 mM, K^+ 4.2 mM, Cl^-139.6 mM, phosphate buffer 9.6 mM [24].

Bovine serum, or natural lubricant, is a synovial fluid extracted from the joint of a young bovine according to current legislation. It was kept in a refrigerator at −20 °C before each test.

Density was calculated by means of the precision balance before described, equipped like a hydrostatic system that follows the Archimede's principle. Each test was performed at room temperature for five minutes. The density (ρ) was evaluated from dry (P_{dry}) and wet (P_{wet}) weight of the sample, before and after the immersion in ethanol, with the following equation:

$$\rho = \frac{P_{dry}}{P_{dry} - P_{wet}} \rho_{eth} \tag{2}$$

where ρ_{eth} is equal to 0.790 g/cm^3. The resulting density value of each sample was the average of n = 3 measurements.

2.2. Fourier Transform Infrared (ATR/FTIR) Spectroscopy

The chemical composition of the different materials was analyzed by attenuated total reflectance Fourier transform infrared (ATR/FTIR) spectroscopy (Shimazu, Tokyo, Japan). The spectra were obtained with the Prestige-21 FTIR spectrometer equipped with an AIM-8800 infrared microscope (Shimadzu, Tokyo, Japan), using the incorporated 3mm diameter Ge attenuated total reflectance (ATR) semicircular prism. Furthermore, the spectra were recorded with an incident angle of 30° with a resolution of 4 cm^{-1} (64 scan) and were in the range of 650–4000 cm^{-1}. The Prestige software (IRsolution, version 1.10, Shimadzu, Tokyo, Japan) was used to further analyze the spectra.

Table 1. Values of reference UP and of the nanocomposites at different immersion times. U, ultra high molecular weight polyethylene (UHMWPE); P, paraffin oil; C, carbon nanofiller.

Sample Code	Density (g/mL)
UP	0.866 ± 0.001
UPC 0.1 wt %	0.864 ± 0.001
UPC 0.5 wt %	0.862 ± 0.002
UPC 1.0 wt %	0.862 ± 0.004
UPC 1.5 wt %	0.863 ± 0.002
UPC 2.0 wt %	0.865 ± 0.002

2.3. Fourier Transform Infrared (ATR/FTIR) Spectroscopy

In order to evaluate the bioactivity, the materials were soaked in simulated body fluid (SBF) with ion concentration nearly equal to those in human blood plasma [27].

Table 2. Body fluid (simulated body fluid, SBF) composition.

Ion	Concentration/mol m^3	
	SBF	Human Blood Plasma
Na^+	142.0	142.0
K^+	5.0	5.0
Mg^{2+}	1.5	1.5
Ca^{2+}	2.5	2.5
Cl^-	147.8	103.0
HCO_3^-	4.2	27.0
HPO_4^{2-}	1.0	1.0
SO_4^{2-}	0.5	0.5

The SBF was prepared from NaCl, $NaHCO_3$, KCl, $MgCl_2$, 1 M HCl, $CaCl_2{\cdot}6H_2O$, and Na_2SO_4 (Sigma-Aldrich, St. Louis, MO, USA) with a concentration that was suggested by Kokubo [27] (Table 2). The pH of the buffer was adjusted to pH 7.4 using 1 M HCl. The solution was exchanged every two days to avoid depletion of the ionic species in the SBF owing to the formation of biominerals. After 21 days of exposure at 37 °C, the samples were removed from the SBF and air-dried in a desiccator.

Fourier transform infrared (ATR/FTIR) spectroscopy (Shimadzu, Tokyo, Japan) was used to observe the characteristic peaks of hydroxyapatite layer on the surface of materials.

2.4. Antibacterial Activity

Escherichia coli, Gram-negative (ATCC 25922) and *Enterococcus faecalis*, Gram-positive (ATCC 29212) were used to evaluate the antibacterial properties of UHMWPE with 2.0 wt % of paraffin oil in function of different amounts of carbon nano-filler.

The bacterial culture was diluted in distilled water to produce a bacterial cell suspension of 10×10^5 CFU/mL (where CFU is for colony forming unit). *E. coli* and *E. faecalis* were inoculated in TBX medium (Tryptone Bile X-Gluc) (Liofilchem, Italy) and in Slanetz Bartley agar base (Liofilchem, Italy), respectively.

The materials were incubated against E. *coli* for 24 h at 36 °C and against E. *faecalis* for 24 h at 44 °C. The microbial growth was evaluated by observing the diameter of the inhibition halo (ID). The obtained values are the mean standard (SD) deviation of the measurements carried out on samples analyzed three times.

2.5. Cytotoxicity

3-(4,5-dimethylthiazol-2-yl)-2,5-diphenyl tetrazolium bromide (MTT) assay was used to determine the metabolic activity on NIH-3T3 murine fibroblast cells. For this purpose, cells were grown in Dulbecco's modified Eagle medium supplemented with 10% fetal bovine serum, 50.0 U/mL penicillin, and 100.0 µg/mL streptomycin, at 37 °C in a humidified atmosphere containing 5% CO_2. When cells were seeded at a density equal to 5.0×10^5 per well, onto six-well plates, they are directly exposed to disks of synthesized material. After 6, 12, and 24 h of incubation, cells were treated with MTT (500 µL; 0.50 mg/mL), previously dissolved in culture media, for 2 h at 37 °C in a 5% CO_2 humidified atmosphere. The MTT solution was then removed and dimethyl sulphoxide (DMSO) was added to dissolve the original formazan. Finally, the absorbance at 570 nm of each well was determined using a Victor3 Perkin Elmer fluorescence and absorbance reader. The cell viability was expressed as a percentage of mitochondrial redox activity of the cells directly exposed to powders, compared with an unexposed control.

3. Results and Discussion

3.1. Wear Behavior

Considering a comparison of the specific wear rate (Ws) of the control sample (UP) and the nanocomposites in different lubricating media reported in Figure 1, we have the following trend of Ws:

Air > Water > Artificial Lubricant > Natural Lubricant

Thus, the natural bovine serum has the best lubricant effect, while the air (or no lubricant) has the worst, as expected. The highest reduction occurs in natural lubricant, for all the samples, regardless of the CNF amount. In particular, the decrease was −78.03% in the UP control sample (from $47.3 \times \cdot 10^{-6}$ mm^3/Nm to $10.39 \cdot \times 10^{-6}$ mm^3/Nm, $p < 0.0001$) and −89.01% (from $40.78 \cdot \times 10^{-6}$ mm^3/Nm to $4.48 \cdot \times 10^{-6}$ mm^3/Nm, $p < 0.0001$) in the UPC 0.5 wt %, and −93.29% (from $37.91 \times \cdot 10^{-6}$ mm^3/Nm to $2.54 \times \cdot 10^{-6}$ mm^3/Nm, $p < 0.0001$) in the UPC 1.0 wt %. These data suggest that UPC 1.0 wt % nanocomposite exhibits the highest wear rate reduction, regardless of the lubricating media. In fact, for higher filler amounts, the wear rate improves again.

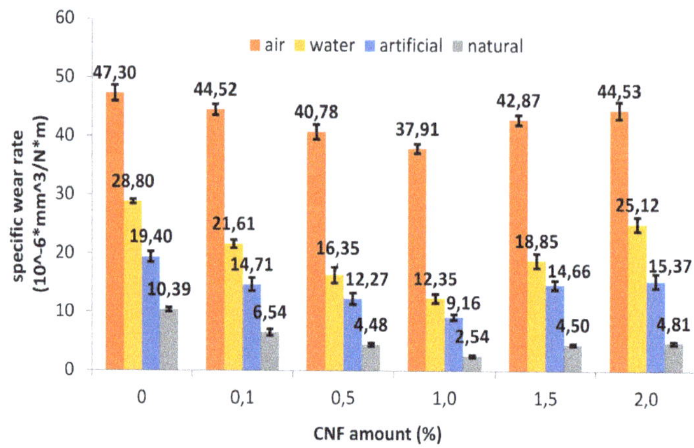

Figure 1. Wear rate of all of the samples vs. the carbon nano-filler (CNF) amount in different media (air, distilled water, and artificial and natural lubricant). The differences between the groups are statistically significant according to two-way analysis of variance (ANOVA) interaction and post hoc Bonferroni test, $p < 0.0001$).

This is because the filler dispersion changes with the different loads; at a low amount, such as 0.1 or 0.5 wt %, the dispersion is poor, while at higher amounts, such as 1.5 or 2.0 wt %, the great filler amount let to a filler agglomeration, which forms weak points that act as localized stress concentration points with an increases of specific wear rate improvement for an improvement of weight loss [28].The inferential analysis revealed statistically significant differences in the effect of the media (air, distilled water, and artificial and natural lubricant) on the nanocomposites ($p < 0.0001$). The statistical analysis carried out also showed that both variables (carbon nanofiller content and type of medium used) have a statistically significant effect on the wear data obtained (interaction, $p < 0.0001$).

3.2. Fourier Transform Infrared (ATR/FTIR) Spectroscopy

The chemical interactions among the components in the materials and its chemical structure were evaluated using ATR/FTIR spectroscopy. In Figure 2 are reported the spectra of all samples with different percentages of carbon nano-filler (curves b, c, d, e) compared with UHMWPE with paraffin oil and without carbon nano-filler (curve a).

In all spectra, the bands at 2930 cm^{-1} and 2847cm^{-1} are observed owing to the –CH–CH asymmetric and symmetric stretching modes [29,30]. In addition, the peak at 1465 cm^{-1} was attributed to –CH asymmetrical bending, while the C–C stretching vibrations could be assigned at 1242 cm^{-1} [29,30]. Furthermore, in the spectrum of UHMWPE with paraffin oil and without carbon nano-filler (curve a), the peaks at 1743 cm^{-1} and 721 cm^{-1} were detectable. The peak at 721 cm^{-1} was reported to be related to the asymmetric angular deformation of CH_2 groups in mineral oil [31]. The band at 1743 cm^{-1} could be attributable to an oxidative degradation process that occurs during the high temperature processing. In fact, the peak at 1743 cm^{-1} is attributed to the absorption of carbonyl species (C=O) [32], while oxygen bearing functionalities such as C–O stretching could be assigned at 1242 cm^{-1}. Observing the spectra of the polymer containing the different percentage of carbon nano-filler, the peak at 1743 cm^{-1} disappeared, and it was weakly detectable in UPC 2.0 wt % CNF spectrum (Figure 2 panel E). This could be because of the ability of carbon nano-filler (CNF)to improve resistance to oxidative degradation. Indeed, the C–O stretching vibration was also more pronounced in UPC 2.0 wt % CNF spectrum. The different change in shape and absorbance of the peak at 1242 cm^{-1} could be the result of CNF C–C stretching modes overlapping [33].

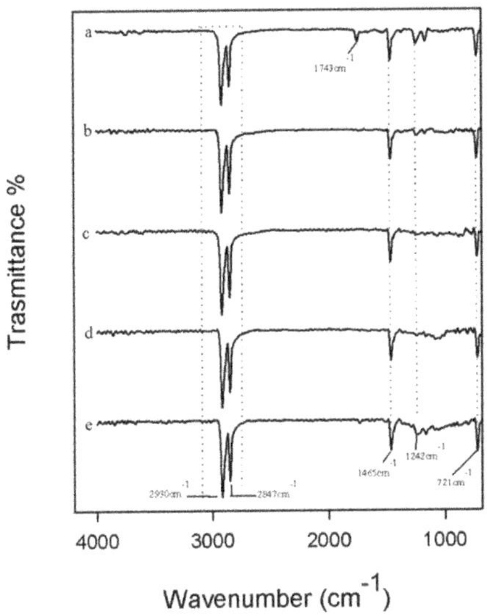

Figure 2. Total reflectance Fourier transform infrared (ATR/FTIR) spectroscopy spectra of (**a**) UP; (**b**) UPC 0.1 wt % CNF; (**c**) UPC 0.5 wt % CNF; (**d**) UPC 1.0 wt % CNF; and (**e**) UPC 2.0 wt % CNF. U, ultra high molecular weight polyethylene (UHMWPE); P, paraffin oil; C, carbon nanofiller.

3.3. *Bioactivity Test*

The bioactivity properties of the different materials were evaluated using Kokubo's test. The materials were soaked in simulated body fluid (SBF) for 21 days, and after this exposure time, the formation of the hydroxyapatite on the surfaces of all samples was detected by ATR/FTIR.

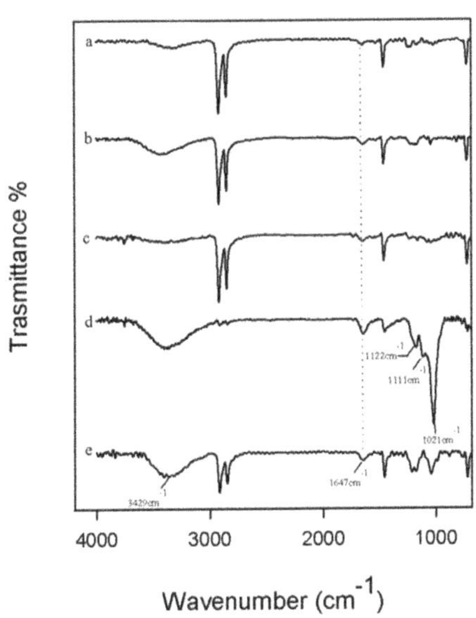

Figure 3. Total reflectance Fourier transform infrared (ATR/FTIR) spectroscopy spectra of (**a**) UP; (**b**) UPC 0.1 wt % CNF; (**c**) UPC 0.5 wt % CNF; (**d**) UPC 1.0 wt % CNF; and (**e**) UPC 2.0 wt % CNF after 21 days in simulated body fluid (SBF).

In Figure 3, the spectra of all materials are reported, in which it is possible to observe the peaks at 3429 and 1647 cm^{-1} that correspond to the –OH groups [34,35]. Comparing all materials, in the spectra of UHMWPE with paraffin oil and with 1.0 wt % of carbon nano-filler (curve d), the typical bands of the hydroxyapatite layer are clearly visible. In fact, the peak at 1021 cm^{-1} is the result of the absorption of the stretching mode of the PO_4^{-3} ions [36–38]. In addition, the P–O bond of the phosphate group appears at 1122 and 1111 cm^{-1} [39]. Furthermore, only in this spectrum, the typical peaks of the polymer are not very intense, which is probably because the material's surface is covered by a hydroxyapatite layer that is very thick compared with the others [40,41]. In conclusion, all materials seem to be bioactive because the hydroxyapatite bands are present in all spectra, but when carbon nano-filler is added to 1.0 wt %, the hydroxyapatite peaks are more intense, suggesting that this material is the most bioactive of all.

According to the wear resistance test result, the optimal wear performance and optimal bioactivity are noticed in the same nanocomposite, containing 1.0 wt % of carbon nano-filler. This result confirms that the filler dispersion is a key parameter to achieve the best mechanical as well as biological response, as discussed in Section 3.1.

3.4. Antibacterial Activity

The materials with the same percentage of paraffin oil and several amounts of carbon nano-filler were incubated with gram-positive and gram-negative bacterial strains. In Figure 4A, the representative images of both bacteria are reported, while the bacterial growth is shown in Figure 4B. The UP material with only 2.0 wt % of paraffin oil was used as control in order to evaluate the effects of carbon nano-filler on bacterial growth. In all plates, after 24 h of incubation against E. coli and E. faecalis, the inhibition halo (ID) was not observed, regardless of the carbon nano-filler content. Therefore, the results show that the materials do not have toxic effects against gram-positive and gram-negative bacterial strains. In addition, the materials during the incubation period did not release any product that could damage the bacteria growth, suggesting that they could be used as optimal materials for joint replacement (JR) in artificial knees and hips prosthesis.

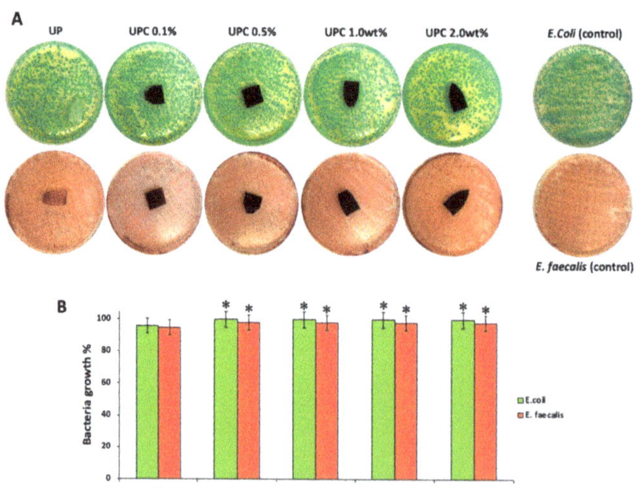

Figure 4. (**A**) Representative image of E. coli and E. faecalis incubated with all materials. (**B**) Bacteria growth (%) of E. coli and E. faecalis incubated with UP, UPC 0.1 wt %, UPC 0.5 wt %, UPC 1.0 wt %, and UPC 2.0 wt %. Values are the mean SD of measurements carried out on samples analyzed three times. The means and S.D. are shown. * $p < 0.05$ vs. the bacteria control treated without materials.

3.5. Cytotoxicity

The nanocomposites were investigated for their biocompatibility through MTT-direct contact test. The data (Figure 5) acquired highlighted that all the samples avoided cytotoxic effects, and that the increase in amount percentage in carbon nano-filler positively affected NIH-3T3 viability. In fact, all the samples appeared to induce a cell growth increase with respect to UP material, reaching its maximum by the UPC1% sample. This latter showed a percentage increase in cell viability equal to 10.2% compared with UP. Thus, the incorporation of CNFs in a small dose level percentage seemed to not exert evident redox mitochondrial modification. Previous studies evidenced that multiwall carbon nanotubes caused a significant time- and dose-dependent decrease in cells' viability [42,43], and that they could be toxic at sufficiently high concentrations. Indeed, it was reported that carbon nanomaterials present significantly different cytotoxicity depending on their physicochemical properties, including size, length, shape, and surface area [44–46]. The lack of filler coalescence, defined as a consequence of CNF incorporation, suggested the constitution of an improved UHMWPE-based material, in which the absence of debris could favour the slowing down of inflammatory and oxidative stress effects commonly ascribed to these materials [47–52].

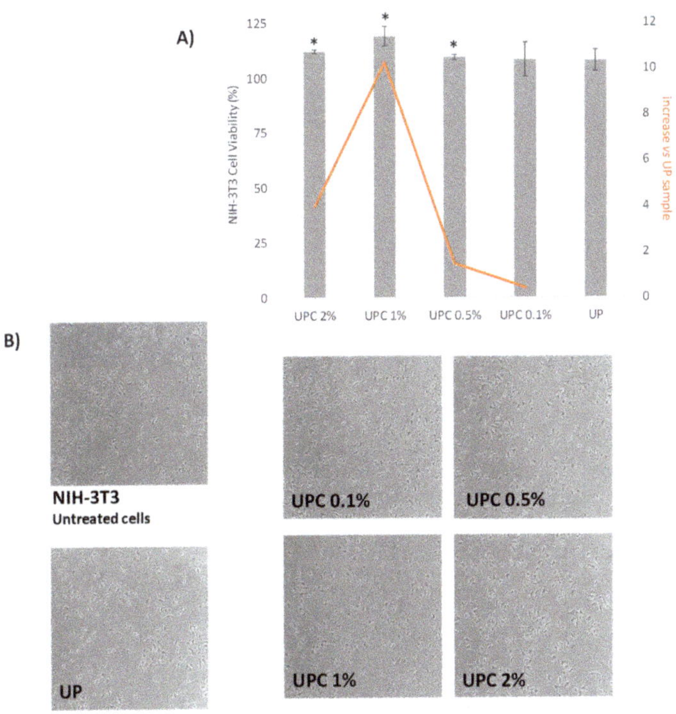

Figure 5. (**A**) Data from MTT assay expressed as NIH-3T3 cells viability (%) vs. untreated cells. Values are the mean ± SD of two independent experiments performed in triplicate. * $p < 0.05$ vs. untreated cells. The increase in cell viability vs. UP sample is also shown. (**B**) Representative cell morphology images of NIH-3T3 cells after treatment with investigated samples. Images were acquired by Inverted Phase Contrast Brightfield Zeiss Primo Vert Microscope.

4. Conclusions

In this paper, we investigated the biological and antibacterial activity of wear resistant nanocomposites based on biomedical grade UHMWPE paraffin oil and carbon nano-filler. Paraffin oil favors the workability of the typical high viscosity UHMWPE and the filler dispersion within the polymer. Carbon nanofiller act as lubricant to lower the friction between the UHMWPE and other counterparts. Furthermore, carbon nano-filler prevents the oxidative degradation of the nanocomposite during its processing, as observed by the FTIR analysis.

The wear test was performed in different lubricating media (no lubricant or air, distilled water, simulated synovial fluid, and bovine serum). The lubricating effect was in the following order: natural lubricant > artificial lubricant > water > air. The more wear resistant nanocomposite resulted it being reinforced with 1.0 wt % of CNF regardless of the lubricating media because of the best filler dispersion within the polymeric matrix. A filler load lower than 1.0 weight percentage, as well as a higher one, are too little and too much, respectively, in order to achieve a homogeneous dispersion and to avoid filler coalescence, respectively. This aspect also reflects the bio-activity response of the materials, which has the best hydroxyapatite production in just the 1.0 wt % sample.

This suggests that it could be applicable in artificial prosthesis. In order to verify in more depth the suitability of these nanocomposites, further studies are underway on the release of components over time into biological fluids.

Author Contributions: For research articles with several authors, a short paragraph specifying their individual contributions must be provided. The following statements should be used "conceptualization, M.C. and A.V.; methodology, M.C., S.P.; software, C.S.; validation, M.C., A.V. and S.P.; formal analysis, C.S., and G.D.P.; investigation, C.S.; resources, A.V.; data curation; M.C.; writing—original draft preparation, A.V.; writing—review and editing, M.C.; visualization, S.P.; supervision, M.C., A.V. All authors have read and agreed to the published version of the manuscript.

Funding: The work was financially supported by V:ALERE 2019 grant support from University of Campania "Luigi Vanvitelli" of CHIMERA.

Acknowledgments: Authors thank dott. Eugenio Pedullà of General surgery and medical-surgical specialties Department A.O.U. Policlinico Vittorio Emanuele-Catania, Italy for his help in conducting statistical analyses.

Conflicts of Interest: The authors declare no conflict of interest.

References

1. Gomez-Barrena, E.; Puertolas, J.-A.; Munuera, L.; Konttinen, Y.T. Update on UHMWPE research from the bench to the bedside. *ActaOrthop* **2008**, *79*, 832–840. [CrossRef] [PubMed]
2. Zhou, J.; Chakravartula, A.; Pruitt, L.; Komvopoulos, K. Tribological and Nanomechanical Properties of Unmodified and Crosslinked Ultra-High Molecular Weight Polyethylene for Total Joint Replacements. *J. Tribol.* **2004**, *126*, 386. [CrossRef]
3. Ge, S.; Kang, X.; Zhao, Y. One-year biodegradation study of UHMWPE as artificial joint materials: Variation of chemical structure and effect on friction and wear behavior. *Wear* **2011**, *271*, 2354–2363. [CrossRef]
4. Yousef, S.; Visco, A.M.; Galtieri, G.; Nocita, D.; Espro, C. Wear behaviour of UHMWPE reinforced by carbon nanofiller and paraffin oil for joint replacement. *Mater. Sci. Eng. C* **2017**, *73*, 234–244. [CrossRef] [PubMed]
5. Visco, A.; Yousef, S.; Galtieri, G.; Nocita, D.; Njuguna, J. Thermal, Mechanical and Rheological Behaviors of Nanocomposites Based on UHMWPE/Paraffin Oil/Carbon Nanofiller Obtained by Using Different Dispersion Techniques. *JOM-Springer* **2016**, *68*, 1078–1089. [CrossRef]
6. Atwood, S.A.; van Citters, D.W.; Patten, E.W.; Furmanski, J.; Ries, M.D.; Pruitt, L.A. Tradeoffs amongst fatigue, wear, and oxidation resistance of cross-linked ultra-high molecular weight polyethylene. *J. Mech. Behav. Biomed. Mater.* **2011**, *4*, 1033–1045. [CrossRef] [PubMed]
7. Lewis, G.; Fencl, R.M.; Carroll, M.; Collins, T. The relative influence of five variables on the in vitro wear rate of uncrosslinked UHMWPE acetabular cup liners. *Biomaterials* **2003**, *24*, 1925–1935. [CrossRef]
8. Bruck, A.L.; KanagaKaruppiah, K.S.; Sundararajan, S.; Wang, J.; Lin, Z. Friction and wear behavior of ultrahigh molecular weight polyethylene as a function of crystallinity in the presence of the phospholipid dipalmitoyl phosphatidylcholine. *J. Biomed. Mater. Res. Part. B Appl. Biomater.* **2010**, *93B*, 351–358. [CrossRef]
9. Visco, A.M.; Campo, N. Properties of nanocomposites based on polyethylene (UHMWPE) and carbon nanotubes mixed by high energy ball milling and UV-source irradiated. *Int. J. Polym. Anal. Charact.* **2012**, *17*, 144–157.

10. Muratoglu, O.K.; Bragdon, C.R.; O'Connor, D.O.; Jasty, M.; Harris, W.H.; Gul, R.; McGarry, F. Unified wear model for highly crosslinked ultra-high molecular weight polyethylenes (UHMWPE). *Biomaterials* **1999**, *20*, 1463–1470. [CrossRef]
11. Visco, A.M.; Campo, N.; Torrisi, L.; Cristani, M.; Trombetta, D.; Saija, A. Electron beam irradiated UHMWPE: Degrading action of air and hyaluronic acid. *Bio-Med. Mater. Eng.* **2008**, *18*, 137–148. [CrossRef]
12. Gul, R.M. The effects of peroxide content on the wear behavior, microstructure and mechanical properties of peroxide cross-linked UHMWPE used in total hip replacement. *J. Mater. Sci. Mater. Med.* **2008**, *19*, 2427–2435. [CrossRef] [PubMed]
13. Atkinson, J.R.; Cicek, R.Z. Silanecrosslinked polyethylene for prosthetic applications Part I. Certain physical and mechanical properties related to the nature of the material. *Biomaterials* **1983**, *4*, 267–275. [CrossRef]
14. Atkinson, J.R.; Cicek, R.Z. Silanecrosslinked polyethylene for prosthetic applications: II. Creep and wear behaviour and a preliminary moulding test. *Biomaterials* **1984**, *5*, 326–335. [CrossRef]
15. Oral, E.; GodleskiBeckos, C.; Malhi, A.S.; Muratoglu, O.K. The effects of high dose irradiation on the cross-linking of vitamin E-blended ultrahigh molecular weight polyethylene. *Biomaterials* **2008**, *29*, 3557–3560. [CrossRef]
16. Visco, A.M.; Campo, N.; Brancato, V.; Trimarchi, M. Influence of thea-tocopherol load and the annealing treatment on the wear resistance of biomedical UHMWPE radiated with electron beam. *Int. J. Polym. Anal. Charact.* **2013**, *18*, 545–556. [CrossRef]
17. Oral, E.; Christensen, S.D.; Malhi, A.S.; Wannomae, K.K.; Muratoglu, O.K. Wear Resistance and Mechanical Properties of Highly Cross-linked, Ultrahigh–Molecular Weight Polyethylene Doped With Vitamin E. *J. Arthroplast.* **2006**, *21*, 580–591. [CrossRef]
18. Oral, E.; GodleskiBeckos, C.; Lozynsky, A.J.; Malhi, A.S.; Muratoglu, O.K. Improved resistance to wear and fatigue fracture in high pressure crystallized vitamin E-containing ultra-high molecular weight polyethylene. *Biomaterials* **2009**, *30*, 1870–1880. [CrossRef]
19. Oral, E.; Doshi, B.N.; Fung, K.; O'Brien, C.; Wannomae, K.K.; Muratoglu, O.K. Chemically Cross-Linked UHMWPE With Superior Toughness. *J. Orthop. Res.* **2019**, *37*, 2182–2188. [CrossRef]
20. Hussain, M.; Naqvi, R.A.; Abbas, N.; Khan, S.M.; Nawaz, S.; Hussain, A.; Zahra, N.; Khalid, M.W. Ultra-High-Molecular-Weight-Polyethylene (UHMWPE) as a Promising Polymer Material for Biomedical Applications: A Concise Review. *Polymers* **2020**, *12*, 323. [CrossRef]
21. Panin, S.V.; Kornienko, L.A.; Nguyen, D.A.; Alexenko, V.O.; Ivanova, L.R. Enhancement of mechanical and tribotechnical properties of polymer composites with thermoplastic UHMWPE and PEEK matrices by loading carbon nanofibers/nanotubes. In Proceedings of the 11th International Conference on Mechanics, Resource and Diagnostics of Materials and Structures, AIP Conference Proceedings Mechanics, Resource and Diagnostics of Materials and Structures (Mrdms-2017), Ekaterinburg, Russia, 11–15 December 2017; Eduard, S., Gorkunov, V.E., Sunder, R., Eds.; American Institute of Physics: College Park, MD, USA, 2017; Volume 1915, p. 030015.
22. Kurtz, S.M. *The UHMWPE Handbook*, 1st ed.; Elsevier Academic Press: San Diego, CA, USA; London, UK, 2004.
23. Puértolas, J.A.; Kurtz, S.M. Evaluation of carbon nanotubes and graphene as reinforcements for UHMWPE based composites in arthroplastic applications: A review. *J. Mech. Behav. Biomed. Mater.* **2014**, *39*, 129–145. [CrossRef] [PubMed]
24. Visco, A.; Yousef, S.; Scolaro, C.; Espro, C.; Cristani, M. Tribological Behavior of Nanocomposites Based on UHMWPE Aged in Simulated Synovial Fluid. *Polymers* **2018**, *10*, 1291. [CrossRef] [PubMed]
25. Fiorito, S.; Serafino, A.; Andreola, F.; Togna, A.; Togna, G. Toxicity and biocompatibilità of carbon nanoparticles. *J. Nanosci. Nanotech.* **2006**, *6*, 1–9. [CrossRef] [PubMed]
26. Agarwal, G.; Patnaik, A.; Sharma, R.K. Parametric optimization and three-body abrasive wear behavior of sic chopped glass fiber reinforced epoxy composites. *Int. J. Compos. Mater.* **2013**, *3*, 32–38.
27. Kokubo, T.; Takadama, H. How useful is SBF in predicting in vivo bone bioactivity? *Biomaterials* **2006**, *27*, 2907–2915. [CrossRef]
28. Harsha, A.P.; Joyce, Y.J. Challenges associated with using bovine serum in wear testing orthopaedic biopolymers. *J. Eng. Med.* **2011**, *225*, 948–958. [CrossRef]
29. Ahmadi, M.; Masoomi, M.; Safi, S.; Zabihi, O. Interfacial evaluation of epoxy/carbon nanofiber nanocomposite reinforced with glycidyl methacrylate treated UHMWPE fiber. *J. Appl. Polym. Sci.* **2016**, *133*. [CrossRef]

30. Barick, A.K.; Tripathy, D.K. Effect of nanofiber on material properties of vapor-grown carbon nanofiber reinforced thermoplastic polyurethane (TPU/CNF) nanocomposites prepared by melt compounding. *Compos. Part. A Appl. Sci. Manuf.* **2010**, *41*, 1471–1482. [CrossRef]
31. Houmard, M.; Nunes, E.H.M.; Vasconcelos, D.C.L.; Berthomé, G.; Joud, J.-C.; Langlet, M.; Vasconcelos, W.L. Correlation between sol–gel reactivity and wettability of silica films deposited on stainless steel. *Appl. Surf. Sci.* **2014**, *289*, 218–223. [CrossRef]
32. Azam, A.M.; Ali, A.; Khan, H.; Yasin, T.; Mehmood, M.S. Analysis of degradation in UHMWPE a comparative study among the various commercial and laboratory grades UHMWPE. In *IOP Conference Series: Materials Science and Engineering*; IOP Publishing: Bristol, UK, 2016.
33. Ma, H.; Zeng, J.; Realff, M.L.; Kumar, S.; Schiraldi, D.A. Processing, structure, and properties of fibers from polyester/carbon nanofiber composites. *Compos. Sci. Technol.* **2003**, *63*, 1617–1628. [CrossRef]
34. Ohtsuki, C.; Kokubo, T.; Yamamuro, T. Mechanism of apatite formation on $CaOSiO_2P_2O_5$ glasses in a simulated body fluid. *J. Non-Cryst. Solids* **1992**, *143*, 84–92. [CrossRef]
35. Catauro, M.; Barrino, F.; Poggetto, G.D.; Crescente, G.; Piccolella, S.; Pacifico, S. New SiO_2/Caffeic acid hybrid materials: Synthesis, spectroscopic characterization, and bioactivity. *Materials* **2020**, *13*, 394. [CrossRef] [PubMed]
36. Koutsopoulos, S. Synthesis and characterization of hydroxyapatite crystals: A review study on the analytical methods. *J. Biomed. Mater. Res.* **2002**, *62*, 600–612. [CrossRef] [PubMed]
37. Hong, L.; Wang, Y.L.; Jia, S.R.; Huang, Y.; Gao, C.; Wan, Y.Z. Hydroxyapatite/bacterial cellulose composites synthesized via a biomimetic route. *Mater. Lett.* **2006**, *60*, 1710–1713.
38. Catauro, M.; Tranquillo, E.; Poggetto, G.D.; Pasquali, M.; Dell'Era, A.; VecchioCiprioti, S. Influence of the heat treatment on the particles size and on the crystalline phase of TiO_2 synthesized by the sol-gel method. *Mataerials* **2018**, *11*, 2364. [CrossRef]
39. Klee, W.E.; Engel, G. IR spectra of the phosphate ions in various apatites. *J. Inorg. Nuclear Chem.* **1970**, *32*, 1837–1843. [CrossRef]
40. Kokubo, T.; Ito, S.; Huang, Z.T.; Hayashi, T.; Sakka, S.; Kitsugi, T.; Yamamuro, T. Ca, P-rich layer formed on high-strength bioactive glass-ceramic A-W. *J. Biomed. Mater. Res.* **1990**, *24*, 331–343. [CrossRef]
41. Kokubo, T.; Kushitani, H.; Sakka, S.; Kitsugi, T.; Yamamuro, T. Solutions able to reproduce in vivo surface-structure changes in bioactive glass-ceramic A-W3. *J. Biomed. Mater. Res.* **1990**, *25*, 721–734. [CrossRef]
42. Patlolla, A.; Knighten, B.; Tchounwou, P. Multi-walled carbon nanotubes induce cytotoxicity, genotoxicity and apoptosis in normal human dermal fibroblast cells. *Ethn Dis.* **2010**, *20* (Suppl. 1), S1–65–72.
43. Catauro, M.; Papale, F.; Bollino, F.; Picolella, S.; Marciano, S.; Nocera, P.; Pacifico, S. Silica/quercetin sol-gel hybrids as antioxidant dental implant materials. *Sci. Technol. Adv. Mater.* **2015**, *16*, 035001. [CrossRef]
44. Yuan, X.; Zhang, X.; Sun, L.; Wei, Y.; Wei, X. Cellular Toxicity and Immunological Effects of Carbon-based Nanomaterials. *Part. Fibre Toxicol.* **2019**, *16*, 18. [CrossRef] [PubMed]
45. Ricciotti, L.; Roviello, G.; Tarallo, O.; Borbone, F.; Ferone, C.; Colangelo, F.; Catauro, M.; Cioffi, R. Synthesis and characterizations of melamine-based epoxy resins. *Int. J. Mol. Sci.* **2013**, *14*, 18200–18214. [CrossRef] [PubMed]
46. Pacifico, S.; Piccolella, S.; Papale, F.; Nocera, P.; Lettieri, A.; Catauro, M. A polyphenol complex from Thymus vulgaris L. plants cultivated in the Campania Region (Italy): New perspectives against neuroblastoma. *J. Funct. Foods* **2016**, *20*, 253–266. [CrossRef]
47. Gazzano, E.; Bracco, P.; Bistolfi, A.; Aldieri, E.; Ghigo, D.; Boffano, M.; Costa, L.; del Prever, E.B. Ultra high molecular weight polyethylene is cytotoxic and causes oxidative stress, even when modified. *Int. J. Immunopathol. Pharmacol.* **2011**, *24* (Suppl. 2), 61–67. [CrossRef] [PubMed]
48. Nguyen, K.; Garcia, A.; Sani, M.-A.; Diaz, D.; Dubey, V.; Clayton, D.; Poggetto, G.D.; Cornelius, F.; Payne, R.J.; Separovic, F.; et al. Interaction of N-terminal peptide analogues of the Na+,K+-ATPase with membranes. *BBA—Biomembranes* **2018**, *1860*, 1282–1291. [CrossRef]
49. Kiventer, J.; Lancellotti, I.; Catauro, M.; Poggetto, F.D.; Leonelli, C.; Illikainen, M. Alkali activation as new option for gold mine tailings inertization. *J. Clean. Prod.* **2018**, *187*, 76–84. [CrossRef]
50. Samal, S.; Kolinova, M.; Rahier, H.; Poggetto, G.D.; Blanco, I. Investigation of the internal structure of fiber reinforced geopolymer composite under mechanical impact: A micro computed tomography (μCT) study. *Appl. Sci.* **2019**, *9*, 516. [CrossRef]

51. Torrisi, L.; Visco, A.; Campo, N. Pulsed laser treatments of polyethylene films. *Nucl. Instrum. Meth. B* **2010**, *268*, 3117–3121. [CrossRef]
52. Cioffi, R.; Aronne, P.P.A.; Catauro, M.; Quattroni, G. Glass-Ceramics from fly ash with added Li_2O. *J. Eur. Ceram. Soc.* **1994**, *13*, 143–148. [CrossRef]

polymers

MDPI

Review

History of Cyclodextrin Nanosponges

Ilona Krabicová [1,†], Silvia Lucia Appleton [2,†], Maria Tannous [2], Gjylije Hoti [2], Fabrizio Caldera [2], Alberto Rubin Pedrazzo [2], Claudio Cecone [2], Roberta Cavalli [3] and Francesco Trotta [2,*]

1 Department of Nanochemistry, Advanced Technologies and Innovation, Technical Institute for Nanomaterials, University of Liberec, Studentská 1402/2, 461 17 Liberec, Czech Republic; ilona.krabicova@tul.cz

2 Department of Chemistry, University of Torino, via P. Giuria 7, 10125 Torino, Italy; silvialucia.appleton@unito.it (S.L.A.); maria_tannous@hotmail.com (M.T.); gjylije.hoti@unito.it (G.H.); fabrizio.caldera@unito.it (F.C.); alberto.rubinpedrazzo@unito.it (A.R.P.); claudio.cecone@unito.it (C.C.)

3 Department of Drug Science and Technology, University of Torino, via P. Giuria 9, 10125 Torino, Italy; roberta.cavalli@unito.it

* Correspondence: francesco.trotta@unito.it

† Both authors contributed equally to this work.

Received: 19 April 2020; Accepted: 7 May 2020; Published: 14 May 2020

Abstract: Nowadays, research in the field of nanotechnology and nanomedicine has become increasingly predominant, focusing on the manipulation and development of materials on a nanometer scale. Polysaccharides have often been used as they are safe, non-toxic, hydrophilic, biodegradable and are low cost. Among them, starch derivatives and, in particular, cyclodextrin-based nanosponges (CD NSs) have recently emerged due to the outstanding properties attributable to their peculiar structure. In fact, alongside the common polysaccharide features, such as the presence of tunable functional groups and their ability to interact with biological tissues, thus giving rise to bioadhesion, which is particularly useful in drug delivery, what makes CD NSs unique is their three-dimensional network made up of crosslinked cyclodextrin units. The name "nanosponge" appeared for the first time in the 1990s due to their nanoporous, sponge-like structure and responded to the need to overcome the limitations of native cyclodextrins (CDs), particularly their water solubility and inability to encapsulate charged and large molecules efficiently. Since CD NSs were introduced, efforts have been made over the years to understand their mechanism of action and their capability to host molecules with low or high molecular weight, charged, hydrophobic or hydrophilic by changing the type of cyclodextrin, crosslinker and degree of crosslinking used. They enabled great advances to be made in various fields such as agroscience, pharmaceutical, biomedical and biotechnological sectors, and NS research is far from reaching its conclusion. This review gives an overview of CD NS research, focusing on the origin and key points of the historical development in the last 50 years, progressing from relatively simple crosslinked networks in the 1960s to today's multifunctional polymers. The approach adopted in writing the present study consisted in exploring the historical evolution of NSs in order to understand their role today, and imagine their future.

Keywords: history; cyclodextrin nanosponge; crosslinked polymer

1. Introduction

Cyclodextrins (CDs) are natural oligosaccharides widely used in numerous fields, including biomedicine, cosmetics, food industry, wastewater remediation and catalysis. The popularity of CDs is mainly attributable to their inclusion capacity and ability to improve desired physico-chemical properties of guest molecules, such as apparent solubility and stability.

However, native CDs have some limitations, among all solubility. Time-consuming and expensive separation techniques would be necessary to recover the CDs from an aqueous environment. When CD polymers came on the scene, this limit was overcome and their solubility could be tuned by changing the degree of crosslinking. Nowadays, CD insoluble polymers are usually called "nanosponges" (NS), referring to their sponge-like structure, which has high porosity and capacity of entrapping various kinds of molecules into the matrix [1].

Being insoluble is not the only advantage of CD NSs. In fact, while internal cavities of CDs can host hydrophobic molecules, the interstitial pores present between crosslinker units and the external walls of CDs make NSs capable of also entrapping hydrophilic molecules [2,3]. This ability has triggered extensive research, and CD NSs have emerged as a promising material in various fields such as environmental, enzymological, agricultural, biomedical, catalytical and pharmaceutical applications, as well as in gas storage, flame retardants, etc. [4–9], which will be discussed in depth in this review.

Today's main field of investigation is nanomedicine due to the target reached by nanoparticles as drug delivery systems [10]. Nanoparticles can be prepared using polymers, which must at least be biocompatible and are better if biodegradable. Use has been made of many different polymeric materials, such as polylactic and polyglycolic acids, polycaprolactone, polyacrylic acids, proteins, polypeptides (gelatin) and polysaccharides, which include CD NSs [10].

The advantages of using polysaccharides are that they can be easily modified chemically due to the presence of derivable groups on the molecular chains, and they are safe, non-toxic, hydrophilic, biodegradable and are low cost as they are readily available in nature. In addition, the hydrophilic groups contained in most natural polysaccharides, such as hydroxyl, carboxyl and amino groups, which may form non-covalent bonds with biological tissues (mainly epithelia and mucous membranes), thus giving rise to bioadhesion, prolong the residence time of the encapsulated drug and solve the bioavailability problems. These properties combined with the capability of NSs to carry a wide variety of drugs make NSs the therapeutic nanocarriers of choice.

A comprehensive overview of the current state of these NS drug delivery systems is provided in the review conducted by Caldera et al. [11], in which they were classified into four generations.

The 1st generation comprises urethane, carbonate, ether and ester NSs synthesized by reacting CDs with a crosslinking agent. In the 2nd generation, there are polymers with specific properties, e.g., fluorescence or charged side chains. The 3rd generation contains stimuli-responsive NSs modifying their behavior according to changes in the environment, such as pH, temperature gradients or oxidative/reducing conditions. The 4th generation includes molecularly imprinted NSs with high selectivity towards specific guest molecules.

NSs have proven capable of keeping up with the advances in nanomedicine, responding positively to the need for targeted treatments, aimed at improving the efficacy and reducing the adverse effects of the drugs. This approach is called "targeted drug delivery" and consists of assisting the drug molecule to preferably reach the desired site (cell/tissue/organ) [12]. Much consideration has, therefore, been given in pharmaceutical research to nanoparticles, including NSs, able to perform such delivery. The potential of nanosponges, in particular, is attributable to the presence of functional groups on their surface, which has led to their being recently used for grafting biological ligands capable of binding to specific receptors on the surface of the target cells, thus increasing cellular uptake of the drug encapsulated inside the NSs [12] and, consequently, the possibility of improving therapeutic efficacy.

These NSs functionalized with biological ligands may be considered as belonging to a new, fifth generation.

Today's NSs are the result of intense research conducted over the years. They have progressed from the relatively simple crosslinked networks of the 1960s to today's multifunctional polymers.

The synthesis of NSs over the years has evolved towards greener processes, starting from the use of organic solvents, which made the washing step of NSs necessary because of the toxicity, passing through water to the most recent solvent-free synthesis [13]. NS formulations have also been optimized by means of mathematical tools to ensure the best solution for the delivery of drugs [14–17]. The number of analytical tools has increased in order to improve the characterization of NSs and better understand their properties and behavior.

The purpose of this review is to provide an overview of research on CD NSs, by exploring their historical development to make the reader aware of how they were developed and used to meet the changing needs of the period, and highlighting their properties and enormous potential for the future.

We believe that studying the history of nanosponges is important because it allows us to understand the past, which, in turn, enables us to understand the present and possibly have a glimpse of the future opportunities of such innovative and promising polymers. As Winston Churchill said, "The farther backward you can look, the farther forward you are likely to see".

2. Origin and Historical Development of Nanosponges Over the Last 50 Years

This review describes in detail the research related to crosslinked insoluble CD polymers from their origin to today's multifunctional polymers.

2.1. From the 1960s to the 1980s: Origin of Insoluble Crosslinked Cyclodextrin Polymers

The history of crosslinked insoluble CD polymers dates back to 1965 when Solms and Egli published a study on the preparation and inclusion properties of novel network polymers made up of CDs crosslinked with epichlorohydrin (EPI). Firstly, CDs dissolved in water were activated with sodium borohydride and hot solution of sodium hydroxide, and then EPI was added as a crosslinker. The authors studied the binding properties of this novel material in comparison to EPI-dextran network polymers. It was capable of including iodine and several organic compounds, aniline, pyridine, benzaldehyde, butyric acid, *p*- and *o*-nitrophenol are a few examples. This study suggested that the inclusion ability could be useful in separation techniques based not only on separating molecules on the basis of their size but also of their shape, an example of which is the separation of *p*- from *o*-nitrophenol and of differences in their inclusion behavior [18]. This idea was further developed in the 1970s with the aim to prepare novel stationary phases for nucleic acids [19] and mandelic acid derivatives [20].

2.2. From the 1980s to the 1990s: Investigation on Polymer Properties and Applications

In the 1980s, research was more focused on exploring and understanding the properties of these new materials, rather than investigating only their binding abilities. In 1980, insoluble porous polymers such as polyurethane-CD network polymers were examined. They were obtained by using hexamethylene diisocyanate (HMDI), 1,3-bis (isocyanatomethyl) cyclohexane and 1,3-bis (isocyanatomethyl) benzene as crosslinkers. The synthesis consisted of stirring CDs with diisocyanate in pyridine subjected to heating. The properties of these new polymers were investigated more in depth compared to previous studies: thermal analysis, Brunauer, Emmett and Teller (BET) measurement of surface area, elemental analysis (detection of unreacted –OH groups in CD) and gas chromatography were employed to study the interactions between CD polymers and organic compounds (e.g., benzene, toluene, cyclohexane, ethanol, methyl ethyl ketone, propanol and others) [21]. Moreover, the influence of the type of crosslinker (1,2-ethanediol diglycidyl ether, 1,4-butanediol diglycidyl ether, 1,6-hexanediol diglycidyl ether, 1,3-butadiene diepoxide, 1,7-octadiene diepoxide, EPI) as well as the degree of crosslinking on guest binding properties of CDs were investigated [22].

In the 1980s and the 1990s, CD polymers were studied for food application for the first time. For example, grapefruit juice was debittered by removing naringin and limonin with the use of polymers, such CDs crosslinked with EPI, hexamethylene diisocyanate and phenyl isocyanate [23]. In addition, their encapsulation capability was also tested on other food components, such as caffeine, vanillin and theobromine [24].

Insoluble crosslinked polymers in the form of gels were examined for the sorption of textile dyes [25], and the same polymers were later used to entrap aromatic pollutants, such as phenol, *p*-nitrophenol, benzoic acid, *p*-nitrobenzoic acid, *β*-naphthol, chlorophenols and 4-*tert*-butylbenzoic acid [26].

2.3. 1999:"In the Beginning Was the Word"

The term "nanosponge" was used for the first time in 1999 by Min Ma and De Quan Li [27,28]. They described novel nanoporous polymers made up of CDs connected with diisocyanate linkers [28]. The preparation was very simple: CDs were mixed with the crosslinker in DMF and heated for 24 h. These NSs exhibited a surprisingly high adsorption capacity despite their low surface area, thus opening new possibilities in the water remediation field. Their properties, such as adsorption capacity, tunability and low cost, overcame the limits of the purification methods used so far, such as reverse osmosis and adsorption on activated carbon, or zeolites. A great advantage was the possibility to modify the properties of these polymers by changing the degree of crosslinking and the kind of crosslinker. The entrapment of organic pollutants from an aqueous medium was strictly dependent on the size of CD cavities, and the NSs were readily available after washing with an organic solvent.

The 20th century came to an end with an innovation in CD chemistry, which will influence a broad range of scientific fields. However, a decade passed before the importance of NSs was widely recognized.

2.4. From 2000 to 2009: New Millennium Came with New Applications

In the new millennium NSs found interesting opportunities in new fields. Nevertheless, already known applications of CD NSs were not neglected.

The problem of water purification was still one of the main points of interest, being a critical issue of daily life. The challenge consisted of improving water purification by developing novel insoluble CD materials with enhanced affinity to organic contaminants. A β-CD polymer highly crosslinked with EPI was prepared to entrap polycyclic aromatic hydrocarbons and pharmaceutical compounds from water [29]. The entrapment behavior showed the dependency on the polarity of the pollutant, being more effective with nonpolar compounds than ionized ones. The binding ability was affected by the presence of alcohols during the pollutant entrapment, as alcohols have a high affinity to CD cavity. Nevertheless, such competitive behavior offered an easy and efficient way to recover the polymer by washing in alcohol. Moreover the trapping potential could be enhanced via functionalization of CD NSs [30]. In 2005 (Italian patent), Trotta and collaborators replaced the potentially toxic diisocyanates used as NS crosslinking agents with carbonate compounds. They found that CDs crosslinked through carbonate bonds were in the form of micro- or nanoporous materials able to remove chlorinated persistent organic pollutants (POPs) from aqueous solutions even at low concentrations (such as a few ppb) and to strongly bind them [11,31,32].

The undesirable effects of some chemicals present in water even at very low concentrations created the need to improve filtration methods in order to discard them. Polyurethane cyclodextrin NSs were investigated for high-efficiency removal of aromatic and chlorinated hydrocarbons, using disinfection by-products with unpleasant odor as model pollutants. All the materials tested showed elimination of the pollutants to undetectable values; therefore, the efficacy obtained was higher than that of activated carbon [33]. Nanosponges were used also for discarding carcinogenic compounds from drinking water supplies, which can be harmful even at low concentrations, such as *N*-nitrosodimethylamine (NDMA). According to the study of Mhlongo et al., cyclodextrin polyurethane NSs showed great potential in NDMA entrapment [34].

CD NSs were not examined only for removing organic pollutants, but also for trapping heavy metals, thus finding application in environmental and pharmaceutical fields. The most common heavy-metal pollutants include arsenic, cadmium, chromium, copper, nickel, lead and mercury. The consumption of contaminated water causes the intake of these heavy-metals, which can accumulate in tissues and, therefore, are extremely harmful to human health. CDs crosslinked with pyromellitic dianhydride (PMDA) were tested for binding metal ions, such as Al (III), Mn (II), Co (II), Ni (II), Cu (II), Cd (II) and others. It was the first time that acidic properties of a CD polymer were investigated by pH-metric titrations with standard NaOH solutions in order to obtain protonation constant values of individual specific protogenic sites. The complexation of heavy metal cations on the polymer was tested at different pH values. In most of the cases the retention was mainly pH dependent and higher than 70% [35]. The idea was further developed by comparing different types of crosslinkers, such as naphthalene dianhydride and diphenyl carbonate (DPC) [36].

In the new millennium, nanotechnology and nanobiotechnology gained increasing importance and will change medicine greatly in the coming decades. Nanomedicine is believed to be a powerful tool to significantly improve the current existing drug delivery systems and treatments, thus improving the quality of life of patients.

Intense research on nanoparticulate systems began in this period, and various kinds of materials were proposed as drug delivery systems with improved solubility, pharmacokinetics, distribution, sustained release and cellular targeting of drugs.

CD NSs came to the scene due to their promising properties, which are attributable to their unique structure. At first, the research was focused on their capacity to encapsulate various kinds of drugs. Attention was also paid to their safety, negligible toxicity and biodegradability, as they were intended for human use. To assure the quality parameters listed above, efforts were made to better understand how nanosponges worked, their interaction with drugs and how they influenced the properties of the drugs. In addition to FTIR and HPLC methods, employed to examine the structure and loading properties, as well as DSC thermal analysis, X-ray diffraction, photon correlation spectroscopy, optical microscopy, TEM, electrophoretic mobility and zeta potential, hemolysis, cytotoxicity and in vitro release experiments were used and became essential to fully characterize these innovative drug delivery systems.

For the first time, Cavalli et al. tested the capacity of carbonate CD NSs to load drugs both lipophilic (e.g., dexamethasone or flurbiprofen) and hydrophilic (e.g., doxorubicin), and a sustained release of the drugs was achieved [2,37]. Carbonate CD NSs also showed the ability to significantly increase the solubility of the antifungal drug itraconazole and, thus, potentially improve its bioavailability. Moreover, the effect could be even enhanced by using additives such as copolyvidonum [38].

In the end of the decade, NSs found also application as a substrate for enzyme immobilization. Carbonate NSs increased the thermostability, pH stability and also storage stability of catechol 1,2-dioxygenase enzyme. NSs with this immobilized enzyme were used to create a small-scale bioreactor for "green" production of *cis, cis*-muconic acid, a precursor for industrially valued adipic acid [39].

2.5. From 2010 to 2015: Focus on Nanosponges as Delivery Systems

In these years, the capability of CD NSs to load molecules of pharmaceutical interest was extensively studied.

Alongside their use as drug carriers, they were found to be also suitable for gases, such as oxygen and carbon dioxide. The former can solve hypoxic conditions of tissues, the latter provide beneficial physiological effects like blood vessel dilation, blood circulation improvement and activation of gastrointestinal movement.

The capacity of cyclodextrins to store gases in their cavity has been known for a long time, dating back to 1987 when the encapsulation of carbon dioxide with CDs was patented in Japan anticipating its uses in cosmetics, cleansing and personal care products [40]. Among all, α-cyclodextrins were of choice because their gases have a low molecular weight and small size, unlike β-cyclodextrins, which have a higher dimension of the inner cavity and therefore do not fit the requirements to host gases. For the first time, Cavalli et al., and then Trotta et al., demonstrated that alongside cyclodextrins, crosslinked cyclodextrins, including β-CDs, were not only able to host gases but also turned out to be particularly advantageous in this field.

In particular, in 2010 Cavalli et al. [41] investigated the capacity of nanosponges synthesized using different kinds of cyclodextrins α, β, γ-CD crosslinked with 1,1′-carbonyldiimidazole (CDI) to deliver oxygen. These nanosponges, especially β-NS, were able to encapsulate, store and release oxygen for a prolonged period, which could be enhanced by means of ultrasound as external stimulus in an in vitro environment. The year after, Trotta et al. loaded inside carbonate β-CD NSs not only oxygen and carbon dioxide, but also 1-methylcyclopropene (1-MCP), opening a new possible application, such as the improvement of the vase life of *Dianthus caryophyllus* cut flowers. The oxygen loading and release confirmed the results of the study conducted by Cavalli et al., carbon dioxide was entrapped at atmospheric pressure and room temperature, and a significant amount of carbon dioxide was retained even at 373 K for 36 h under vacuum. 1-MCP included in β-CD NS showed a superior antiethylenic effect in long-lasting cut flowers in contrast to commercially available products [40]. Moreover, Seglie et al. demonstrated that 1-MCP encapsulated in β-CD NS was even more effective than the 1-MCP gaseous application at different concentrations, preventing pigment degradation in petals and reducing endogenous ethylene production [42]. Later, the effectiveness of the non-volatile formulation of 1-MCP complex in controlling *Botrytis cinerea* damage on carnation cut flowers was able to control fungal diseases of cut flowers in the postharvest environment [43].

A number of NSs-based drug delivery systems with different types of CDs and crosslinkers have been developed in these years. They contributed to improve the solubility, stability, sustained release, enhancement of permeability and bioavailability and activity of drugs. Moreover, they enabled alternative routes of administration to be chosen, thus favoring patient compliance and reducing side effects; ocular and transdermal delivery are a few examples.

Cancer drugs were extensively studied in order to find the best nanoparticulate delivery system capable of improving their efficacy and reducing their well-known side effects. CD NSs were proposed as a promising solution in the studies reviewed below.

Ansari et al. developed various β-CD NS crosslinked with diphenylcarbonate (DPH) with ratios 1:2, 1:4 and 1:8 to find the best carrier for loading paclitaxel [44]. Another attempt was made by Mognetti et al. who found an alternative to classical paclitaxel formulation in Cremophor EL: fluorescent NSs were synthesized and tested in vitro on cancer cells. As the anticancer activity of paclitaxel was enhanced, it was believed that the nanosponges adhering to/interacting with the cell membrane promoted the release of the drug [45].

Camptothecin, used for hematological and solid tumors, was encapsulated in DPH-linked NSs and tested on human prostate cancer cells. β-CD NS carriers were able to overcome chemical disadvantages of the drug and improve in vitro anti-tumor efficacy in androgen refractory models of prostate cancer DU145 and PC-3 [46].

Carboxylated β-CD NSs were effective nanocarriers for oral delivery of tamoxifen [47], for delivery of calcium in hyperphosphatemia [48], for curcumin in cancer treatment [49], naphthaleneacetic acid on rhizogenesis of globe artichoke [50] and acyclovir [51].

Lembo et al. evidenced another extremely powerful property of NSs: the possibility to make them fluorescent, which is particularly useful for cellular trafficking studies. The method consisted of adding a pre-formed carbonate NS to a fluorescein isothiocyanate solution in DMSO and incubating at 90 °C for 3 h. After the solid was recovered by filtration, it was reacted with succinic anhydride to obtain fluorescent NSs bearing carboxylic groups [51].

β-CD NS prepared with DPH as a cross-linker were successfully used for ocular delivery of dexamethasone. The drug was retained for a longer time in the eye, thus increasing its corneal permeability [52].

Nanosponges, as mentioned before, were employed also in drug delivery through the skin. Imiquimod used in the prevention and treatment of post-burn hypertrophic scars was loaded in β-CD/PMDA nanosponges [53]. The same kind of nanosponge was documented for the first time in Conte et al.'s study as a multifunctional ingredient in semisolid formulations for drug delivery to the skin [54]. The role of the NS in the solubilization and stabilization of benzoporphyrin-derivative monoacid ring A (BPDMA), all-trans retinoic acid (atRA) and on skin permeation of diclofenac (DIC) was tested. The nanosponge, being able to stabilize light-sensitive drugs and to localize the action of highly penetrating drugs in the external layers of skin, proved to be particularly useful in topical formulations.

Again, β-CD/PMDA nanosponges were investigated by Shende and co-workers to deliver meloxicam in order to improve its solubility and bioavailability as well as to prolong its release for anti-inflammatory and analgesic effects. Physical mixing, kneading and sonication were used to extend the duration of the drug release [55].

Positive results were achieved also with another non-steroidal anti-inflammatory drug (NSAID) such as ibuprofen. This drug was loaded inside β-CD/ethylenediaminetetraacetic acid (EDTA) dianhydride nanosponges. The dynamic properties of ibuprofen, especially in the gel state, were studied for advanced formulations using high-resolution magic angle spinning (HRMAS) NMR spectroscopy. The polymeric network was believed to affect the diffusive regimes of ibuprofen [56].

The administration of antioxidants is challenging due to their low solubility and physico-chemical stability. A number of studies were carried out to assess the suitability of NSs for the delivery of these molecules. CDI-linked β-CD NSs were used to deliver resveratrol and polyphenols found in apples (rutin, phloridzin and chlorogenic acid) [57]. The latter were successfully loaded in β-CD NS crosslinked with hexamethylene diisocyanate. Having significantly enhanced the photostability and antioxidant activities, NSs offered a potential drug delivery system for oral and topical delivery [58,59]. Moreover, gamma-oryzanol (GO), a mixture of ferulic acid esters, usually employed to stabilize food and pharmaceutical raw materials as sunscreen in cosmetic formulations, were loaded in NSs. The photodegradation of GO after exposure to UVA or UVB irradiation was slowed down, and the antioxidant effect was still present when included in nanosponges. In addition, a certain accumulation of GO was found in in vitro experiments on porcine ear skin [60].

Drug delivery is not the only field in which CD NSs found interesting opportunities: in these years agricultural and flame retardancy applications can be found.

Efforts were made to make agriculture more efficient, not only in terms of fertilization, but also to protect crops from pests and increase food shelf-life. Functionalized β-CD NSs were investigated for the growth, conservation, protection and disinfection of vegetable organisms [61]. NSs were used to increase the efficacy and bioavailability of nutrients and, thus, enhance the plant growth. Moreover, NSs may decrease toxic side-effects and enhance biodegradability of commercially available fertilizers. The positive effect of NSs as fertilizer carriers on plant growth was confirmed in a work focused on the development of a new iron fertilizer using a β-CD/PMDA nanosponge. The Fe-NS had a positive effect on the growth and re-greening in sweet corn as well as tomato plants [62].

As mentioned above, it is necessary not only to grow more plants, but also to protect the quality and safety of crops. The agricultural commodities might be devaluated by a number of pests, including fungi, after which contamination with harmful mycotoxins remains. Polyurethane NSs were studied to reduce the concentration of mycotoxins (Ochratoxin A) in aqueous solutions with positive results. CD NSs showed good potential for further use in decontamination of beverages produced from the affected commodities [63].

More and more emphasis is placed on the safety of materials in all fields, including electronics or construction. Polymers and textiles are very useful for preparation of cables, insulation, etc. However, the flammability of such materials is a serious problem limiting the application. To overcome this issue, materials are enriched with flame-retardant additives, such as phosphates, inorganic hydroxides, halogen and metal–halogen derivatives. These additives prevent ignition or slow down its development in three ways: endothermic degradation of retardant cooling down the substrate, creation of a protective char surface layer and/or emission of non-combustible gases. However, it is necessary to use a high amount of flame retardants (20–60 wt %) [64], which might affect the final properties of the materials. Development of efficient new fire retardant agents may lower the effective dose as well as preserve human health by using more eco-friendly compounds and eliminating the emission of toxic smoke [64–66].

Nanosponges were studied as novel flame retardant systems [64–66]. The first study was conducted by Alongi et al., consisting of nanosponges loaded with phosphorus derivatives, which were entrapped into the internal cavities of cyclodextrins and/or interstitial space between cyclodextrins and the crosslinker by mechanical grinding. Unlike traditional systems, this complex, which was stable in processing conditions, had the advantage of having NS acting as both carbon sources and foam forming agents able to generate phosphoric acid in situ directly, thus protecting the copolymer against combustion [64].

2.6. From 2016 to Present: State-of-the-Art and Future Prospects of Nanosponges

The increasing number of publications on CD-based NSs over the years shows that they have emerged during the last decade (Figure 1) and have attracted researchers' attention worldwide in numerous fields (Figure 2).

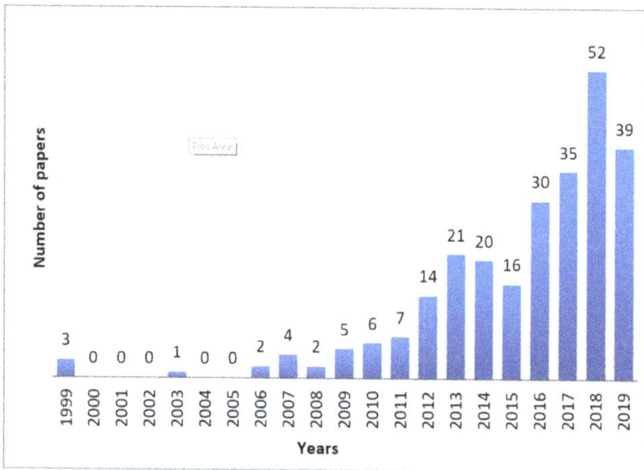

Figure 1. Number of papers on cyclodextrin-based nanosponges (CD NSs) published over the years from 1999 until 2019.

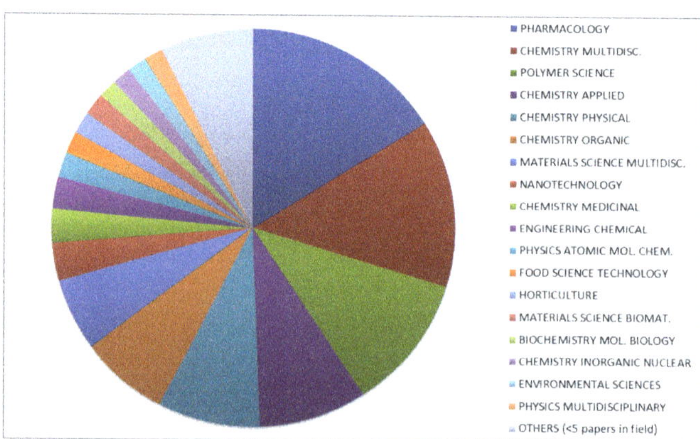

Figure 2. Research areas in which CD NSs are involved.

The main area of investigation is nanomedicine [67], in which CD NSs have been employed mainly as drug delivery systems [68]. NSs gained great attention due to the ability to host various kinds of drugs, thus improving their bioavailability and, in addition to this, due to their lack of toxicity as demonstrated by Shende et al. in 2015 in their acute and repeated dose toxicity study [69].

In the last few years (2016–2019), all four CD-based nanosponge generations have been investigated in the pharmaceutical field [11]. β-CD crosslinked with DMC, CDI, DPC, PMDA and CA (citric acid), belonging to the first generation, have been employed in pharmaceutical research (Table 1). Some examples are provided below.

Table 1. Historical excursus of publications on CD-based NSs employed in pharmaceutical research.

Year	Nanosponge/Cross-l	Drug	Indication	References
2010–2015	β-CD/DPC	Camptothecin	Pharmaceutical application	[46,70]
	β-CD/CDI	Resveratrol	Pharmaceutical application	[58]
	Fluorescent NS	Paclitaxel	Pharmaceutical application	[45]
	Carboxylated nanosponges	Acyclovir	Antivirial efficacy	[51]
	β-CD/CDI	Rutin, phloridzin and chlorogenic acid	Anti-cancer, anti-diabetic, antiobesity, neuronal protective properties etc.	[57]
	β-CD/DPC	Gabapentin	Treatment of partial seizures in pediatric	[71]
2016	β-CD/PMDA	Lansoprazole	Gastric ulcers	[72]
	β-CD/DPC	Quercetin	Cancer treatment	[73]
	β-CD /CDI	Erlotinib hydrochloride (ERL)	Cancer treatment	[74]
	GSH–NS (β-CD/PMDA)	Doxorubicin	Anticancer activity	[75]
2017	β-CD/EDTA	Ibuprofen	diseases	[76]
	β-CD/DPC	2-(3,4-dimethoxyphenyl)-3-phenyl-4H-pyrido [1,2-a] pyrimidin-4-one (DB103)	Cardiovascular diseases, drug-eluting stents (DES)	[77]
	β-CD/DPC	Chrysin	Optimal anti-oxidant and anti-tumorous properties	[78]
	β-CD/DPC	Nifedipine	Treatment of angina pectoris and hypertension	[79]
	β-CD/CDI	Camptothecin (CPT)	Antitumor efficacy, inhibition effect on prostate cancer	[80,81]
	β-CD/DPC	Cefadroxil (CFD)	Against variety of Gram-positive and Gram-negative bacteria	[82]
	β-CD/DPC	Efavirenz and Rilpivirine HCl	HIV	[83,84]
	β-CD/PMDA	Erlotinib glutathione	Cancer	[85]
2018	β-CD/DMC	Ellagic acid	Cancer	[86]
	β-CD/DPC	Mebendazole	Lymphatic worm infestations	[16]
	β-CD/TDI	Naproxen	Inflammation	[87]
	Electrospun β-CD/PMDA	*N,N*-diethyl-3-toluamide	Infectious diseases	[88]
	β-CD/CDI; PMDA	Rilpivrine	HIV	[89]
	β-CD/CDI	Atorvastatin calcium	Dyslipidemia	[90]
	β-CD/DPC	Norfloxacin	Urinary tract infections	[91]
	β-CD/CDI	Doxorubicin	Cancer	[92]
2019	β-CD/NDCA	Sage essential oil	Diabetes	[93]
	β-CD/DPC; PMDA	Curcumin and Resveratrol	Cancer	[15,94,95]
	β-CD / DPC	Ferulic acid and Imatinib Mesylate	Cancer	[17,96]
	β-CD/EPI	Curcumin	Cancer	[97,98]
	β-CD/CDI	Kynurenic acid and Paliperidone	Neurological disorders and Schizophrenia	[99,100]
	β-CD/PMDA	Imiquimod	Topical diseases	[101]

Patel and co-workers (2016) investigated the ability of β-cyclodextrin/DPC NSs to host both hydrophilic drugs, such as gemcitabine, and lipophilic drugs, such as bicalutamide, paclitaxel and letrozole. Lipophilic drugs had a higher drug loading capacity than hydrophilic ones because of their large number of lipophilic sites available for drug complexation. Moreover, the drug loading capacity seemed to be dependent on the synthetic route selected: it was lower if the NS was obtained by stirring than when a more powerful and intense method, like sonication, was used, a larger particle size being obtained in the former case. In the same year, β-CD/PMDA was used to encapsulate lansoprazole. The positive results obtained made this kind of NS promising for the treatment of gastric ulcers [72].

NSs were also used as multifunctional direct compression excipients for tablet designing without adding any binder, lubricant or disintegrant. This kind of NS was synthesized by crosslinking β-CD with CA [102].

In 2017, β-CD/CDI NSs loaded with anti-cancer drugs, such as erlotinib (an epidermal growth-factor receptor tyrosine kinase inhibitor) and camptothecin (an inhibitor of DNA Topoisomerase-I), increased their oral bioavailability, solubility and dissolution, minimizing the dose-related adverse effects [74,80,81]. Celecoxib, possessing analgesic and anti-inflammatory actions, was loaded in a β-CD/NN-methylene bisacrylamide nanosponge, which was incorporated in a hydrogel for topical application. The drug solubility and bioavailability were therefore enhanced [103].

β-CD/DPC NS was used for the delivery of drugs for the treatment of HIV, such as efavirenz and rilpivirine, in order to enhance their solubility and the bioavailability [83,84]. An improvement in the aqueous solubility was again achieved by using the same kind of NS (β-CD/DPC) for the delivery of chrysin, having antioxidant and anti-tumorous properties [78].

Curcumin, resveratrol and a combination of the two were loaded in β-CD NSs and tested [15,94,98]. Curcumin and resveratrol were combined in 2019 by Pushpalatha et al. in order to exploit their synergistic effect against breast cancer through transdermal delivery [15]. The nanosponges not only enhanced the release in vitro of both curcumin and resveratrol 10 and 2.5 times, respectively, but also the combination showed a synergistic cytotoxic effect on the breast cancer cell line selected. The year before, Pushpalatha and co-workers explored the effect of different kinds of crosslinked cyclodextrins (β-CD, DPC and β-CD, PMDA) for the delivery of resveratrol. Having selected the best crosslinker ratio, the NSs were compared in terms of physico-chemical characterization together with photodegradation, in vitro drug release, in vitro cytotoxicity and in vivo tests in order to find the most suitable choice for the selected drug [95]. Resveratrol and oxyresveratrol were also loaded in carbonate CD NSs to improve the solubility, release profile, photostability, antioxidant and cytotoxicity activity [104].

Alongside resveratrol, other natural antioxidants, such as ellagic and ferulic acids, were studied [86]. When loaded in NSs, their solubility was enhanced, making low solubility no longer a limit for their application in the food and pharmaceutical industries. Dhakar and co-workers studied the enhancement in aqueous solubility, antioxidant activity and in vitro cell toxicity of kynurenic acid (KYNA) loaded in β-CD/CDI NS. As a high solubilization and drug loading of KYNA were obtained, CD NSs proved to be suitable for biological applications [99].

Alongside drugs, the capability of NSs of reacting with fluorescent molecules was also explored, as this property would be particularly useful in vivo as they would act as fluorescent probes. This capability was confirmed by Ncube et al.'s study in which the isothiocyanate form of fluorescent dyes (rhodamine and fluorescein) was used to react with an epichlorohydrin β-CD polymer, found useful in cancer therapy [105]. Fluorescent NSs belong to the second generation of NSs, as reviewed by Caldera et al. [11].

The need to improve drug efficacy and minimize dosage in order to reduce side effects has stimulated the development of novel drug delivery systems capable of controlled release triggered by stimulating signals. These NSs, which belong to the third generation, have been developed recently. In 2016, Trotta et al. developed glutathione-responsive NSs capable of releasing entrapped anticancer drugs in response to intracellular stimuli, such as intracellular glutathione [75]. The synthesis consisted of a single-step reaction between β-CD and 2-hydroxyethyl disulfide in the presence of PMDA as a cross-linker [106]. GSH-NSs loaded with doxorubicin showed remarkably higher effectiveness than free drug in cancer cells characterized by high GSH content both in vitro and in vivo. In 2017, erlotinib hydrochloride, which was associated with severe toxicity when administered in a systemic and uncontrolled way, was loaded in GSH-NSs [85].

Biodistribution and in-vivo tumor growth inhibition studies revealed drug release to the cancer cell, thus preventing unnecessary drug exposure, and exhibited extended drug release proportional to the external GSH concentration. In 2018, plant hormones (strigolactones) were loaded into GSH/pH-sensitive NSs in order to investigate whether this stimuli-responsive nanocarrier reduced the viability of prostate cancer cells in vitro. The cytotoxic effects on prostate cancer cells have been enhanced after incorporation of SLs into GSH/pH-sensitive NSs [107].

Other types of pH-sensitive NSs were also explored. In 2019, cyclodextrins and calixarenes were combined to develop pH-sensitive NSs (CyCaNS), the adsorption and release capabilities of which were tested using as a model drug for tetracycline antibiotics [108]. pH sensitivity, according to Fontana et al., may be due to the Coulomb interactions between the positively charged guest molecule and the average charge density on the polymeric framework (due to the presence of ionizable amine or carboxyl groups deriving from chemical post-modification). Interesting improvements of the biocidal activity were achieved. These improvements may be attributable once again to the presence of particularly favorable Coulomb interactions occurring between the NS carrier and the bacterial cell wall. Further studies are ongoing in order to better understand the possible mechanisms implicated in the CyCaNS-bioactive molecule-bacterial cell interaction.

NSs obtained via interaction between template molecules and polymers by covalent, semi-covalent or non-covalent bonding are included within the group of Molecularly Imprinted Polymers (MIPs) and belong to the fourth generation [11]. The product is a heteropolymer matrix with specific recognition elements for the template molecule, even though it has been removed. MIPs find applications in several fields alongside drug delivery, such as separation sciences and purification, biological antibodies receptors systems and catalysis. Deshmukh and co-workers developed biomimetics for glucose estimation using molecularly and non-molecularly imprinted polymers of β-CD/PMDA NSs. The rapid extraction of the template from the samples, high adsorption efficiency and non-toxicity are some of the properties of MIP-NSs. They have several advantages over non-molecularly imprinted polymers (NIP NSs) due to the diffusion of the template in the formed cavity and its high surface area.

Trotta and co-workers synthesized new MIP-NSs by cross-linking β-CD with CDI in DMF in the presence of L-DOPA as a template molecule (a pro-drug for the treatment of Parkinson's disease). It has been confirmed that MIP-NSs show a slower and more prolonged release profile making them a promising alternative for storage and controlled delivery of L-DOPA [109]. MIPs could be also an effective approach for the synthesis of catalysts capable of overcoming limitations for processes requiring an optimal arrangement of several reacting molecules or of converting bulky molecules in two-phase systems [110–114].

In 2018, attempts were made to overcome the inherent lack of cellular binding ability of NSs, which has limited their application in drug delivery. In fact, medical research has recently invested heavily on drug delivery systems capable of reaching the target site (cell/tissue/organ) in order to improve the efficacy of the drug and limit its adverse effects. This therapeutic approach called "targeted drug delivery" has been experimented especially on anticancer drugs to limit their toxicity. Various approaches have been proposed, including the functionalization of nanoparticles with natural ligands, which bind to specific receptors on the surface of the target cells, thus increasing cellular drug uptake [12]. CD-based NSs, even if still at an early stage, have been studied in this field of application as they can be easily functionalized with ligands as mentioned in the introduction of this review. A method explored successfully by Singh P. and co-workers was the functionalization of the surface of CD-based NSs with cholesterol, which is a ubiquitous endogenous molecule, responsible for cell interactions and protein binding. Doxorubicin was selected as a model drug, and its cellular uptake revealed an enhanced effect of doxorubicin when loaded in this innovative carrier [92].

This approach was also used for theranostic nanomedicines, which have attracted huge interest for imaging-guided drug/gene delivery in cancer treatment because of the combination of diagnostic and therapeutic functions. For example, folic acid was used as a targeting agent in a novel theranostic system based on a CD NS polymer anchored on the surface of magnetite nanoparticles, which was then decorated with folic acid [97]. Facile condensation polymerization with carboxyl-functionalized nitrogen-doped carbon-quantum dots (CQDs) and β-cyclodextrin as multifunctional monomers has been used to synthesize fluorescent hyper-crosslinked β-CD–CQD hybrid NSs for tumor theranostic application. Doxorubicin (DOX) was loaded to β-CD–CQD via host–guest complexation, showing a pH-responsive, controlled release in the stimulated tumor microenvironment. Their potential application in tumor imaging has been demonstrated by easy synthesis, excellent biocompatibility, strong bright blue fluorescence emission, pH-responsive sustained release and enhanced anticancer activity [115].

Alongside drugs, mosquito repellents were also encapsulated inside nanosponges to improve the performance of *N*,*N*-diethyl-meta-toluamide (DEET) commonly used to prevent dangerous infections transmitted by insects. DEET itself shows repellence against mosquitoes, bugs and ticks but is not effective for a long time, whereas when encapsulated inside CD NSs it has a prolonged residence time. DEET-loaded NSs were synthesized using CDI and PMDA. DEET was complexed with the β-CD cavities before or after cross-linking. The encapsulation efficiency, loading capacity and washing durability were studied. The persistence of DEET on polyester fabrics was observed in CDI-NSs preloaded with 5:1 molar ratio of β-CD due to stronger DEET–NS interaction [116]. Moreover, DEET was loaded in β-CD/PMDA NS microfibers, and its release over time was monitored. Cecone and co-workers produced these fibers using an electrospinning process, and from their pyrolysis a novel carbon material was obtained. There has been tremendous progress with electrospun fibers because of their various applications such as filtration, tissue engineering, biosensors, drug delivery, wound dressings and enzyme immobilization [117,118].

Alongside pharmaceutical application, in the last few years NSs have been studied also in other fields, of which the food industry, environment, textile industry, solid-phase extraction and catalysis (Table 2) are a few examples.

Table 2. Historical excursus of publications on CD-based NSs employed in food, textile, environmental fields, etc.

Years	Nanosponge/Crosslinker	Adsorbate	Field	References
1980–2000	CDs/HMDI, phenyl isocyanate, EPI	Naringin, limonin, debittering of grapefruit juice,	Food	[23]
	CDs/EPI	Caffeine, vanillin and theobromine		[24]
	CDs / EPI, HMDI and Phenyl isocyanate	Textile dyes	Textile	[25]
		Aromatic pollutants, such as phenol, *p*-nitrophenol, benzoic acid, *p*-nitrobenzoic acid, *β*-naphthol, chlorophenols and 4-*tert*-butylbenzoic acid	Environment	[26]
	CDs/HMDI or toluene 2,6-diisocyanate	Organic pollutants	Water Treatment (Environment)	[119]
2000–2009	β-CD/EPI	Naphthalene, 2-naphthol and naproxen	Water Treatment (Environment)	[29]
	CDs/CDI, DMC	Chlorinated persistent organic pollutants (POPs)	Water Treatment (Environment)	[32]
	CD/HMDI or toluene-2,4-diisocyanate	p-nitrophenol and pentachlorophenol,		[30]
	β-CD/PMDA	Heavy metals (Al (III), Mn (II), Co (II), Ni (II), Cu (II), Cd (II) etc.)	Water Treatment (Environment)	[35]
		Carcinogenic N-nitrosodimethylamine (NDMA)	Water Treatment (Environment)	[34]
	Carbonate NSs	Increase the thermostability, pH stability and storage stability of catechol 1,2-dioxygenase	As a substrate for enzyme immobilization	[39]
2010–2019	β-CD/DPC	Phosphorus derivatives	Flame retardancy	[64]
	α, β, γ-CD / CDI; PMDA	Oxygen	Hypoxia/Reoxygenation and Biomedical application	[41,120]
	β-CD /CDI	Oxygen, carbon dioxide and 1-methylcyclopropene 1-MCP carriers.	biomedical, environmental and floriculture application	[40]
	β-CD/PMDA	Iron fertilizer	Agricultural application	[62]
	CD/DPC	Xanthene derivatives	Catalytic Activity	[121]
	CD/CDI	Coriander essential oil	food packaging	[122]

The increased consumption of packaged food and consumer demand for safer and minimally processed food have driven research towards natural alternatives to synthetic antimicrobials and active/intelligent packaging capable of releasing these antimicrobials. NSs have been investigated as feasible candidates due to their encapsulation ability, sustained release, low cost and lack of toxicity. In the last few years the antimicrobial properties of cinnamon and coriander essential oils have been exploited by incorporating them into CD NSs to successfully create stable controlled release systems [123,124]. In the study conducted by Silva et al., various kinds of NSs (synthesized using CD monomers, such as α-CD, hydroxypropyl-β-CD and maltodextrin) were investigated in order to find the most suitable ones for both essential oils. The results demonstrated that loading was dependent on the type of solvent used and that NSs had different affinities for the oils tested as well as for each of their components [122].

The problem of water pollution is still a significant concern, although a lot of effort has been made by the industrial sector over the last 30 years. In 2017, Salgin et al. tested the capability of NSs (β-CD/HDMI) to remove organic pollutants such as p-nitrophenol (p-NP) from water [125].

Still, in 2019 the traditional use of NSs is proposed again for the removal of pollutants such as heavy metals [126], boric acid and organic micropollutants [127]. Rubin Pedrazzo and co-workers demonstrated that nature-friendly NSs obtained by crosslinking β-CD with CA in water were also capable of adsorbing a high amount of heavy metal ions. Particularly new (there is little information about it in the literature) is the use of NSs as advanced pharmaceutical carriers combined with textile materials, namely bio-functional textiles [128]. They are promising as wearable platforms for controlled drug release through skin [128,129]. In the study conducted by Mihailiasa et al., cotton fibers were functionalized with melatonin-loaded carbonate NSs (β-CD/CDI) to prepare a bioactive functional fabric. Melatonin formed a molecular dispersion in the NS cavities, and its loading efficiency was estimated at 8 wt %. An in vitro release study was conducted and showed that the melatonin release rate from the functionalized fabrics could be described by a zero-order law [129].

NSs have also aroused great interest for green extractions through solid-phase extraction. They have already been studied for extracting steroids from a complex matrix like urine, using both β-CD-EPI NS and β-CD urethane NSs [130]. The latter has also been used as a sorbent of carcinogenic aromatic amines from water [130]. These NSs have a huge potential because they are inexpensive, versatile and have evolved in order to make the extraction procedure greener and have a better performance. Thus, efforts are being made not only to prepare non-polluting and re-utilizable NSs, but also functionalized ones with both ionic and neutral groups to adsorb organic compounds, metals and ions simultaneously, as well as amphoteric NSs to be used over a wide pH range.

Another research field in which NSs, especially the carbonate ones, have been investigated is catalysis. Sadjadi and co-workers suggested CD NS-based catalysts as potential candidates for promoting chemical reactions. Heteropolyacids (HPAs) are considered a non-toxic class of catalysts and have been widely used for their strong Bronsted acidity and redox potential. The NS was amine-functionalized followed by HPA immobilization. Synthesis of a variety of xanthenes in aqueous media has been carried out using this catalyst. The merits of this novel procedure were the green and mild reaction conditions as well as excellent yield. Another efficient hybrid catalyst was developed combining HPAs with ionic liquids (ILs). HPA-IL immobilized in the cavities of CD NS was used to promote the cascade reaction of hydrazine hydrate, ethyl acetoacetate, α- or β-naphthol and benzaldehyde. In terms of reactivity and reusability, both hybrid catalysts showed superior catalytic performance [121,131].

Considering the great variety of applications, researchers are increasingly aware of NS potential and need to fully characterize them in order to understand their structure and mechanism of action. To study the interaction of drugs loaded with NS and to emphasize the process of their fabrication, synthesis and design, NS characterization using suitable analytical techniques is the need of the hour. Full analytical characterization of NSs can be useful to select the most appropriate polymer, crosslinker and increase the pharmaceutical applications of these systems along with possible patenting and marketing implications [132].

Knowledge of the analytical tools can significantly improve the assessment of the quality parameters of NSs, i.e., their safety, negligible toxicity, superior inclusion capability, marked swelling behavior and biodegradability, which are fundamental for their use in drug delivery, drug targeting, tissue engineering and regenerative medicine. Therefore, with the growing importance of CD NSs, many characterization techniques have been used. Fourier transform infrared spectroscopy in attenuated total reflectance geometry (FTIR-ATR), CHNS/O analysis, scanning electron microscopy (SEM), energy dispersive X-ray spectroscopy (EDX) analysis, light scattering analysis [133], low-frequency Raman scattering [134], high-resolution magic angle spinning (HRMAS) NMR spectroscopy, carbon-13 nuclear magnetic resonance (^{13}C NMR), proton nuclear magnetic resonance (^{1}H NMR), thermogravimetric analysis (TGA), potentiometric titration, Brunauer–Emmett–Teller (BET) analysis, transmission electron microscopy (TEM), differential scanning calorimetry (DSC) and X-ray diffraction (XRD) have been extensively employed in the studies over the years. In addition, the nano-formulations developed need to be evaluated as far as encapsulation efficiency and loading capacity, stability, in vitro drug release, in vivo pharmacokinetic release, efficacy and toxicity are concerned in order to ensure quality and efficacy [35,37,53,56,57,73,76,78,98,109,122,125,135–140].

In recent studies, special attention has been given to optimizing formulations in order to find the best solution for the active molecule to be delivered, also by means of mathematical tools, such as a design approach. Experimental design helps to minimize the experiments, develop the process and improve the product quality. It consists of exploring the behavior of NSs in various experimental conditions in a limited number of tests and straightforward identification of the variables that affect the system the most, taking into account the synthesis [141] or the formulation process [14–16]. For example, this approach was adopted by Kamble and co-workers to synthesize NSs considering the effect that various levels of crosslinking agents and βCD concentrations had on porosity, drug encapsulation, zeta potential and drug release [17]. Furthermore, Pushpalatha and co-workers optimized the curcumin-resveratrol loaded NS hydrogel formulation using a factor 3-level Box–Behnken design. The concentrations of the two ingredients used for the hydrogel, together with the pH, were selected as independent variables at three levels. Transdermal flux of curcumin, transdermal flux of resveratrol and spreadability of gel were the dependent variables evaluated for optimization [15]. Singireddy and co-workers optimized the synthesis of CD NSs to avoid low yields and batch-to-batch variations due to differences in experimental conditions (reaction temperature in °C, reaction time in min and stirring speed in rpm). The reaction conditions for the synthesis of NSs were optimized by using central composite design and response surface methodology [141].

Pushpalatha and co-workers also used the hierarchy analysis approach to choose the best cross-linker among various kinds for obtaining NSs for drug delivery [14]. The most common cross-linkers were investigated. The selection was made on the basis of the process, materials and physicochemical characteristics of the output following the schematic procedure present in Figure 3. From this study, it emerged that this kind of analysis helped to reduce the number of experiments, shorten the development process and improve product quality.

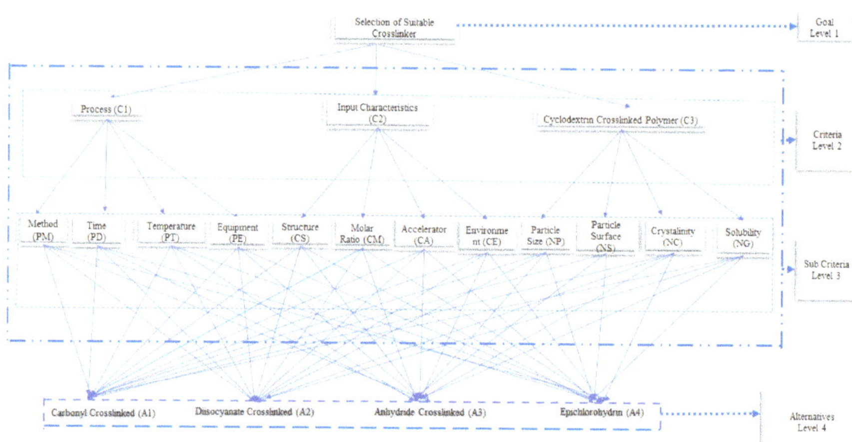

Figure 3. Hierarchy analysis approach to choose the best cross-linker.

3. Conclusions and Future Perspectives

The historical description of CD NSs shows how they emerged over the years (Figure 4, Tables 1 and 2).

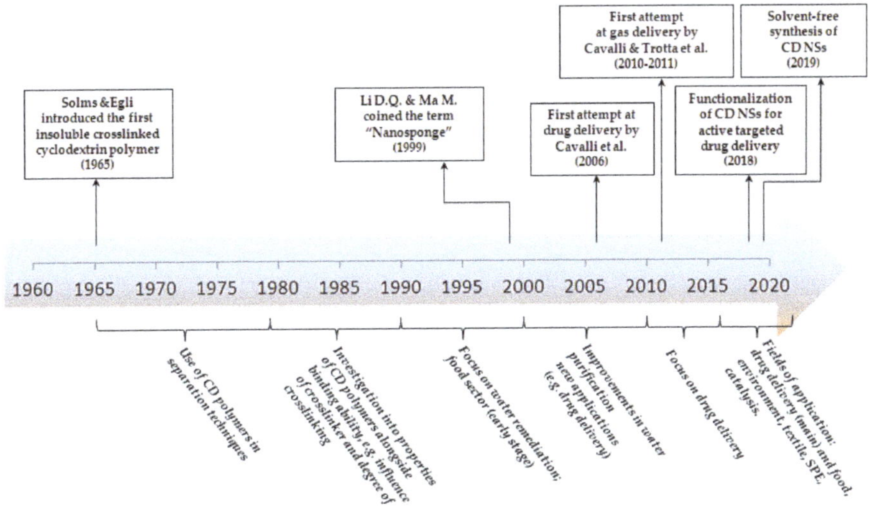

Figure 4. Timeline of the historical development of cyclodextrin-based nanosponges.

It all began in the 1960s when simple network polymers made up of crosslinked CDs were introduced for the first time. Their binding properties tested on organic compounds suggested a possible application in separation techniques, which was further developed in the 1970s with the production of stationary phases for nucleic acids, etc.

In the 1980s, research explored new polymers and made efforts to understand their properties alongside their binding ability. The influence of the crosslinker and the degree of crosslinking on guest binding properties of CDs were investigated for the first time.

In the 1990s, CD polymers found application as debittering agents and food component carriers (e.g., caffeine, vanillin and theobromine). Moreover, in the water remediation field, they overcame the limits of purification methods used up to then due to their high adsorption capacity, tunability and low cost. At the end of this decade they were called "nanosponges" for their nanoporous sponge-like structure.

In the new millennium, new opportunities for nanosponges were studied without neglecting the known applications, such as water purification, with efforts made to improve NS trapping potential by functionalizing them or by replacing potentially toxic crosslinking agents with carbonate compounds, improving the removal of organic pollutants even at a few ppb.

For the first time, nanosponges were investigated as drug delivery systems. Different kinds of drugs were successfully loaded, and a sustained release was achieved. In addition, their safety, negligible toxicity and biodegradability became a matter of concern as they were intended for human use.

This field of application was extensively studied in the following years (2010–2015) lengthening the list of drugs delivered, e.g., anticancer drugs, polyphenols, L-Dopa, NSAIDs and gases of pharmaceutical interest (i.e., oxygen and carbon dioxide).

The period 2016–2019 has seen pharmaceutics as the main field of application. All NS generations are present, including innovative smart nanosponges capable of releasing drugs triggered by external stimuli (i.e., pH and GSH) and the most recent ones having natural ligands grafted on the surface able to perform active targeting.

In these years, alongside the pharmaceutical field, the great versatility of NSs has been confirmed by studies conducted in the food industry identifying new applications, such as active/intelligent packaging. Other applications have been found in the environmental field, the textile industry, solid-phase extraction and catalysis.

Much attention is being paid to fully characterize NSs in terms of structure and mechanism of action with a view to selecting the most appropriate polymer also with the aid of mathematical tools, such as a design approach to rationalize experimentation and improve the product quality.

The success of cyclodextrin-based nanosponges certainly is due to their ability to keep up with the times while retaining their initial features, i.e., low cost, environmental compatibility, non-toxicity and the ability to host various kinds of molecules. Their synthesis has evolved in the direction of greener processes culminating in the most recent solvent-free synthesis. All of these advantages would make NSs suitable for future industrial scale up.

For the reasons discussed above, NS research has not yet reached its conclusion. On the contrary, the potential advantages that could be obtained from their use certainly justify further studies aimed, on the one hand, at investigating in greater depth their existing fields of application in which appropriately optimized NSs are useful as carriers and, on the other, at exploring new fields in which their potential could be exploited to the full as a promising, safe innovation for human health and activities.

Author Contributions: I.K., S.L.A., M.T. and G.H. contributed equally in collecting the historical data. I.K. drafted the text together with S.L.A., who also rearranged and finalized the manuscript. F.C. checked the correctness of the contents of the entire paper. A.R.P. and C.C. revised the contents from the chemical point of view. R.C. coordinated the activities and F.T. conceived the idea and proofread the manuscript. All authors have read and agreed to the published version of the manuscript.

Funding: The work of Ilona Krabicová was supported by the Ministry of Education, Youth and Sports of the Czech Republic and the European Union—European Structural and Investment Funds in the frames of Operational Programme Research, Development and Education—project Hybrid Materials for Hierarchical Structures (HyHi, Reg. No. CZ.02.1.01/0.0/0.0/16_019/0000843).

Conflicts of Interest: The authors declare no conflict of interest. The funders had no role in the design of the study; in the collection, analyses, or interpretation of data; in the writing of the manuscript, or in the decision to publish the results.

References

1. Torne, S.J.; Ansari, K.A.; Vavia, P.R.; Trotta, F.; Cavalli, R. Enhanced oral paclitaxel bioavailability after administration of paclitaxel-loaded nanosponges. *Drug Deliv.* **2010**, *17*, 419–425. [CrossRef]
2. Cavalli, R.; Trotta, F.; Tumiatti, W. Cyclodextrin-based nanosponges for drug delivery. *J. Incl. Phenom. Macrocycl. Chem.* **2006**, *56*, 209–213. [CrossRef]
3. Vyas, A.; Saraf, S.; Saraf, S. Cyclodextrin based novel drug delivery systems. *J. Incl. Phenom. Macrocycl. Chem.* **2008**, *62*, 23–42. [CrossRef]
4. Trotta, F. *Cyclodextrin Nanosponges and Their Applications*; Bilensoy, E., Ed.; John Wiley & Sons, Inc.: Hoboken, NJ, USA, 2011.
5. Subramanian, S.; Singireddy, A.; Krishnamoorthy, K.; Rajappan, M. Nanosponges: A novel class of drug delivery system—Review. *J. Pharm. Pharm. Sci.* **2012**, *15*, 103–111. [CrossRef]
6. Shringirishi, M.; Prajapati, S.K.; Mahor, A.; Alok, S.; Yadav, P.; Verma, A. Nanosponges: A potential nanocarrier for novel drug delivery-a review. *Asian Pac. J. Trop. Dis.* **2014**, *4*, S519–S526. [CrossRef]
7. Chilajwar, S.V.; Pednekar, P.P.; Jadhav, K.R.; Gupta, G.J.; Kadam, V.J. Cyclodextrin-based nanosponges: A propitious platform for enhancing drug delivery. *Expert Opin. Drug Deliv.* **2014**, *11*, 111–120. [CrossRef] [PubMed]
8. Trotta, F.; Dianzani, C.; Caldera, F.; Mognetti, B.; Cavalli, R. The application of nanosponges to cancer drug delivery. *Expert Opin. Drug Deliv.* **2014**, *11*, 931–941. [CrossRef]
9. Trotta, F.; Zanetti, M.; Cavalli, R. Cyclodextrin-based nanosponges as drug carriers. *Beilstein J. Org. Chem.* **2012**, *8*, 2091–2099. [CrossRef]
10. Liu, Z.; Jiao, Y.; Wang, Y.; Zhou, C.; Zhang, Z. Polysaccharides-based nanoparticles as drug delivery systems. *Adv. Drug Deliv. Rev.* **2008**, *60*, 1650–1662. [CrossRef]
11. Caldera, F.; Tannous, M.; Cavalli, R.; Zanetti, M.; Trotta, F. Evolution of Cyclodextrin Nanosponges. *Int. J. Pharm.* **2017**, *531*, 470–479. [CrossRef]
12. Kutova, O.M.; Guryev, E.L.; Sokolova, E.A.; Alzeibak, R.; Balalaeva, I.V. Targeted delivery to tumors: Multidirectional strategies to improve treatment efficiency. *Cancers* **2019**, *11*, 68. [CrossRef] [PubMed]

13. Trotta, F.; Rubin Pedrazzo, A. Processo Per La Preparazione Di Una Nanospugna. IT Patent P021499IT-01, 15 November 2019.
14. Pushpalatha, R.; Selvamuthukumar, S.; Kilimozhi, D. Hierarchy analysis of different cross-linkers used for the preparation of cross-linked cyclodextrin as drug nanocarriers. *Chem. Eng. Commun.* **2018**, *205*, 759–771. [CrossRef]
15. Pushpalatha, R.; Selvamuthukumar, S.; Kilimozhi, D. Cyclodextrin nanosponge based hydrogel for the transdermal co-delivery of curcumin and resveratrol: Development, optimization, in vitro and ex vivo evaluation. *J. Drug Deliv. Sci. Technol.* **2019**, *52*, 55–64. [CrossRef]
16. Rao, M.R.P.; Raut, S.P.; Shirsath, C.T.; Jadhav, M.B.; Chandanshive, P.A. Self-nanoemulsifying Drug Delivery System of Mebendazole for Treatment of Lymphatic Filariasis. *Indian J. Pharm. Sci.* **2018**, *80*, 1057–1068. [CrossRef]
17. Kamble, M.; Zaheer, Z.; Mokale, S.; Zainuddin, R. Formulation Optimization and Biopharmaceutical Evaluation of Imatinib Mesylate Loaded β-Cyclodextrin Nanosponges. *Pharm. Nanotechnol.* **2019**, *7*, 343–361. [CrossRef]
18. Solms, J.; Egli, R.H. Harze mit Einschlusshohlräumen von Cyclodextrin-Struktur. *Helv. Chim. Acta* **1965**, *48*, 1225–1228. [CrossRef]
19. Hoffman, J.L. Chromatography of nucleic acids on cross-linked cyclodextrin gels having inclusion-forming capacity. *J. Macromol. Sci. Part A—Chem.* **1973**, *7*, 1147–1157. [CrossRef]
20. Harada, A.; Furue, M.; Nozakura, S.-I. Optical resolution of mandelic acid derivatives by column chromatography on crosslinked cyclodextrin gels. *J. Polym. Sci. Polym. Chem. Ed.* **1978**, *16*, 189–196. [CrossRef]
21. Mizobuchi, Y.; Tanaka, M.; Shono, T. Preparation and sorption behaviour of cyclodextrin polyurethane resins. *J. Chromatogr. A* **1980**, *194*, 153–161. [CrossRef]
22. Sugiura, I.; Komiyama, M.; Toshima, N.; Hirai, H. Immobilized β-Cyclodextrins. Preparation with Various Crosslinking Reagents and the Guest Binding Properties. *Bull. Chem. Soc. Jpn.* **1989**, *62*, 1643–1651. [CrossRef]
23. Shaw, P.E.; Buslig, B.S. Selective removal of bitter compounds from grapefruit juice and from aqueous solution with cyclodextrin polymers and with Amberlite XAD-4. *J. Agric. Food Chem.* **1986**, *34*, 837–840. [CrossRef]
24. Su, C.-S.; Yang, C.-P. Partial removal of various food components from aqueous solution using crosslinked polymers of cyclodextrins with epichlorohydrin. *J. Sci. Food Agric.* **1991**, *54*, 635–643. [CrossRef]
25. Shao, Y.; Martel, B.; Morcellet, M.; Weltrowski, M.; Crini, G. Sorption of textile dyes on β-Cyclodextrin-epichlorhydrin gels. *J. Incl. Phenom. Mol. Recognit. Chem.* **1996**, *25*, 209–212. [CrossRef]
26. Crini, G.; Bertini, S.; Torri, G.; Naggi, A.; Sforzini, D.; Vecchi, C.; Janus, L.; Lekchiri, Y.; Morcellet, M. Sorption of aromatic compounds in water using insoluble cyclodextrin polymers. *J. Appl. Polym. Sci.* **1998**, *68*, 1973–1978. [CrossRef]
27. Li, D.Q.; Ma, M. Nanosponges: From inclusion chemistry to water purifying technology. *Chemtech* **1999**, *29*, 31–37.
28. Li, D.Q.; Ma, M. Nanoporous polymers: New nanosponge absorbent media. *Filtr. Sep.* **1999**, *36*, 26–28. [CrossRef]
29. Orprecio, R.; Evans, C.H. Polymer-immobilized cyclodextrin trapping of model organic pollutants in flowing water streams. *J. Appl. Polym. Sci.* **2003**, *90*, 2103–2110. [CrossRef]
30. Mamba, B.B.; Krause, R.W.; Malefetse, T.J.; Nxumalo, E.N. Monofunctionalized cyclodextrin polymers for the removal of organic pollutants from water. *Environ. Chem. Lett.* **2007**, *5*, 79–84. [CrossRef]
31. Li, D.; Ma, M. Nanosponges for water purification. *Clean Prod. Process.* **2000**, *2*, 112–116. [CrossRef]
32. Trotta, F.; Tumiatti, W. Cross-Linked Polymers Based on Cyclodextrins for Removing Polluting Agents. U.S. Patent US20050154198A1, 14 July 2005.
33. Mhlanga, S.D.; Mamba, B.B.; Krause, R.W.; Malefetse, T.J. Removal of organic contaminants from water using nanosponge cyclodextrin polyurethanes. *J. Chem. Technol. Biotechnol.* **2007**, *82*, 382–388. [CrossRef]
34. Mhlongo, S.H.; Mamba, B.B.; Krause, R.W. Monitoring the prevalence of nitrosamines in South African waters and their removal using cyclodextrin polyurethanes. *Phys. Chem. Earth* **2009**, *34*, 819–824. [CrossRef]
35. Berto, S.; Bruzzoniti, M.C.; Cavalli, R.; Perrachon, D.; Prenesti, E.; Sarzanini, C.; Trotta, F.; Tumiatti, W. Synthesis of new ionic β-cyclodextrin polymers and characterization of their heavy metals retention. *J. Incl. Phenom. Macrocycl. Chem.* **2007**, *57*, 631–636. [CrossRef]

36. Berto, S.; Bruzzoniti, M.C.; Cavalli, R.; Perrachon, D.; Prenesti, E.; Sarzanini, C.; Trotta, F.; Tumiatti, W. Highly crosslinked ionic β-cyclodextrin polymers and their interaction with heavy metals. *J. Incl. Phenom. Macrocycl. Chem.* **2007**, *57*, 637–643. [CrossRef]
37. Trotta, F.; Cavalli, R. Characterization and applications of new hyper-cross-linked cyclodextrins. *Compos. Interfaces* **2009**, *16*, 39–48. [CrossRef]
38. Swaminathan, S.; Vavia, P.R.; Trotta, F.; Torne, S. Formulation of betacyclodextrin based nanosponges of itraconazole. *J. Incl. Phenom. Macrocycl. Chem.* **2007**, *57*, 89–94. [CrossRef]
39. Di Nardo, G.; Roggero, C.; Campolongo, S.; Valetti, F.; Trotta, F.; Gilardi, G. Catalytic properties of catechol 1,2-dioxygenase from Acinetobacter radioresistens S13 immobilized on nanosponges. *Dalt. Trans.* **2009**, 6507–6512. [CrossRef] [PubMed]
40. Trotta, F.; Cavalli, R.; Martina, K.; Biasizzo, M.; Vitillo, J.; Bordiga, S.; Vavia, P.; Ansari, K. Cyclodextrin nanosponges as effective gas carriers. *J. Incl. Phenom. Macrocycl. Chem.* **2011**, *71*, 189–194. [CrossRef]
41. Cavalli, R.; Akhter, A.K.; Bisazza, A.; Giustetto, P.; Trotta, F.; Vavia, P. Nanosponge formulations as oxygen delivery systems. *Int. J. Pharm.* **2010**, *402*, 254–257. [CrossRef]
42. Seglie, L.; Martina, K.; Devecchi, M.; Roggero, C.; Trotta, F.; Scariot, V. The effects of 1-MCP in cyclodextrin-based nanosponges to improve the vase life of Dianthus caryophyllus cut flowers. *Postharvest Biol. Technol.* **2011**, *59*, 200–205. [CrossRef]
43. Seglie, L.; Spadaro, D.; Trotta, F.; Devecchi, M.; Gullino, M.L.; Scariot, V. Use of 1-methylcylopropene in cyclodextrin-based nanosponges to control grey mould caused by Botrytis cinerea on Dianthus caryophyllus cut flowers. *Postharvest Biol. Technol.* **2012**, *64*, 55–57. [CrossRef]
44. Ansari, K.A.; Torne, S.J.; Vavia, P.R.; Trotta, F.; Cavalli, R. Paclitaxel Loaded Nanosponges: In-Vitro Characterization and Cytotoxicity Study on MCF-7 Cell Line Culture. *Curr. Drug Deliv.* **2011**, *8*, 194–202. [CrossRef] [PubMed]
45. Mognetti, B.; Barberis, A.; Marino, S.; Berta, G.; De Francia, S.; Trotta, F.; Cavalli, R. In vitro enhancement of anticancer activity of paclitaxel by a Cremophor free cyclodextrin-based nanosponge formulation. *J. Incl. Phenom. Macrocycl. Chem.* **2012**, *74*, 201–210. [CrossRef]
46. Minelli, R.; Cavalli, R.; Ellis, L.; Pettazzoni, P.; Trotta, F.; Ciamporcero, E.; Barrera, G.; Fantozzi, R.; Dianzani, C.; Pili, R. Nanosponge-encapsulated camptothecin exerts anti-tumor activity in human prostate cancer cells. *Eur. J. Pharm. Sci.* **2012**, *47*, 686–694. [CrossRef] [PubMed]
47. Torne, S.; Darandale, S.; Vavia, P.; Trotta, F.; Cavalli, R. Cyclodextrin-based nanosponges: Effective nanocarrier for Tamoxifen delivery. *Pharm. Dev. Technol.* **2013**, *18*, 619–625. [CrossRef] [PubMed]
48. Shende, P.; Deshmukh, K.; Trotta, F.; Caldera, F. Novel cyclodextrin nanosponges for delivery of calcium in hyperphosphatemia. *Int. J. Pharm.* **2013**, *456*, 95–100. [CrossRef] [PubMed]
49. Darandale, S.S.; Vavia, P.R. Cyclodextrin-based nanosponges of curcumin: Formulation and physicochemical characterization. *J. Incl. Phenom. Macrocycl. Chem.* **2013**, *75*, 315–322. [CrossRef]
50. Cavallaro, V.; Trotta, F.; Gennari, M.; Di Silvestro, I.; Pellegrino, A.; Barbera, A.C. Effects of the complex nanosponges-naphthaleneacetic acid and β cyclodextrins on in vitro rhizogenesis of globe artichoke. *Acta Hortic.* **2013**, *983*, 369–372. [CrossRef]
51. Lembo, D.; Swaminathan, S.; Donalisio, M.; Civra, A.; Pastero, L.; Aquilano, D.; Vavia, P.; Trotta, F.; Cavalli, R. Encapsulation of Acyclovir in new carboxylated cyclodextrin-based nanosponges improves the agent's antiviral efficacy. *Int. J. Pharm.* **2013**, *443*, 262–272. [CrossRef]
52. Swaminathan, S.; Vavia, P.R.; Trotta, F.; Cavalli, R. Nanosponges encapsulating dexamethasone for ocular delivery: Formulation design, physicochemical characterization, safety and corneal permeability assessment. *J. Biomed. Nanotechnol.* **2013**, *9*, 998–1007. [CrossRef]
53. Bastiancich, C.; Scutera, S.; Alotto, D.; Cambieri, I.; Fumagalli, M.; Casarin, S.; Rossi, S.; Trotta, F.; Stella, M.; Cavalli, R.; et al. Cyclodextrin-Based Nanosponges as a Nanotechnology Strategy for Imiquimod Delivery in Pathological Scarring Prevention and Treatment. *J. Nanopharm. Drug Deliv.* **2015**, *2*, 311–324. [CrossRef]
54. Conte, C.; Caldera, F.; Catanzano, O.; D'Angelo, I.; Ungaro, F.; Miro, A.; Pellosi, D.S.; Trotta, F.; Quaglia, F. β-cyclodextrin nanosponges as multifunctional ingredient in water-containing semisolid formulations for skin delivery. *J. Pharm. Sci.* **2014**, *103*, 3941–3949. [CrossRef]
55. Shende, P.K.; Gaud, R.S.; Bakal, R.; Patil, D. Effect of inclusion complexation of meloxicam with β-cyclodextrin- and β-cyclodextrin-based nanosponges on solubility, in vitro release and stability studies. *Colloids Surf. B Biointerfaces* **2015**, *136*, 105–110. [CrossRef] [PubMed]

56. Ferro, M.; Castiglione, F.; Punta, C.; Melone, L.; Panzeri, W.; Rossi, B.; Trotta, F.; Mele, A. Anomalous diffusion of ibuprofen in cyclodextrin nanosponge hydrogels: An HRMAS NMR study. *Beilstein J. Org. Chem.* **2014**, *10*, 2715–2723. [CrossRef] [PubMed]
57. Ramírez-Ambrosi, M.; Caldera, F.; Trotta, F.; Berrueta, L.; Gallo, B. Encapsulation of apple polyphenols in β-CD nanosponges. *J. Incl. Phenom. Macrocycl. Chem.* **2014**, *80*, 85–92. [CrossRef]
58. Ansari, K.A.; Vavia, P.R.; Trotta, F.; Cavalli, R. Cyclodextrin-based nanosponges for delivery of resveratrol: In vitro characterisation, stability, cytotoxicity and permeation study. *AAPS PharmSciTech* **2011**, *12*, 279–286. [CrossRef] [PubMed]
59. Anandam, S.; Selvamuthukumar, S. Fabrication of cyclodextrin nanosponges for quercetin delivery: Physicochemical characterization, photostability, and antioxidant effects. *J. Mater. Sci.* **2014**, *49*, 8140–8153. [CrossRef]
60. Sapino, S.; Carlotti, M.E.; Cavalli, R.; Ugazio, E.; Berlier, G.; Gastaldi, L.; Morel, S. Photochemical and antioxidant properties of gamma-oryzanol in beta-cyclodextrin-based nanosponges. *J. Incl. Phenom. Macrocycl. Chem.* **2013**, *75*, 69–76. [CrossRef]
61. Roggero, C.M.; Di Carlo, S.; Tumiatti, V.; Tumiatti, M.; Vecchi, M.; Scariot, V.; Kapila, S. Use of Functionalised Nanosponges for the Growth, Conservation, Protection and Disinfection of Vegetable Organisms. WO Patent 2013046165, A1, 4 April 2013.
62. Vercelli, M.; Gaino, W.; Contartese, V.; Gallo, L.; Di Carlo, S.; Tumiatti, V.; Larcher, F.; Scariot, V. Preliminary studies on the effect of Fe-nanosponge complex in horticulture. *Acta Sci. Pol. Hortorum Cultus* **2015**, *14*, 51–58.
63. Appell, M.; Jackson, M.A. Sorption of ochratoxin A from aqueous solutions using β-cyclodextrin-polyurethane polymer. *Toxins* **2012**, *4*, 98–109. [CrossRef]
64. Alongi, J.; Pošković, M.; Frache, A.; Trotta, F. Novel flame retardants containing cyclodextrin nanosponges and phosphorus compounds to enhance EVA combustion properties. *Polym. Degrad. Stab.* **2010**, *95*, 2093–2100. [CrossRef]
65. Alongi, J.; Poskovic, M.; Visakh, P.M.; Frache, A.; Malucelli, G. Cyclodextrin nanosponges as novel green flame retardants for PP, LLDPE and PA6. *Carbohydr. Polym.* **2012**, *88*, 1387–1394. [CrossRef]
66. Lai, X.; Zeng, X.; Li, H.; Yin, C.; Zhang, H.; Liao, F. Synergistic effect of phosphorus-containing nanosponges on intumescent flame-retardant polypropylene. *J. Appl. Polym. Sci.* **2012**, *125*, 1758–1765. [CrossRef]
67. Fulekar, M.H.; Pathak, B. *Environmental Nanotechnology*; CRC Press: Boca Raton, FL, USA, 2017; Volume 1, ISBN 9781498726245.
68. Swaminathan, S.; Cavalli, R.; Trotta, F. Cyclodextrin-based nanosponges: A versatile platform for cancer nanotherapeutics development. *Wiley Interdiscip. Rev. Nanomed. Nanobiotechnol.* **2016**, *8*, 579–601. [CrossRef]
69. Shende, P.; Kulkarni, Y.A.; Gaud, R.S.; Deshmukh, K.; Cavalli, R.; Trotta, F.; Caldera, F. Acute and Repeated Dose Toxicity Studies of Different β-Cyclodextrin-Based Nanosponge Formulations. *J. Pharm. Sci.* **2015**, *104*, 1856–1863. [CrossRef] [PubMed]
70. Swaminathan, S.; Pastero, L.; Serpe, L.; Trotta, F.; Vavia, P.; Aquilano, D.; Trotta, M.; Zara, G.P.; Cavalli, R. Cyclodextrin-based nanosponges encapsulating camptothecin: Physicochemical characterization, stability and cytotoxicity. *Eur. J. Pharm. Biopharm.* **2010**, *74*, 193–201. [CrossRef] [PubMed]
71. Rao, M.R.P.; Bhingole, R.C. Nanosponge-based pediatric-controlled release dry suspension of Gabapentin for reconstitution. *Drug Dev. Ind. Pharm.* **2015**, *41*, 2029–2036. [CrossRef] [PubMed]
72. Shende, P.; Chaphalkar, R.; Deshmukh, K.; Gaud, R.S. Physicochemical Investigation of Engineered Nanosuspensions Containing Model Drug, Lansoprazole. *J. Dispers. Sci. Technol.* **2016**, *37*, 504–511. [CrossRef]
73. Singireddy, A.; Subramanian, S. Cyclodextrin nanosponges to enhance the dissolution profile of quercetin by inclusion complex formation. *Part. Sci. Technol.* **2016**, *34*, 341–346. [CrossRef]
74. Dora, C.P.; Trotta, F.; Kushwah, V.; Devasari, N.; Singh, C.; Suresh, S.; Jain, S. Potential of erlotinib cyclodextrin nanosponge complex to enhance solubility, dissolution rate, in vitro cytotoxicity and oral bioavailability. *Carbohydr. Polym.* **2016**, *137*, 339–349. [CrossRef]
75. Daga, M.; Ulllio, C.; Argenziano, M.; Dianzani, C.; Cavalli, R.; Trotta, F.; Ferretti, C.; Zara, G.P.; Gigliotti, C.L.; Ciamporcero, E.S.; et al. GSH-targeted nanosponges increase doxorubicin-induced toxicity "in vitro" and "in vivo" in cancer cells with high antioxidant defenses. *Free Radic. Biol. Med.* **2016**, *97*, 24–37. [CrossRef]
76. Ferro, M.; Castiglione, F.; Pastori, N.; Punta, C.; Melone, L.; Panzeri, W.; Rossi, B.; Trotta, F.; Mele, A. Dynamics and interactions of ibuprofen in cyclodextrin nanosponges by solid-state NMR spectroscopy. *Beilstein J. Org. Chem.* **2017**, *13*, 182–194. [CrossRef] [PubMed]

77. Coviello, V.; Sartini, S.; Quattrini, L.; Baraldi, C.; Gamberini, M.C.; La Motta, C. Cyclodextrin-based nanosponges for the targeted delivery of the anti-restenotic agent DB103: A novel opportunity for the local therapy of vessels wall subjected to percutaneous intervention. *Eur. J. Pharm. Biopharm.* **2017**, *117*, 276–285. [CrossRef] [PubMed]
78. Sundararajan, M.; Thomas, P.A.; Venkadeswaran, K.; Jeganathan, K.; Geraldine, P. Synthesis and characterization of chrysin-loaded β-cyclodextrin-based nanosponges to enhance in-vitro solubility, photostability, drug release, antioxidant effects and antitumorous efficacy. *J. Nanosci. Nanotechnol.* **2017**, *17*, 8742–8751. [CrossRef]
79. Shringirishi, M.; Mahor, A.; Gupta, R.; Prajapati, S.K.; Bansal, K.; Kesharwani, P. Fabrication and characterization of nifedipine loaded β-cyclodextrin nanosponges: An in vitro and in vivo evaluation. *J. Drug Deliv. Sci. Technol.* **2017**, *41*, 344–350. [CrossRef]
80. Gigliotti, C.L.; Minelli, R.; Cavalli, R.; Occhipinti, S.; Barrera, G.; Pizzimenti, S.; Cappellano, G.; Boggio, E.; Conti, L.; Fantozzi, R.; et al. In vitro and in vivo therapeutic evaluation of camptothecin-encapsulated β-cyclodextrin nanosponges in prostate cancer. *J. Biomed. Nanotechnol.* **2016**, *12*, 114–127. [CrossRef]
81. Gigliotti, C.L.; Ferrara, B.; Occhipinti, S.; Boggio, E.; Barrera, G.; Pizzimenti, S.; Giovarelli, M.; Fantozzi, R.; Chiocchetti, A.; Argenziano, M.; et al. Enhanced cytotoxic effect of camptothecin nanosponges in anaplastic thyroid cancer cells in vitro and in vivo on orthotopic xenograft tumors. *Drug Deliv.* **2017**, *24*, 670–680. [CrossRef]
82. Dubey, P.; Sharma, H.K.; Shah, S.; Tyagi, C.K.; Chandekar, A.R.; Jadon, R.S. Formulations and evaluation of Cyclodextrin complexed Ceadroxil loaded nanosponges. *Int. J. Drug Deliv.* **2017**, *9*, 84. [CrossRef]
83. Rao, M.R.P.; Shirsath, C. Enhancement of Bioavailability of Non-nucleoside Reverse Transciptase Inhibitor Using Nanosponges. *AAPS PharmSciTech* **2017**, *18*, 1728–1738. [CrossRef]
84. Rana, Z.; Zahid, Z.; Jaiprakash, N. Sangshetti Mufassir, M. Enhancement of oral bioavailability of anti-HIV drug rilpivirine HCl through nanosponge formulation. *Drug Dev. Ind. Pharm.* **2017**, *43*, 2076–2084.
85. Momin, M.M.; Zaheer, Z.; Zainuddin, R.; Jaiprakash, N. Extended release delivery of erlotinib glutathione nanosponge for targeting lung cancer. *Artif. Cells Nanomed. Biotechnol.* **2017**, *46*, 1064–1075. [CrossRef]
86. Mady, F.M.; Ragab, S.; Ibrahim, M. Cyclodextrin-based nanosponge for improvement of solubility and oral bioavailability of Ellagic acid. *Pak. J. Pharm. Sci.* **2018**, *31*, 2069–2076.
87. Deshmukh, K.; Shende, P. Toluene diisocyanate cross-linked β-cyclodextrin nanosponges as a pH-sensitive carrier for naproxen. *Mater. Res. Express* **2018**, *5*. [CrossRef]
88. Cecone, C.; Caldera, F.; Trotta, F.; Bracco, P.; Zanetti, M. Controlled Release of DEET Loaded on Fibrous Mats from Electrospun PMDA/Cyclodextrin Polymer. *Molecules* **2018**, *23*, 1694. [CrossRef]
89. Rao, M.R.P.; Chaudhari, J.; Trotta, F.; Caldera, F. Investigation of Cyclodextrin-Based Nanosponges for Solubility and Bioavailability Enhancement of Rilpivirine. *AAPS PharmSciTech* **2018**, *19*, 2358–2369. [CrossRef] [PubMed]
90. Zidan, M.F.; Ibrahim, H.M.; Afouna, M.I.; Ibrahim, E.A. In vitro and in vivo evaluation of cyclodextrin-based nanosponges for enhancing oral bioavailability of atorvastatin calcium. *Drug Dev. Ind. Pharm.* **2018**, *44*, 1243–1253. [CrossRef]
91. Mendes, C.; Meirelles, G.C.; Germano, C.; Assreuy, J.; Silva, M.A.S.; Ponchel, G. Cyclodextrin based nanosponge of norfloxacin: Intestinal permeation enhancement and improved antibacterial activity. *Carbohydr. Polym.* **2018**, *195*, 586–592. [CrossRef] [PubMed]
92. Singh, P.; Ren, X.; Guo, T.; Wu, L.; Shakya, S.; He, Y. Biofunctionalization of β -cyclodextrin nanosponges using cholesterol. *Carbohydr. Polym.* **2018**, *190*, 23–30. [CrossRef] [PubMed]
93. Nait Bachir, Y.; Nait Bachir, R.; Hadj-Ziane-Zafour, A. Nanodispersions stabilized by β-cyclodextrin nanosponges: Application for simultaneous enhancement of bioactivity and stability of sage essential oil. *Drug Dev. Ind. Pharm.* **2019**, *45*, 333–347. [CrossRef]
94. Pushpalatha, R.; Selvamuthukumar, S.; Kilimozhi, D. Cross-linked, cyclodextrin-based nanosponges for curcumin delivery—Physicochemical characterization, drug release, stability and cytotoxicity. *J. Drug Deliv. Sci. Technol.* **2018**, *45*, 45–53. [CrossRef]
95. Pushpalatha, R.; Selvamuthukumar, S.; Kilimozhi, D. Carbonyl and carboxylate crosslinked cyclodextrin as a nanocarrier for resveratrol: In silico, in vitro and in vivo evaluation. *J. Incl. Phenom. Macrocycl. Chem.* **2018**, *92*, 261–272. [CrossRef]

96. Rezaei, A.; Varshosaz, J.; Fesharaki, M.; Farhang, A.; Jafari, S.M. Improving the solubility and in vitro cytotoxicity (anticancer activity) of ferulic acid by loading it into cyclodextrin nanosponges. *Int. J. Nanomed.* **2019**, *14*, 4589–4599. [CrossRef] [PubMed]
97. Gholibegloo, E.; Mortezazadeh, T.; Salehian, F.; Forootanfar, H.; Firoozpour, L.; Foroumadi, A.; Ramazani, A.; Khoobi, M. Folic acid decorated magnetic nanosponge: An efficient nanosystem for targeted curcumin delivery and magnetic resonance imaging. *J. Colloid Interface Sci.* **2019**, *556*, 128–139. [CrossRef] [PubMed]
98. Gholibegloo, E.; Mortezazadeh, T.; Salehian, F.; Ramazani, A.; Amanlou, M.; Khoobi, M. Improved curcumin loading, release, solubility and toxicity by tuning the molar ratio of cross-linker to β-cyclodextrin. *Carbohydr. Polym.* **2019**, *213*, 70–78. [CrossRef] [PubMed]
99. Dhakar, N.K.; Caldera, F.; Bessone, F.; Cecone, C.; Rubin, A.; Cavalli, R.; Dianzani, C.; Trotta, F. Evaluation of solubility enhancement, antioxidant activity, and cytotoxicity studies of kynurenic acid loaded cyclodextrin nanosponge. *Carbohydr. Polym.* **2019**, *224*. [CrossRef]
100. Sherje, A.P.; Surve, A.; Shende, P. CDI cross-linked β-cyclodextrin nanosponges of paliperidone: Synthesis and physicochemical characterization. *J. Mater. Sci. Med.* **2019**, *30*. [CrossRef]
101. Argenziano, M.; Haimhoffer, A.; Bastiancich, C.; Caldera, F.; Trotta, F.; Scutera, S.; Alotto, D.; Fumagalli, M.; Musso, T.; Castagnoli, C.; et al. In Vitro Enhanced Skin Permeation and Retention of Imiquimod Loaded in β -Cyclodextrin Nanosponge Hydrogel. *Pharmaceutics* **2019**, *11*, 138. [CrossRef]
102. Garcia-Fernandez, M.J.; Tabary, N.; Chai, F.; Cazaux, F.; Blanchemain, N.; Flament, M.P.; Martel, B. New multifunctional pharmaceutical excipient in tablet formulation based on citric acid-cyclodextrin polymer. *Int. J. Pharm.* **2016**, *511*, 913–920. [CrossRef]
103. Gangadharappa, H.V.; Chandra Prasad, S.M.; Singh, R.P. Formulation, in vitro and in vivo evaluation of celecoxib nanosponge hydrogels for topical application. *J. Drug Deliv. Sci. Technol.* **2017**, *41*, 488–501. [CrossRef]
104. Dhakar, N.K.; Matencio, A.; Caldera, F.; Argenziano, M.; Cavalli, R.; Dianzani, C.; Zanetti, M.; López-Nicolás, J.M.; Trotta, F. Comparative evaluation of solubility, cytotoxicity and photostability studies of resveratrol and oxyresveratrol loaded nanosponges. *Pharmaceutics* **2019**, *11*, 545. [CrossRef]
105. Ncube, P.; Krause, R.W.M.; Mamba, B.B. Detection of chloroform in water using an azo dye-modified β-cyclodextrin—Epichlorohydrin copolymer as a fluorescent probe. *Phys. Chem. Earth* **2014**, *67–69*, 79–85. [CrossRef]
106. Trotta, F.; Caldera, F.; Dianzani, C.; Argenziano, M.; Barrera, G.; Cavalli, R. Glutathione Bioresponsive Cyclodextrin Nanosponges. *Chempluschem* **2016**, *81*, 439–443. [CrossRef] [PubMed]
107. Argenziano, M.; Lombardi, C.; Ferrara, B.; Trotta, F.; Caldera, F.; Blangetti, M.; Koltai, H.; Kapulnik, Y.; Yarden, R.; Gigliotti, L.; et al. Glutathione/pH-responsive nanosponges enhance strigolactone delivery to prostate cancer cells. *Oncotarget* **2018**, *9*, 35813–35829. [CrossRef] [PubMed]
108. Fontana, R.M.; Milano, N.; Barbara, L.; Di Vincenzo, A.; Gallo, G.; Meo, P. Lo Cyclodextrin-Calixarene Nanosponges as Potential Platforms for pH-Dependent Delivery of Tetracycline. *ChemistrySelect* **2019**, *4*, 9743–9747. [CrossRef]
109. Trotta, F.; Caldera, F.; Cavalli, R.; Soster, M.; Riedo, C.; Biasizzo, M.; Uccello Barretta, G.; Balzano, F.; Brunella, V. Molecularly imprinted cyclodextrin nanosponges for the controlled delivery of L-DOPA: Perspectives for the treatment of Parkinson's disease. *Expert Opin. Drug Deliv.* **2016**, *13*, 1671–1680. [CrossRef]
110. Byrne, M.E.; Park, K.; Peppas, N.A. Molecular imprinting within hydrogels. *Adv. Drug Deliv. Rev.* **2002**, *54*, 149–161. [CrossRef]
111. Seong, H.; Lee, H.B.; Park, K. Glucose binding to molecularly imprinted polymers. *J. Biomater. Sci. Polym. Ed.* **2002**, *13*, 637–649. [CrossRef]
112. Deshmukh, K.; Tanwar, Y.S.; Shende, P.; Cavalli, R. Biomimetic estimation of glucose using non-molecular and molecular imprinted polymer nanosponges. *Int. J. Pharm.* **2015**, *494*, 244–248. [CrossRef]
113. Rousseau, J.; Menuel, S.; Rousseau, C.; Hapiot, F.; Monflier, E. Cyclodextrins as Porous Material for Catalysis. In *Organic Nanoreactors: From Molecular to Supramolecular Organic Compounds*; Elsevier Inc.: Amsterdam, The Netherlands, 2016; pp. 15–42. ISBN 9780128018101.
114. Zaidi, S.A. Molecular imprinting: A useful approach for drug delivery. *Mater. Sci. Energy Technol.* **2020**, *3*, 72–77. [CrossRef]
115. Pei, M.; Pai, J.Y.; Du, P.; Liu, P. Facile Synthesis of Fluorescent Hyper-Cross-Linked β-Cyclodextrin-Carbon Quantum Dot Hybrid Nanosponges for Tumor Theranostic Application with Enhanced Antitumor Efficacy. *Mol. Pharm.* **2018**, *15*, 4084–4091. [CrossRef]

116. Peila, R.; Scordino, P.; Shanko, D.B.; Caldera, F.; Trotta, F.; Ferri, A. Synthesis and characterization of β-cyclodextrin nanosponges for N,N-diethyl-meta-toluamide complexation and their application on polyester fabrics. *React. Funct. Polym.* **2017**, *119*, 87–94. [CrossRef]
117. Cecone, C.; Caldera, F.; Anceschi, A.; Scalarone, D.; Trotta, F.; Bracco, P.; Zanetti, M. One-step facile process to obtain insoluble polysaccharides fibrous mats from electrospinning of water-soluble PMDA/cyclodextrin polymer. *J. Appl. Polym. Sci.* **2018**, *135*, 46490. [CrossRef]
118. Cecone, C.; Zanetti, M.; Anceschi, A.; Caldera, F.; Trotta, F.; Bracco, P. Microfibers of microporous carbon obtained from the pyrolysis of electrospun β-cyclodextrin/pyromellitic dianhydride nanosponges. *Polym. Degrad. Stab.* **2019**, *161*, 277–282. [CrossRef]
119. Ma, M.; Li, D.Q. New organic nanoporous polymers and their inclusion complexes. *Chem. Mater.* **1999**, *11*, 872–874. [CrossRef]
120. Femminò, S.; Penna, C.; Bessone, F.; Caldera, F.; Dhakar, N.; Cau, D.; Pagliaro, P.; Cavalli, R.; Trotta, F. α-Cyclodextrin and α-Cyclodextrin Polymers as Oxygen Nanocarriers to Limit Hypoxia/Reoxygenation Injury: Implications from an In Vitro Model. *Polymers* **2018**, *10*, 211. [CrossRef]
121. Sadjadi, S.; Heravi, M.M.; Daraie, M. Cyclodextrin nanosponges: A potential catalyst and catalyst support for synthesis of xanthenes. *Res. Chem. Intermed.* **2017**, *43*, 843–857. [CrossRef]
122. Silva, F.; Caldera, F.; Trotta, F.; Nerín, C.; Domingues, F.C. Encapsulation of coriander essential oil in cyclodextrin nanosponges: A new strategy to promote its use in controlled-release active packaging. *Innov. Food Sci. Emerg. Technol.* **2019**, *56*. [CrossRef]
123. Simionato, I.; Domingues, F.C.; Nerín, C.; Silva, F. Encapsulation of cinnamon oil in cyclodextrin nanosponges and their potential use for antimicrobial food packaging. *Food Chem. Toxicol.* **2019**, *132*. [CrossRef]
124. Silva, F.; Simionato, I.; Domeño, C.; Domingues, F.C. Cyclodextrin nanosponges as a new encapsulating agent for essential oils and their effectiveness against foodborne pathogens. *Facta Univ.* **2018**, *16*, 42.
125. Salgin, S.; Salgin, U.; Vatansever, Ö. Synthesis and Characterization of β-Cyclodextrin Nanosponge and Its Application for the Removal of p-Nitrophenol from Water. *Clean Soil Air Water* **2017**, *45*. [CrossRef]
126. Pedrazzo, A.R.; Smarra, A.; Caldera, F.; Musso, G.; Dhakar, N.K.; Cecone, C.; Hamedi, A.; Corsi, I.; Trotta, F. Eco-Friendly beta-cyclodextrin and Linecaps Polymers for the Removal of Heavy Metals. *Polymers* **2019**, *11*, 1658. [CrossRef]
127. Liao, X.; Zhang, Q. Mesoporous Polymer Nanosponges Immobilized with Functional Polyols for Rapid Removal of Boric Acid and Organic Micropollutants. *ACS Appl. Polym. Mater.* **2019**, *1*, 2089–2098. [CrossRef]
128. Massella, D.; Argenziano, M.; Ferri, A.; Guan, J.; Giraud, S.; Cavalli, R.; Barresi, A.A.; Salaün, F. Bio-Functional Textiles: Combining Pharmaceutical Nanocarriers with Fibrous Materials for Innovative Dermatological Therapies. *Pharmaceutics* **2019**, *11*, 403. [CrossRef] [PubMed]
129. Mihailiasa, M.; Caldera, F.; Li, J.; Peila, R.; Ferri, A.; Trotta, F. Preparation of functionalized cotton fabrics by means of melatonin loaded β-cyclodextrin nanosponges. *Carbohydr. Polym.* **2016**, *142*, 24–30. [CrossRef] [PubMed]
130. Gentili, A. Cyclodextrin-based sorbents for solid phase extraction. *J. Chromatogr. A* **2019**, 460654. [CrossRef]
131. Sadjadi, S.; Heravi, M.M.; Daraie, M. A novel hybrid catalytic system based on immobilization of phosphomolybdic acid on ionic liquid decorated cyclodextrin-nanosponges: Efficient catalyst for the green synthesis of benzochromeno-pyrazole through cascade reaction: Triply green. *J. Mol. Liq.* **2017**, *231*, 98–105. [CrossRef]
132. Kumar, S.; Rao, R. Analytical tools for cyclodextrin nanosponges in pharmaceutical field: A review. *J. Incl. Phenom. Macrocycl. Chem.* **2019**, *94*, 11–30. [CrossRef]
133. Zhao, F.; Repo, E.; Yin, D.; Meng, Y.; Jafari, S.; Sillanpää, M. EDTA-Cross-Linked β-Cyclodextrin: An Environmentally Friendly Bifunctional Adsorbent for Simultaneous Adsorption of Metals and Cationic Dyes. *Environ. Sci. Technol.* **2015**, *49*, 10570–10580. [CrossRef]
134. Venuti, V.; Rossi, B.; D'Amico, F.; Mele, A.; Castiglione, F.; Punta, C.; Melone, L.; Crupi, V.; Majolino, D.; Trotta, F.; et al. Combining Raman and infrared spectroscopy as a powerful tool for the structural elucidation of cyclodextrin-based polymeric hydrogels. *Phys. Chem. Chem. Phys.* **2015**, *17*, 10274–10282. [CrossRef]
135. Hayiyana, Z.; Choonara, Y.; Makgotloe, A.; Toit, L.; Kumar, P.; Pillay, V. Ester-Based Hydrophilic Cyclodextrin Nanosponges for Topical Ocular Drug Delivery. *Curr. Pharm. Des.* **2017**, *22*, 6988–6997. [CrossRef]
136. Pawar, S.; Shende, P.; Trotta, F. Diversity of β -cyclodextrin-based nanosponges for transformation of actives. *Int. J. Pharm.* **2019**, *565*, 333–350. [CrossRef]

137. Morin-Crini, N.; Winterton, P.; Fourmentin, S.; Wilson, L.D.; Fenyvesi, É.; Crini, G. Water-insoluble β-cyclodextrin–epichlorohydrin polymers for removal of pollutants from aqueous solutions by sorption processes using batch studies: A review of inclusion mechanisms. *Prog. Polym. Sci.* **2018**, *78*, 1–23. [CrossRef]
138. Deshmukh, K.; Tanwar, Y.S.; Sharma, S.; Shende, P.; Cavalli, R. Functionalized nanosponges for controlled antibacterial and antihypocalcemic actions. *Biomed. Pharmacother.* **2016**, *84*, 485–494. [CrossRef] [PubMed]
139. Russo, M.; Saladino, M.L.; Chillura Martino, D.; Lo Meo, P.; Noto, R. Polyaminocyclodextrin nanosponges: Synthesis, characterization and pH-responsive sequestration abilities. *RSC Adv.* **2016**, *6*, 49941–49953. [CrossRef]
140. Junthip, J.; Promma, W.; Sonsupap, S.; Boonyanusith, C. Adsorption of paraquat from water by insoluble cyclodextrin polymer crosslinked with 1,2,3,4-butanetetracarboxylic acid. *Iran. Polym. J.* **2019**, *28*, 213–223. [CrossRef]
141. Singireddy, A.; Pedireddi, S.R.; Subramanian, S. Optimization of reaction parameters for synthesis of Cyclodextrin nanosponges in controlled nanoscopic size dimensions. *J. Polym. Res.* **2019**, *26*. [CrossRef]

polymers

MDPI

Article

Delayed Addition of Template Molecules Enhances the Binding Properties of Diclofenac-Imprinted Polymers

Laura Anfossi, Simone Cavalera, Fabio Di Nardo, Giulia Spano, Cristina Giovannoli and Claudio Baggiani *

Department of Chemistry, University of Torino, 10125 Torino, Italy; laura.anfossi@unito.it (L.A.); simone.cavalera@unito.it (S.C.); fabio.dinardo@unito.it (F.D.N.); giulia.spano@unito.it (G.S.); cristina.giovannoli@unito.it (C.G.)

* Correspondence: claudio.baggiani@unito.it

Received: 24 April 2020; Accepted: 18 May 2020; Published: 21 May 2020

Abstract: It has been reported that in the molecular imprinting technique, the use of preformed oligomers instead of functional monomers increases the stability of the non-covalent interactions with the template molecule, providing a sharp gain in terms of binding properties for the resulting imprinted polymer. Based on this theory, we assumed that the delayed addition of template molecules to a polymerization mixture enhances the binding properties of the resulting polymer. To verify this hypothesis, we imprinted several mixtures of 4-vinylpyridine/ethylene dimethacrylate (1:6 mol/mol) in acetonitrile by adding diclofenac progressively later from the beginning of the polymerization process. After polymerization, the binding isotherms of imprinted and non-imprinted materials were measured in acetonitrile by partition equilibrium experiments. Binding data confirm our hypothesis, as imprinted polymers prepared by delayed addition, with delay times of 5 and 10 min, showed higher binding affinity (K_{eq} = 1.37 × 10^4 L mol^{-1} and 1.80 × 10^4 L mol^{-1}) than the polymer obtained in the presence of template at the beginning (K_{eq} = 5.30 × 10^3 L mol^{-1}). Similarly, an increase in the imprinting factor measured vs. the non-imprinted polymer in the binding selectivity with respect to mefenamic acid was observed. We believe that the delayed addition approach could be useful in prepar imprinted polymers with higher binding affinity and increased binding selectivity in cases of difficult imprinting polymerization.

Keywords: diclofenac; mefenamic acid; molecular imprinting; delayed addition; binding isotherm; binding selectivity; imprinting factor

1. Introduction

Molecularly imprinted polymers (MIPs) are obtained by polymerizing a mixture of cross-linkers and functional monomers in the presence of a template molecule (i.e., the pre-polymerization mixture) [1–3]. The current molecular imprinting concept assumes that the origin of imprinted binding sites is the non-covalent complexes between the template molecules and functional monomers in the pre-polymerization mixture, and that the polymerization process stabilizes these complexes through the formation of a dense three-dimensional polymeric network around the template molecules [4,5]. Typically, the resulting MIP shows high affinity for the template molecule, with binding equilibrium constants (K_{eq}) on the order of 10^3–10^5 L mol^{-1}; whereas in pre-polymerization mixtures, the strength of the template–functional monomer interaction is usually much weaker, with K_{eq} values between 1 and 10 L mol^{-1} [6,7]. This binding enhancement is attributed to the growth of a three-dimensional polymer network that progressively embeds the template molecules. Because of the oligomeric chains' flexibility, conformational changes maximizing the interactions with the template are allowed. The result is a net entropic gain, but while polymerization is proceeding, the newly formed cross-links rapidly stiffen the polymer structure around the template and freeze the binding pocket in its definitive shape.

The role of the oligomeric chains as functional "macromonomers" has been shown in several ways. Efficient target rebinding has been reported when linear pre-polymerized oligomeric chains that were synthesized onto the surface of silica beads were used to imprint cholic acid [8], uric acid [9], and the alkaloid cytisine [10]. In this approach, macromonomers are formed in situ before the beginning of the polymerization process, and cross-linking is achieved by the addition of a proper bifunctional reagent after the addition of the template. Porogen-soluble linear oligomers have been synthesized and directly used as functional macromonomers to imprint (−)-cinchonidine [11], theophylline [12], bisphenol A, atrazine, and ketoprofen [13]. Several other proteins have been imprinted through a similar approach in the so-called "assisted recognition by polymeric chain" (ARPC) method [14–20]. The main drawbacks associated with the use of pre-polymerized oligomeric chains as functional macromonomers consist of complexity in their preparation (requiring long, multistep synthetic methods), the difficulty of their isolation from the reaction mixture, and difficulty in their manipulation due to their viscous nature. A possible alternative to overcome these drawbacks could consist of avoiding the isolation of the pre-polymerized functional macromonomers, by directly using oligomeric chains formed early in situ at the beginning of the bulk polymerization process, before reaching the gelling stage.

Since diclofenac is a proof-of-concept template molecule, the goal of this paper is to show that an enhanced imprinting effect can be obtained by delaying the addition of the template molecule because of the presence of a large number of growing oligomeric chains that are able to settle multiple non-covalent interactions with the template itself. We have experimentally verified this hypothesis by studying a series of diclofenac-imprinted polymers that were prepared by adding the template molecule at fixed times after the start of polymerization and comparing their binding properties with those of a similar MIP that was prepared using the usual approach, that is, by adding the template at the start of the polymerization process.

2. Materials and Methods

2.1. Materials

2,2′-Azobis-(2-methylpropionitrile), diclofenac sodium (DIC), ethylene dimethacrylate, mefenamic acid (MEF), and 4-vinylpyridine were purchased from Sigma-Aldrich-Fluka (Milan, Italy). Polymerization inhibitors were removed from the monomer solutions by clean-up on activated alumina. Acetic acid, acetonitrile, and methanol were purchased from VWR International (Milan, Italy). All the reagents used were of analytical grade. Diclofenac was transformed in its free acid form as follows: 250 mg of sodium salt was dissolved in 40 mL of deionized water. Then, the pH was adjusted to 1 using aqueous hydrogen chloride 1 M, and the turbid suspension was extracted three times using 15 mL of chloroform. The organic fractions were dried over anhydrous sodium sulphate and evaporated to dryness under reduced pressure, yielding the free acid quantitatively. Diclofenac and mefenamic acid stock solutions were prepared by dissolving 25 mg of the substance in 25 mL of acetonitrile then stored in the dark at −20 °C.

2.2. Polymers Preparation

A 300 μg/mL diclofenac solution in acetonitrile was prepared by dissolving the template under sonication and maintaining it at 60 °C under a nitrogen atmosphere. In 4-mL glass vials equipped with Mininert valves, pre-polymerization mixtures were prepared by dissolving 108 μL of 4-vinylpyridine (1.0 mmol), 1.13 mL of ethylene dimethacrylate (6 mmol), and 2 mg of 2,2′-azobis-(2-methylpropionitrile) in 400 μL of acetonitrile. The vials were purged with nitrogen, sealed, and left to polymerize at 60 °C. The template was added at 0, 5, 10, 15, 20, and 30 min from the start of polymerization by adding 245 μL of diclofenac solution (0.25 mmol) with an HPLC syringe and rapidly vortexing the vials. After overnight polymerization, the obtained bulk polymers were broken with a steel spatula, mechanically ground in a mechanical ball mill, and wet-sieved to 15–38 μm particle size. The particulate was packed in 5-mL polypropylene SPE cartridges and exhaustively washed with methanol–acetic acid 9:1 (*v/v*) until the polymers were deemed free from diclofenac by HPLC analysis of the eluates. Then, the particulate was washed with abundant acetonitrile, dried in the oven at 80 °C, and stored at room temperature. A blank polymer (NIP) was prepared and treated in the same manner, but without the template.

2.3. Light Scattering of Polymerization Mixtures

Pre-polymerization mixtures were polymerized for 5, 10, 15, and 20 min, then diluted 1:1 (*v/v*) with 5 mol L^{-1} hydroquinone in acetonitrile to stop the polymerization process. The size distribution of the macromonomers was determined at 25 °C using an ALV/NIBS-HPPS particle sizer equipped with an ALV-5000 multiple tau digital correlator (ALV, Langen, Germany). The measured autocorrelation function of the scattered light was processed by ALV Correlator software (ver.3.0), obtaining the number-weighted size distribution of the macromonomers.

2.4. HPLC Analysis

Reverse-phase HPLC analysis was performed on the Nova-Pak C18 (125 × 3.9 mm, 5 μm) from VWR (Milano, Italy). The HPLC apparatus was a LaChrom *Elite* (L-2130 constant-flow quaternary pump, L-2400 UV-Vis detector, L-2200 autosampler, and data acquisition system EZChrom *Elite* 3.1) from VWR Hitachi (Milan, Italy). The mobile phase was a 55:45 (*v/v*) acetonitrile–0.02 mole L^{-1} sodium phosphate buffer with a pH of 4.6. The flow rate was set at 1.0 mL/min, the injection volume was 5 μL, and the detection wavelength was 278 nm.

Diclofenac and mefenamic acid standard solutions at concentrations ranging from 2 to 200 μg/mL were prepared in the mobile phase immediately before use. The standards were analyzed three times consecutively and peak areas were plotted against the concentration. The calibration plot was drawn by using a weighted linear regression (weight = 1/conc, r^2 = 0.998). The limit of quantification (LOQ_{DIC} = 2.8 μg/mL, LOQ_{MEF} = 2.6 μg/mL) was calculated as LOQ = 10 Sy/b, where Sy is the standard error of the response and b is the slope of the calibration plot.

2.5. Equilibrium Batch Rebinding

About 10 mg of the polymer was exactly weighed in 2-mL flat-bottom amber glass vials. Then, 500 μL of acetonitrile solutions containing increasing amounts of diclofenac or mefenamic acid ranging from 4 to 200 μg were added and sonicated for 10 min. The vials were incubated overnight at room temperature under continuous agitation on a horizontal rocking table. The solutions were filtered on 0.22 μm nylon membranes, and free amounts of ligand were measured by HPLC analysis. Each experimental point was assessed as the average of three repeated measurements. The binding isotherms were calculated using the SigmaPlot 12 (Systat Software Inc., Richmond, CA, USA). Non-linear least-square fitting was applied to the averaged experimental data, using a simple Langmuir isotherm model [21,22]:

$$B = \frac{K_{eq} B_{max} F}{1 + K_{eq} F}, \quad (1)$$

where B is the ligand bound to the polymer, F is the ligand not bound to the polymer, K_{eq} is the equilibrium binding constant, and B_{max} is the binding site concentration. To ensure robust results, weighted (1/y) Pearson VII limit minimization was chosen as the minimization method. To avoid being trapped in local minima, which might give incorrect results, the fitting was carried out several times by using different initial guess values for the isotherm parameters.

The imprinting factor, IF, was calculated as the ratio between the equilibrium binding constants relative to the imprinted and non-imprinted polymers:

$$IF = \frac{K_{\text{eq}}^{MIP}}{K_{\text{eq}}^{NIP}}. \tag{2}$$

The binding selectivity, α, was calculated as the ratio between the equilibrium binding constants relative to mefenamic acid and diclofenac:

$$\alpha = \frac{K_{\text{eq}}^{MEF}}{K_{\text{eq}}^{DIC}}. \tag{3}$$

3. Results and Discussion

3.1. Synthesis of Imprinted Polymers

A preliminary study on the gelation time of the pre-polymerization mixtures was performed in the same experimental conditions used to prepare diclofenac-imprinted polymers. They were based on a previously reported formulation producing an imprinted polymer with good binding properties towards the chosen template molecule [23]. These experiments show that turbidity was clearly appreciable with the naked eye after about 40 min from the start of the thermal polymerization process. Early formation of nanogel particles in the pre-polymerization mixture was confirmed by dynamic light scattering measurements performed 5, 10, 15, and 20 min from the start of the polymerization (Figure S1). The presence of nanoparticles became experimentally measurable from 10 min onwards and the corresponding diameters were 15.5 ± 2.9 (10 min), 18.8 ± 4.4 (15 min), and 24.5 ± 8.3 nm (20 min), with a polydispersity index (PDI) of 0.61, 0.68, and 0.82, respectively.

On the basis of these preliminary studies, it was decided to carry on with the delayed addition experiments by adding preheated diclofenac solutions to the polymerization mixtures at 5, 10, 15, 20, and 30 min from the start of the polymerization process. Immediately after the addition of the template, a clear change in the solution color from yellow to dark orange was observed for all the considered mixtures. In addition, after the gelling phase started, the color of bulk polymers progressively changed from dark orange to deep red. On the other hand, the addition of acetonitrile only to the control mixture polymerized without the template did not affect the solution's color at all, although a change from yellow to pink after the gelation of the mixture was observed (Figure 1). In fact, as previously reported in the literature [24], when 4-vinylpyridine is used as the functional monomer, the change in the adsorption spectra after template addition would qualitatively confirm the presence of non-covalent interactions between the template molecules and pre-polymerization mixtures, without giving a direct insight on the extent of such interactions. However, it is necessary to add that the color of the polymerization mixture in the presence of the template can be attributed with good approximation to the presence of complexes between the oligomers in the formation, the template itself, and traces of metal present and deriving from the alumina used to remove the inhibitor radicals from the monomers [25]. Since these complexes are known and reported in the literature [26,27], their contribution cannot be ruled out a priori.

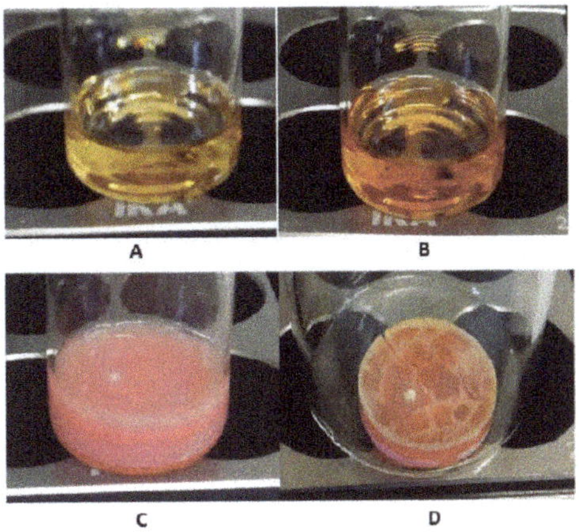

Figure 1. (**A**) The MIP-10 polymerization mixture immediately before the addition of diclofenac; (**B**) The MIP-10 polymerization mixture after the addition of diclofenac; (**C**) The non-imprinted polymer (NIP) polymerization mixture immediately after gelation; (**D**) The MIP-10 polymerization mixture after complete gelation. MIP-10: molecularly imprinted polymers (MIPs) prepared by template addition 10 min from the start of the polymerization.

3.2. Binding Properties of Imprinted Polymers

The binding properties of bulk polymers were quantitatively evaluated by measuring the binding isotherms for the template diclofenac and the related mefenamic acid (Figure 2), respectively.

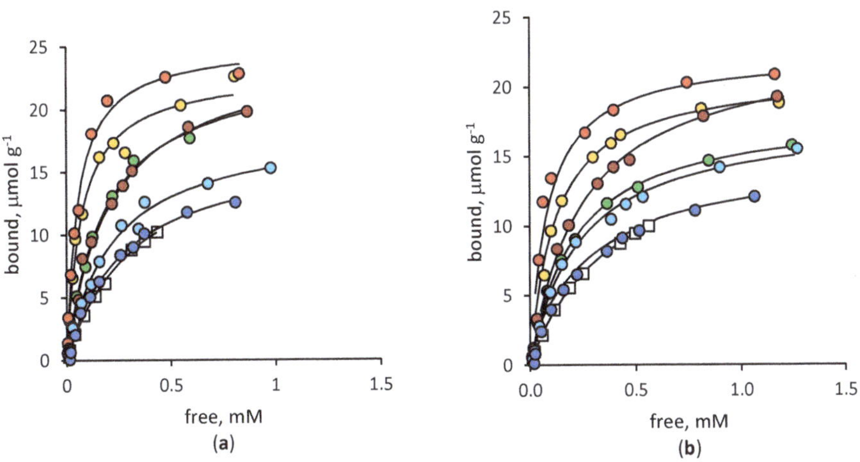

Figure 2. Binding isotherms for NIP (open squares), MIP-0 (green circles), MIP-5 (yellow circles), MIP-10 (red circles), MIP-15 (dark red circles), MIP-20 (cyan circles), MIP-30 (blue circles). (**a**) Diclofenac; (**b**) Mefenamic acid.

Considering the binding equilibrium constants reported in Table 1 (the corresponding statistical evaluation is reported as Tables S1–S3), the imprinted polymer prepared in the presence of the template at the beginning of the polymerization process (MIP-0) showed a statistically significant ($\alpha = 0.05$, $n = 11$, $t = 3.427$) affinity that was higher than that of NIP for diclofenac. This is obvious, as it must happen in the case of a successful imprinting effect. It is worth noting that this difference between MIP and NIP progressively increased in the case of MIPs prepared by template addition at 5 and 10 min from the start of the polymerization (MIP-5, MIP-10), whereas it showed a sharp decrease, eventually becoming statistically indistinguishable from NIP, when the template was added 30 min from the start ($\alpha = 0.05$, $n = 11$, $t = 1.925$). The same behavior could be observed in the case of mefenamic acid, where the affinity increased from MIP-0 to MIP-10, then dropped rapidly and became indistinguishable from the NIP for MIP-30 ($\alpha = 0.05$, $n = 11$, $t = 2.052$).

Regarding the binding site concentration, MIP-0 showed a statistically significant difference ($p = 0.05$, $n = 11$, $t = 4.075$) with respect to NIP in the presence of diclofenac as a ligand. This difference slightly increased from MIP-5 to MIP-10, whereupon binding site concentration values started to decrease until they became statistically indistinguishable from NIP ($p = 0.05$, $n = 11$, MIP-20, $t = 0.247$, MIP-30, $t = 1.448$). Concerning mefenamic acid as a ligand representative of diclofenac-analogous molecules, the binding site concentration showed the same trend, although the values were slightly lower. It must be noted that this difference was statistically significant only for NIP and MIP from MIP-0 to MIP-10, while it was not for the remaining polymers.

Table 1. Calculated binding parameters (±standard error) for diclofenac and mefenamic acid measured on non-imprinted (NIP) and imprinted (MIP) polymers prepared by adding the template at 0, 5, 10, 15, 20, and 30 min from the start of polymerization.

	Diclofenac		Mefenamic Acid	
Polymer	**K_{eq}, 10^3 L mol^{-1}**	**B_{max}, µmol g^{-1}**	**K_{eq}, 10^3 L mol^{-1}**	**B_{max}, µmol g^{-1}**
NIP	2.86 ± 0.18	18.4 ± 0.7	2.72 ± 0.11	16.4 ± 0.3
MIP-0	5.30 ± 0.62	24.1 ± 1.1	3.86 ± 0.12	18.5 ± 0.2
MIP-5	13.70 ± 2.04	23.0 ± 1.0	8.58 ± 0.20	20.9 ± 0.1
MIP-10	17.98 ± 1.68	25.1 ± 0.7	12.03 ± 1.41	22.4 ± 1.3
MIP-15	4.75 ± 0.47	24.8 ± 1.0	4.07 ± 0.26	23.2 ± 0.6
MIP-20	4.55 ± 0.52	18.7 ± 0.3	4.30 ± 0.33	17.8 ± 0.5
MIP-30	3.70 ± 0.36	17.0 ± 0.3	3.39 ± 0.28	15.3 ± 0.5

The effect of the delayed template addition can be further highlighted by considering the imprinting factors, as reported in Figure 3. When the template was present in the polymerization mixture from the start of the process, the resulting polymer (MIP-0) showed a relatively small but statistically significant imprinting effect for both diclofenac ($\alpha = 0.05$, $n = 11$, $t = 3.509$) and mefenamic acid ($\alpha = 0.05$, $n = 10$, $t = 6.003$). Meanwhile, in conditions where delayed addition was employed, the imprinting effect markedly increased when the template was added after 5 and 10 min (MIP-5, MIP-10), but did not when the template was added later (MIP-15, MIP-20). Likewise, the imprinting effect was completely suppressed in the case of MIP-30.

As a consequence of the changing binding properties of the MIP, the binding selectivity was also clearly affected by the delayed template addition in the polymerization mixture. As reported in Figure 4, the NIP did not show any binding selectivity between the template diclofenac and the related mefenamic acid ($\alpha = 0.95 \pm 0.07$), while the polymer prepared in the presence of the template from the beginning of the polymerization process (MIP-0) showed a moderate degree of binding selectivity ($\alpha = 0.73 \pm 0.09$). As in the case of the imprinting factor, in the presence of delayed addition conditions, the binding selectivity markedly increased when the template was added after 5 and 10 min (MIP-5, $\alpha = 0.63 \pm 0.10$; MIP-10, $\alpha = 0.67 \pm 0.10$), but not when the template was added 15 min from the beginning of the polymerization process (MIP-15, $\alpha = 0.86 \pm 0.10$). Furthermore, the binding selectivity was completely lost in the case of polymers prepared with even later addition of the template (MIP-20, $\alpha = 0.95 \pm 0.13$; MIP-30, $\alpha = 0.92 \pm 0.12$).

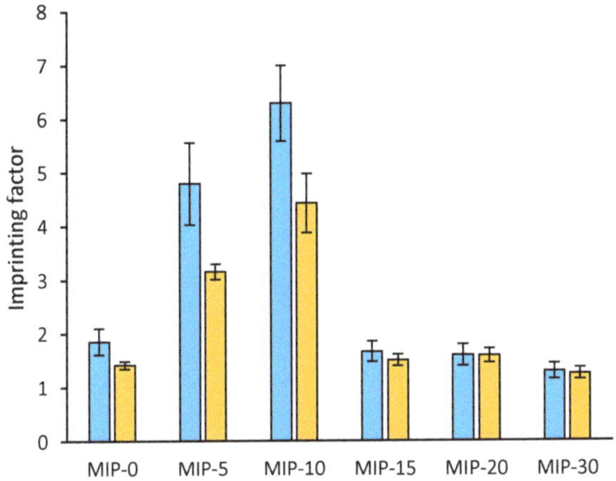

Figure 3. Imprinting factors (±standard error) for diclofenac (cyan bars) and mefenamic acid (yellow bars), calculated as the ratio between the equilibrium binding constants relative to the ligand for the imprinted and non-imprinted polymers.

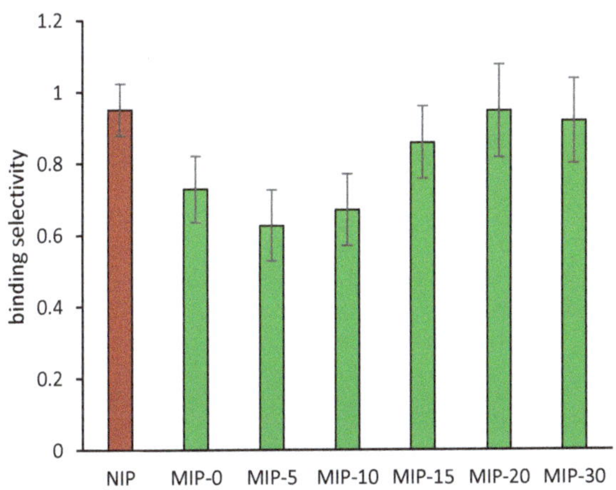

Figure 4. Binding selectivity (±standard error), calculated as the ratio between the equilibrium binding constants relative to mefenamic acid and diclofenac.

4. Discussion

From the experimental data obtained, it is worth highlighting that the addition of template molecules soon after the start of the polymerization process (5–10 min) enhanced the imprinting effect and binding selectivity, mainly by increasing the binding affinity constant of the resulting polymer. On the contrary, when template molecules were added later (15–30 min), they no longer seemed able to imprint the polymer effectively. To explain this behavior, it is useful to consider the way a bulk polymer is formed from its polymerization mixture. It is common knowledge that at the start of the polymerization process, monomers rapidly form oligomeric chains. At first, chains are poorly cross-linked and therefore highly flexible and soluble in the polymerization medium. When they grow further, cross-linking increases until they exceed their solubility limit, giving rise to nanogel particles, which may further grow and aggregate to finally form the bulk polymer. As recently demonstrated in the case of imprinted polymers prepared by solid-phase synthesis [28], the interactions between the template molecules and the pre-polymerization mixture can lead to three different outcomes:

(a) When template molecules are present in the pre-polymerization mixture from the start, they will find a reaction medium that is very rich in functional monomers, but with no oligomers. Because the polymerization process is fast and dominated by cross-linking steps, weak complexes between template molecules and functional monomers will prevail by producing an MIP with a relatively low imprinting effect.

(b) When template molecules are added after the start of the polymerization process, they will find a reaction medium enriched with functional macromonomers and only a few remaining monomers. Because the functional macromonomers are highly flexible, they will freely rearrange around the template. Chelate effects and chain desolvation greatly stabilize these complexes, producing an MIP with an enhanced imprinting effect.

(c) When template molecules are added very late in the polymerization process, they will find a reaction medium mainly composed of preformed nanogel particles that will form a continuous cross-linked macrogel. In this condition, the interactions between the template molecules and the growing polymer particles become increasingly disadvantaged, since cross-linked nanogel particles are stiffer than functional macromonomers. As a consequence, an MIP with a low or negligible imprinting effect is the more predictable outcome.

5. Conclusions

Our experimental data confirm that an enhanced imprinting effect occurred when a delayed addition of diclofenac as the template molecule in a polymerization mixture was performed. This behavior can be explained by the presence of growing oligomeric chains capable of settling multiple non-covalent interactions with the template itself, as well as the possible presence of metal ions. The binding isotherms highlight that the imprinting effect was due to the increase in binding affinity. We hypothesize that the experimental results described here can be related to a general behavior of any imprinted polymer prepared by radical polymerization, regardless of the nature of the template molecule, its final morphology, and the resulting structure. We speculate that this effect could be controlled by the nature of the polymerization process. Furthermore, we believe that the delayed addition approach could be useful whenever difficult imprinting polymerization takes place, as in the case of template molecules that are unable to establish strong interactions with functional monomers, and thus lead to the formation of imprinted polymers with high imprinting factor.

Supplementary Materials: The following are available online at http://www.mdpi.com/2073-4360/12/5/1178/s1. Figure S1: Dynamic light scattering of polymerization mixtures. Tables S1 and S2: Statistical evaluation of ligand binding to polymers. Table S3: Statistical evaluation of imprinting factors.

Author Contributions: Conceptualization, C.B.; methodology, C.B. and C.G.; investigation, L.A., S.C. and F.D.N.; data curation, G.S.; writing—original draft preparation, C.G.; writing—review and editing, C.B. All authors have read and agreed to the published version of the manuscript.

Funding: This research received no external funding.

Acknowledgments: The authors are grateful to Marco Minella (Department of Chemistry, University of Torino) for providing the technical assistance in performing the DLS measurements.

Conflicts of Interest: The authors declare no conflict of interest.

References

1. Haupt, K.; Linares, A.V.; Bompart, M.; Tse Sum Bui, B. Molecularly imprinted polymers. *Top. Curr. Chem.* **2012**, *325*, 1–28.
2. Chen, L.; Xu, W.; Li, J. Recent advances in molecular imprinting technology: Current status, challenges and highlighted applications. *Chem. Soc. Rev.* **2011**, *40*, 2922–2942. [CrossRef]
3. Chen, L.; Wang, X.; Lu, W.; Wu, X.; Li, J. Molecular imprinting: Perspectives and applications. *Chem. Soc. Rev.* **2016**, *45*, 2137–2211. [CrossRef]
4. Biffis, A.; Graham, N.B.; Siedlaczek, G.; Stalberg, S.; Wulff, G. The synthesis, characterization and molecular recognition properties of imprinted microgels. *Macromol. Chem. Phys.* **2001**, *202*, 163–171. [CrossRef]

5. Salian, V.D.; Byrne, M.E. Living radical polymerization and molecular imprinting: Improving polymer morphology in imprinted polymers. *Macromol. Mater. Eng.* **2013**, *298*, 379–390. [CrossRef]
6. O'Mahony, J.; Molinelli, A.; Nolan, K.; Smyth, M.R.; Mizaikoff, B. Towards the rational development of molecularly imprinted polymers: 1H NMR studies on hydrophobicity and ion-pair interactions as driving forces for selectivity. *Biosens. Bioelectron.* **2005**, *20*, 1884–1893. [CrossRef] [PubMed]
7. O'Mahony, J.; Molinelli, A.; Nolan, K.; Smyth, M.R.; Mizaikoff, B. Anatomy of a successful imprint: Analysing the recognition mechanisms of a molecularly imprinted polymer for quercetin. *Biosens. Bioelectron.* **2006**, *21*, 1383–1392. [CrossRef] [PubMed]
8. Gao, B.; Lu, J.; Chen, Z.; Guo, J. Preparation and recognition performance of cholic acid-imprinted material prepared with novel surface-imprinting technique. *Polymer* **2009**, *50*, 3275–3284. [CrossRef]
9. Gao, B.; Liu, S.; Li, Y. Preparation and recognition performance of uric acid-imprinted material prepared with novel surface imprinting technique. *J. Chromatogr. A* **2010**, *1217*, 2226–2236. [CrossRef] [PubMed]
10. Gao, B.; Qinyuan, N.; Ruikui, D. Preparation and recognition performance of cytisine alkaloid-imprinted material prepared using novel surface molecular imprinting technique. *J. Sep. Sci.* **2010**, *33*, 1338–1348. [CrossRef] [PubMed]
11. Matsui, J.; Tamaki, K.; Sugimoto, N. Molecular imprinting in alcohols: Utility of a pre-polymer based strategy for synthesizing stereoselective artificial receptor polymers in hydrophilic media. *Anal. Chim. Acta* **2002**, *466*, 11–15. [CrossRef]
12. Li, Z.; Day, M.; Ding, J.; Faid, K. Synthesis and characterization of functional methacrylate copolymers and their application in molecular imprinting. *Macromolecules* **2005**, *38*, 2620–2625. [CrossRef]
13. Kitayama, Y.; Yoshikawa, K.; Takeuchi, T. Post-cross-linked molecular imprinting with functional polymers as a universal building block for artificial polymeric receptors. *Macromolecules* **2017**, *50*, 7526–7534. [CrossRef]
14. Long, Y.; Xing, X.; Han, R.; Sun, Y.; Wang, Y.; Zhao, Z.; Mi, H. Two-step purification of low-content cellular protein using protein-imprinted polymers. *Anal. Biochem.* **2008**, *380*, 268–275. [CrossRef] [PubMed]
15. Guo, M.; Zhao, Z.; Fan, Y.; Wang, C.; Shi, L.; Xia, J.; Long, Y.; Mi, H. Protein-imprinted polymer with immobilized assistant recognition polymer chains. *Biomaterials* **2006**, *27*, 4381–4387. [CrossRef] [PubMed]
16. Han, R.; Xing, X.; Wang, Y.; Long, Y.; Sun, Y.; Zhao, Z.; Mi, H. Separation/enrichment of active natural low content protein using protein imprinted polymer. *J. Chromatogr. B* **2008**, *873*, 113–118. [CrossRef]
17. Gao, J.; Tian, H.; Wang, Y.; Yang, Q.; Liu, D.; Wang, Y.; Mi, H. The design of protein-imprinted polymers as antibody substitutes for investigating proteineprotein interactions. *Biomaterials* **2012**, *33*, 3344–3352. [CrossRef]
18. Liu, D.; Yang, Q.; Jin, S.; Song, Y.; Gao, J.; Wang, Y.; Mi, H. Core-shell molecularly imprinted polymer nanoparticles with assistant recognition polymer chains for effective recognition and enrichment of natural low-abundance protein. *Acta Biomater.* **2014**, *10*, 769–775. [CrossRef]
19. Zhang, X.; Liu, S.; Pan, J.; Jia, H.; Chen, Z.; Guo, T. Multifunctional oligomer immobilized on quartz crystal microbalance: A facile and stabilized molecular imprinting strategy for glycoprotein detection. *Anal. Bioanal. Chem.* **2019**, *411*, 3941–3949. [CrossRef]
20. Zhao, W.; Li, B.; Xu, S.; Huang, X.; Luo, J.; Zhu, Y.; Liu, X. Electrochemical protein recognition based on macromolecular self-assembly of molecularly imprinted polymer: A new strategy to mimic antibody for label-free biosensing. *J. Mater. Chem. B* **2019**, *7*, 2311–2319. [CrossRef]
21. Garcia-Calzon, J.A.; Diaz-Garcia, M.E. Characterization of binding sites in molecularly imprinted polymers. *Sens. Actuat. B* **2007**, *123*, 1180–1194. [CrossRef]
22. Kumar, K.V.; Gadipelli, S.; Wood, B.; Ramisetty, K.A.; Stewart, A.A.; Howard, C.A.; Brett, D.J.L.; Rodriguez-Reinoso, F. Characterization of the adsorption site energies and heterogeneous surfaces of porous materials. *J. Mater. Chem. A* **2019**, *7*, 10104–10137. [CrossRef]
23. Baggiani, C.; Giovannoli, C.; Anfossi, L.; Passini, C.; Baravalle, P.; Giraudi, G. A connection between the binding properties of imprinted and non-imprinted polymers: A change of perspective in molecular imprinting. *J. Am. Chem. Soc.* **2012**, *134*, 1513–1518. [CrossRef] [PubMed]
24. Muhammad, T.; Cui, L.; Wang, J.D.; Piletska, E.V.; Guerreiro, A.R.; Piletsky, S.A. Rational design and synthesis of water-compatible molecularly imprinted polymers for selective solid phase extraction of amiodarone. *Anal. Chim. Acta* **2012**, *709*, 98–104. [CrossRef]
25. Khvostikov, V.A.; Karandashev, V.K.; Burmii, Z.P. Analysis of α-alumina samples by mass spectrometry with inductively coupled plasma and laser ablation. *Inorg. Mater.* **2015**, *51*, 1410–1415. [CrossRef]

26. Li, X.Y.; Ma, L.; Huang, Y.P.; Liu, Z.S.; Aisa, H.A. Preparation of metallic pivot-based imprinted monoliths with a hydrophilic macromonomer. *RSC Adv.* **2015**, *5*, 36753–36761. [CrossRef]
27. Feng, J.; Li, F.; Ran, R.X.; Huang, Y.P.; Liu, Z.S. Synergistic effect of metal ions pivot and macromolecular crowding reagents on affinity of molecularly imprinted polymer. *Eur. Polym. J.* **2019**, *120*, 109242. [CrossRef]
28. Cowen, T.; Stefanucci, E.; Piletska, E.; Marrazza, G.; Canfarotta, F.; Piletsky, S.A. Synthetic mechanism of molecular imprinting at the solid phase. *Macromolecules* **2020**, *53*, 1435–1442. [CrossRef]

polymers

MDPI

Review

Life-Cycle Assessment in the Polymeric Sector: A Comprehensive Review of Application Experiences on the Italian Scale

Ignazio Blanco [1,*], Carlo Ingrao [2] and Valentina Siracusa [2]

1 Department of Civil Engineering and Architecture, University of Catania an UdR-Catania Consorzio INSTM, Viale Andrea Doria 6, 95125 Catania, Italy

2 Department of Chemical Sciences, University of Catania, Viale Andrea Doria 6, 95125 Catania, Italy; ing.carloingrao@gmail.com (C.I.); vsiracus@dmfci.unict.it (V.S.)

* Correspondence: iblanco@unict.it

Received: 2 May 2020; Accepted: 23 May 2020; Published: 26 May 2020

Abstract: In recent years, a growing media campaign has demonized the use of plastic tout court, as solely responsible for environmental problems. Behind what is now vulgarly called plastic there are actually many applications and uses without which our daily life would be greatly penalized in the most common and routine actions. Our belief, in the role of researchers who have made polymers and their derivatives their main research object, is that sustainable use of polymeric materials is not only possible but is above all necessary. For this reason, in this review which is part of the Special Issue "State-of-the-Art Polymer Science and Technology in Italy", we offer a rundown of life-cycle assessment (LCA) studies on polymers used in the most important production and commercial sectors carried out in the last few years by Italians researchers.

Keywords: polymers; life cycle assessment; polymeric materials; sustainability; plastic; food packaging; LCA

---

1. Introduction

'Plastic' is now accepted as a universal term encompassing the huge array of polymers that are currently available in the market and are characterized by widely diverse chemical composition, mechanical properties, manufacturing methods and raw material feedstock (e.g., petroleum vs. renewable resources) [1]. This is what has made it possible to utilize them in almost every human activity [2].

Plastics present chemico-physical, mechanical, optical, and biological properties that, being adjustable for a huge number of disparate tasks, have contributed significantly to the expanding utilization of those plastics, making them increasingly prevalent in daily life [2,3].

It is estimated that since 1950 8.3 billion tons of plastics have been manufactured worldwide [1]. In 2018, global plastics production almost reached 360 million tons, with Europe contributing about 17% of that amount. The vast majority of the polymers utilized in the plastic industry are fossil-derived, which means that they are produced from crude oil and other fuels like natural gas and coal [1,3]. Most of them fall into the following categories, as defined by the Plastics Industry Society: poly(ethylene terephthalate) (PET), high-density polyethylene (HDPE), poly(vinyl chloride) (PVC), low-density polyethylene (LDPE), polypropylene (PP), polystyrene (PS) and 'others' [1].

Polyolefins, mainly PP and polyethylene (PE), are the leading polymers, covering around 50% of all possible applications [4], as shown in Figure 1.

Figure 1. Distribution of plastics demand by resin types in the year 2018. Reprinted from [4], © 2020 with permission from PlasticsEurope.

The packaging sector is hugely demanding of those polyolefins and of other polymers, like PET whose core application is the packaging of water and other beverages [5], and PS for the production of fresh-food packaging trays [6]. The huge consumption of polymers in the packaging sector should be attributed to a combination of benefits including light weight, flexibility, strength, transparency, impermeability, and ease of sterilization [7,8].

Other highly demanding sectors are buildings, mainly because of the consumption of "hard" plastic materials like HDPE and PVC, and automotive, due to utilization of PP and other thermoplastics. Additionally, agriculture is a big consumer of PP and LDPE, which are utilized mostly for the production of protected cultivation films, nets, pipes and irrigation/drainage systems, and packaging [9]. The wide range of these applications have led to the elaboration of the new term 'plasticulture', to strengthen the importance that plastic commodities and related manufacturing industries have in agriculture [2,10].

All over Europe, consumption of plastic films for covering tunnels in A-shaped greenhouses and in tunnel-greenhouses is equal to an average of 72,000 and 75,000 tons per year, respectively [11,12]. A large part of this consumption comes from the Mediterranean region, especially from those countries like Italy where protected crops are largely cultivated [12]. In Italy, agricultural plastic consumption was estimated to be 350,000–400,000 tons per year, with 30% in the field of greenhouse claddings, low tunnels, soil mulching, vineyards nets and so on [11,13].

That said, current plastics-demanding sectors are depicted in Figure 2, as extrapolated from the latest PlasticsEurope's report [4].

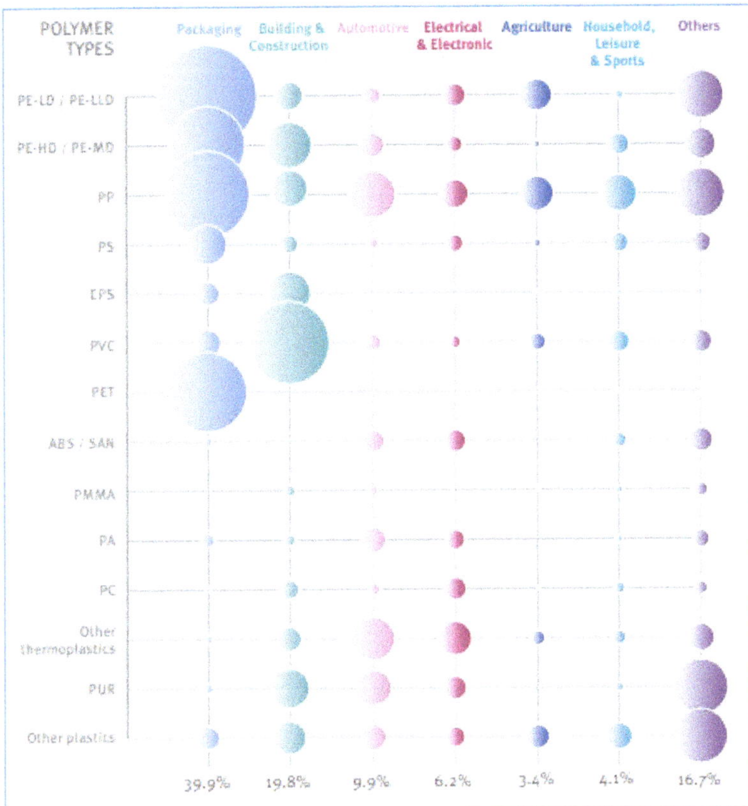

Figure 2. Plastics demand by segments and polymer types in 2018. Reprinted from [4], © 2020 with permission from PlasticsEurope.

The environmental impacts caused by the plastic industry are increasingly being understood, and so are increasingly gaining the attention of scientists, producers, company managers and owners, as well as policy- and decision-makers, at the international level [1,14].

To perform the functions that they are designed for, plastics cause a set of environmental impacts, in almost all phases of their life cycles, mainly deriving from exploitation of primary-energy resources and from the emission of greenhouse gases (GHGs) and of other polluting substances [6,15]. Furthermore, the massive consumption of those products results in consistent waste generation, with environmental and socio-economic consequences harmful to the heath of humans, animals, and plants and to the quality of the global ecosystem [16]. Parlato et al. reported that 26 million tons of plastic waste get generated every year in Europe and that just 30% of those is recycled, so representing a great loss from an economic and environmental viewpoint [17]. Still, the environmental benefits of recycling are well documented, with lots of studies available in the literature. Recently, in the field of protected crop cultivation, Cascone et al. [12] explored relevant energy and environmental issues associated with the recycling of post-use greenhouse-covering film into LDPE granules usable, as an intermediate product, in a wide range of downstream systems. The study highlighted that recycling is highly energy and water demanding, as well as highly GHG emitting. Despite this, the delivered (recycled) LDPE granulate was far more sustainable than—and so was validated as a feasible substitute for—the virgin counterpart.

Plastics are mainly produced for durable scopes, which makes them able to persist undegraded for decades in the environment where they are disposed of, with a persistence time going well beyond the predictable one [3,16,18–21]. It is this persistence making plastics accumulate in increasing quantities in the marine environment [22]. This should be attributed to marine-water bodies, like oceans and seas, being downhill and downstream virtually from every human living place, and to around half of the world population living within 50 miles (around 80 km) of those bodies. Consequently, lightweight plastic trash, like bags, bottles and cutlery [16], because of the lack of effective recovery systems, blows and runs off into the sea [22]. That is when it starts moving to a huge number of habitats, with a set of significantly damaging impacts to humans and to wildlife [22].

These and other related concerns emphasize the need to push towards sustainable, disposal scenarios that are oriented to recycling and so to the production of quality secondary raw materials [12]; and to alternatives to petroleum mineral-based products [16].

In this regard, attention is being increasingly focused upon the development of eco-friendly, innovative materials: chief among them, polymers derived from biological feedstocks, known as *biopolymers*, have been documented as effective to replace unsustainable, fossil-derived plastics [2,23].

They represent, in fact, a renewable substitute to the traditional petroleum-based plastics, and can be derived from different biomass types, including agricultural products, such as corn or soybeans, as well as algae or food waste [2]. Both recycled-polymers and biopolymers can contribute to greening the plastics sector, and accelerating the transition towards sustainable, equitable, post-fossil carbon societies [24].

The increasing utilization of polymers in the multitude of sectors shown in Figure 2 is one good reason for development of environmental assessments as the essential base to find the improvements that can be made in the life cycle of plastic products. Life-cycle assessment (LCA) is one of the tools that are currently available in the market and can be used for such a purpose. Walker and Rothman recommended its use to compare the environmental impact of bio-based and fossil-based polymers and, so, to account for the important balances and trade-offs between polymers and between impacts, which is essential in order to make sensible material selection [1].

LCA of products and processes is widely recognized, with two main beneficial features that are related to the 'cradle-to grave' approach, and to the use of a functional unit (FU) that enables comparative evaluations [25]. LCA is based upon a clearly structured methodology that is ruled by International Standards 14040 and 14044 [26,27]. It is a systematic tool that enables qualifying and quantifying the relevant environmental loads and impacts that are associated with the life cycle of a product or service [6,25]. Preparation of raw and auxiliary materials from resource extraction, transportation, manufacture of the product, end use and ultimately disposal of the product itself are accounted for in LCAs [25,28].

LCA has been significantly improved over the past decades, mainly thanks to the development of activities for: improvement of currently used databases; integration of quality assurance; improvement of completeness, transparency and consistency of assessments; and harmonization of methodological aspects and applications [25,29,30]. This has contributed to its emergence as a valuable decision-support tool for both policy makers and company owners in assessing the cradle-to-grave impacts of a product or process.

Over the years, practitioners worldwide have used LCA to explore relevant environmental aspects in a wide range of sectors including automotive, buildings and construction, energy, electronics, agriculture, and food production and packaging [6,11,31].

In this context, this study was aimed at overviewing the state-of-the-art of LCAs in the plastics sector at the Italian level. This has been done to complement the Special Issue about the "State-of-the-Art of Polymer Science and Technology in Italy", by providing information on the relevant environmental issues of plastic material applications in the Italian region (Figure 3).

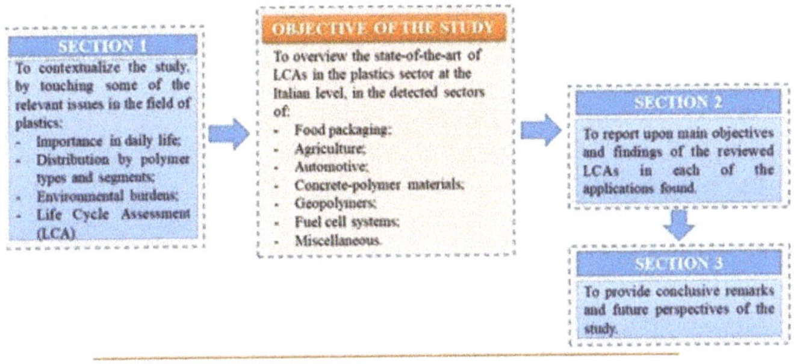

Figure 3. Review's flowchart.

The study was motivated by the importance of making polymer supply chain actors aware that a sustainable use of polymeric materials is not only possible but is above all necessary. In this way, there is greening not only of life cycles of those materials but also of the downstream systems in which they are utilized as such and/or are processed into finished, more complex commodities. This review offers a rundown of LCAs on polymers used in the most important production and commercial sectors, carried out in the last few years by Italian researchers.

2. Applications

The authors of this paper reviewed the findings of relevant articles exploring the application of LCA in the polymeric sector. The bibliographical search was conducted by browsing scientifically-recognized databases, like 'Scopus', 'Web of Science', 'Science Direct', and 'Springer Link'. Combinations of key-words connected with the research field investigated were used by the authors, and included: 'polymers', 'plastics', 'environmental sustainability', 'carbon footprint', and 'life-cycle assessment'. To be consistent with the objectives of this study, papers conducted at the Italian level were then selected and were extrapolated from the ensemble of papers found. A total of 58 LCAs were detected by the authors in several research areas, including food packaging, agriculture, automotive, admixed concretes, geopolymers, and fuel cells. Therefore, it can be highlighted that Italian research touched on most of the applications shown in Figure 2. The biggest number of LCAs (19) were found in the food-packaging sector, and regarded films, trays and bottles.

Finally, many of the reviewed papers were authored by team members belonging to different universities and research institutions, so remarking that environmental research on polymers presents such a complexity that it requires investigation through a multidisciplinary approach.

Each of the papers are discussed in the following sections in terms of key objectives and findings: additional information may be gathered by the reader from the original papers.

2.1. Food Packaging

As reported by Siracusa et al. [32] and as can be extensively read on the European Bioplastics website [33], the largest field of application of extracted crude oil is for plastics, with a current global production of 200 million tons and an annual grow of about 5%, with a consequent concern about ecological problems due to their total non-biodegradability [34]. As mentioned in the Introduction, within the petrochemical-based plastics materials, PET, PVC, PE, PP, PS and polyamide (PA) are the most common and used in the food-packaging field because their large availability, low cost and good performance such as tensile and tear mechanical strength; being a good barrier to oxygen, carbon dioxide and aroma compounds; heat sealability; and so on. Considering that when they are used as food-packaging materials they are not totally recyclable, also because they are often contaminated by foodstuff and biological substance, several thousands of tons of goods are every year landfilled, increasing the problem of municipal waste disposal [35]. The growing environmental interest imposes on packaging polymers as well as the technological process both user-friendly and eco-friendly attributes. In order to take into consideration this concern, LCA is a useful technique to evaluate the environmental impact of such materials, considering two fields of interest. The first one is the final destination of post-consumer plastics, comparing recycling with other options such as source reduction, incineration and landfilling as well as energy conversion and chemical recycling options. The second one is to take into consideration alternative materials such standard glass and/or metal packaging and/or biodegradable polymers that could be recycled or composted. These options must be considered in their entire life cycle and can be summarized in Figure 4, using the usual scheme of the LCA methodology:

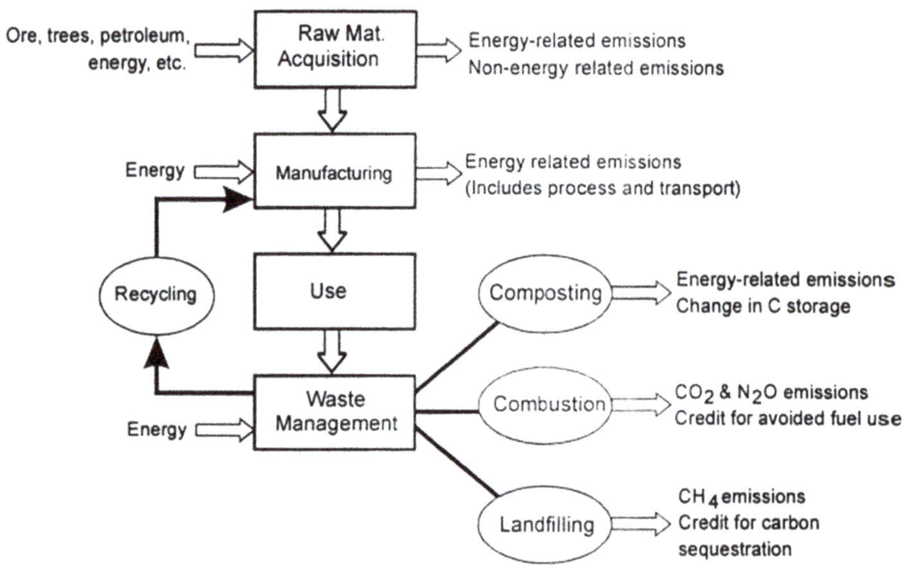

Figure 4. Municipal solid-waste life cycle. Reprinted from [36], © 2020 with permission from Wiley.

As reported before, for many years, polymers from fossil fuel resources have supplied most common packaging materials such as glass and steel, thanks to their great features especially softness, lightness and transparency. Due to their non-biodegradability, plastics use has led to serious ecological problems. Despite their total substitution with eco-friendly materials being just impossible to achieve in a near future, especially for specific applications such as food packaging, the use of bioplastics (biodegradable and/or compostable materials) should be considered. In this context, biodegradable polymers as well as bio-based polymers are being promoted as a good environmental alternative to the conventional plastics, derived from petroleum resources. They can be obtained from renewable resources or from waste, as food and plants, and can be recycled [32]. If landfilled, they become carbon dioxide and water, through composting. The main biodegradable polymer, compared with the conventional ones, is poly(lactic acid) (PLA), whose properties and performance make it suitable for food packaging application but not with success in every field, such as for example not for carbonate beverage bottles. LCA is a good tool that could be used to assess the overall environmental burden of a product and of the system used for manufacturing, use and disposal of such product. System boundaries have to be perfectly defined and are needed in order to clearly separate the system from the environment. As a basis for this analysis, the system boundaries for a biodegradable polymer such as PLA and conventional plastics such as PP, PE, PET, PS and so on, are illustrated in Figure 5, as an example of the most common polymers employed for packaging application.

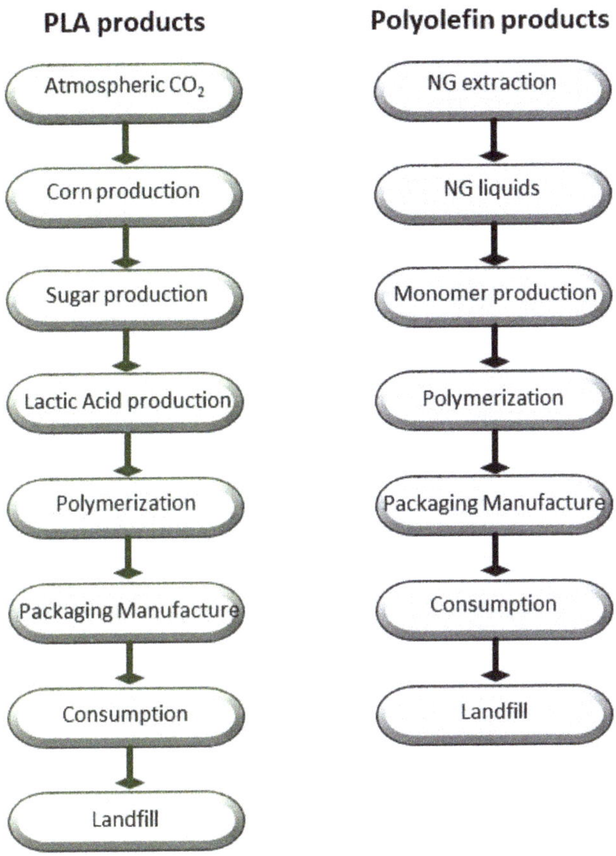

Figure 5. System boundaries for biodegradable poly(lactic acid) (PLA) and synthetic polyolefins.

After use, package is discarded and, most of the time, is landfilled. In many industrialized countries other treatments could be considered and are used, with important implications for the LCA results. As reported by Costa, Quintero and Dias [37], until December 2018, 144 articles written in English appeared in international literature dealing with the LCA application in the field of plastics materials; of those, 49 articles were devoted to real case studies and the remaining devoted to reviews, viewpoint, mixed approaches and methodological development articles, presenting the LCA foundations, main methods, current operation state and challenges. Several articles have focused their attention on energy consumption, global warming potential (GWP), LCA of mechanical recycling technologies and on processes. The major numbers of articles are from Germany (22), followed by the United States of America (16), Italy (14) and China (13) and then The Netherlands (10), Spain (6) and France (5). While the USA and Germany reported each one 11 articles on real LCA cases study, Germany and Italy reported each one 11 contributions as reviews and viewpoints on the development of the LCA methodology. Choosing Italy as geographically representative situation, in the knowledge of the authors until 2020, the manuscripts dedicated to the food-packaging plastics materials total 20, discussed in this review, in order to understand the use of such methodology in this important field of interest.

Mainly, two kind of plastic materials were the object of such LCA studies: materials obtained from non-renewable fossil fuel resources, as mentioned in the manuscript, and biodegradable polymers, obtained from renewable resources, such as PLA. It must be considered that also other materials are fully used in the foods packaging field, such as mixed materials, multilayer films, bio-based materials and polymers derived from microorganisms such as polyhydroxyalkanoates (PHAs), but in lower amounts due to the high price and lower performance.

Among the series of synthetic polymers present in the plastic market and especially in the food-packaging field, one of the most used is PET, with a global production of 30.3×10^6 tons in 2017 [38], and consequently the most present one in plastics urban waste, followed by PE and PP. Its performance, characteristics and environmental impact were compared with biodegradable polymers coming from renewable resources. Among the biodegradable polymers, PLA is surely the most used in the packaging field and studied in its technological characteristics and performances. One of the first manuscript dealing with a LCA study on such a polymer, but not in Italy, was that of Bohlmann [36], who carried out a LCA comparing PLA and PP, used as food packaging materials. In this first work, four energy main categories were considered, including fuel production (e.g., electricity consumption), fuel use for processing (e.g., steam consumption), transportation energy and feedstock energy. The selected functional unit was one metric ton (1000 kg) of yogurt packed in plastics, purchased by the consumer. In order to permit comparison with other LCA manuscripts, the energy consumption per kg of polymer used as packaging materials was also considered. Stanford Research Institute (SRI) Consulting's Process Economics Program (PEP) data were used, as good source of information for mass and energy balances. It was assumed that filling, distribution and use of PLA and PP yogurt cups was exactly the same. A cradle-to-grave analysis was performed, focusing on greenhouse gas emissions as the most relevant impact category due to the fact that for SETAC (Society of Environmental Toxicology and Chemistry) it is the only indicator with great relevance for biodegradable polymers such as PLA. Consequently, emissions of greenhouse gases are converted to metric tons of carbon equivalent (MTCE) and normalized for kg of polymer and for one ton of packed yogurt FU. As reported by Bohlmann, due to the fact that PLA consumes almost no feedstock energy, this polymer is more energy efficient than PP for food packaging application. Furthermore, PLA and PP greenhouse gas emissions become equivalent if it is considered that the carbon embodied in PLA is fully landfilled. Considering that the biodegradation of PLA in landfill could have an important effect on greenhouse gas emissions, uncertainties could be present in the LCA results. In addition, data analysis can be affected by uncertainty due to the fact that many plants in the world produce PP with differences in processes, energy usage and environmental loads from plant to plant. For PLA instead, there was only one commercial plant [39] thus making unavailable a sufficient number of data. To make the comparison of data the most consistent possible, Bohlmann chose the PEP database. Subsequently, several authors applied the LCA methodology on plastic materials. In this review we have focused our attention on Italian case studies.

In 2010, Girone and Piemonte [40] analyzed two different disposal scenarios, such as composting and recycling processes, for PLA (made from Nature Works) and Mater-Bi (made from Novamont) considering the land-use change (LUC) emissions, comparing also cost and benefits of the bioplastic disposal versus conventional plastics. In 2011, the same authors [41] realized a cradle-to-grave LCA study of PET compared with PLA bottles for drinking water, assuming the same waste scenarios like landfill and incineration. The results obtained highlighted that, despite the benefits related to the use of renewable resources for the production of biodegradable bottles in respect of non-renewable resources for PET production, higher impact value for the categories human health and ecosystem quality were recorded, due to the use of pesticides, consumption of land and water for the production of row PLA. However, it must be considered that LCAs' unfavorable results were related to the fact that in 2011 the technology for PET production and recycling was much more mature, tested and optimized both for production and recycling processes in respect to that of PLA, which was at an early stage of development, with a very limited and localized production. The results were interesting when different final dispositions were compared: recycling for PET and composting and recycling, in different percentages, for PLA. Authors reported that the best benefit in the use of PLA could be achieved only if it could be totally recycled because composting and incineration showed a higher environmental impact than recycling. Quite similar results have been obtained when recycling of 100% PET bottles was compared with incineration/recycling, in the amount of at least 20%–80% of PLA bottles. The same mixed scenarios as final bottles destinations have been also considered. In particular for PET 70% recycling and 30% landfilling, 70% recycling and 30% incineration and finally 50% recycling, 30% incineration and 20% landfilling; for PLA 50% recycling and 50% landfill, 50% recycling and 50% composting, 40% recycling 30% composting, 15% landfilling, and 15% incineration. Comparable results were obtained when the macro category "resources" was considered.

In the same year, 2011, Piemonte [42] carried out an LCA cradle-to-grave analysis comparing the results obtained choosing as functional units 1000 shells of PLA, polymer produced from Nature Works (USA), and 1000 shells of Mater-Bi, a starch based bioplastics, made from Novamont (Italy), both used as food packaging materials. The scope of their work was to assess the energy and GHGs emissions coming from the production of bioplastics, compared with the standard materials, as well as the best waste final disposition, in order to maximize energy saving. The Intergovernmental Panel on Climate Change (IPCC) and Global Warming and Cumulative Energy Demand (CED) methods were used, with an end point approach, considering three "Damage" categories such as human health, the ecosystem quality and the non-renewable resources depletion as the most high-impact categories. PET and PE data, used for comparison, were based on the Eco-profiles of the European plastics industry. Piemonte found that the replacement of the petroleum-based plastics with bioplastics could be considered an interesting strategy towards the sustainable development, for energy and GHG emissions savings. As a final destination of bioplastics, mechanical recycling could be considered a good alternative, maximizing the energy saving and reducing the renewable resources consumption. In addition, Piemonte underlined that the results reported are interesting in terms of the environmental reliability of bioplastics, but cannot be taken as guidelines for the market use of PET or PE rather than PLA or Mater-Bi, and vice versa.

In 2013, Toniolo et al. [43] analyzed with the LCA methodology two kinds of plastic product: a multilayer plastic tray and a PET tray, used for food packaging. For the first product an end-of-life scenario that includes land-filling and incineration was chosen, whilst for the second product an end-of-life scenario comprising recycling, land-filling and incineration was considered. They studied how an innovative and recyclable package material could be environmentally preferable in respect to a non-recyclable one, explaining how the utilization of recycled materials represents a great effort to reduce the environmental burdens. The package produced employing a recyclable mono-material film is more environmentally advisable than the multilayer ones, for all of the impact categories analyzed. The results were also tested by a sensitivity analysis and an uncertainty analysis in order to confirm the results of the life-cycle impact assessment. The study was used to demonstrate that the life-cycle approach is an important tool to assess how a prevention activity to reduce waste production is actually an environmentally sustainable alternative, providing also great decision-making support in the field of packaging waste management. In 2017 they completed their research considering allocation strategies for implementation in the modeling of the burdens of input waste and end-of-life disposal of such food-packaging materials [44].

During 2014, three papers appeared on LCA studies: one dealing with an environmental assessment study of a multilayer material bag used for food packaging [45], the second one dealing with an economic and environmental assessment study performed on reusable plastic containers used for a food catering supply chain [46], and the third one based on the environmental assessment of two packaging materials for poultry products, such as a polystyrene-based tray and an aluminum-based trays (70 wt% primary and 30 wt% secondary aluminum) [47].

In the first paper, Siracusa et al. reported an LCA case study on a multilayer film bag made of PA and LDPE, used for vacuum or modified atmosphere packaging (MAP) technology food preservation. Following the ISO standards 14040:2006 and 14044:2006 rules, a functional unit of 1 m^2 of plastic film was chosen, with a cradle to factory-gate approach, including the phases of the raw materials for bag production and processing and the bag's delivery to both the food production and packaging plant.

Due to the use of crude oil and natural gas, the most impacting phases were the production of PA and LDPE granules. The most affected damage category was "Resources", followed by "Climate Change", "Human Health", and "Ecosystem Quality" respectively. Film thickness reduction and the use of recycled PA granules were proposed to the firm in order to reduce the total damage. The assessment results showed that the two proposals allowed a reduction of about 25% and 15% (respectively) of the assessed damage.

In the second paper, Accorsi et al. reported that during the last few decades, Europeans have drastically modified their food consumption habits through the food catering sector, and chosen to eat out or buy take-away foods, thus increasing drastically the amounts of package waste, exacerbated as the food catering chain service does not currently employ reusable packaging systems. They compared a multi-use system to traditional single-use packaging (e.g., wooden boxes, disposable plastic crates and cardboard boxes), used for a fresh fruit and vegetable Italian catering chain, from vendors to final customers, to quantify the economic returns and environmental impacts of the reusable plastic container (RPC). A carbon footprint (CF) analysis was performed for the environmental assessment. As plastic materials, 1 kg of PP for the production of 1 kg of single-use plastic crates and RPCs was considered, while if an amount of 2.8% of scrap is added during the production phase, 0.925 kg of polypropylene granulate can be expected to yield one kilogram of product. The LCA results demonstrated that the environmental impact associated with the single-use network was mainly due to the manufacturing phase, due to the great volume of the packages required over the year. It was observed that the transportation phase, as well as the different disposal scenarios, significantly affected the sustainability and the environmental impacts of the RPC system. From the study it was observed that the RPC system will lead to a reduced environmental impact in terms of CO_{2eq} emissions while the overall economic return, in the opinion of the authors, is projected to be negative, due to a cost increasing of about 0.06 € per kilogram of handled food product.

In the third paper, Zampori and Dotelli described a cradle-to-grave case study, considering the production, use phase (such as cooking), and different end-of-life scenarios. Greenhouse Gas Protocol (GGP), Cumulative Energy Demand and International Reference Life-Cycle Data System (ILCD) midpoint method were used for the life-cycle impact assessment (LCIA) measurements. The aluminum bowl was designed for use during the cooking stage of the poultry product and to reduce the cooking time allowing consequently a reduction in electric energy equal to 0.21 kWh (1.38 kWh instead of 1.59 kWh), that is of primary importance to reduce overall emissions, expressed as CO_{2eq} emissions, especially in those countries, such as Italy, where the use of fossil fuels is predominant for electric energy production. As regards the production stage of the two trays, it was found that the PS tray was more sustainable than the aluminum one. Furthermore, in order to reduce the environmental impact, the trays were designed to lower the overall emissions, determined by GGP and CED. In particular, it was found that the Al tray was less sustainable than the PS one, according to a "recycled content" scenario, but it was more sustainable when a "substitution" scenario was applied. In particular, the authors found that if an Al tray made of totally recycled material is used, also if this is less sustainable than a PS one, the choice of the Al tray becomes the most sustainable option thanks to the lower impacts recorded during the cooking stage that were calculated to be the most impactful over the entire life cycle of the two alternatives by a "cradle-to-grave" approach.

In 2015, Ingrao et al. [6] performed an LCA study of foamy polystyrene trays used for fresh meat packaging. The FU was identified with 1 kg of packed-trays delivered to food production and packaging firms, with a maximum capacity amounting to almost 800 cm^3 of food. The tray size was chosen as the object of the study because it was the most commercialized one amongst the other different types produced by the Italian firm and so representing perfectly its core-business. The study highlighted that the highest environmental impacts come from PS-granule production and electricity consumption. The authors underlined as there were no margins for improvement in the production of PS granules, in the transport of materials and of the trays' inputs. On the contrary, changing the energy source into a renewable one (by installing, for instance, a wind power plant) would enable a 14% damage reduction.

In 2015, Conte et al. [48] focused their attention on 24 scenarios of packaging used for cheese obtained from sheep. The goal of the study was to analyze the environmental implications of different packaging systems, considering the potential food loss, in order to obtain an eco-indicator able to quantify the environmental effect of different packaging materials. The geographical context was referred to two area of the Basilicata region: Matera and Potenza (South of Italy). The LCA methodology was applied following two approaches: an attributional approach (AA), in order to consider the impacts related to the adoption of different packaging solutions, by taking into account only the phase of packaging and including the production and disposal phases of packaging materials; and a consequential approach (CA), with the aim of obtaining information about the consequences of the decisions in different packaging choices. In this last case, the investigation was oriented to the environmental implications in using different plastic films and gaseous atmospheres, together with the inside effects such as the production and disposal of plastic materials, and outside effects such as shelf-life and food losses. In this context, four types of plastic films were studied: a multi-layer high-barrier film of PET (13%) co-extruded with PE (78%), exposed to anti-fog treatment (9% AF); a common multi-layer film of polyamide (21% nylon), PE (74%) and ethylene vinyl acetate (EVA) (5%); oriented polypropylene (OPP) (100%) of 60 micron thickness; OPP (100%) of 35 micron thickness. For each heat-sealed bag, composed as described above, six gaseous atmospheres were considered: air, vacuum and four MAP obtained by mixing N_2 and CO_2 at various percentage. Two scenarios were considered for packaging disposal: the multilayer systems recovered for energy use (incineration) whereas the PP was completely recycled. The authors found that, if an attributional LCA approach was followed, thinner and recyclable packaging materials resulted in being more sustainable while if consequential type approach was followed, packaging able to guarantee a longer shelf life, thus reducing the food loss become the most sustainable. Furthermore, considering that the environmental impact caused by cheese production is generally high if it is compared with the other stages of the life cycle, the packaging possibility to reduce food losses was a key factor from an environmentally proper choice point of view, more than packaging production and/or packaging disposal.

Accorsi et al. [49], in 2015, reported a LCA case study on extra-virgin olive oil (EVOO) bottled in glass, PET and recyclable PET (R-PET), quantifying the impacts of global warming potential, ozone layer depletion, non-renewable energy use, acidification, eutrophication and photochemical smog. They focused their attention on the EVOO processing and bottling, on the EVOO and packaging supply, on the distribution to customers and on the management of end-of-life (EoL) treatments (recycling and disposal of wastes). The functional unit was a 1 L bottle of EVOO (i.e., equal to 0.916 kg·L), supplied by an Italian olive oil firm. The LCA analysis was performed according to an attributional and a consequential approach, in order to quantify the impact of the different technological approach. The authors found that if high fractions of package waste is devoted to recycling (over 40%), EVOO bottled in glass becomes less impactful than PET. Finally, LCA highlighted the potentiality of R-PET bottles in reducing the environmental impact of EVOO supply chains.

In 2017 three papers appeared on LCA case studies. The first one was performed on cheesecake normally packaged in low-density polyethylene film [50], with a limited shelf-life due to fast microbial growth. The authors decided to study the effect of different packaging conditions like use multilayer gas barrier trays, subsequently wrapped with a multilayer gas and water barrier film (i.e., AerPack packaging), in order to extend the food shelf life. As control batches, food was packaged in gas barrier recycled PET (named XrPet) trays and wrapped with XrPet films. Samples were characterized not only from a chemico-physical, microbiological and sensory parameters point of view, but an LCA study was also carried out, considering those new packaging solutions a possible way to minimize transport costs and to generate economies in manufacturing. The chosen FU was a unit of food sold, corresponding to a tray containing two cheesecakes, with a total weight of approximately 300 g. The ReCiPe endpoint method was used for this LCA study. Metal depletion, followed by human toxicity, freshwater and marine ecotoxicity were the impact categories that varied the most, relating to a high environmental impact of packaging. The ReCiPe normalized results confirmed that the AerPack materials showed a lower environmental impact than the XrPet. Moreover, considering the economic sustainability approach, the authors underlined as a change in the packaging technology to keep the product fresh longer, could allow the cheesecake producer to minimize transport costs, generating economies in manufacturing. By contrast with what was considered by Conte et al. [48] that provided an eco-indicator index able to quantify the environmental indirect effects related to the different choices in the food packaging, Gutierrez et al. [50] utilized an economic model considering the minimum amount of product delivered on consignment that must to be sold, defining the impact of shelf life on the food losses.

The second paper, published by Ingrao et al. [16], reports the evaluation of life-cycle environmental impacts for fresh-food packaging trays constituted by expanded-PLA (EPLA), compared with standard trays used for fresh meat, constituted of expanded PS (EPS). The purpose was to understand if the biodegradable polymers could be an interesting and environmentally valid alternatives to the synthetic ones, by assessing the global environmental impact but considering more damage and impact categories (DCs and ICs) than those reported in previous studies by the same authors [3,6]. In particular, the comparison was carried out using an end-point approach level and the results were shown to be associated with the life-cycle of the trays (Figure 6) and the DCs and ICs (Figure 7). In both figures, values were expressed as pt kg^{-1}.

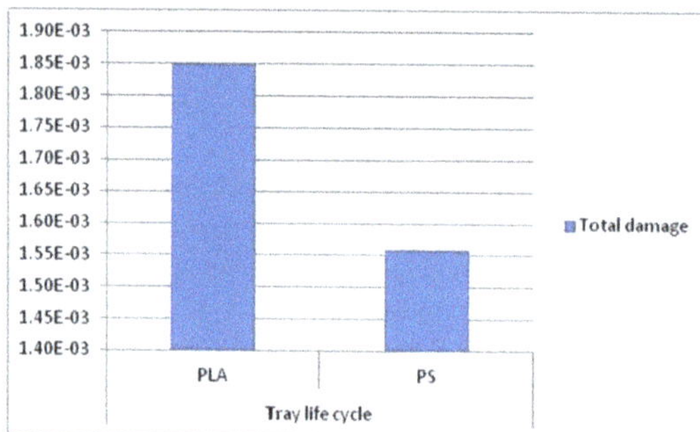

Figure 6. A comparison of the (total) damages associated with the life cycles of PLA and polystyrene (PS) trays. Values are expressed as pt.kg$_{tray}$$^{-1}$. Reprinted from [16], © 2020 with permission from Elsevier.

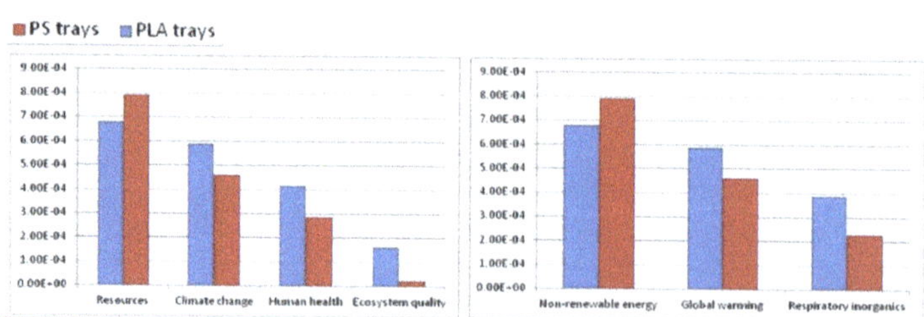

Figure 7. A comparison of the damages associated with both DCs and ICs in the life cycles of PLA and PS trays. Values are expressed as pt.kg$_{tray}$$^{-1}$. Reprinted from [16], © 2020 with permission from Elsevier.

The authors concluded that the most impacting phase were represented by the production of PLA granules, due to the corn-cultivation phase, by the energy consumed in the processing of those granules and, mostly, by their transportation from America to Italy, and in Italy direct to the tray manufacturing factory. In this context it was observed that the transport issue was responsible for the PLA-trays' higher impacts, compared with the PS ones.

The third paper was reported by Cinelli et al. [51], evaluating the degradability in compost of some representative examples of materials prepared with pristine PLA, PLA with chitin nano-fibrils (PLA/NC), PLA with NC and polyethylene glycol 400 (PLA/NC/PEG400) used as plasticizing additive, suitable to be used for production of rigid or flexible packaging. Due to the fact that a biodegradable polymer was blended with an active compound that might hinder the compostability of such materials, for authors it was important to assess the test of degradability on those innovative materials, bio-based and containing biodegradable components. The functional units were respectively 1 Kg of nano chitin fibrils, and 1 Kg of PLA/NC/PEG400 pellets. The materials based on PLA and NC showed disintegration in compost, evidencing no toxicity, while the production process of NC was affected by a high consumption of water and chemicals. LCA analysis measured the environmental benefits for PLA/NC/PEG400 products, mainly related to the renewable sources of most of the used components such as chitin, chitosan, PLA, glycerol, as well as the compostability of the obtained product. The LCA study was aimed at suggesting improvements for the production of NC based materials in terms of sustainability and future application in food packaging market.

A year later, Petrucci et al. [52] reported a comparison between limonene plasticized PLA films containing cellulose nanocrystals (CNC) and an acetyl tributyl citrate (ATBC) plasticized PLA films, having equivalent mechanical properties, containing organo-modified montmorillonite (OMMT). A cradle-to-gate approach was followed, using the Eco-Indicator 99 as the main method to assess the environmental impacts and 1 kg of plastic film as the functional unit in order to identify the most critical inputs. Physical and geographical limits of the investigated PLA whole life cycle, including both nanocomposite manufacturing plant and the composting end of life, were considered as system boundaries. The film service life and transports were excluded from the study. The authors found that the chemical extraction of CNC caused a large environmental footprint and a relatively relevant energy expense, despite it being a biobased filler. Furthermore, the environmental impact of PLA/limonene film reinforced with 1% in weight of CNC (PLA/CNC/limonene) was comparable to the environmental impact of PLA films reinforced with OMMT and plasticized with a petroleum based plasticizer (ATBC) (PLA/OMTT/ATBC).

In 2019, Blanc et al. [53] performed an LCA case study of raspberry supply chains from an environmental, economic, and social point of view, using bio-based plastics packaging. LCA, life-cycle costing (LCC), and externality assessment (ExA) methodologies were applied to assess the impacts. As traditional materials they considered PET and PLA as innovative packaging. A cradle-to-grave approach was followed, with the supply chain framework reported in Figure 8.

Figure 8. Raspberry supply chain framework. Reprinted from [53].

As a functional unit a standard sale of 125 g flow pack was chosen. The results, reported on Figure 9 showed that bio-based plastics have lower environmental and social impacts than the conventional one, whereas the latter is the best choice according to a classic economic approach. The authors underlined as a success the use of bio-based material, instead of the traditional plastics (such as PE and PET), is linked to the ability to operate in a systemic way along the whole supply chain.

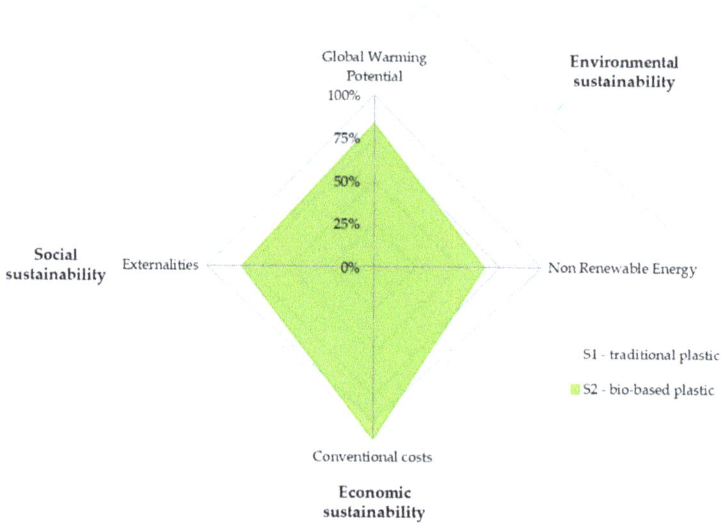

Figure 9. Sustainable performance results according to the triple bottom line (TBL) approach. Reprinted from [53].

In the same year, Giovenzana et al. [54] reported the results of a case study on the environmental impact analysis made on a new packaging solution used to oil quality preservation. This new packaging solution, a three layers packaging made from two bio-based polymers such as PLA treated with metallization and Bio-PE was compared with the traditional packages such as PE, PET and aluminum. An LCA cradle-to-grave approach was used, choosing as functional unit a single-use packaging olive oil content equal to 10 mL. With respect to traditional packaging, the innovative one obtained better performance in the human health impact categories but the fermentation process of sugar cane, used to obtain the PLA raw material, showed higher impacts in the ecosystem quality impact categories related to water resource depletion, freshwater eco-toxicity and land use. The improvement of environmental sustainability using an innovative packaging solution was not confirmed underline as the use of the bio-based product does not always represent the better way in terms of environmental sustainability, especially when recycling becomes an important aspect to be considered.

Finally, in 2020 Ferrara et al. [55] performed an LCA case study comparing the environmental performance of Italian wine packed in traditional single-use glass bottle and wine packed using four different packaging alternatives such as aseptic carton, bag-in-box, refillable glass bottle and multilayer PET bottle. All packages were chosen by the authors because they considered it suitable to contain wine while protecting it from light and oxygen during the storage time (considered within one year) and preserving its quality level. However, for a considerable length of time before opening (over one year), they reported clearly that the aseptic carton, bag-in-box and multilayer PET bottle cannot substitute glass bottles for high-quality aged wine [56]. All primary data were obtained by the packaging companies for weight, size and composition, for all packaging components (primary, secondary and tertiary packages), comprising also transport, distribution and final disposal scenarios. Secondary data were obtained from literature and technical documents. ReCiPe 2016 H was used as evaluation method, with both midpoint and endpoint approaches. From the results it was observed that the most environmental alternative was the bag- in-box packaging, slightly better than the aseptic carton, due to the composition of the containers, lower packaging weight and greater palletizing efficiency. Compared with single-use glass bottles, the impact of the bag-in-box was 60% to 90% lower. Considering as alternative scenarios the weight of containers, the wine distribution distance and the packaging disposal scenario, it was observed that decreasing the distribution distance, the environmental performances of refillable glass bottles could be comparable to those of the other two alternatives, pointing out as glass bottle reuse in Italy is a convenient alternative only when the distribution is performed at distances less than 100 km (local market). This study provided also useful indications for industry and government stakeholders for new strategies in order to enhance the sustainability of the wine life cycle, as well as countering the prejudice that glass packaging would be more sustainable than plastic or multilayer packaging.

2.2. Agriculture

Degradability and renewability have become one of the guiding principles for sustainable production and consumption of plastic commodities [2]. These two aspects have been documented to be largely satisfied by biopolymers: the latter have, indeed, been expanding mainly in response to the growing concerns about the sustainability of conventional plastics and to the environmental impacts caused by the uncontrolled disposal of plastic wastes [3]. Over the years, such polymers have been developed and improved in a way to become valid substitutes to the conventional ones, and so to find applications in a wide range of sectors.

In this regard, Shen et al. [57] identified the following biopolymers as utilizable in agriculture: starches; PLA; PHA; poly(butylene succinate) (PBS); PE and PVC produced from ethanol. Riggi et al. [2] reviewed their applications in crop production, and came to the conclusion that, by replacing their petroleum-based counterparts, these polymers may contribute to the greening of plasticulture. Furthermore, in their review study, Riggi et al. detected several research advances and patents, mainly based upon application of methodologies to expand the usage of biopolymers. This is the case of biodegradable spray mulching coatings from hydrolyzed proteins that resulted as enabling several benefits like: reducing mulch application costs; adding plant adjuvants in the sprayed mixture; and making spray mulching more versatile and utilizable where the conventional, LDPE-film based one is not [2,58].

Innovation in the mulch application field was researched also by Gorgitano and Pirilli [59], who explored the feasibility and environmental impact of using biopolymers as alternative film materials to the fossil counterparts. In particular, the authors' attention was focused upon comparing LDPE to a Mater-bi (MBi)-based biodegradable polymer as mulch film in the growing of different greenhouse crops, namely lettuce, cantaloupe melon, and kohlrabi. For the study development, the authors carried out an LCA considering a set of impact categories to be consistent with the system investigated and best describe its environmental profile, including the potentials of global warming, acidification, and eutrophication, as extrapolated from the CML1 baseline. The Functional Unit chosen by the authors was 1 ha land where cultivation was carried out under greenhouse conditions using traditional and innovative mulching film, alternatively. The system boundaries were of the gate-to-gate type, and so encompassed all material and energy inputs, as well as all the activities carried out, from production to harvest. The study allowed the authors to understand that, on average, for all crops and impact categories investigated by the authors, MBi-based films perform better than the LDPE ones, but results are significantly affected by the mulch film dimensions and features, as well as by the given cultivation practices.

In another study, Russo and Scarascia Mugnozza [60] carried out a LCA of table tomato production, comparing traditional with hydroponic farming. In both cases, cultivation took place in greenhouse systems characterized by equal size (10×100 m^2) but with different materials and structures. The FU was represented by 1 kg table tomatoes with the system boundaries going from production to harvesting and later distribution, including greenhouse maintenance and component replacement. With regard to the greenhouse system, for all impact categories considered according to CML 2 2000, a pitched-roof woody structure with an LDPE film cover was far more environmentally sustainable than a pitched-roof structure in zinc-coated steel with a glass cover and being more sustainable than a vaulted roof structure in a zinc-coated structure with LDPE as covering film. Therefore, it can be asserted that the combined utilization of wood and plastics enable more sustainable greenhouse systems but this solution was not feasible for a series of surrounding conditions and the steel-plastic mix system may be considered as the environmentally viable trade-off: as a matter fact, it was chosen by the authors to compare the conventional vs. the hydroponic cultivation. Related results from this part of the study have not been discussed here because they were considered by the authors as falling outside the scope of this review paper.

The studies reviewed found ways for the greening of plasticulture also through incremental innovation solutions, but there are still challenges to face, as related to background conditions, materials utilized and production processes, and sustainability assessment methodologies. Agreeing with Riggi et al. [2], those challenges can be solved through a multidisciplinary approach that combines the efforts of scientists and technicians with expertise in subjects like chemistry, engineering, agronomy and biology.

2.3. Automotive

The automotive industry has an important multiplier effect on the economy, because it is important not only for downstream industries but in particular for upstream industries such as chemical ones. Despite the fact that 80% of the growth in the sector is expected to occur outside the European Union (EU), it remains crucial for the economy of the old continent. 13.8 million Europeans (2.6 million directly) are involved by the car industry (6.1% of total European employment). But overall, the EU is at the forefront of investment in research and development (R&D) [61]. Thus the sector, in order to maintain and/or strengthen this leadership, in the last two decades has been searching for a sustainable approach in the design, manufacture and recycle/disposal of the materials employed in the automotive industry. The use of polymers and composite materials in the automotive industry has been increasing in recent years (Figure 10) both because legislative and consumer demands for lighter weight and fuel-efficient vehicles and because the design and complex parts of cars are more easily mouldable [62].

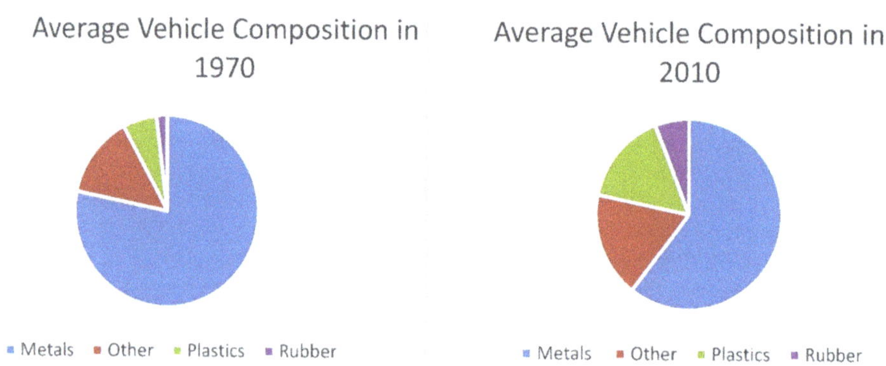

Figure 10. Change in vehicle composition from 1970 to 2010. Reprinted from [62].

Based upon the subject literature, the sustainable development of supply processes consist of four levels, starting from the normative up to the monitoring of those processes, with their early detection and implementation in between [63]. Generally speaking, it is obvious that prevention is the key point to be taken into account for achieving a more sustainable production procedure, thus we cannot leave aside an analysis tool like LCA.

Passarini et al. considered the increase in light material content such as polymers in new vehicle production, and hoping for an improvement of polymer separation according to an ecodesign-oriented strategy, applied the LCA to the end-of-life treatments [64]. The research group at Bologna "Alma

Mater Studiorum" University, by considering the residual waste after shredding end-of-life vehicles (ELVs) aimed at proposing the trend of the environmental effects of treating an automotive shredder residue (ASR). They analyzed, according to the Ecoindicator 99 method (Pré Consultant, 2001) and using the SimaPro 7.1 software, integrated with Ecoinvent 2.0 database (Ecoinvent, 2008), the potential benefits from the recycling of polyolefins from ELVs, for the Italian situation. In fact, polymers' growing share in the composition of new cars obliged to reconsider the EoL stage, trying to find effective recycling methods for polymers, which initially were sent to landfill due to the difficulty in being sorted and recycled with respect to steel and light metals. In particular, Passarini et al. evaluated how the different treatments (gasification, incineration, landfill, mechanical recycling) worked as a function of the time-dependent waste composition, showing the importance of LCA application as a tool for identifying a suitable waste management hierarchy, in accordance with the Directive 2008/98/EC. They concluded that recycling plants, modeling mechanical and chemical recycling options, achieved the lowest impacts due to the combination of material and energy recovery, with a consequent decrease in the residual amount of waste disposed of in landfill. Since the presence of recyclable materials and the net calorific value are fundamental parameters for the mass and energy balance during the Life Cycle Inventory (LCI) phase, they highlighted the need to consider their long-term evolving composition (ASR mix composition) in order to choose the most environmentally suitable technologies, especially when an investment in new technologies and plants, planned to be in operation for many years, is made.

D'Errico and Ranza applied LCA to determine the environmental impact of an Eco-floor panel for cars in comparison with two other conventional manufacturing process cycles [65]. Namely, they compared a stamped steel floor panel, a polyacrylonitrile (PAN) carbon fiber blended with epoxy resin panel and a panel addressed by recycled eco-magnesium. They assumed a double scenario and reported the 3 process cycles and relative GWP per kg of material produced (Figure 11):

(i) 30% in situ recycled mixed with 70% of fresh material;
(ii) 30% in situ recycled mixed with 70% of recycled chips.

The purpose of the Milanese researchers was the evaluation of lightweight solutions in a cradle-to-exit gate stage modality, thus including the so called dirty-phase that is different for each considered material. The result of this approach was that, considering a car a lifespan of 200,000 km, carbon fiber reinforced polymer (CRFP) panels showed the best performance. On the other hand, novel Mg-based panels have not any technology limit in using 100% of recycled feedstock materials, thus allowing to align final GW values of low-impacting Eco-Mg produced by chips to those low values typical of a CRFP panel.

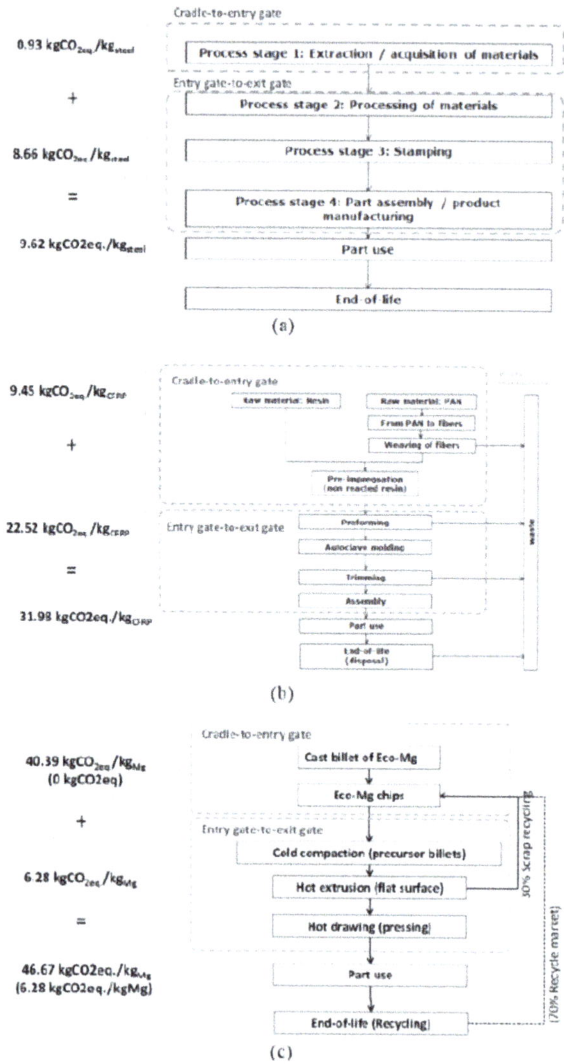

Figure 11. Process cycles and GWP results by Carbon Calculation over the Life Cycle (CCalC) software calculation: (**a**) LCA of conventional steel floor pan; (**b**) LCA of CFRP floor pan; (**c**) LCA for novel process route of extrusion of in-situ recycled Ecochips mixed with: (i) 70% of fresh material provided (cast billet and ecochips pathway); (ii) (in parenthesis) 70% of raw material entirely provided by recycling market. Reprinted from [65], © 2020 with permission from Springer Nature.

Ingarao et al. evaluated the environmental impact of some materials in means of transportation, by considering all their life cycle stages [66]. Their contribution aiming at improving the reliability of modeling the material (including recycling) and the manufacturing steps of aluminum-based components in comparison with those of glass-fiber-reinforced polymers (GFRP), considering light-weight benefits without neglecting the material production impact. Taking advantage of an industrial case study, coupling experimental measurements with industrial data and literature analysis, they carried out a life-cycle analysis of window panels to be assembled on the Italian high-speed trains. In particular, they focused on an innovative recycling strategy and manufacturing inventory issues. Concerning the functional unit, the life cycle of a single train window panel was analyzed. Each train, covering an annual distance of 500,000 km and an expected lifespan of 25 years, was considered for 112 panels, setting the FU as one of this panel. The system boundaries were settled from cradle-to-grave adopting the recycling as EoL strategy.

The impact of material production, product manufacturing, transportation, product use and recycling were evaluated, analyzing primary energy and CO_2 emissions flows throughout the panel life cycle. As shown in Figure 12, GFRP performed worst, both for CO_2 emissions and energy demand, with the use that has proved, by far, the most impactful of the various phases.

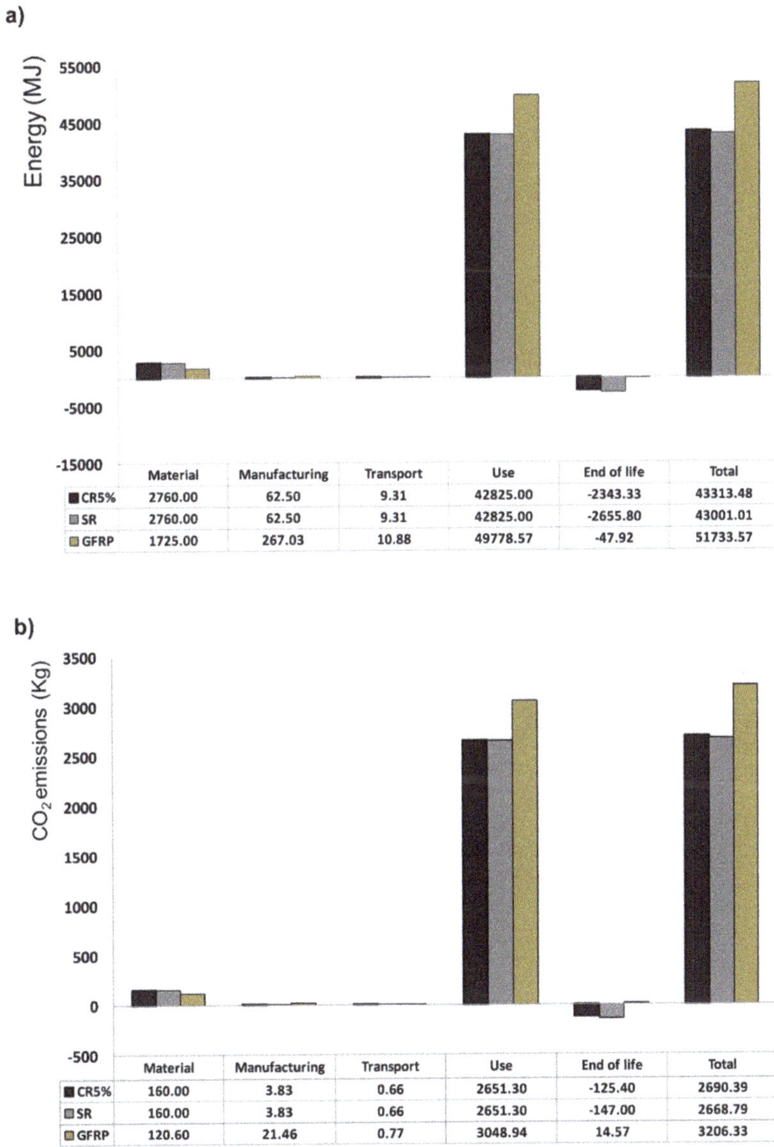

	Material	Manufacturing	Transport	Use	End of life	Total
CR5%	2760.00	62.50	9.31	42825.00	-2343.33	43313.48
SR	2760.00	62.50	9.31	42825.00	-2655.80	43001.01
GFRP	1725.00	267.03	10.88	49778.57	-47.92	51733.57

	Material	Manufacturing	Transport	Use	End of life	Total
CR5%	160.00	3.83	0.66	2651.30	-125.40	2690.39
SR	160.00	3.83	0.66	2651.30	-147.00	2668.79
GFRP	120.60	21.46	0.77	3048.94	14.57	3206.33

Figure 12. Comparative life cycle inventory results: (**a**) primary energy, (**b**) CO_2 emissions. Reprinted from [66], © 2020 with permission from Elsevier.

As regards the other phases, GRFP performed better during the material production (1725 vs. 2760 MJ and 120.6 vs. 160 kg of CO_2), but subtracting the credit arising for recycling, the aluminum resulted in being by far the best solution for both the considered metrics. Finally, they concluded that for high-speed trains and in general for long service life transportation means light-weighting is definitely the strategy to follow, considering also that the long traveling distance allows for the possible worse environmental performances in material or manufacturing steps to be easily compensated.

Also Zanchi et al. evaluated, according to the LCA and LCC [67] methods, a light weighting strategies aiming at reducing impact during the use phase of a vehicle. They compared two different composite-based solutions for a dashboard panel manufactured by Magneti Marelli®, the first one reinforced with talc whilst the second based on the use of hollow glass microspheres (HGM), considering the environmental assessment and the economic sustainability [68]. The economic and environmental analyses were based on a detailed data collection (comprising primary data), including and comparing, in the first case, the production cost with the cost saving during the use phase, and, in the second case, different indicators aiming at highlighting the trade-off among manufacturing and use and possible EoL scenarios. They studied three different PP layers with two solutions differing in the bottom one (Figure 13), the first reinforced with 25% talc and the second one reinforced with 23% HGM.

Figure 13. Automotive dashboard panel: finalized component (a); bottom layer (b); upper mantle (c). Reprinted from [68], © 2020 with permission from Elsevier.

They settled as FU a dashboard panel of an Italian car (Alfa Romeo Mito 955 diesel engine) having a lifespan of 150,000 km/10 years. The system boundaries included materials production, component manufacturing, transport of materials and of the finalized dashboard to plant for its assembly to the vehicle, use phase and EoL treatments. Results showed that HGM-reinforced composite was likely better from an environmental point of view for those impact categories where the use phase is more involved. The increase of material processing impact did not compromise benefits in terms of GWP and primary energy demand due to weight reduction, nevertheless it affected resource depletion and ecotoxicity indicators negatively, but overall, the EoL phase was not affected significantly. They thus concluded that, despite a higher material cost, the use of HGM reinforced composites was found economically preferable as demonstrated also by the breakeven point and it was suitable for mass customization.

Zanchi et al. continued their LCA investigation extending the previous methodological approach in the design of other Magneti Marelli® components by using GaBi software [69]. The various automotive components were sorted by three vehicle systems: drivetrain (namely: air intake manifold, throttle body, muffler, fuel tank); interior (dashboard, pedal box support, brake pedal); Suspensions and chassis (Crossmember, Suspension arm). Each component was analyzed in terms of elementary flows and products/waste flows having as goal and scope the determination of the environmental performances along the entire life cycle of innovative lightweight solutions when compared with the current heavier designs. They pointed out the difficult of benefits prediction when lightweight polymeric materials were used in the place of heavy metal ones, again for the trade-off among manufacturing and use. Thus, they concluded their efforts in this field with a series of take-home messages:

- A closer synergy between the various experts involved in the design phase is necessary to select materials that actually affect the reduction of impacts, thus enhancing increasingly accurate data collection and environmental modelling.

- The introduction of different impact categories, then GWP could make the results interpretation more difficult. They recommended ensuring an updating process of internal LCA expertise concerning continuous scientific progress in the LCA fields regarding methods to calculate and interpret the impacts.
- The assumptions on the final treatment of ASR (landfill or incineration) and recycling rate values become significant for a complete substitution of heavy metals with polymer-based materials.
- They recommended including secondary effects in the LCA comparison to obtain more precise results concerning benefits from lightweight design solutions.

The production of light and high-performance components for the automotive was also the object of the study of Vita and his colleagues [70]. The researchers of the Universities of Ancona and Viterbo, by the means of LCA, compared the environmental impacts of three alternative resin transfer molding (RTM) manufacturing processes for the production of CFRP car hoods: namely Low Pressure-RTM (LP-RTM), High Pressure-RTM (HP-RTM) and Compression-RTM (C-RTM) (Figure 14).

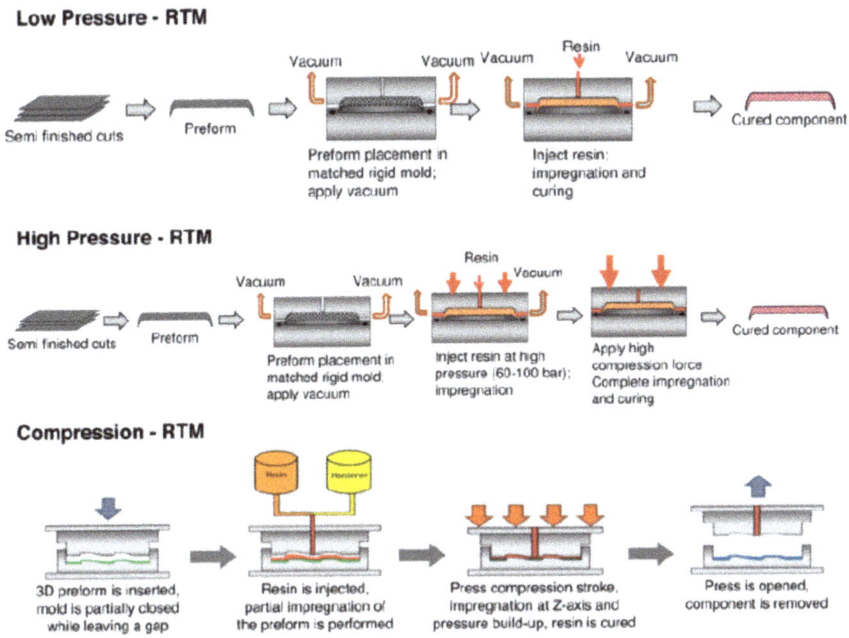

Figure 14. Resin transfer molding (RTM) variants analyzed. Reprinted from [70], © 2020 with permission from Elsevier.

A car hood, mounted in an Italian luxury car, was taken as reference part and the functional unit was defined as the production of 1000 CFRP car hoods through the three processes under investigation. The system boundaries included the material extraction, manufacturing and EoL phases for those flows directly involved in processes. The LCA analysis was carried out using SimaPro 8.05.13 as software and the database EcoInvent 3.1 as the supporting inventory database. They observed that the mold size directly influenced the energy consumption of the heating and cooling systems, thus it was evident that moving towards low-pressure methods, such as C-RTM or LP-RTM, led to a drastic reduction of impacts caused by the manufacturing of CFRP components (Figure 15). Anyhow, a reduction of the electricity in these systems could be desirable, for example by using more efficient heating methods and materials.

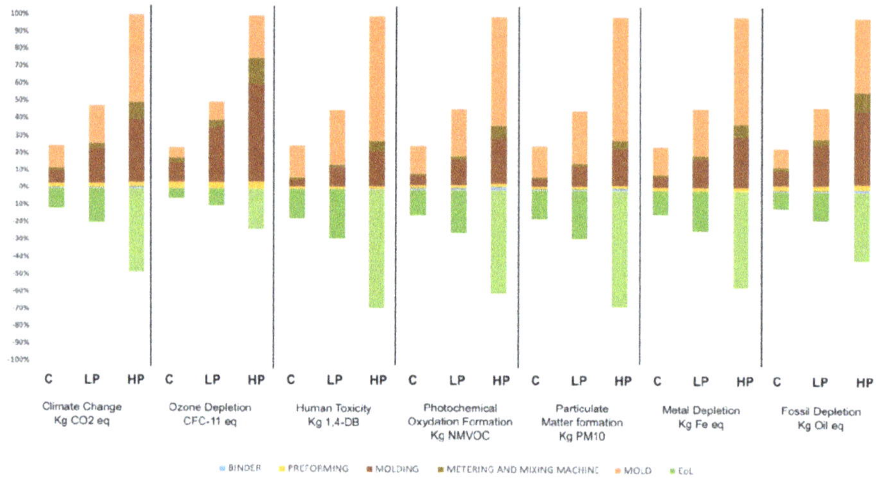

Figure 15. Comparison of the RTM variants in terms of ReCiPe mid-points. Reprinted from [70], © 2020 with permission from Elsevier.

Ingarao and some foreign colleagues, combining the generalized rule-of-mixture (ROM) model and the Ashby material selection one, compared flax-fiber reinforced polymers (FRPs) and GFRPs as lightweight components for automotives [71]. The purpose of this research group was to assess, by the means of LCA, the cradle-to-grave environmental impact of flax FRPs with different design specifications (fiber format, volume fraction, manufacturing technique, and load-bearing capacity) and comparing to those of the conventional GFRPs. The functional unit was designed as 200,000 km driving distance during the lifespan. Two types of flax FRPs, injection molded short flax fiber reinforced polypropylene (short flax-PP) and compound/compression molded flax mat polypropylene (flax mat-PP) film, were investigated and compared with conventional glass mat-PP and short glass fiber-PP. They evaluated multiple impact categories and incineration with energy recovery was chosen as the EoL scenario for both the composites. The flax mat-PP showed, under the equal stiffness criterion, a lower environmental impact than a glass fiber-mat in most of the impact categories (Figure 16), due to its lightweight structure, however, flax mat-PP composites suffered from environmental issues common to agroproducts. By contrast, considering the dominant impact categories, the total life cycle impact of the short flax fiber-PP was very similar to that of the short glass fiber-PP. Finally, since the fiber volume fraction can be considered the critical parameter they evaluated its influence on the impacts, showing the best volume fractions (28–32% v/v) for the flax mat-PP composite differently than flax fiber-PP which exhibited steady decreased life-cycle CO_2 emissions.

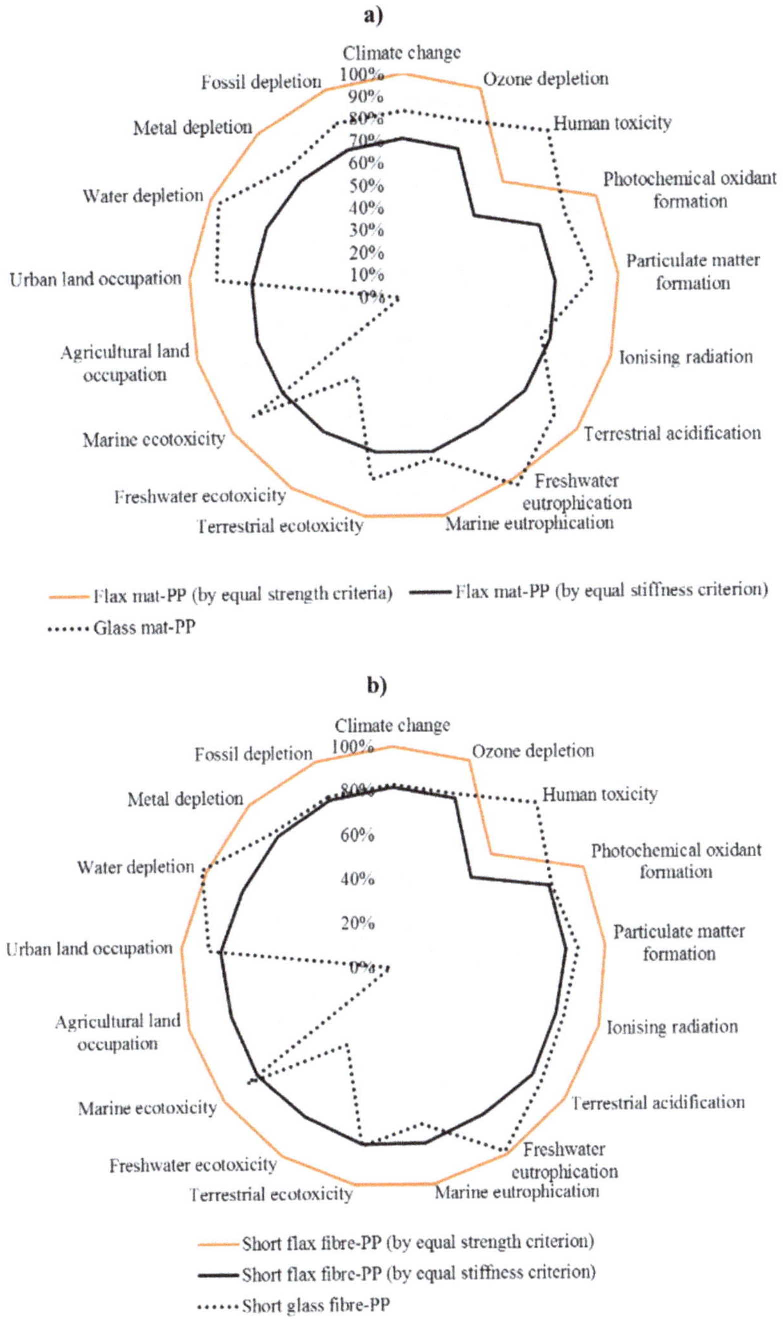

Figure 16. Life cycle impact comparison between flax fiber-reinforced polymers (FRPs) and glass fiber-reinforced polymers (GFRPs) with equal stiffness and equal strength design criteria. (**a**) Flax mat-PP. (**b**) Short flax fiber-PP. Reprinted from [71], © 2020 with permission from Wiley.

2.4. Concrete-Polymeric Materials

Concrete-polymer materials are extensively used due to their unique properties leading to an extended service life, in particular in the structural applications. The major causes of deterioration of reinforced concrete structures, such as the corrosion of the steel reinforcement, industrial chemicals, deicing salts, and moisture, can be overcome by using FRP bars as an alternative type of reinforcement [72,73].

Redaelli et al. carried out a LCA sustainability analysis on a prototype culvert made of different type of concrete and reinforcement (carbon steel, austenitic and duplex stainless steels, and glass fiber–reinforced polymer) providing a comparison among the different typologies [74]. The analysis was firstly carried out from cradle to the gate (i.e., up to culvert installation) by assuming a 5 m culvert segment as FU, for the LCI and GFRP were considered primary data by the industry partners, whilst for the remaining materials and processes were used secondary data from the Ecoinvent database. Among the different concrete formulations, that containing carbon steel (CS) resulted the cheapest to the gate. They successively extended their study to the operational and EoL stages of the infrastructure, by performing cradle-to-the-grave analysis over a period of 100 years. In these new boundaries, the formulation containing CS resulted the worst one, both in terms of environmental impact and cost, because of the additional required maintenance. The formulation used glass fiber–reinforced polymer reinforcements proved to be the cheapest solution and, as regards the environmental impacts, the most sustainable solution, followed by the stainless steel. The LCA results, among the different formulations, did not change when the study period was decreased to 50 years.

A LCA and also a LCC analysis on four non-corrosive reinforcement materials for concrete structures were carried out by Dotelli et al. taking into account stainless-steel (SS); GFRP rebars, CFRP strands and epoxy-coated steel (ECS) reinforcing bars [75,76]. SIMAPRO (version 8.5.2.0, 2018) software and Ecoinvent database, as source of secondary data, were used firstly in a cradle-to-gate (raw materials, transportation, and construction) investigation and then in a cradle-to-grave (raw materials, transportation, construction, maintenance, and EOL) one. The selected impact categories were GWP 100, ozone-depletion potential (ODP), acidification potential (AP), photochemical oxidant creation potential (POCP) and eutrophication potential (EP). CFRPs reinforced formulations outperformed in four out of five categories, resulting in not being environmentally competitive only for the ozone depletion. Concrete reinforced with SS well performing in three out of five categories, resulted in being the first alternative to the fiber-reinforced polymer. Anyway, due to the large contribution in terms of global warming of the SS's reinforced concrete (plus 35,508 kg CO_2 eq.) the former formulation remained the largely preferable choice. Their results indicated the ECS reinforcing bars as the least environmental- friendly solution, since there was not a single category for which it was an ideal alternative. Also as regards the demolition and landfill activities at EoL, the results of Dotelli and his colleagues indicated the FRP formulations as the best option, because did not require major operations (such as cathodic protections or reinforcement replacement), and, thus, did not require traffic disruptions or traffic diversions.

The comparison between the use of FRP and steel in reinforcing the masonry buildings was also the subject of the study carried out by Russo et al. at the University of Venice [77]. They evaluated the LCA of the two options when using as retrofit work to increase the overall strength of masonry structures against seismic actions. They observed that, in terms of sustainability, costs and environmental impact, the weight of steel and the need for maintenance can become a critical problem, thus proposing as a sustainable solution the use of FRP profiles instead of the steel ones.

Puccini et al. approaching to a sustainable design and management of transport infrastructures applied the LCA tool to study different solutions of low noise pavements. Their idea was based on replacing, in the bituminous mixtures, the classic virgin non-renewable materials with secondary materials to lower landfill pressure, maintaining adequate functional properties [78]. Thus, the classic hot mix asphalt (HMA) was compared with crumb rubber (CR) from EoL tires and reclaimed asphalt pavement (RAP). They carried out a cradle to grave study, from production to EoL stages, including laying and maintenance in an Italian urban road located in the municipality of Massarosa in the province of Lucca (Tuscany). They settled on 400 m and 464.5 m of pavement of the road as FU and the data collected were analyzed by SimaPro 8.3.0 software. The researchers of the University of Pisa demonstrated how the use of recycled materials leads to a 50% reduction of the environmental burdens, confirming the high impacts associated with the use of the classic HMA mixtures (Figure 17).

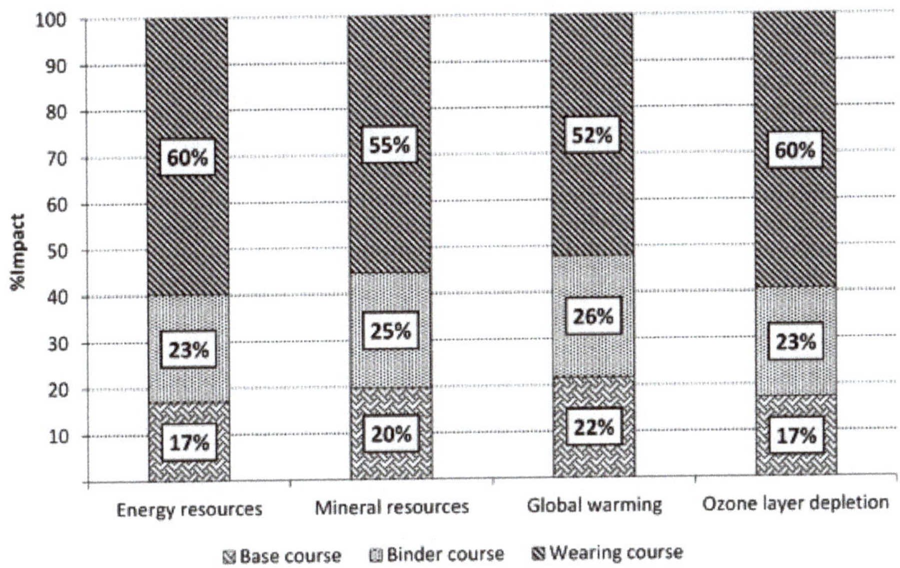

Figure 17. Contribution of the pavement layers to selected impact categories, scenario S. Reprinted from [78].

Some of the observed reduction (material depletion and energy consumption) was attributed to the different level of maintenance requested. As regards the reduction in raw materials and hydrocarbons emission was attributed to the use of reclaimed materials. Thus, the combined use of recycled materials and warm technologies was indicated as a methodological approach to improve the environmental sustainability of low noise pavements.

2.5. Geopolymers

In the previous paragraph we got to see how, considering concrete the most commonly used construction material worldwide, many efforts were carried out in recent years to improve their functional properties trying to increase, in the meantime, its environmental performance. In fact it is well known that this huge amount leads to a relevant environmental impact, concrete occupy large volumes in landfills, and its manufacturing is highly energy intensive. Cement industry can be considered among the largest CO_2 emission sources [79]. For the above cited reason both sector industry and academy devoted their studies to eco-sustainable materials, such as geopolymers, as components of a new class of low-energy materials maintaining the specific properties of cements [80]. As we reported for concrete and also for geopolymers LCA represents the most common approach to assess potential environmental impacts, showing that the inclusion of industrial waste provides the enhancement of its environmental performance and the reduction of costs. In particular, the recycling of waste, by using in their formulation slags, calcined clays and coal fly ashes [81–84], leads to a reduction in energy and natural resource consumption.

A LCA investigation, concerning the environmental burdens related to the use of wastes from recycling construction and demolition treatment plant as raw materials in the geopolymer's production, was carried out by Petrillo et al. regarding the production of concrete blocks in plants located in the South of Italy [85]. Their study referred to different plants located in the Campania Region (South of Italy), starting from raw materials acquisition and processing up to the production and packing of concrete paving blocks, thus reporting results in a cradle-to-gate approach. The ordinary Portland cement (OPC) block were obtained by mixing cement with recycled clay soil as solid precursors and blast furnace slag, designing as functional unit 1 m^2 of paver blocks. The various life-cycle stages of brick products were modelled and assessed with SimaPro VC LCA software, by using Eco-indicator 99 and CML 2000 baseline.

In both the methods used, they indicated the binder and concrete production processes as the hot spot for energy consumption and thus the main responsible for CO_2 emissions (Figure 18).

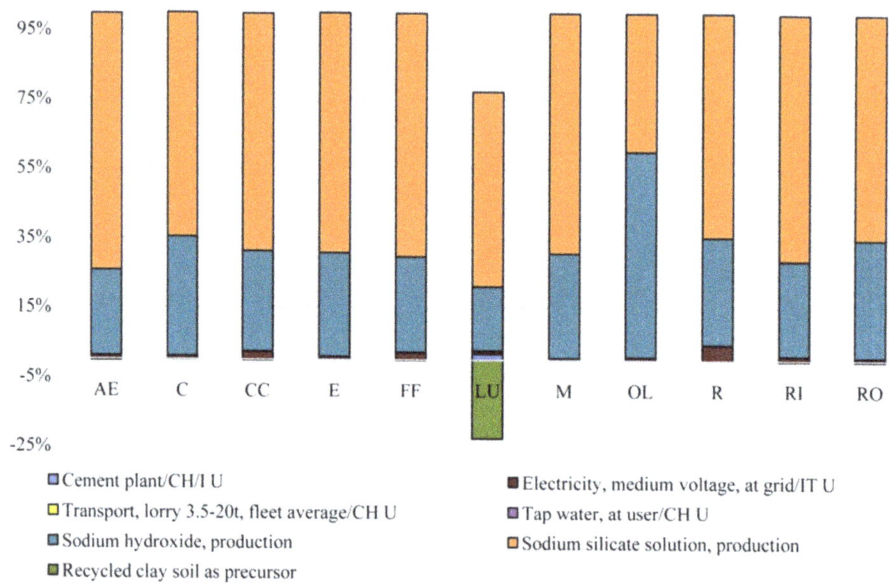

Figure 18. ECO 99 characterization values of 1212 kg of geopolymeric binder (AE: acidification/ eutrophication; C: carcinogenics; CC: climatic change; E: ecotoxicity; FF: fossil fuels; LU: land use; M: minerals; OL: ozone layer; R: radiation; RI: respiratory inorganics; RO: respiratory organics). Reprinted from [85], © 2020 with permission from Wiley.

Other relevant impact categories, involving the use and production of alkali activators, were the ozone layer depletion and eutrophication ones for the CML method and respiratory inorganics and ozone layer depletion for Eco-indicator 99 (Figure 19). They concluded that the use of recycled clay soil or waste for making paving blocks would address sustainable issues such as the resource conservation and conversion of wastes and byproducts to useful and valuable products, although an LCC analysis could be necessary.

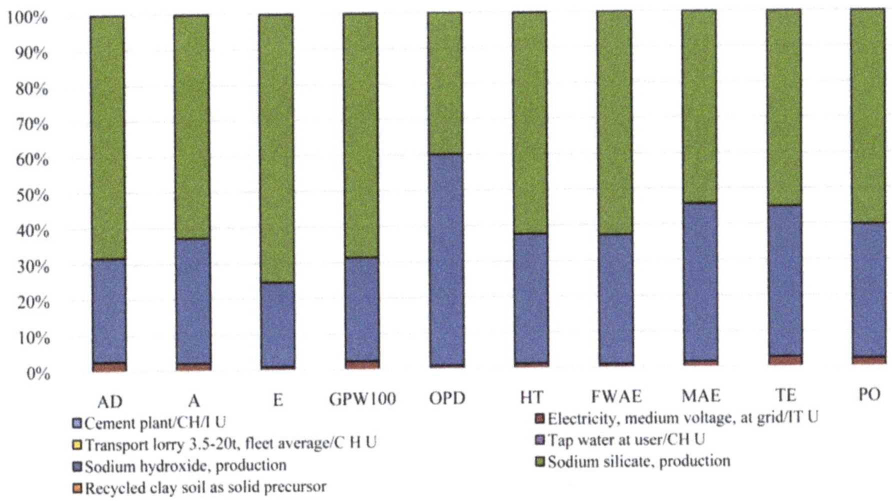

Figure 19. CML characterization values of 1212 kg of geopolymeric binder (AD: abiotic depletion; A: acidification; E: eutrophication; FWAE: fresh water aquatic ecotox; GWP: global warming potential; HT: human toxicity; MAE: marine aquatic ecotoxicity; ODP: ozone layer depletion; PO: photochemical oxidation; TE: terrestrial ecotoxicity). Reprinted from [85], © 2020 with permission from Wiley.

Tugnoli et al. carried out a LCA analysis aiming at evaluating the replacement of the conventional cementitious binders in the passive fire protection (PFP) systems with geopolymer-based matrices [86], because of their high thermal stability and low thermal conductivity [83,87]. They compared the environmental and economic performance of a PFP geopolymer with a commercial cement, having similar fireproofing properties, that was used as reference. The research group of the University of Bologna divided the LCA study into two cases, A and B respectively. In the former they investigated a geopolymer-based OPC replacement block, by designing a geopolymer concrete block with 1 m^3 as FU, and comparing it with data obtained in previous studies. In the second case, they investigated a geopolymer-based mixture (waste fly ash from the flue gas treatment of a coal-fired power plant as a precursor and Na-based alkali activators) for PFP application with the purpose to define its environmental footprint. To this aim and to make possible the comparison, they designed as FU an insulating PFP layer of 66.5 m^2. The analyses were carried out from cradle-to-grave with secondary data coming from the Ecoinvent database 2.0 and the thinkstep free GaBi LCI datasets. The main unit processes for both cases are reported in Figure 20, where the considered transportation distances are reported as well.

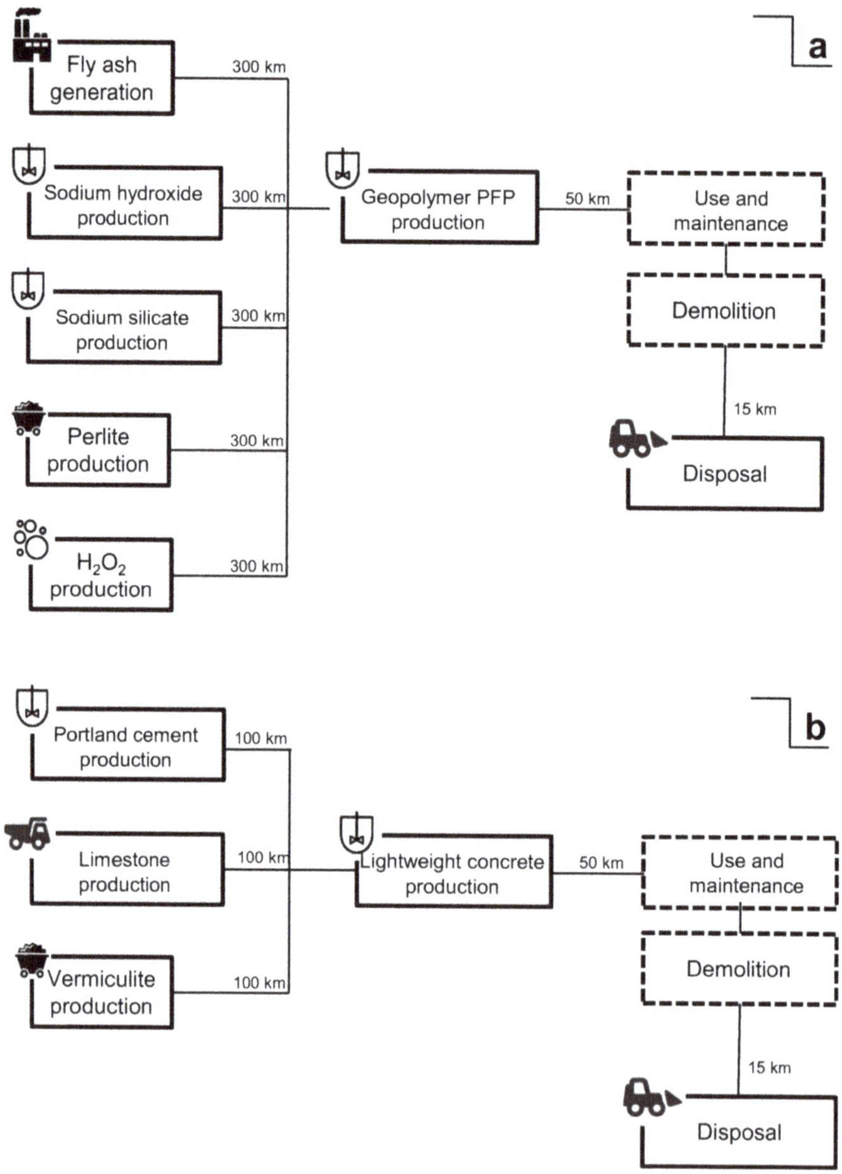

Figure 20. Main unit processes considered for (**a**) geopolymer based passive fire protection (PFP) mix and (**b**) lightweight concrete PFP mix in case B. Dashed box: processes outside system boundaries. Reprinted from [86], © 2020 with permission from Springer Nature.

By making a balance among the avoiding of CO_2 emissions from the clinker process in cement manufacturing and the alkali activation of the geopolymers (affecting relevant parameters like AP, EP) the proposed new geopoloymeric systems performed a life-cycle impact which was 27% lower than that of the reference in terms of GW. These results confirmed the overall sustainability and the economic viability (Figure 21) of geopolymers for PFP applications with the authors' recommendation in employing sodium silicate produced via the hydrothermal route and in partial substitution of sodium silicate with waste-derived alkali activators.

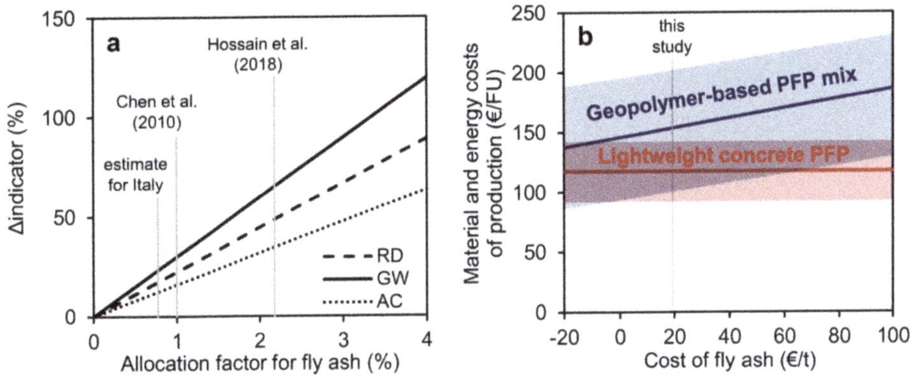

Figure 21. (**a**) Variation of the main environmental indicators for the geopolymer-based PFP mixture as a function of allocation of the coal life cycle impacts to the fly ash. (**b**) Estimated material and energy costs for the geopolymer-based mixture and for the lightweight concrete alternative as a function of the cost of fly ash. Reprinted from [87], © 2020 with permission from Springer Nature.

Aiming at demonstrating the feasibility of the production of aerated systems based on geopolymer hybrid foams (GHF), derived from fly ash (GHF-FA), Roviello et al. applied LCA to evaluate the impacts of this process in comparison with GHF based on metakaolin (GHF-MK) and commercial aerated autoclaved concrete (AAC) systems [88]. In the authors' intentions the data deriving from LCA must allowed to overcome the issues that, at that time, limited the geopolymers commercialization. To reach this goal they emphasized the inexpensiveness of the raw material and processes.

In particular, they aimed at developing an eco-friendly process that, using waste raw materials, adopted the circular economy philosophy. They enclosed in their study raw materials, transport, manufacturing and disposal, thus considering a cradle-to-gate approach (Figure 22). LCA was carried out by the means of the SimaPro© software v.8, designing as FU 1 m^3 of GHF and by using for primary data available datasets and literature data whilst, for secondary data, the Ecoinvent v.3 database, literature review and expert judgment. As expected, Roviello and her colleagues found that showed the best environmental performance paying, however, in terms of mechanical properties with respect to the systems. Thus, they demonstrated the feasibility of GHF-FA industrial production highlighting, however, the need for further studies to make comparable its usage performance to that of GHF-MK.

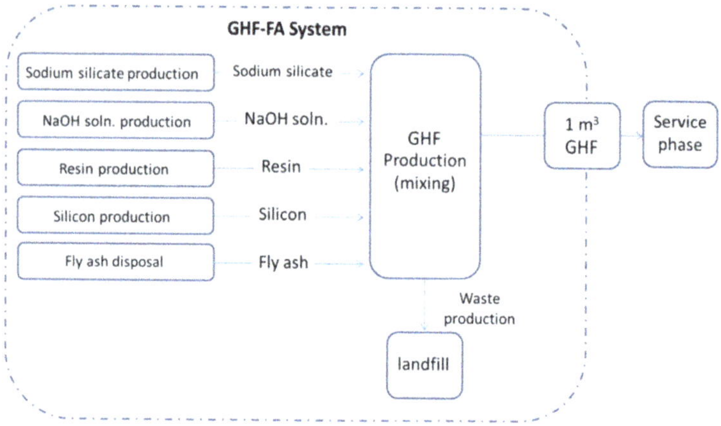

Figure 22. Description of geopolymer hybrid foams derived from fly ash (GHF-FA) life-cycle production. Reprinted from [88], © 2020 with permission from Elsevier.

2.6. Polymeric Materials in Fuel Cell

It is well known that automotive sector, from production to the road, contributes substantially to GHG emissions and air pollution [89,90]. In one of the previous paragraphs, we dealt with the production assessment, whilst in this one we face the problems related with the automobile's use. The GHG emissions from road transport accounted for 18 and 28% in EU and USA, respectively [91,92] and in the last decade battery electric vehicles (BEVs) were proposed as environmental alternative to the fossil-based internal combustion engine vehicles (ICEVs) showing, however, disadvantages as regards short driving range and long refueling time. For this reason in the last years a technology, already developed in the 1960 s [93,94], like that of fuel cells (FC), has been reconsidered [95]. Various LCA analyses were carried out focusing mainly on the fuel cycle and accounting for roughly on the production and disposal of the cars [90].

Thus, in moving towards the understanding of the real environmental impacts of producing FC vehicles, in comparison to the classic or alternative green cars, and overcoming the limitations of existing studies, Lettieri et al. focused, by the means of LCA tools, on the manufacturing stages of a polymer electrolyte membrane (PEM) FC and compared the results obtained with those relating the production of BEV and ICEV [96]. They carried out a LCA analysis of FCV, ICEV and a BEV, respectively, considering their whole life cycle, including the manufacturing of the vehicles, their use phase and end-of-life (Figure 23), assuming a lifetime of 150,000 km for all three vehicles and a functional unit of 1 km driven by one car.

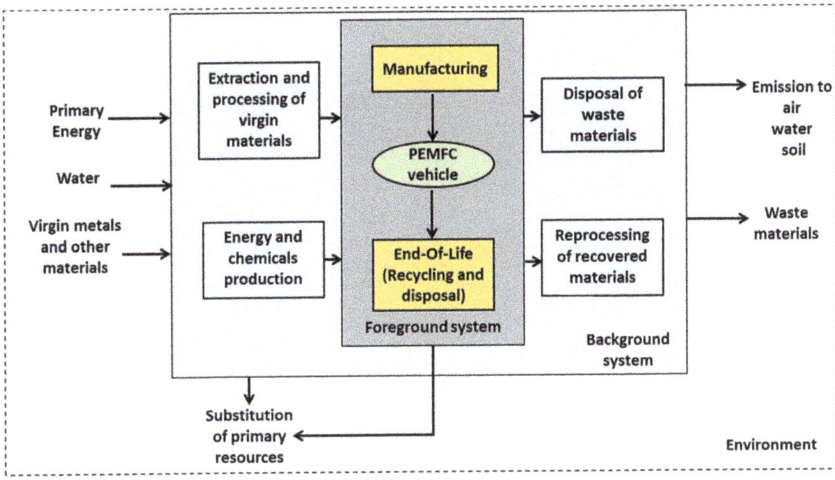

Figure 23. System boundary of the whole life cycle assessment of a polymer electrolyte membrane (PEM) fuel cell passenger vehicle. Reprinted from [96], © 2020 with permission from Elsevier.

The results obtained showed different environmental impacts of the three different vehicles as a function of the stages considered. FCV performed worse in the production process because of the hydrogen tank and the fuel cell stack, despite the scenario proposed suggested a 25% potential reduction in the climate change impact category. By contrast, analyzing the use stage FCV performed well in respect to ICEV, showing the highest GWP due to the fossil carbon emissions associated to the use of diesel (Figure 24).

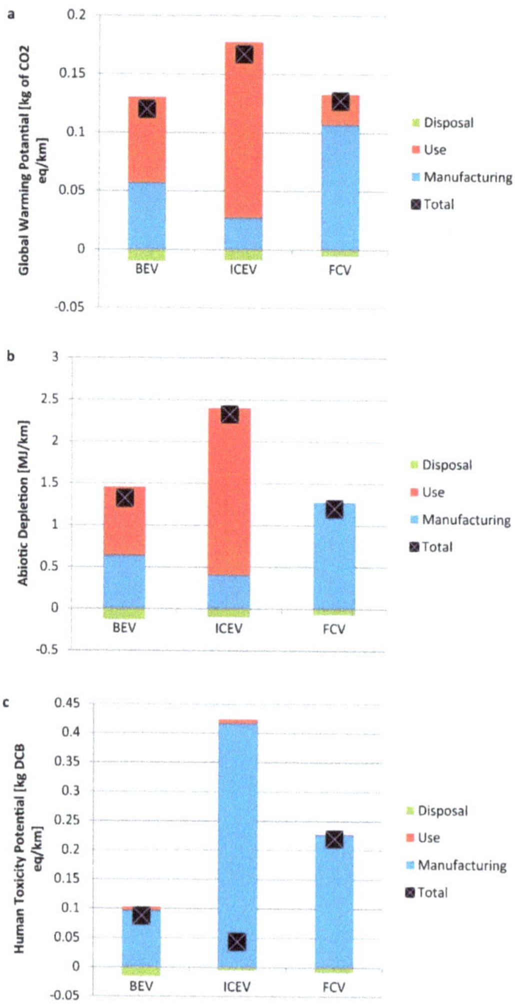

Figure 24. (**a**) Global warming potential, (**b**) abiotic depletion, (**c**) human toxicity potential of fuel cell vehicle (FCV), battery electric vehicle (BEV) and internal combustion engine vehicle (ICEV) for the whole life cycle. Reprinted from [96], © 2020 with permission from Elsevier.

The GWP of the disposal stage was shown negligible for all three technologies, thus according to the authors FCV showed a significant improvement in the use stage but reducing the environmental impact associated with its manufacture still represents an important challenge that need to be addressed in future years.

Di Marcoberardino et al. after designed and tested a novel micro-combined heat and power (CHP) system based on membrane reactor and PEM fuel cell [97,98] proceeded with their study evaluating the environmental performance of the system, by using LCA, and the economic assessment to make it sustainable and cost-effective over its lifetime [99].

In particular, they focused their analysis on aspects such as process intensification and optimized thermal integration thus trying to reach a reduction of the material requirements, considering also aspects beside greenhouse gas emissions: water withdrawal, resources depletion, human health. They compared their innovative FC with a commercial PEM for various residential applications, supplied with conventional natural gas from electric grids in two different countries (Italy and Germany), thus making the assessment as generic as possible. Italy and Germany were chosen to compare different loads because of the different climate, and different natural gas and electricity costs. The evaluation was carried out, taking into account the full life cycle needed to provide heat and electricity (Figure 25), and assigning the functional unit as "provide one, two or 10 German or Italian dwellings with useful heat and electricity produced by a 5 kWel PEMFC micro-CHP over one year". The LCA software used was IMPACT 2002+ v2.2 considering the following indicators: carbon footprint, resources, impacts on ecosystem quality, water withdrawal and impacts on human health. Among the results, whilst the impacts on the water withdrawal and human health can be positive or negative depending on the case, the innovative system undoubtedly reduced the carbon footprint impact, thus becoming competitive, from an economic point of view.

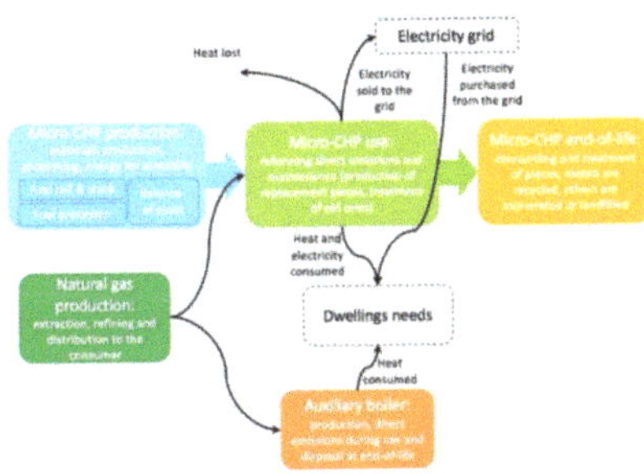

Figure 25. System boundaries considered for the LCA for the assessment of the steam reformer (SR) and autothermal reformer membrane reactor (ATR-MR) systems. Reprinted from [99], © 2020 with permission from Elsevier.

2.7. Various Life-Cycle Assessment (LCA) Applications on Polymers

Given the specific applications we have seen in the above reported sections, in general in the last decade LCA has been applied to explore new design routes, processes and final treatment of polymer or polymer-based materials. The main goals that the authors addressed in the last paragraph was the reduction of the environmental impacts, proposing a different strategy, by the help of LCA, compared with the standard one. Bio-based polymeric materials in particular have attracted the attention of those researchers aiming at improving their environmental performance.

La Rosa et al., by means of LCA, explored the possibility of improving the eco efficiency of glass fibre composite materials to be applied on a pipe system used to transport cooling sea water in a Sicilian petrochemical company [100]. In order to replace part of the conventional fibres with hemp mats a comparative LCA was performed on two different elbow-fittings made of glass fibre/thermoset composite and hybrid (glass fibre-hemp)/thermoset composite, respectively. The study was carried out from cradle to grave by considering as FU one elbow fitting used in the chemical plant pipeline, with an estimated life of 20 years and using data collected from the process as primary and the Ecoinvent v2.2 database for the secondary ones. They found an increase of the eco-efficiency for the hemp mat-reinforced composites, particularly during the production phase due to the 'green' origin of the hemp mat. The same authors, by using the Simapro 7.2 software, carried out a cradle-to-manufacture study in order to compare the production of an eco-sandwich panel containing cork, hemp and bio-based epoxy resin as natural materials with respect to a traditional sandwich made of epoxy resin and polyurethane/glass-fibers [101]. Aiming at evaluating the environmental impacts related to their production, they settled on an eco-sandwich panel sized (0.400 × 0.400 × 0.02 m) as FU (Figure 26).

Figure 26. Eco-sandwich with core in granulated cork and skins in hemp/bio-resin. Reprinted from [101], © 2020 with permission from Elsevier.

LCA results showed a good improvement in terms of environmental impacts and energy use compared with the petroleum-based products, despite the major contributions to the impact (in both cases, eco-sandwich and traditional sandwich) being due to the use of epoxy resin (environmental impact up to >85%).

Lettieri et al., in the direction of a greening of the epoxy based materials, carried out a cradle to gate study comparing bacterial cellulose (BC)- and nanofibrillated cellulose (NFC)-reinforced epoxy composites with PLA and 30 wt.-% randomly oriented GFR/PP composites [102].

They assumed 1 kg of NFC-reinforced epoxy composite as FU and used version 6 of the GaBi software. The adopted model, developed by the Centre for Environmental Science in Leiden University [103], in order to minimize uncertainties provided midpoint indicators to model the effects of substances on the environment at an early stage (also known as the problem-oriented approach). Their research highlighted good mechanical properties for the proposed systems with respect to neat PLA, but in the meantime a higher GWP and abiotic depletion potential of fossil fuels of BC- and NFC-reinforced epoxy composites. They observed that by enlarging the boundaries to the grave (Figure 27) and increasing the nanocellulose content up to 60%, the GWP and abiotic depletion potential of the innovative composites could be lower than neat PLA, suggesting higher nanocellulose loading to reach the desired "greener credentials".

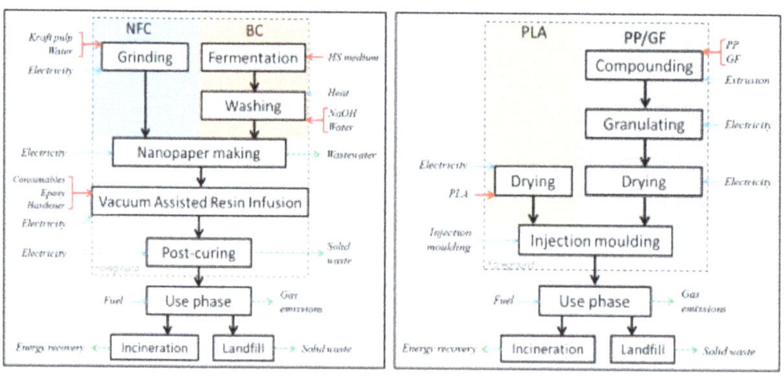

Figure 27. Schematic diagram showing the system boundaries of the model representing the life cycle of bacterial cellulose (BC-) and nanofibrillated cellulose (NFC)-reinforced polymer composites (**left**), and PLA and glass-fibre/polypropylene (GF/PP) composite (**right**). The red, blue and green arrows represent consumables or raw materials required, energy input and waste (materials and energy), respectively. Reprinted from [102], © 2020 with permission from Elsevier.

A couple of years since their first studies, La Rosa et al., continuing LCA investigations on epoxies, proposed the idea of the recyclable composites production by engineering a commercial amine-based epoxy curing agent. After the manufacturing and the chemical characterization of this CF/epoxy composites, they evaluated the environmental impacts associated with the recyclability of the thermoset composite [104]. Thus aiming at highlighting the benefits of using Recyclamines® in the production of structural fibre reinforced composites. The Ecoinvent database was used for the data and a CF/thermoset panel, manufactured through high-pressure (HP) RTM and chemically treated in order to recover clean CFs as well as an epoxy thermoplastic, was designed as FU. The first LCA results showed the environmental advantages in recovering clean CFs as well as a thermoplastic polymer that can be reused and re-processed. In a further study, they tested the efficacy of the recycling process, developed by Connora Technologies (Hayward, CA, USA), at the plant level, highlighting the great innovation introduced by Connora Technologies that induced the conversion of thermosets into thermoplastics [105]. The research group at the University of Catania obtained a series of LCA results, demonstrating a very low environmental impact associated with the process and the avoiding of furthermore impacts due to the recovery of the epoxy-composite constituents (fibres and matrix).

People working in manufacturing are always searching for new materials and process optimization to guarantee quality and safety standards. Gagliardi et al. focused their research on the evaluation of the impacts associated with the different joining processes (i.e., mechanical, thermal and chemical) involved in the assembling of materials, with particular regard to the hybrid ones [106]. They carried out cradle-to-gate and cradle-to-grave analyses but not considering the use phase and focusing mainly in the recycling aspects. In the first part of the study, the energy consumption and CO_2 emissions of the investigated processes, showed the best environmental performance for the mechanical fastening performed by bolts. When they enlarged the boundaries, considering the recycling, the adhesive bonding and ultrasonic spot welding became environmentally friendly as well.

The lowering of the environmental impacts and toxicity for human health in the design and production of materials for X-ray shielding applications was explored by Milanesio et al. [107], who proposed a new epoxy based composites filled with barium sulfate and bismuth oxide instead of the traditional screens made of lead and steel. The low-cost lightweight shielding composite materials was tested by LCA, from cradle to gate, by using openLCA 1.9 and the Ecoinvent 3.5 database. The proposed new epoxy-based panel performed better than both the traditional panels, in terms of GW and particulate formation with respect to the lead one and in terms of human health with respect to the steel one (Figure 28).

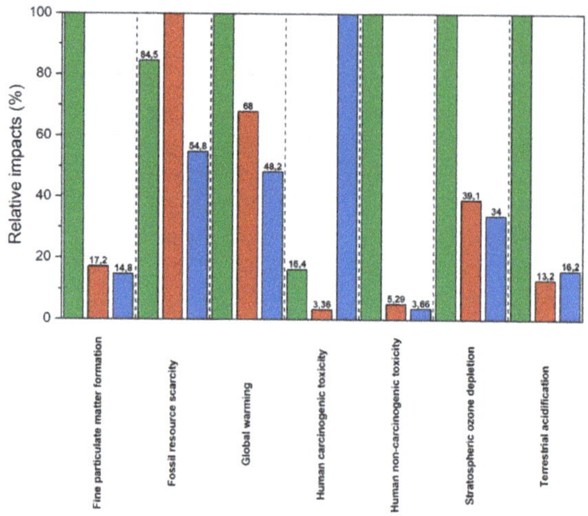

Figure 28. The histogram showing the relative impacts for the main and most reliable factors studied. Data were normalized to the maximum value for a rapid evaluation of the advantages and disadvantages of each material. Green bars refer to the impacts calculated for lead, red bars for the composite samples and blue bars are referred to steel impacts. Reprinted from [107].

3. Future Outlooks

The fair amount of works analyzed in this review allowed the authors to understand not only that in Italy are LCA applications held in high regard, but, also, to draw considerations and look at the multitude of perspectives that LCA can offer to different sectors dealing with the design, production, marketing and final treatment of polymeric materials. As regards food packaging, the most investigated sector in Italy for number of manuscripts, application of LCA was double addressed. On the one hand, towards the aspects inherent the final treatment of the plastic materials to mitigate, or completely remove, the dispersion of the packaging in the environment, and on the other hand to replace the raw material to make it free from oil derivation. In the next few years, the authors expect a growing interest in the first of the two aspects, aiming at reducing the persistence of packaging in the environment. As far as they are concerned, the authors will continue working in this field of research by evaluating the relevant environmental impacts, and the consequential damages, that are associated with naturally derived polymers with high barrier properties for optimal food conservation levels.

Furthermore, agriculture is a very sensible research area, especially in Mediterranean regions, like Sicily, where there is a massive consumption of plastic materials for different purposes, from mulching to greenhouse covers, through irrigation systems. Despite this, the authors found that very little has been done thus far at the environmental level compared with the other areas reviewed. This stimulates expansion of research on the subject to contribute to greening not only plastics' life cycles but also the agricultural systems where those plastics' are utilized. With regard to this point, this paper's team of authors intend to co-work in the direction of developing and testing innovative plastic formulations that are compatible with the soil and/or enable improving its properties, in order to find the truly feasible ones. Experimental research will then be complemented with LCA to validate those formulations from an environmental perspective.

Similarly, in the building sector, new concrete formulations can be designed using recycled plastics, without compromising the structural properties of the artefact.

The review highlighted LCA to be an irreplaceable tool being able to couple economic, environmental and performance considerations thus leading to discovery of the most high-performing solution or, at least, the viable trade-off.

Last but not least the automotive sector is historically a very important sector of the Italian economy. Here, LCA practitioners can indulge themselves, both for the innumerable polymeric components to be studied and for the possibility of extending or restricting the boundaries, consequently focusing upon an impact category rather than another one. In this regard it is worth making a consideration to pay due attention to comparing scenarios that are likely to be similar but which can, instead, show big differences just by expanding the investigated field.

4. Conclusions

In the framework of the environmental impact assessments, in the last years, LCA has emerged as a valuable decision-support tool for decision-makers, in both political and industrial areas. Consequently, LCA has assumed over the years a key role as a tool for identifying cradle-to-grave impacts of products in a multitude of sectors: among them, polymers is one of the most investigated, considering the uncountable number of disparate applications.

Through this literature review, the authors achieved their goal of reviewing and building upon LCAs in the Italian polymeric sector, to highlight the most relevant environmental issues.

It was found that LCA was used in the majority of those sectors where plastic products are utilized, namely:

- agriculture;
- food production and packaging;
- automotive;
- concrete-polymeric materials;
- geopolymers;
- polymeric materials in fuel cells.

In each sector and for each application, the Italian researchers proposed, on the basis of LCA results, sometimes from the cradle to the grave and at other times from the cradle to the gate, the best solution to minimize the environmental impact and ensure a viable choice also from an economic point of view. Food production and packaging resulted the most studied sector with LCA, but in all the other sectors we recorded increasing attention on the whole polymer's life cycle and the need to show its sustainability.

Author Contributions: I.B., C.I. and V.S. planned, organized and wrote the paper. All authors have read and agreed to the published version of the manuscript.

Funding: Ignazio Blanco is grateful to the University of Catania for supporting this research within the "Bando-CHANCE" and to the Department of Civil Engineering and Architecture within "Piano per la Ricerca 2016–2018–Linea Intervento 1 and 2".

Conflicts of Interest: The authors declare no conflict of interest.

References

1. Walker, S.; Rothman, R. Life cycle assessment of bio-based and fossil-based plastic: A review. *J. Clean. Prod.* **2020**, *261*, 121158. [CrossRef]
2. Riggi, E.; Santagata, G.; Malinconico, M. Bio-based and biodegradable plastics for use in crop production. *Recent Patents Food, Nutr. Agric.* **2011**, *3*, 49–63. [CrossRef] [PubMed]
3. Ingrao, C.; Tricase, C.; Cholewa-Wójcik, A.; Kawecka, A.; Rana, R.L.; Siracusa, V. Polylactic acid trays for fresh-food packaging: A Carbon Footprint assessment. *Sci. Total Environ.* **2015**, *537*, 385–398. [CrossRef] [PubMed]
4. PlasticsEurope, 2019. Plastics—The Facts 2019. An Analysis of European Plastics Production, Demand and Waste Data. Available online: https://www.plasticseurope.org/application/files/9715/7129/9584/FINAL_web_version_Plastics_the_facts2019_14102019.pdf (accessed on 20 April 2020).
5. Licciardello, F. Packaging, blessing in disguise. Review on its diverse contribution to food sustainability. *Trends Food Sci. Technol.* **2017**, *65*, 32–39. [CrossRef]

6. Ingrao, C.; Giudice, A.L.; Bacenetti, J.; Khaneghah, A.M.; Sant'Ana, A.S.; Rana, R.L.; Siracusa, V.; Sant'Ana, A.D.S. Foamy polystyrene trays for fresh-meat packaging: Life-cycle inventory data collection and environmental impact assessment. *Food Res. Int.* **2015**, *76*, 418–426. [CrossRef]
7. Siracusa, V.; Blanco, I.; Romani, S.; Tylewicz, U.; Dalla Rosa, M. Gas Permeability and Thermal Behavior of Polypropylene Films Used for Packaging Minimally Processed Fresh-Cut Potatoes: A Case Study. *J. Food Sci.* **2012**, *77*, E264–E272. [CrossRef]
8. Ingrao, C.; Siracusa, V. Quality—And sustainability—Related issues associated with biopolymers for food packaging applications: A comprehensive review. In *Biodegradable and Biocompatible Polymer Composites—Processing, Properties and Applications. Woodhead Publishing*; Shimpi, N.G., Ed.; Elsevier: Amsterdam, The Netherlands, 2017; pp. 401–418; ISBN 9780081009703.
9. Scarascia-Mugnozza, G.; Sica, C.; Russo, G. Plastic materials in european agriculture: actual use and perspectives. *J. Agric. Eng.* **2012**, *42*, 15–28. [CrossRef]
10. Kijchavengkul, T.; Auras, R.; Rubino, M.; Ngouajio, M.; Fernandez, R.T. Assessment of aliphatic–aromatic copolyester biodegradable mulch films. Part I: Field study. *Chemosphere* **2008**, *71*, 942–953. [CrossRef]
11. Picuno, P.; Sica, C.; Laviano, R.; Dimitrijevic, A.; Scarascia-Mugnozza, G. Experimental tests and technical characteristics of regenerated films from agricultural plastics. *Polym. Degrad. Stab.* **2012**, *97*, 1654–1661. [CrossRef]
12. Cascone, S.; Ingrao, C.; Valenti, F.; Porto, S.M. Energy and environmental assessment of plastic granule production from recycled greenhouse covering films in a circular economy perspective. *J. Environ. Manag.* **2019**, *254*, 109796. [CrossRef]
13. Scarascia-Mugnozza, G.; Sica, C.; Picuno, P. The optimisation of the management of agricultural plastic waste in italy using a geographical information system. *Acta Hortic.* **2008**, *801*, 219–226. [CrossRef]
14. Leceta, I.; Guerrero, P.; Cabezudo, S.; De La Caba, K. Environmental assessment of chitosan-based films. *J. Clean. Prod.* **2013**, *41*, 312–318. [CrossRef]
15. Bertolini, M.; Bottani, E.; Vignali, G.; Volpi, A. Comparative Life Cycle Assessment of Packaging Systems for Extended Shelf Life Milk. *Packag. Technol. Sci.* **2016**, *29*, 525–546. [CrossRef]
16. Ingrao, C.; Gigli, M.; Siracusa, V. An attributional Life Cycle Assessment application experience to highlight environmental hotspots in the production of foamy polylactic acid trays for fresh-food packaging usage. *J. Clean. Prod.* **2017**, *150*, 93–103. [CrossRef]
17. Parlato, M.C.M.; Valenti, F.; Porto, S.M.C. Covering plastic films in greenhouses system: A GIS-based model to improve post use sustainable management. *J. Environ. Manag.* **2020**, *263*, 110389. [CrossRef] [PubMed]
18. Blanco, I.; Abate, L.; Antonelli, M.L. The regression of isothermal thermogravimetric data to evaluate degradation Ea values of polymers: A comparison with literature methods and an evaluation of lifetime prediction reliability. *Polym. Degrad. Stab.* **2011**, *96*, 1947–1954. [CrossRef]
19. Blanco, I. End-life prediction of commercial PLA used for food packaging through short term TGA experiments: Real chance or low reliability? *Chin. J. Polym. Sci.* **2014**, *32*, 681–689. [CrossRef]
20. Blanco, I. Lifetime prediction of food and beverage packaging wastes. *J. Therm. Anal. Calorim.* **2015**, *125*, 809–816. [CrossRef]
21. Blanco, I. Lifetime Prediction of Polymers: To Bet, or Not to Bet—Is This the Question? *Materials* **2018**, *11*, 1383. [CrossRef]
22. Moore, C.J. Synthetic polymers in the marine environment: A rapidly increasing, long-term threat. *Environ. Res.* **2008**, *108*, 131–139. [CrossRef]
23. Sukan, A.; Roy, I.; Keshavarz, T. Dual production of biopolymers from bacteria. *Carbohydr. Polym.* **2015**, *126*, 47–51. [CrossRef] [PubMed]
24. Ingrao, C.; Bacenetti, J.; Bezama, A.; Blok, V.; Goglio, P.; Koukios, E.G.; Lindner, M.; Nemecek, T.; Siracusa, V.; Zabaniotou, A.; et al. The potential roles of bio-economy in the transition to equitable, sustainable, post fossil-carbon societies: Findings from this virtual special issue. *J. Clean. Prod.* **2018**, *204*, 471–488. [CrossRef]
25. Dassisti, M.; Intini, F.; Chimienti, M.; Starace, G. Thermography-enhanced LCA (Life Cycle Assessment) for manufacturing sustainability assessment. The case study of an HDPE (High Density Polyethylene) net company in Italy. *Energy* **2016**, *108*, 7–18. [CrossRef]
26. ISO (International Organization for Standardization). 2006. ISO 14040:2006–Environmental Management—Life Cycle Assessment—Principles and Framework (2006).

27. ISO (International Organization for Standardization), 2006. ISO 14044: 2006–Environmental Management—Life Cycle Assessment—Requirements and Guidelines (2006).
28. Henton, D.; Gruber, P.; Lunt, J.; Randall, J. Polylactic Acid Technology. In *Natural Fibers, Biopolymers, and Biocomposites*; Mohanty, A.K., Misra, M., Drzal, L.T., Eds.; CRC Press: Boca Raton, FL, USA, 2005; pp. 527–577.
29. Jeswani, H.K.; Azapagic, A.; Schepelmann, P.; Ritthoff, M. Options for broadening and deepening the LCA approaches. *J. Clean. Prod.* **2010**, *18*, 120–127. [CrossRef]
30. European Commission (EC)—Joint Research Centre (JRC)—Institute for Environment and Sustainability (IES). *International Reference Life Cycle Data System (ILCD) Handbook—General guide for Life Cycle Assessment—Detailed Guidance*, 1st ed.; Publications Office of the European Union: Luxembourg, 2010. [CrossRef]
31. Madival, S.; Auras, R.; Singh, S.P.; Narayan, R. Assessment of the environmental profile of PLA, PET and PS clamshell containers using LCA methodology. *J. Clean. Prod.* **2009**, *17*, 1183–1194. [CrossRef]
32. Siracusa, V.; Rocculi, P.; Romani, S.; Rosa, M.D. Biodegradable polymers for food packaging: A review. *Trends Food Sci. Technol.* **2008**, *19*, 634–643. [CrossRef]
33. European Bioplastics. Available online: www.european-bioplastics.org (accessed on 20 April 2020).
34. Sorrentino, A.; Gorrasi, G.; Vittoria, V. Potential perspectives of bio-nanocomposites for food packaging applications. *Trends Food Sci. Technol.* **2007**, *18*, 84–95. [CrossRef]
35. Kirwan, M.J.; Plant, S.; Strawbridge, J.W. Plastics in Food Packaging. *Food Beverage Packag. Technol.* **2011**, *7*, 157–212. [CrossRef]
36. Bohlmann, G.M. Biodegradable packaging life-cycle assessment. *Environ. Prog.* **2004**, *23*, 342–346. [CrossRef]
37. Costa, D.; Quinteiro, P.; Dias, A. A systematic review of life cycle sustainability assessment: Current state, methodological challenges, and implementation issues. *Sci. Total Environ.* **2019**, *686*, 774–787. [CrossRef]
38. Gomes, T.S.; Visconte, L.L.Y.; Pacheco, E.B.A.V. Life Cycle Assessment of Polyethylene Terephthalate Packaging: An Overview. *J. Polym. Environ.* **2019**, *27*, 533–548. [CrossRef]
39. Vink, E.T.; Rábago, K.R.; Glassner, D.; Gruber, P.R. Applications of life cycle assessment to NatureWorks™ polylactide (PLA) production. *Polym. Degrad. Stab.* **2003**, *80*, 403–419. [CrossRef]
40. Gironi, F.; Piemonte, V. Bioplastics disposal: How to manage it. *Waste Manag. 2010* **2010**, *140*, 261–271. [CrossRef]
41. Gironi, F.; Piemonte, V. Life cycle assessment of polylactic acid and polyethylene terephthalate bottles for drinking water. *Environ. Prog. Sustain. Energy* **2010**, *30*, 459–468. [CrossRef]
42. Piemonte, V. Bioplastic Wastes: The Best Final Disposition for Energy Saving. *J. Polym. Environ.* **2011**, *19*, 988–994. [CrossRef]
43. Toniolo, S.; Mazzi, A.; Niero, M.; Zuliani, F.; Scipioni, A. Comparative LCA to evaluate how much recycling is environmentally favourable for food packaging. *Resour. Conserv. Recycl.* **2013**, *77*, 61–68. [CrossRef]
44. Toniolo, S.; Mazzi, A.; Pieretto, C.; Scipioni, A. Allocation strategies in comparative life cycle assessment for recycling: Considerations from case studies. *Resour. Conserv. Recycl.* **2017**, *117*, 249–261. [CrossRef]
45. Siracusa, V.; Ingrao, C.; Giudice, A.L.; Mbohwa, C.; Rosa, M.D. Environmental assessment of a multilayer polymer bag for food packaging and preservation: An LCA approach. *Food Res. Int.* **2014**, *62*, 151–161. [CrossRef]
46. Accorsi, R.; Cascini, A.; Cholette, S.; Manzini, R.; Mora, C. Economic and environmental assessment of reusable plastic containers: A food catering supply chain case study. *Int. J. Prod. Econ.* **2014**, *152*, 88–101. [CrossRef]
47. Zampori, L.; Dotelli, G. Design of a sustainable packaging in the food sector by applying LCA. *Int. J. Life Cycle Assess.* **2013**, *19*, 206–217. [CrossRef]
48. Conte, A.; Cappelletti, G.M.; Nicoletti, G.; Russo, C.; Del Nobile, M. Environmental implications of food loss probability in packaging design. *Food Res. Int.* **2015**, *78*, 11–17. [CrossRef] [PubMed]
49. Accorsi, R.; Versari, L.; Manzini, R. Glass vs. Plastic: Life Cycle Assessment of Extra-Virgin Olive Oil Bottles across Global Supply Chains. *Sustainability* **2015**, *7*, 2818–2840. [CrossRef]
50. Gutierrez, M.M.; Meleddu, M.; Piga, A. Food losses, shelf life extension and environmental impact of a packaged cheesecake: A life cycle assessment. *Food Res. Int.* **2017**, *91*, 124–132. [CrossRef] [PubMed]
51. Cinelli, P.; Coltelli, M.B.; Mallegni, N.; Morganti, P.; Lazzeri, A. Degradability and Sustainability of Nanocomposites Based on Polylactic Acid and Chitin Nano Fibrils. *Chem. Eng. Trans.* **2017**, *60*, 115–120.

52. Petrucci, R.; Fortunati, E.; Puglia, D.; Luzi, F.; Kenny, J.M.; Torre, L. Life Cycle Analysis of Extruded Films Based on Poly(lactic acid)/Cellulose Nanocrystal/Limonene: A Comparative Study with ATBC Plasticized PLA/OMMT Systems. *J. Polym. Environ.* **2017**, *26*, 1891–1902. [CrossRef]
53. Blanc, S.; Massaglia, S.; Brun, F.; Peano, C.; Mosso, A.; Giuggioli, N.R. Use of Bio-Based Plastics in the Fruit Supply Chain: An Integrated Approach to Assess Environmental, Economic, and Social Sustainability. *Sustainability* **2019**, *11*, 2475. [CrossRef]
54. Giovenzana, V.; Casson, A.; Beghi, R.; Tugnolo, A.; Grassi, S.; Alamprese, C.; Casiraghi, E.; Farris, S.; Fiorindo, I.; Guidetti, R. Environmental Benefits: Traditional Vs Innovative Packaging for Olive Oil. *Chem. Eng. Trans.* **2019**, *75*, 193–198.
55. Ferrara, C.; De Feo, G. Comparative life cycle assessment of alternative systems for wine packaging in Italy. *J. Cleaner Prod.* **2020**, *259*, 120888. [CrossRef]
56. Cleary, J. Life cycle assessments of wine and spirit packaging at the product and the municipal scale: A Toronto, Canada case study. *J. Clean. Prod.* **2013**, *44*, 143–151. [CrossRef]
57. Shen, L.; Haufe, J.; Patel, M.K. *Product Overview and Market Projection of Emerging Bio-Based Plastics*; Pro-Bip 2009 Final Report; Utrecht University: Utrecht, The Netherlands, 2009; Volume 245.
58. Schettini, E.; Sartore, L.; Barbaglio, M.; Vox, G. Hydrolyzed protein based materials for biodegradable spray mulching coatings. *Acta Hortic.* **2012**, *17*, 359–366. [CrossRef]
59. Gorgitano, M.T.; Pirilli, M. Life cycle economic and environmental assessment for a greening agriculture. *Qual. Access Success* **2016**, *17*, 181–185.
60. Russo, G.; Scarascia-Mugnozza, G. lca methodology applied to various typology of greenhouses. *Acta Hortic.* **2005**, 837–844. [CrossRef]
61. Report of the European Commission on Internal Market, Industry, Entrepreneurship and SMEs. Available online: https://ec.europa.eu/growth/sectors/automotive_en (accessed on 3 April 2020).
62. Miller, L.; Soulliere, K.; Sawyer-Beaulieu, S.; Tseng, S.; Tam, E.K.L. Challenges and Alternatives to Plastics Recycling in the Automotive Sector. *Materials* **2014**, *7*, 5883–5902. [CrossRef] [PubMed]
63. Koplin, J.; Seuring, S.; Mesterharm, M. Incorporating sustainability into supply management in the automotive industry e the case of the Volkswagen AG. *J. Clean. Prod.* **2007**, *15*, 1053–1062. [CrossRef]
64. Passarini, F.; Ciacci, L.; Santini, A.; Vassura, I.; Morselli, L. Auto shredder residue LCA: Implications of ASR composition evolution. *J. Clean. Prod.* **2012**, *23*, 28–36. [CrossRef]
65. D'Errico, F.; Ranza, L. Comparative Environmental Benefits of Lightweight Design in the Automotive Sector: The Case Study of Recycled Magnesium Against CFRP and Steel. In *Magnesium Technology 2015*; Manuel, M.V., Singh, A., Alderman, M., Neelameggham, N.R., Eds.; Springer: Berlin/Heidelberg, Germany, 2015.
66. Ingarao, G.; Deng, Y.; Marino, R.; Di Lorenzo, R.; Franco, A.L. Energy and CO 2 life cycle inventory issues for aluminum based components: The case study of a high speed train window panel. *J. Clean. Prod.* **2016**, *126*, 493–503. [CrossRef]
67. Hunkeler, D.; Lichtenvort, K.; Rebitzer, G.; Ciroth, A.; SETAC-Europe (Eds.) Environmental Life Cycle Costing. In *ISO 14040, 2006. Environmental Management e Life Cycle Assessment e Principles and Framework*; SETAC; CRC Press: Boca Raton, FL, USA, 2008; ISBN 978-1-880611-83-8.
68. Delogu, M.; Zanchi, L.; Maltese, S.; Bonoli, A.; Pierini, M. Environmental and economic life cycle assessment of a lightweight solution for an automotive component: A comparison between talc-filled and hollow glass microspheres-reinforced polymer composites. *J. Clean. Prod.* **2016**, *139*, 548–560. [CrossRef]
69. Delogu, M.; Zanchi, L.; Dattilo, C.A.; Maltese, S.; Riccomagno, R.; Pierini, M. Take-Home Messages from the Applications of Life Cycle Assessment on Lightweight Automotive Components. *SAE Tech. Paper Ser.* **2018**, *80*, 352–357. [CrossRef]
70. Vita, A.; Castorani, V.; Germani, M.; Marconi, M. Comparative life cycle assessment of low-pressure RTM, compression RTM and high-pressure RTM manufacturing processes to produce CFRP car hoods. *Procedia CIRP* **2019**, *80*, 352–357. [CrossRef]
71. Deng, Y.; Guo, Y.; Wu, P.; Ingarao, G. Optimal design of flax fiber reinforced polymer composite as a lightweight component for automobiles from a life cycle assessment perspective. *J. Ind. Ecol.* **2019**, *23*, 986–997. [CrossRef]
72. Lee, J.-Y.; Kim, T.-Y.; Kim, T.-J.; Yi, C.-K.; Park, J.-S.; You, Y.-C.; Park, Y.-H. Interfacial bond strength of glass fiber reinforced polymer bars in high-strength concrete. *Compos. Part B Eng.* **2008**, *39*, 258–270. [CrossRef]

73. Pour, S.M.; Alam, M.; Milani, A.S. Improved Bond Equations for Fiber-Reinforced Polymer Bars in Concrete. *Materials* **2016**, *9*, 737. [CrossRef] [PubMed]
74. Redaelli, E.; Arrigoni, A.; Carsana, M.; Dotelli, G.; Gastaldi, M.; Lollini, F.; Bertola, F.; Canonico, F.; Nanni, A. Culvert Prototype Made with Seawater Concrete: Materials Characterization, Monitoring, and Environmental Impact. *Adv. Civ. Eng. Mater.* **2019**, *8*, 787–802. [CrossRef]
75. Cadenazzi, T.; Dotelli, G.; Rossini, M.; Nolan, S.; Nanni, A. Cost and environmental analyses of reinforcement alternatives for a concrete bridge. *Struct. Infrastruct. Eng.* **2019**, *16*, 787–802. [CrossRef]
76. Cadenazzi, T.; Dotelli, G.; Rossini, M.; Nolan, S.; Nanni, A. Life-Cycle Cost and Life-Cycle Assessment Analysis at the Design Stage of a Fiber-Reinforced Polymer-Reinforced Concrete Bridge in Florida. *Adv. Civ. Eng. Mater.* **2019**, *8*, 89–94. [CrossRef]
77. Russo, S.; Ippolito, I.; Bergamo, C. FRP Pultruded Material as Reinforcement for Masonry: Expected Interaction in the Medium and Long Time. *Key Eng. Mater.* **2019**, *817*, 89–94. [CrossRef]
78. Puccini, M.; Leandri, P.; Tasca, A.L.; Pistonesi, L.; Losa, M. Improving the Environmental Sustainability of Low Noise Pavements: Comparative Life Cycle Assessment of Reclaimed Asphalt and Crumb Rubber Based Warm Mix Technologies. *Coatings* **2019**, *9*, 343. [CrossRef]
79. Benhelal, E.; Zahedi, G.; Shamsaei, E.; Bahadori, A. Global strategies and potentials to curb CO2 emissions in cement industry. *J. Clean. Prod.* **2013**, *51*, 142–161. [CrossRef]
80. Davidovits, J. Geopolymers. *J. Therm. Anal.* **1991**, *37*, 1633–1656. [CrossRef]
81. Messina, F.; Ferone, C.; Colangelo, F.; Cioffi, R. Low temperature alkaline activation of weathered fly ash: Influence of mineral admixtures on early age performance. *Constr. Build. Mater.* **2015**, *86*, 169–177. [CrossRef]
82. Molino, B.; Vincenzo, A.; Ferone, C.; Messina, F.; Colangelo, F.; Cioffi, R. Recycling of Clay Sediments for Geopolymer Binder Production. A New Perspective for Reservoir Management in the Framework of Italian Legislation: The Occhito Reservoir Case Study. *Materials* **2014**, *7*, 5603–5616. [CrossRef] [PubMed]
83. Catauro, M.; Tranquillo, E.; Barrino, F.; Poggetto, G.D.; Blanco, I.; Cicala, G.; Ognibene, G.; Recca, G. Mechanical and thermal properties of fly ash-filled geopolymers. *J. Therm. Anal. Calorim.* **2019**, *138*, 3267–3276. [CrossRef]
84. Blanco, I.; Poggetto, G.D.; Morrone, B.; Tranquillo, E.; Barrino, F.; Catauro, M. Fly Ash Filled Geopolymers: Preparation and Thermal Study. *Macromol. Symp.* **2020**, *389*, 1699–1708. [CrossRef]
85. Petrillo, A.; Cioffi, R.; De Felice, F.; Colangelo, F.; Borrelli, C. An environmental evaluation: A comparison between geopolymer and OPC concrete paving blocks manufacturing process in Italy. *Environ. Prog. Sustain. Energy* **2016**, *35*, 1699–1708. [CrossRef]
86. Pozzo, A.D.; Carabba, L.; Bignozzi, M.C.; Tugnoli, A. Life cycle assessment of a geopolymer mixture for fireproofing applications. *Int. J. Life Cycle Assess.* **2019**, *24*, 1743–1757. [CrossRef]
87. Kamseu, E.; Ceron, B.; Tobias, H.; Leonelli, E.; Bignozzi, M.; Muscio, A.; Libbra, A. Insulating behavior of metakaolin-based geopolymer materials assess with heat flux meter and laser flash techniques. *J. Therm. Anal. Calorim.* **2011**, *108*, 1189–1199. [CrossRef]
88. Ricciotti, L.; Occhicone, A.; Petrillo, A.; Ferone, C.; Cioffi, R.; Roviello, G. Geopolymer-based hybrid foams: Lightweight materials from a sustainable production process. *J. Clean. Prod.* **2020**, *250*, 119588. [CrossRef]
89. Pollet, B.G.; Staffell, I.; Shang, J. Current status of hybrid, battery and fuel cell electric vehicles: From electrochemistry to market prospects. *Electrochim. Acta* **2012**, *84*, 235–249. [CrossRef]
90. Miotti, M.; Hofer, J.; Bauer, C. Integrated environmental and economic assessment of current and future fuel cell vehicles. *Int. J. Life Cycle Assess.* **2015**, *22*, 94–110. [CrossRef]
91. Hill, N.; Brannigan, C.; Wynn, D.; Milnes, R.; van Essen, H.; den Boer, E.; van Grinsven, A.; Ligthart, T.; van Gijlswijk, R. The Role of GHG Emissions from Infrastructure Construction, Vehicle Manufacturing, and ELVs in Overall Transport Sector Emissions. Task 2 Paper Produced as Part of a Contract between European Commission Directorate-General Climate Action and AEA Technology plc. Available online: www.eutransportghg2050.eu (accessed on 20 April 2020).
92. EPA Inventory of U.S. Greenhouse Gas Emissions and Sinks: 1990–2012. Available online: https://www.epa.gov/ghgemissions/inventory-us-greenhouse-gas-emissions-and-sinks-1990-2012 (accessed on 20 April 2020).
93. Broers, G.H.J.; Ketelaar, J.A.A. High Temperature Fuel Cells. *Ind. Eng. Chem.* **1960**, *52*, 303–306. [CrossRef]
94. Cairns, E.J.; Douglas, D.L.; Niedrach, L.W. Performance of fractional watt ion exchange membrane fuel cells. *AIChE J.* **1961**, *7*, 551–558. [CrossRef]

95. Thomas, C. Fuel cell and battery electric vehicles compared. *Int. J. Hydrogen Energy* **2009**, *34*, 6005–6020. [CrossRef]
96. Evangelisti, S.; Tagliaferri, C.; Brett, D.J.; Lettieri, P. Life cycle assessment of a polymer electrolyte membrane fuel cell system for passenger vehicles. *J. Clean. Prod.* **2017**, *142*, 4339–4355. [CrossRef]
97. Di Marcoberardino, G.; Roses, L.; Manzolini, G. Technical assessment of a micro-cogeneration system based on polymer electrolyte membrane fuel cell and fluidized bed autothermal reformer. *Appl. Energy* **2016**, *162*, 231–244. [CrossRef]
98. Di Marcoberardino, G.; Manzolini, G. Investigation of a 5 kW micro-CHP PEM fuel cell based system integrated with membrane reactor under diverse EU natural gas quality. *Int. J. Hydrogen Energy* **2017**, *42*, 13988–14002. [CrossRef]
99. Di Marcoberardino, G.; Manzolini, G.; Guignard, C.; Magaud, V. Optimization of a micro-CHP system based on polymer electrolyte membrane fuel cell and membrane reactor from economic and life cycle assessment point of view. *Chem. Eng. Process. Process. Intensif.* **2018**, *131*, 70–83. [CrossRef]
100. La Rosa, A.D.; Cozzo, G.; Latteri, A.; Recca, A.; Bjorklund, A.; Parrinello, E.; Cicala, G. Life cycle assessment of a novel hybrid glass-hemp/thermoset composite. *J. Clean. Prod.* **2013**, *44*, 69–76. [CrossRef]
101. La Rosa, A.D.; Recca, G.; Summerscales, J.; Latteri, A.; Cozzo, G.; Cicala, G. Bio-based versus traditional polymer composites. A life cycle assessment perspective. *J. Clean. Prod.* **2014**, *74*, 135–144. [CrossRef]
102. Hervy, M.; Evangelisti, S.; Lettieri, P.; Lee, K.-Y. Life cycle assessment of nanocellulose-reinforced advanced fibre composites. *Compos. Sci. Technol.* **2015**, *118*, 154–162. [CrossRef]
103. Guinée, J. Handbook on life cycle assessment—Operational guide to the ISO standards. *Int. J. Life Cycle Assess.* **2001**, *6*, 255. [CrossRef]
104. La Rosa, A.D.; Banatao, D.; Pastine, S.; Latteri, A.; Cicala, G. Recycling treatment of carbon fibre/epoxy composites: Materials recovery and characterization and environmental impacts through life cycle assessment. *Compos. Part B Eng.* **2016**, *104*, 17–25. [CrossRef]
105. La Rosa, A.D.; Blanco, I.; Banatao, D.R.; Pastine, S.J.; Björklund, A.; Cicala, G. Innovative Chemical Process for Recycling Thermosets Cured with Recyclamines® by Converting Bio-Epoxy Composites in Reusable Thermoplastic—An LCA Study. *Materials* **2018**, *11*, 353. [CrossRef] [PubMed]
106. Gagliardi, F.; Palaia, D.; Ambrogio, G. Energy consumption and CO2 emissions of joining processes for manufacturing hybrid structures. *J. Clean. Prod.* **2019**, *228*, 425–436. [CrossRef]
107. Lopresti, M.; Alberto, G.; Cantamessa, S.; Cantino, G.; Conterosito, E.; Palin, L.; Milanesio, M. Light Weight, Easy Formable and Non-Toxic Polymer-Based Composites for Hard X-ray Shielding: A Theoretical and Experimental Study. *Int. J. Mol. Sci.* **2020**, *21*, 833. [CrossRef]

polymers

MDPI

Article

Epoxy Based Blends for Additive Manufacturing by Liquid Crystal Display (LCD) Printing: The Effect of Blending and Dual Curing on Daylight Curable Resins

Claudio Tosto [1], Eugenio Pergolizzi [1], Ignazio Blanco [1], Antonella Patti [1], Paul Holt [2], Sarah Karmel [2] and Gianluca Cicala [1,*]

1. Department of Civil Engineering and Architecture (DICAr), University of Catania, Viale Andrea Doria 6, 95125 Catania, Italy; claudio.tosto@unict.it (C.T.); euper@hotmail.it (E.P.); iblanco@unict.it (I.B.); antonella.patti@unict.it (A.P.)
2. Photocentric Ltd., Cambridge House, Oxney Road, Peterborough PE1 5YW, UK; paul.Holt@photocentric.co.uk (P.H.); sarah.karmel@photocentric.co.uk (S.K.)

* Correspondence: gianluca.cicala@unict.it; Tel.: +39-095-738-2760

Received: 22 June 2020; Accepted: 15 July 2020; Published: 18 July 2020

Abstract: Epoxy-based blends printable in a Liquid Crystal Display (LCD) printer were studied. Diglycidyl ether of bisphenol A (DGEBA) mixed with Diethyltoluene diamine (DETDA) was used due to the easy processing in liquid form at room temperature and slower reactivity until heated over 150 °C. The DGEBA/DETDA resin was mixed with a commercial daylight photocurable resin used for LCD screen 3D printing. Calorimetric, dynamic mechanical and rheology testing were carried out on the resulting blends. The daylight resins showed to be thermally curable. Resin's processability in the LCD printer was evaluated for all the blends by rheology and by 3D printing trials. The best printing conditions were determined by a speed cure test. The use of a thermal post-curing cycle after the standard photocuring in the LCD printer enhanced the glass transition temperature T_g of the daylight resin from 45 to 137 °C when post-curing temperatures up to 180 °C were used. The T_g reached a value of 174 °C mixing 50 wt% of DGEBA/DETDA resin with the photocurable resin when high temperature cure cycle was used.

Keywords: additive manufacturing; LCD printing; epoxy; polymer blends; thermomechanical properties

1. Introduction

Liquid crystal display (LCD) 3D printing is based on the use of LCD displays as imaging system. The key property of LCD printing is that the light shines through the flat LCD panels directly onto the uncured resin. This allows light not to expand and, thus, pixel distortion is less of an issue with LCD printing as it happens with DLP printing. In addition, a full layer can be exposed at the same time and there is no need to scan the photopolymer point-by-point like in SLA. This had the advantage of a faster 3D printing process. In many LCD printers the light comes from an array of UV or LED lamps. Other photocuring 3D printing technologies are based on different principles: digital light processing (DLP) uses a projector to project the image of the cross section of the object to be printed; stereolithography apparatus (SLA) uses a laser beam to scan and cure the layer of the photocuring resins point by point; multi jet printing (MJP) prints the models spraying liquid photosensitive resins from several nozzles and then curing the layer by using an UV lamp [1]. Back in 2015, the company Carbon3D reported the process named CLIP (Continuous Liquid Interface Production) which is based on an oxygen permeable membrane leading to the consecutive printing of each layer and therefore speeding up part's production. The CLIP technology is also named DLS (Digital Light Synthesis).

The key aspects is the use of oxygen permeation to inhibit the radical polymerization allowing fast continuous printing [2]. Carbon3D technology is not based only on the use of oxygen permeation membrane, but also on the availability of different resin chemistries ranging from acrylates to epoxies, cyanates and polyurethanes. Some of the best performing resins offered by Carbon3D are based on dual curing approach in which photocuring is followed by thermal curing [3]. Obst et al. [4] discussed the influence of different exposure times on the dual curing reaction of a polyurethane-based resin used in the DLS technology. The resin used in their study was a blend of acrylate and polyurethane. The mechanical characteristics such as tensile strength and elongation at break were shown to be strongly dependent on the degree of UV crosslinking. Redmann et al. [5] focused on an epoxy-based system processable in the Carbon3D printer. The authors addressed the problem of thermal curing optimization for this system. Reducing the thermal curing time while avoiding a negative influence on the final mechanical properties was the main goal. The original curing time with a duration of 750 min was shortened to 200 min with, at the same time, an improvement of glass temperature transition from 146 to 154 °C. Dual curing for acrylate-based systems has been widely studied for several applications: shape memory polymers, optical materials, photolithography, protective coatings, structured surface topologies, and holographic materials [6]. A dual-curing process is defined as a combination of two curing reactions taking place simultaneously or sequentially. Epoxies mixed with a photocurable acrylic resin cured combining photocuring with thermal curing allowed the 3D printing of carbon fiber reinforced composites [7]. Lantean et al. [8] proposed similar approach based on epoxy/acrylate blends for DLP printing. In this study, epoxy was used but no specific thermally activated curing agent was added. The maximum T_g achieved was limited to 108.6 °C for those blends. Daylight resins cure at 460 nm while most of the other photocurable resins cure in the UV region using wavelengths between 325–420 nm. According to Photocentric, "the use of light at longer wave length proved to allow for a deeper penetration depth into the photopolymer material and therefore a more uniform and accurate 3D printing process" [9,10]. The combination of visible light with the use of LCD screens for 3D printing, allows for a faster, more efficient and economical manufacturing process. This is a relevant point for application. In the literature there are increasing examples of the use of additive manufacturing in several fields ranging from microfluidics [11–13], tooling for composites [14,15] and injection molding [16]. Mixing epoxy blends with photocurable resins can open the field to novel formulation too if complex toughened epoxy blends area used [17,18] because this can be a suitable method to obtain the impact resistance of the printed specimens. All these applications might benefit of an accurate control of cavity control [19,20]. The aim of this work was to improve the thermal properties of daylight resins by blending with epoxy thermally curable resins without altering the working principle of the printer. To achieve this goal first we selected a commercial daylight curing resin and we studied its thermal curing properties after LCD printing. The effect of epoxy blending and the use of dual curing as a mean to improve resin's thermal properties was analyzed.

2. Experimental

2.1. Materials and Method

2.1.1. Materials

Two resin systems were mixed: the commercial daylight resin (Cream Hard) obtained by Photocentric Ltd. (Peterborough, UK) and the thermally curable epoxy resin formulated by mixing diglycidyl ether of bisphenol A (DGEBA) and diethyltoluene diamine (DETDA. In the following, we refer to the Cream Hard resin as cream or C while the mixture DGEBA/DETDA is refereed as E. The diglycidyl ether of bisphenol A selected was the Araldite GY 240 kindly donated by Huntsman (Basel, Switzerland). The diethyltoluene diamine (DETDA) was the LonzaCure DETDA kindly donated by Lonza (Basel, Switzerland). DGEBA and DETDA were mixed at room temperature with a 1:1 stoichiometric ratio.

2.1.2. Epoxy-Based Blend Preparation

LonzaCure DETDA is a liquid aromatic amine that can be mixed at room temperature with liquid epoxy. The liquid epoxy formulation obtained is miscible with daylight resins and it is not reactive under daylight. The uncured epoxy formulation was obtained mixing the two monomers (i.e., DGEBA and DETDA) in a centrifugal planetary mixer (ARV-310 by Thinky, Laguna Hills, CA, USA). The conditions used for mixing were: speed 2000 rpm, vacuum pressure of 0.3 kPa and mixing time of 5 min. The epoxy-based blends were prepared by mixing at room temperature the uncured epoxy formulation with the cream resin with different weight ratios (Table 1). The blends were mixed in the centrifugal planetary mixer with the same mixing procedure used for the uncured epoxy formulation. The prepared blends were stored in opaque glass containers avoiding daylight exposure until used. No sign of demixing or incompatibility for the samples stored at room temperature for 3 months was observed. The centrifugal mixer allowed to mix the two resins in 5 min only and, at the same time, to apply deareation avoiding air voids in the cured systems.

Table 1. Resin formulations studied.

ID Sample	Cream [wt%]	Epoxy [wt%]
C	100	0
CE7030	70	30
CE5050	50	50
E	0	100

2.1.3. Resin LCD Printing and Thermal Curing

The LCD printing was carried out on a LC-HR2 LCD printer by Photocentric (Peterborough, UK). The LC-HR2 is a LCD printing machine that uses an iPad screen as display for resin's photocuring. The initial conditions used for printing the cream resin were recommended by Photocentric. Following the printing trials on the epoxy- based blends the printing conditions were tailored to obtain good quality printing (Table 2). The methods used to define the optimized printing parameters for the epoxy blends will be explained in the results and discussion section. The printed samples were first shaped by LCD printing and then thermally post-cured in a standard ventilated oven with two types of cure cycles: isothermal cure for hold time of 3 h; isothermal cure at 140 °C for 2 h followed by a ramp at 2 °C/min to 180 °C and hold at this temperature for 2 h. For the isothermal curing of the cream resin three temperatures were tested: 100, 140 and 150 °C.

Table 2. LCD printing conditions used for specimens manufacturing.

ID Sample	Exposure Time [s]	Z Lift Distance [mm]	Z Lift Speed [mm/min]	Z Retract Speed [mm/min]	Top Time [s]
C	25	3	15	50	5
CE7030	30	10	10	10	10
CE5050	30	10	10	10	10

2.2. *Characterization*

2.2.1. Dynamic Mechanical Analysis (DMA)

Cured specimens with dimensions (10 × 8 × 2) mm^3 were tested using a dynamic mechanical analysis (DMA) machine (Tritech by Triton Ltd., Wrexham, UK) to measure the storage modulus (E') and the $Tan\delta$. The experiments were carried out in single cantilever mode with a 20 μm amplitude and 1 Hz frequency. Two types of experiments were carried out: isothermal testing by heating the DMA chamber at 10 °C/min followed by an isothermal hold for 120 min at different temperatures (i.e., 100, 140 and 150 °C); ramp test at 2 °C/min from 25 to 180 °C. The isothermal test was carried out on the cream sample after LCD printing to evaluate the effect of the thermal crosslinking induced by postcuring. The ramp rate test was used to characterize the viscoelastic properties of the samples after completing the thermal curing cycle. The latter test was used to measure the glass transition temperature (T_g) of the resin that was determined as the *Tanδ* peak.

2.2.2. Differential Scanning Calorimetry (DSC)

Calorimetric measurements were performed using a Shimadzu DSC-60 (Shimadzu, Kyoto, Japan). Resin samples in liquid (uncured resin) and solid (photo- and thermal- cured resin) form were placed in a 40-μL sealed aluminum crucibles. Samples of the average weight of 5–7 mg have been prepared. The samples were heated from room temperature up to 300 °C at a rate of 5 °C/min (first scan). After the first scan, they were cooled to room temperature and reheated to 300 °C, at a heating rate of 5 °C/min (s scan). The glass transition temperature (T_g) was measured as the midpoint of the heat capacity increment Δcp associated with the glass–rubber transition. The specific heat increment Δcp was calculated from the vertical distance between the two extrapolated baselines at the glass transition temperature. The total exothermic heat released was calculated from the area of the DSC exothermic peak.

2.2.3. Rheology Testing

Parallel plate rheometry was carried out on uncured resin blends using an ARES rheometer (TA Instruments, Sesto San Giovanni (Mi), Italy). Disc-disc configuration was used with plates of 25 mm diameter and 1.2 mm gap between plates. Isothermal test at room temperature with strain of 5% and oscillatory frequency of 10 Hz were carried out on the blends.

3. Results and Discussion

3.1. DSC Testing Results

The uncured cream resin was tested before LCD printing by a dynamic temperature scan using differential calorimetry. The result is reported in Figure 1. The graph shows the presence of an exothermic peak centered at 180 °C starting at 150 °C (black curve). The total exothermic heat released was equal to 168.35 J/g. The exothermic peak reveals the presence of some thermally reactive groups in the daylight formulation. The cream resin is a (meth)acrylate-based resin that can be cured by photopolymerization and by heat [21]. The cream resin was analyzed back after the first scan and no exothermic peak was observed (Figure 1, red curve). This finding confirms that all reactive moieties were reacted in the first scan. The thermogram clearly shows a glass transition temperature at 135.6 °C in the second scan (red curve) that is higher than the T_g reported for the cream resin after LCD curing only (i.e., ≈60 °C). Such an improvement can be explained as the result of the increased crosslink density caused by the thermal curing occurring in the first scan.

The unreacted epoxy-based blends were also characterized in their uncured state by DSC. The results are summarized in Figure 2. The addition of the thermally curable resin resulted in the increase of the released heat during the heating ramp (Table 3) as it can be expected because of the presence of a higher content of thermally curable monomers.

Table 3. Data obtained from the DSC analysis of the uncured blends.

ID Sample	Uncured	Photocured	Postcured in Oven	Isothermal [°C]	$Peak_1$ [°C]	$Onset_1$ [°C]	$Heat_1$ [J/g]	$Peak_2$ [°C]	$Onset_2$ [°C]	$Heat_2$ [J/g]
E	•				171.79	129.59	264.52	-	-	-
CE5050	•				183.23	148.69	174.53	259.3	253	9.88
CE7030	•				186.56	143.7	227.98	261.06	243.93	9.31
C	•				181	150.24	168.35	264.83	245.33	41.4
C		•			185.56	183.25	115.71	-	-	-
C		•	•	100	188.22	184.52	55.16	-	-	-
C		•	•	140	189.54	185.71	12.05	-	-	-
C		•	•	150	-	-	-	-	-	-

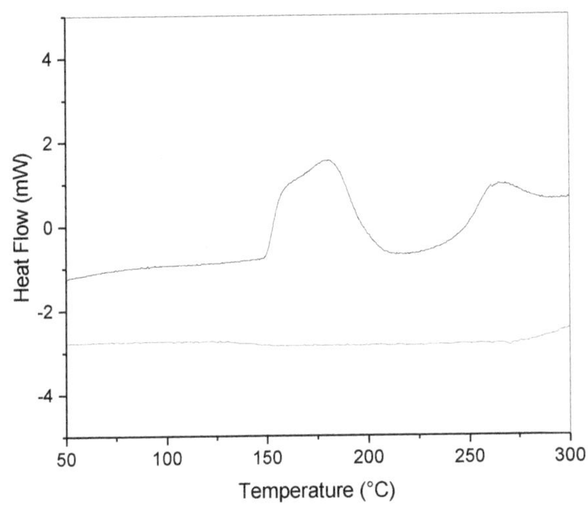

Figure 1. DSC analysis on the hard cream resin: first scan (black); second scan (red).

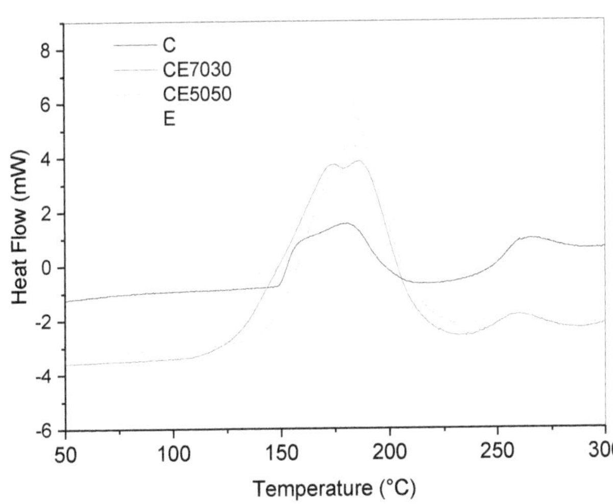

Figure 2. DSC analysis on the epoxy-based blends in their uncured state.

3.2. Rheology Testing

Resin viscosity is a parameter of paramount importance for LCD printing. The resin must flow during platform movement up and down in order to create a homogenous uncured resin layer between the bottom of the vat and the printing part. Photocentric suggests a maximum viscosity (η^*) of 100 Poise to enable resin printing on the LC-HR2 printer. The commercial cream resin used in this study presented a room temperature viscosity of about 1.8 Poise. Adding the epoxy formulation resulted in a viscosity increase with increasing epoxy content. The blend with the highest content of epoxy (i.e., CE5050) showed a room temperature viscosity of 12 Poise (Figure 3). The blend's viscosity largely depends on the epoxy prepolymer molecular weight and, as such, could be tailored by the addition of reactive diluents. We limited blend's viscosity at values within printer specifications and at the same time we increased the T_g values. Notably, the T_g values achieved are higher then the values obtained by most of the current commercially available systems by Photocentric and they are also better that other commercial systems (Table S1). The resins which displayed similar performance where those sold by Carbon3D and based on the dual curing processing. If we compare the T_g results and the full processing times we obtained with the results presented on Carbon3D technology in the open literature [2,4,5] we can notice that proposed approach is fast and lead to comparable T_g performance. The viscosity of the blends studied in this paper are higher than those of the systems commercially available. However, being the values well below the suggested limit of 100 Poise the blends were all regarded as being processable and this was confirmed by our printing trials described in the following section.

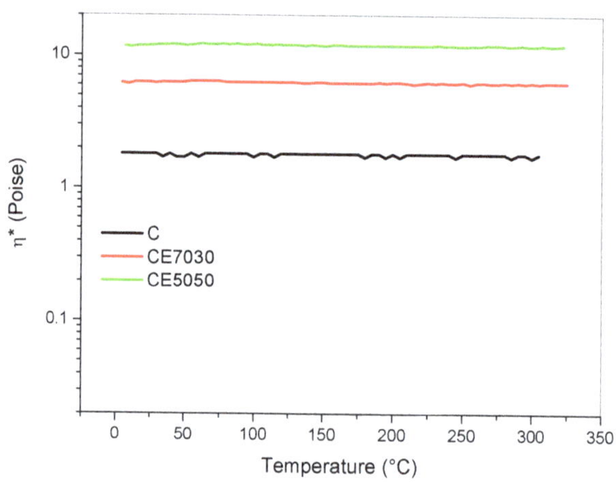

Figure 3. Complex viscosity of the uncured epoxy-based blends: C (blue); CE7030 (red); CE5050 (green).

3.3. Printing Test

Once tested blend's viscosity, printing trials to obtain specimens for thermomechanical testing were carried out. The first trials on the CE7030 blend resulted in unsuccessful printing (Figure 4). The printing failed completely as no shape was formed and only some cured layers deposited on the vat films were found. The same printing parameters used for the cream resin were applied in such trials (see Table 2). The short exposure times seemed not long enough to allow full curing of resin's layers properly. This result is the effect of two parameters: the presence in the blends of the epoxy resin, which is designed for thermal curing only, that reduce the amount of photocurable moieties; the stoichiometric unbalance of the photointiator due to the epoxy resin addition in the blend formulation.

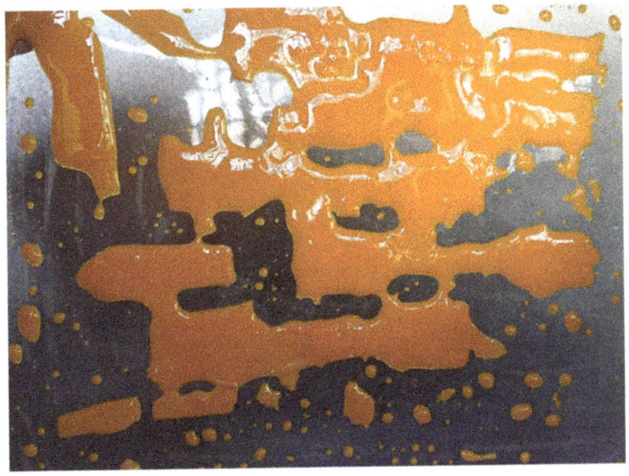

Figure 4. Printing trials for the CE7030 blend with the same printing conditions used for the cream resin (see Table 2).

To optimize the exposure times of the formulated blends the photocuring of one single layer was studied in greater detail using the following procedure. The printer was set to print a single layer prism with a square base of 10 mm. One single layer for the LC-HR2 printer would results in a layer thickness of 100 μm when the cream resin is used. The uncured blend was dropped on a plastic foil and suddenly exposed to the LCD light with different exposure times setting the display to project one layer only for the prism. The LC-HR2 printer works with fixed light intensity therefore this parameter was not investigated. The uncured resin was gently rinsed away with a cloth wetted with isopropyl alcohol (IPA) after light irradiation. The resulting thickness of the printed prism was measured by a caliper. The side of the resulting square prism was measured by a caliper and the deviation from the fixed side (i.e., 10 mm) reported as width overcure (Figure S2). The obtained results are summarized in see Table 4. The results shown in the table demonstrate that adding the epoxy resin to the photocurable resin there is a slightly reduced photo-reactivity for the blend. Therefore, to account for such effect the exposure times increased from 25 s of the cream resin to 30 s for the CE7030 and CE5050 (Table 2). However, the printing results is also affected by the resin flow during the up and down movement of the platform. To account for the increased viscosity of the blends (Figure 3) the platform's controlled movement were modified as reported in Table 2 for the blends. In particular, the platform was lifted higher to 10 mm distance instead of the 3 mm used for the cream and, to ensure the correct timing for resin flow, the Z lift and retract were slower (i.e., 10 mm/min) compared to the cream resin's printing.

Table 4. Average mean values and standard deviations (Std Dev) of the printing trials on the square prism for the LCD printing optimization with the epoxy-based blends

Exposure Time [s]	Thickness [μm]						Width Overcure [%]					
	C		CE7030		CE5050		C		CE7030		CE5050	
	Mean	Std Dev	Mean	Std Dev	Mean	Std Dev	Mean	Std Dev	Mean	Std Dev	Mean	Std Dev
5	-	-	-	-	-	-	-	-	-	-	-	-
10	32	1.0	-	-	-	-	−9.4%	0.14%	-	-	-	-
15	46	0.7	41	1.0	-	-	−8.1%	0.15%	−9.0%	0.17%	-	-
20	82	1.0	73	1.2	-	-	−4.7%	0.19%	−5.3%	0.20%	-	-
25	105	0.9	95	1.3	88	0.8	0.0%	0.14%	−2.0%	0.15%	−4.0%	0.13%
30	117	1.0	108	1.2	102	1.0	1.1%	0.20%	1.5%	0.12%	1.0%	0.19%
35	137	0.7	125	1.2	118	0.9	1.1%	0.16%	1.6%	0.16%	1.5%	0.12%

Following these findings new settings were used Table 2 that yielded good quality specimens as all the shapes drawn were printed (Figure 5) and all the samples had the dimension planned.

Figure 5. Printing trials for the CE7030 blend with the optimized printing conditions after photocuring optimization (see Table 2).

3.4. DMA Testing

The results of the DSC tests showed the presence of an exothermic peak for the uncured cream resin outlining the presence of reactive moieties leading to the thermal curing of this system (see Figure 1). DMA is a suitable technique for the study of the thermal curing of solid systems after vitrification [22]. LCD photo-cured samples were analyzed by running isothermal DMA testing at three different temperatures (i.e., 100, 140 and 150 °C). This test allowed to measure viscoelastic properties changes during the isothermal curing. Storage modulus (E') versus time was monitored in detail. The cream resin tested at 100 °C for 3 h showed only a slight variation of the storage modulus while, when the isothermal test was carried out at 140 or 150 °C, the storage modulus increased of one order of magnitude within the testing time (Figure 6). The increase of the storage modulus is the result of the crosslinking reactions which starts, as demonstrated by the DSC tests (Figure 1), at 150 °C.

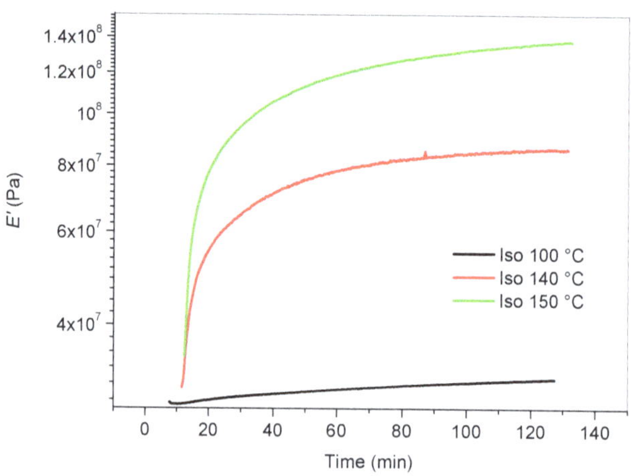

Figure 6. Isothermal DMA curing test on LCD printed Hard Cream specimens: Storage modulus versus time.

After being tested isothermally, the same samples were further analyzed by a ramp rate DMA test (Figure 7). The system cured at 140 °C displayed two $Tan\delta$ peaks at 65 and 126 °C. The resin cured at 150 °C displayed one peak at 134 °C and a shoulder at about 67 °C. The specimen cured at 100 °C showed a slight shift of the $Tan\delta$ peak from 45 °C, measured for the LCD printed sample, to 50 °C.

The shift of T_g peaks, which corresponds to the glass transition temperature (T_g), is the result of the increased crosslink density due to the crosslinking reactions occurring because of the thermal curing of the cream resin. The epoxy DGEBA/DETDA system showed an exothermic peak centered at 172 °C (Figure 2). Therefore, to ensure the full conversion of the epoxy moieties after LCD printing the following post-cure cycle for epoxy modified blends was used: isothermal cure at 140 °C for 2 h, ramp at 2 °C/min up to 180 °C and hold at 180 °C for 2 h. The post- cured samples obtained were analyzed by a ramp test (Figure 8).

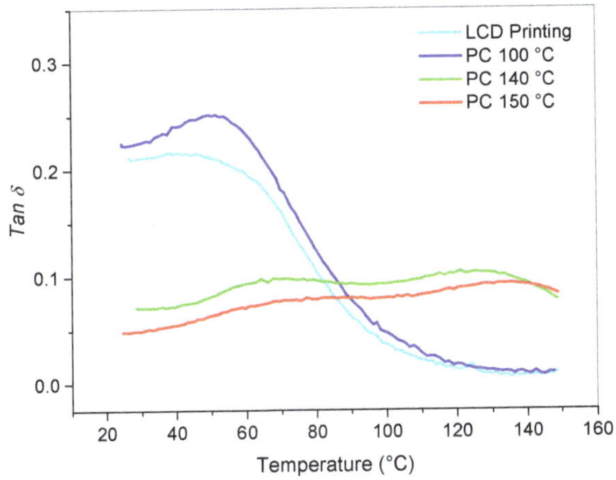

Figure 7. Ramp rate DMA test on isothermally cured Hard Cream resin.

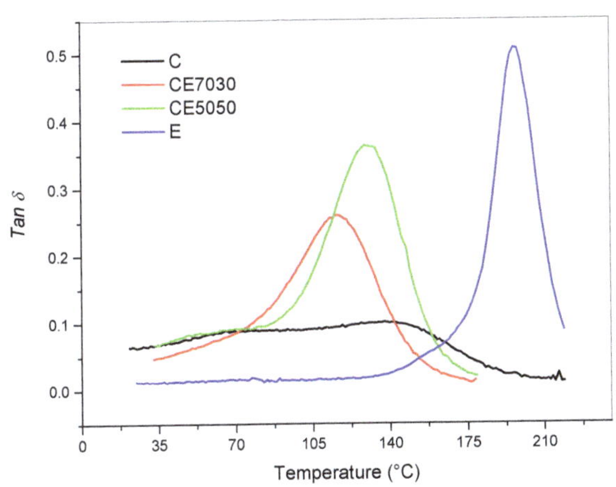

Figure 8. DMA analysis for the LCD printed sample after the post-curing cycle (isothermal cure at 140 °C for 2 h, ramp at 2°C/min to 180 °C and hold at 180 °C for 2 h).

As with the viscosity, if we can consider commercial analogous systems (Table S1) it can be said that with dual curing, resin with higher T_g were obtained. The cream unmodified resin (C) showed two glass transition temperatures (T_g) at 70 and 137 °C while the pure epoxy formulation (E) a T_g of 197 °C. Epoxy blending resulted in one single broad *Tanδ* peak reaching, for the 50:50 blend, with a T_g value of 174 °C. The blends obtained mixing the two formulations showed an increasing T_g

with higher epoxy content (Figure 9). The T_g vs epoxy content graph shows a quasi linear correlation considering only the high T_g for the cream resin. This trend is typically observed for blends forming the so-called interpenetrated network [23].

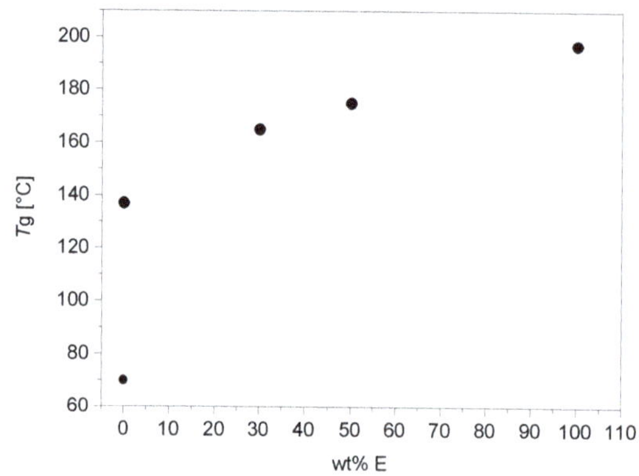

Figure 9. Glass transition temperatures (T_g) obtained as *Tanδ* peaks measured by DMA plotted versus the epoxy content.

4. Conclusions

Daylight curable (meth)acrylate-epoxy blends were studied. Calorimetric analysis demonstrated the presence of thermally reactive moieties for the daylight resin used. Dynamic mechanical analysis showed that LCD printed parts do not develop fully crosslinked network coherently with the information gained from DSC analysis. Treating with thermal postcure the LCD printed daylight resins increased the glass transition temperature from 45 to 137 °C. This T_g increase was obtained with a post treatment that required about 4 h and 20 min of additional thermal curing time. This additional treatment time is not impacting dramatically on overall production times because LC-HR2 printer can print 100 μm every 25 s thus leading to a printing time of 7 h for a complex part with a volume of 78 cm^3 (Figure S1). Regarding this point, we should notice that increasing the printing area on LCD printer do not change the printing time. Therefore, if many similar objects are printed in bigger printers (like for example the Magna model) the printing time remains approximately the same while the curing time can be the same on all the parts. This means that the time to make parts can be overall reduced because of post-curing is carried out at once on all parts. These concepts let Photocentric to achieve high volume throughputs with their big area machines. Therefore, thermal post-curing of daylight resin is an efficient approach for enhancing the final properties of LCD printed parts. The benefits of thermal post-curing were further enhanced when DGEBA/DETDA resin was added to the formulation. The T_g increased up to 165 and 174 °C for blends with 30 and 50 wt% of DEGBA/DETDA, respectively. However, these remarkable T_g increases showed some drawbacks due to the increasing photocuring times to achieve good quality LCD printing. The exposure time needed to be adjusted from 25 s for hard cream resin to 30 s for the epoxy-based blends and the platform to be moved higher and slower to allow resin flow. The changes in these printing parameters lead to longer printing time. For example, the part (Figure S1) with a volume of 78 cm^3 would be of 16 h of printing time. These results offer some important insights for the further development of similar blends. To counteract the reduced photocuring reactivity of the modified blends some adjustments of the type and content of the photoinitiator could a viable solution while, blend's viscosity could be tailored by using reactive diluents. Such developments can be interesting for future applications in the field of tooling for polymers and composites or for specific application in severe environment. In the field of tooling for polymers and composites the need for high temperature resistance is due to the high processing temperatures. For example, many advanced prepregs needed in automotive are cured with processing cycles at 120 °C thus requiring materials with high temperature stability. Other fields which might benefit for the use of high Tg resins and for the high resolution offered by LCD printing are the valves working with hot fluids. Similar applications for dual curable epoxies have been presented by Carbon3D but, according to the result of this paper, LCD printing can be used too if the epoxy blending approach is pursued.

Supplementary Materials: The following are available at http://www.mdpi.com/2073-4360/12/7/1594/s1.

Author Contributions: Conceptualization, G.C., P.H. and S.K. and C.T.; investigation, E.P., C.T., A.P. and I.B.; resources, G.C.; writing–original draft preparation, G.C.; writing–review and editing, I.B. and S.K. All authors have read and agreed to the published version of the manuscript.

Funding: This research was funded by DICAr with the grant PolyADv (59725022078), of University of Catania with the grant Change, and by MIUR with the grant MAMMA (CUP E64I17000260001, Prot. 20179SWLKA). The PhD project was fundend by MIUR within the PON Ricerca e Innovazione 2014–2020 Asse I "Investimenti in Capitale Umano"—Azione I.1 "Dottorati Innovativi Con Caratterizzazione Industriale" Project Title "Advanced Materials by Additive manufacturing" (AMA) (CUP E67I18001120006).

Acknowledgments: Gianluca Cicala acknowledges the grant support of DICAr with the grant PolyADv, of University of Catania with the grant Change and the Italian MIUR for the grant. Claudio Tosto acknowledges Italian MIUR for the funding of his PhD by MIUR.

Conflicts of Interest: The authors declare no conflict of interest. The funders had no role in the design of the study; in the collection, analyses, or interpretation of data; in the writing of the manuscript, or in the decision to publish the results.

Abbreviations

The following abbreviations are used in this manuscript:

η^*	Complex viscosity
DETDA	Diethyltoluene diamine
DGEBA	Diglycidyl ether of bisphenol A
DMA	Dynamic Mechanical Analysis
DSC	Differential Scanning Calorimetry
E'	Storage modulus
Δcp	Heat capacity increment
LCD	Liquid Crystal Display
T_g	Glass-transition temperature

References

1. Quan, H.; Zhang, T.; Xu, H.; Luo, S.; Nie, J.; Zhu, X. Photo-curing 3D printing technique and its challenges. *Bioact. Mater.* **2020**, *5*, 110–115. [CrossRef] [PubMed]
2. Tumbleston, J.R.; Shirvanyants, D.; Ermoshkin, N.; Janusziewicz, R.; Johnson, A.R.; Kelly, D.; Chen, K.; Pinschmidt, R.; Rolland, J.P.; Ermoshkin, A.; et al. Continuous liquid interface production of 3D objects. *Science* **2015**, *347*, 1349–1352. [CrossRef] [PubMed]
3. Xinyu, G.U.; Poelma, J.; Rolland, J.P. *Methods Of Making Three Dimensional Objects From Dual Cure Resins with Supported Second Cure*; WO 2017/040890 A1; World Intellectual Property Organization: Geneva, Switzerland, 2017.
4. Obst, P.; Riedelbauch, J.; Oehlmann, P.; Rietzel, D.; Launhardt, M.; Schmölzer, S.; Osswald, T.A.; Witt, G. Investigation of the influence of exposure time on the dual-curing reaction of RPU 70 during the DLS process and the resulting mechanical part properties. *Addit. Manuf.* **2020**, *32*, 101002. [CrossRef]
5. Redmann, A.; Oehlmann, P.; Scheffler, T.; Kagermeier, L.; Osswald, T.A. Thermal curing kinetics optimization of epoxy resin in Digital Light Synthesis. *Addit. Manuf.* **2020**, *32*, 101018. [CrossRef]
6. Konuray, O.; Fernández-Francos, X.; Ramis, X.; Serra, À. State of the Art in Dual-Curing Acrylate Systems. *Polymers* **2018**, *10*, 178. [CrossRef] [PubMed]
7. Griffini, G.; Invernizzi, M.; Levi, M.; Natale, G.; Postiglione, G.; Turri, S. 3D-printable CFR polymer composites with dual-cure sequential IPNs. *Polymer* **2016**, *91*, 174–179. [CrossRef]

8. Lantean, S.; Roppolo, I.; Sangermano, M.; Pirri, C.F.; Chiappone, A. Development of New Hybrid Acrylic/Epoxy DLP-3D Printable Materials. *Inventions* **2018**, *3*, 29. [CrossRef]
9. Holt, P.; Zaregadeh, H.; Karmel S. *Methods for Making a Metal, Sand Or Ceramic Object by Additive Manufacture and Formulations for Use in Said Methods*; US 2020/0079966 A1; United States Patent Application Publication: Peterborough, Great Britain, UK, 2017.
10. Zhu, J.; Zhang, Q.; Yang, T.; Liu, Y.; Liu, R. 3D printing of multi-scalable structures via high penetration near-infrared photopolymerization. *Nat. Commun.* **2020**, *11*, 3462. [CrossRef] [PubMed]
11. Cairone, F.; Davi, S.; Stella, G.; Guarino, F.; Recca, G.; Cicala, G.; Bucolo, M. 3D-Printed micro-optofluidic device for chemical fluids and cells detection. *Biomed. Microdev.* **2020**, *22*, 37. [CrossRef]
12. Richard, C.; Neild, A.; Cadarso, V.J. The emerging role of microfluidics in multi-material 3D bioprinting. *Lab Chip.* **2020**, *20*, 2044–2056. [CrossRef] [PubMed]
13. Waheed, S.; Cabot, J.M.; Macdonald, N.P.; Lewis, T.; Guijt, R.M.; Paull, B.; Breadmore, M.C. 3D printed microfluidic devices: Enablers and barriers. *Lab Chip* **2016**, *16*, 1993–2013. [CrossRef] [PubMed]
14. Tosto, C.; Latteri, A.; Pergolizzi, E.; Giordano, D.; Abramo, G.; Catenaro, R.; Pignotti, N.; Cicala, G. Additive Manufacturing of Plastics: An Efficient Approach for Composite Tooling. *Macromol. Symp.* **2020**, *389*. [CrossRef]
15. Li, H.; Taylor, G.; Bheemreddy, V.; Iyibilgin, O.; Leu, M.; Chandrashekhara, K. Modeling and characterization of fused deposition modeling tooling for vacuum assisted resin transfer molding process. *Addit. Manuf.* **2015**, *7*, 64–72. [CrossRef]
16. Kampker, A.; Ayvaz, P.; Lukas, G. Direct Polymer Additive Tooling—Economic Analysis of Additive Manufacturing Technologies for Fabrication of Polymer Tools for Injection Molding. *Key Eng. Mater.* **2020**, *843*, 9–18. [CrossRef]
17. Cicala, G.; Recca, G.; Carciotto, S.; Restuccia, C. Development of epoxy/hyperbranched blends for resin transfer molding and vacuum assisted resin transfer molding applications: Effect of a reactive diluent. *Polym. Eng. Sci.* **2009**, *49*, 577–584. [CrossRef]
18. Cicala, G.; Mamo, A.; Recca, G.; Restuccia, C.L. Study on epoxy/thermoplastic blends based on the addition of a novel aromatic block copolymer. *Polym. Eng. Sci.* **2007**, *47*, 2027–2033. [CrossRef]
19. Sapuppo, F.; Bucolo, M.; Intaglietta, M.; Fortuna, L.; Arena, P. A cellular nonlinear network: Real-time technology for the analysis of microfluidic phenomena in blood vessels. *Nanotechnology* **2006**, *17*, S54–S63. [CrossRef]
20. Cairone, F.; Gagliano, S.; Bucolo, M. Experimental study on the slug flow in a serpentine microchannel. *Exp. Ther. Fluid Sci.* **2016**, *76*, 34–44. [CrossRef]
21. Maffezzoli, A.; Terzi, R.; Nicolais, L. Cure behaviour of visible light activated dental composites. *J. Mater. Sci. Mater. Med.* **1995**, *6*, 155–160. [CrossRef]
22. Menard, H.P. *Dynamic Mechanical Analysis: A Practical Introduction*, 2nd ed.; CRC Press: Boca Raton, FL, USA, 2008. [CrossRef]
23. Campbell, J.A.; Inglis, H.; Ng WeiLong, E.; McKinley, C.; Lewis, D.A. Morphology Control in a Dual-Cure System for Potential Applications in Additive Manufacturing. *Polymers* **2019**, *11*, 420. [CrossRef] [PubMed]

MDPI
St. Alban-Anlage 66
4052 Basel
Switzerland
Tel. +41 61 683 77 34
Fax +41 61 302 89 18
www.mdpi.com

Polymers Editorial Office
E-mail: polymers@mdpi.com
www.mdpi.com/journal/polymers

www.ingramcontent.com/pod-product-compliance
Lightning Source LLC
Chambersburg PA
CBHW061552220926
51850CB00005B/317
* 9 7 8 3 0 3 9 4 3 1 8 0 9 *